AF713176

Extracellular Vesicles and their Importance in Human Health

Extracellular Vesicles and their Importance in Human Health

Editor: Erina Jimenez

MURPHY & MOORE
www.murphy-moorepublishing.com

www.murphy-moorepublishing.com

MURPHY & MOORE

Cataloging-in-Publication Data

Extracellular vesicles and their importance in human health / edited by Erina Jimenez.
p. cm.
Includes bibliographical references and index.
ISBN 978-1-63987-783-6
1. Extracellular space. 2. Coated vesicles. 3. Cell organelles. 4. Cell membranes. 5. Cytology. I. Jimenez, Erina.
QH603.E93 E98 2023
571.65--dc23

© Murphy & Moore Publishing, 2023

Murphy & Moore Publishing
1 Rockefeller Plaza,
New York City,
NY 10020, USA

ISBN 978-1-63987-783-6

This book contains information obtained from authentic and highly regarded sources. Copyright for all individual chapters remain with the respective authors as indicated. All chapters are published with permission under the Creative Commons Attribution License or equivalent. A wide variety of references are listed. Permission and sources are indicated; for detailed attributions, please refer to the permissions page and list of contributors. Reasonable efforts have been made to publish reliable data and information, but the authors, editors and publisher cannot assume any responsibility for the validity of all materials or the consequences of their use.

Trademark Notice: Registered trademark of products or corporate names are used only for explanation and identification without intent to infringe.

Contents

Preface

Extracellular vesicles (EVs) are elements enfolded in a lipid bilayer membrane. Similar to exosomes, they are released by all living cells. They comprise a variety of bioactive molecules such as lipids, nucleic acids, metabolites and proteins. EVs have potential as a target for liquid biopsies because they can be detected in nearly all types of body fluids, including breast milk, blood, saliva, ascites and urine. Tumor-secreted EVs are important mediators for intercellular communication in both local and remote microenvironments. They are naturally taken up by the target cells and could be utilized to transfer therapeutic agents. It is crucial to recognize EV subtypes in order to determine which pathways are influenced by EVs and which subtypes can be targeted in liquid biopsies or therapeutic methods. This book provides comprehensive insights on extracellular vesicles. It is compiled in such a manner, that it will provide an in-depth knowledge about the role of extracellular vesicles in human health. This book will serve as a reference to a broad spectrum of readers.

The researches compiled throughout the book are authentic and of high quality, combining several disciplines and from very diverse regions from around the world. Drawing on the contributions of many researchers from diverse countries, the book's objective is to provide the readers with the latest achievements in the area of research. This book will surely be a source of knowledge to all interested and researching the field.

In the end, I would like to express my deep sense of gratitude to all the authors for meeting the set deadlines in completing and submitting their research chapters. I would also like to thank the publisher for the support offered to us throughout the course of the book. Finally, I extend my sincere thanks to my family for being a constant source of inspiration and encouragement.

Editor

1

Extracellular Vesicles in Non-Small-Cell Lung Cancer: Functional Role and Involvement in Resistance to Targeted Treatment and Immunotherapy

Luigi Pasini and **Paola Ulivi** *

Biosciences Laboratory, Istituto Scientifico Romagnolo per lo Studio e la Cura dei Tumori (IRST) IRCCS, 47014 Meldola, Italy; luigi.pasini@irst.emr.it

* Correspondence: paola.ulivi@irst.emr.it

Abstract: Targeted and immunological therapies have become the gold standard for a large portion of non-small cell lung cancer (NSCLC) patients by improving significantly clinical prognosis. However, resistance mechanisms inevitably develop after a first response, and almost all patients undergo progression. The knowledge of such a resistance mechanism is crucial to improving the efficacy of therapies. So far, monitoring therapy responses through liquid biopsy has been carried out mainly in terms of circulating tumor (ctDNA) analysis. However, other particles of tumor origin, such as extracellular vehicles (EVs) represent an emerging tool for the studying and monitoring of resistance mechanisms. EVs are now considered to be ubiquitous mediators of cell-to-cell communication, allowing cells to exchange biologically active cargoes that vary in response to the microenvironment and include proteins, metabolites, RNA species, and nucleic acids. Novel findings on the biogenesis and fate of these vesicles reveal their fundamental role in cancer progression, with foreseeable and not-far-to-come clinical applications in NSCLC.

Keywords: extracellular vesicles; non-small cell lung cancer; resistance; liquid biopsy; diagnostic biomarkers

---

1. Introduction

In the last ten years, treatment of non-small cell lung cancer (NSCLC) has been revolutionized by the introduction of targeted and immunological therapies in clinical practice [1–3]. Currently, the use of tyrosine kinase inhibitors (TKIs) against the epidermal growth factor receptor (EGFR) and the anaplastic lymphoma kinase (ALK) are the first treatment of choice for patients carrying alterations in the *EGFR* and *ALK* genes, as well as for patients with alterations in the proto-oncogene tyrosine-protein kinase ROS-1 (*ROS1*) [4]. On the other side, immune-checkpoint inhibitors represent the first treatment of choice for patients with high expression of the programmed cell death ligand 1 (PD-L1), as monotherapy (pembrolizumab), or in combination with chemotherapy for those patients presenting with no targeted alterations (nivolumab or atezolizumab) [4]. Although both types of treatment prolong progression free survival and improve patient prognosis, resistance mechanisms inevitably arise, leading to disease progression in almost all patients [5]. Currently, molecular profile of NSCLC is assessed by analyzing nucleic acids derived from the primary tumor tissue, but advancements in the ability to detect by longitudinal liquid biopsies the presence of tumor-released particles, including circulating tumor DNA (ctDNA), miRNA, and extracellular vesicles (EVs), has enabled clinicians to better understand the dynamic evolution of this disease [6].

EVs are defined as a heterogeneous group of membrane-delimited nanosized particles actively released by any cell type, including cancer cells [7]. A definite nomenclature consensus has not been established yet, and the various subtypes of cell-released vesicles are classified into exosomes, microvesicles, microparticles, ectosomes, oncosomes, apoptotic bodies (apoEVs), mostly based on their cellular origin and mechanism of biogenesis. However, the improvements of EV isolation methodologies and the emergence on specific markers of EV subtypes have supported the establishment of common guidelines to define experimentally the biological function of different EVs [8]. EVs can also be partially characterized by their size, which can range from 40–50 nm to over 1000 nm, and their molecular content, which can comprise various nucleic acid species, proteins, metabolites, or activated signaling molecules [8].

In this review, we will give particular attention to EVs of endosomal origin (defined as exosomes) and plasma membrane-derived microparticles (microvesicles) [9]. Exosomes and microvesicles share common sorting machineries, although they generate from different cellular compartments and it is the nature and the cellular abundance of the cargoes that determines the fate of a particular vesicle [10]. Several studies have demonstrated the functional relevance of EV release into circulation, indicating their major role in intercellular communication via transfer of their biological material to the recipient cells [10,11]. This evidence has led to the notion that EVs may support tumor growth as direct contributors to the initiation of oncogenesis, immunomodulation, metastatization, and resistance to therapy [12]. Many efforts have been made to understand EV function during cancer progression and, although current protocols of EV isolation are not effective in discriminating normal EVs from cancer-cell-derived EVs, thanks to most recent technical advances, we can envisage the introduction of EV-based liquid biopsy in the near future for the clinical practice of some cancers, including NSCLC.

2. Extracellular Vesicles Biogenesis and Fate

EVs are a group of membranous structures that can be actively released by any cell, which are highly heterogeneous for their origins and content, and are considered to be ubiquitous mediators of intercellular communication [9]. Typically, EVs are classified based on their biological functions and the cellular compartment they originated from, rather than their size and membrane markers, which are normally overlapping between the different types of EVs [7]. In fact, the same marker can be displayed on different EV subtypes, although some markers can be enriched on the surface of particular vesicles, depending on the cell of origin [13]. The most established subtypes of EVs are exosomes, the size of which vary between 40 and 150 nm, and microvesicles that can be up to 1 μm [8,9]. In contrast to microvesicles, which fall from the cell surface due to budding of the plasma membrane, exosomes are derived from the endolysosomal pathway through the intraluminal budding of multivesicular bodies that eventually fuse with the cell membrane [10]. The molecular machineries involved in the different steps of EV biogenesis are for the most part common to exosomes and microvesicles, and the transmembrane proteins sorted on both types of EVs normally reflects the same topology as on the cell membrane (Figure 1).

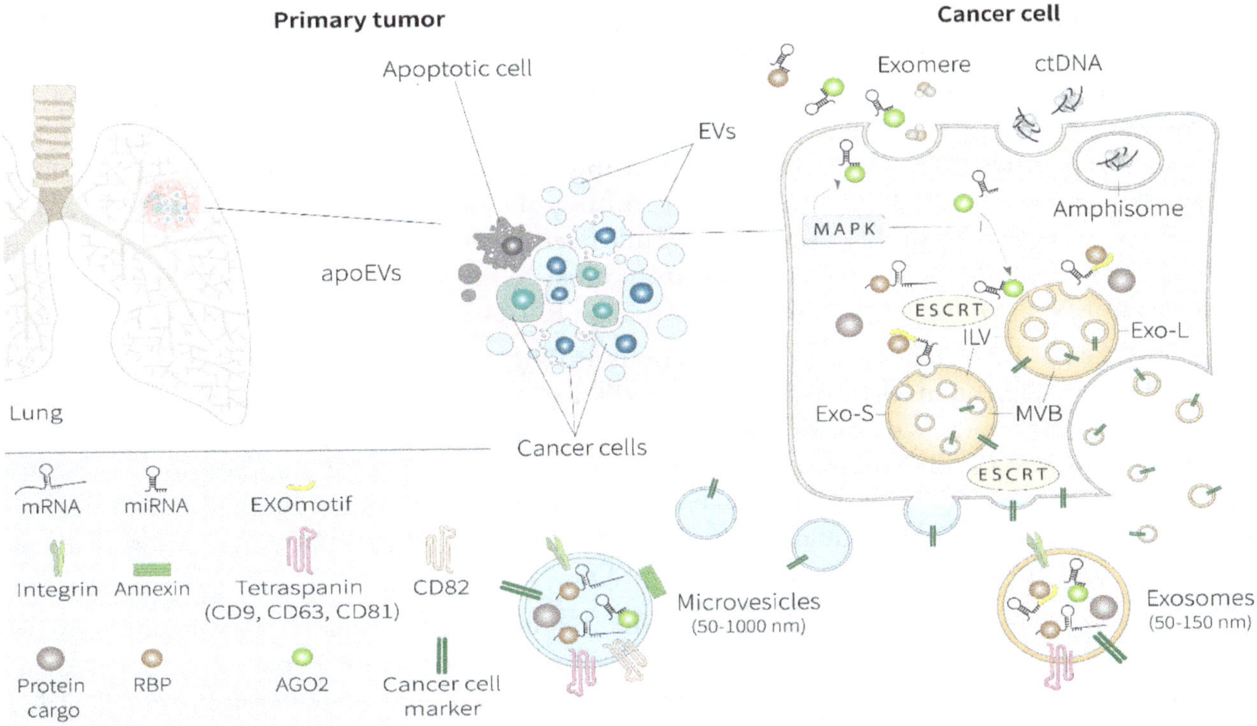

Figure 1. Biogenesis and heterogeneity of tumor-derived extracellular vesicles. A tumor mass is constituted of a heterogeneous population of cancer cells, constantly releasing a variety of extracellular vesicles (EVs), apoptotic bodies (apoEVs), and non-vesicular particles, including exomeres, all involved in the transport of nucleic acids and proteins that can have an active role in promoting tumor survival and aggressiveness. Cargo sorting into EVs involves the endosomal sorting complex required for transport (ESCRT) and partner proteins that include annexins (annexin A1) and tetraspanins (CD9, CD63, CD81, CD82). By reaching their destination exosome cargoes are sorted through the intraluminal vesicles (ILV), which form following the inward of endosomal membrane of the multivesicular bodies (MVBs) and are ultimately secreted upon fusion with the cell surface. Exosomes, similarly to microvesicles, display the same topology of membrane as the cell of origin and can potentially express cancer cell-specific markers. Microvesicles, however, generate from the outward budding of the plasma membrane. Different subsets of exosomes (large and small) with distinct molecular composition have been identified, along with the discovery of a population of abundant non-membranous nanoparticles involved in tumor growth, called exomeres. Exosomes are preferentially enriched with specific populations of miRNA, destined for exportation thanks to the presence of an EXOmotif, recognized by specialized RNA binding proteins (RBPs). During oncogenesis, activation of the mitogen-activated protein kinase (MAPK) pathway suppresses AGO2-mediated sorting of miRNAs to EVs, while favoring AGO2-independent EV loading of miRNAs and non-EV-associated miRNA release. Potentially, microvesicles can accommodate larger RNA species, including transcribing mRNAs. Nucleosomal double strand (dsDNA), although previously reported to be associated with exosomes, might not be directed to the ILVs, and it is instead released through EV-independent mechanisms involving intermediate organelles termed amphisomes. miRNA, microRNA; mRNA, messenger RNA; AGO-2, Argonaute 2; Exo-S, small exosome vesicles; Exo-L, large exosome vesicles.

Exosome cargoes reach their destination either from the Golgi or by internalization from the plasma membrane, before being sorted to the intraluminal vesicles (ILV) that are formed by the inward of endosomal membrane of the multivesicular bodies (MVBs) and are ultimately secreted upon fusion with the cell surface [14]. Thus, exosomes practically are ILVs with distinct biological functions related to their being secreted in the extracellular space [15]. In this context, syntenin is a key adaptor protein that links MVBs to the cytoskeleton, by regulating exosome biogenesis in concert with the ALG-2 interacting protein X (Alix) and the endosomal sorting complex required for transport (ESCRT), while acting as a potential regulator of endosomal targeting of exosome cargoes [16]. For example, selective recruitment of functional receptor proteins on exosomes, including integrin β1 or the major histocompatibility complex (MHC) class I molecules, is regulated at this level of exosome biogenesis [16]. Exosomes can also form in an ESCRT-independent manner that requires the involvement of tetraspanins, specifically CD63, the results of which are particularly enriched on the membrane of exosomes [17]. Tetraspanins, including also other members of the family, CD81 and CD9, directly regulate the selective intracellular routing of cytoplasmic molecules to ILVs, which can result in the co-sorting with other cytosolic proteins, such as the chaperone heat shock 70 kDa protein (HSP70), typically found in exosomes derived from most cell types [18]. Various RNA species, mostly including microRNAs (miRNAs) and small non-coding RNAs (ncRNAs), are transported by exosomes.

The presence of DNA inside EVs depends on methodological drawbacks that are not completely clear [19]. Short DNA sequences traditionally found after exosome isolation could be derived from the co-purification with these EVs during standard isolation methodologies. In fact, double stranded DNA might not be associated with exosomes or with any small EVs at all. DNA apparently is not directed to the ILVs, and it is instead released by the cell through independent mechanisms via intermediate organelles, originating from the nucleus and termed amphisomes, to be liberated, alongside EVs, in the extracellular as free nucleosomes [20].

After loading of their cargo, ILVs are targeted by the plasma membrane to be released as exosomes through the specific involvement of the ESCRT-1 component tumor susceptibility gene 101 protein (TSG101) and RAS-related protein (RAB) GTPases, while fusion with the cell membrane is guided by the soluble NSF attachment protein receptor (SNARE) complex [21,22].

Microvesicles follow a different route of biogenesis, which takes place at the level of discrete microdomains of lipids and membrane-associated proteins that cluster within the plasma membrane after the induction of local changes in Ca2+ concentration [23]. These events are accompanied by modifications of actin and myosin microfilaments that induce the lipid flipping while microvesicles are pinched off the plasma membrane [24]. Release of microvesicles involves TSG101 and ATP-dependent contraction. These molecular rearrangements of the plasma membrane is specifically driven by annexins, such as the annexin A1, along with the action of the tetraspanins CD9, CD81, and CD82 [10]. Microvesicles can be loaded with the same typology of cargos as exosomes through the ESCRT machinery, although microvesicles, given their larger size, might easily accommodate long mRNAs [10,20,25].

The fate of EVs depends on the specific composition of surface receptors that will target them to the right cell type. EVs can remain bound to the plasma membrane through tetraspanin-guided interaction with integrins while waiting for their cargo to be released, initiate signaling transduction, or engage the antigen presentation pathways [26]. EV fusion with recipient cells requires docking at the plasma membrane, followed by activation of extracellular receptors and vesicles internalization, which can be mediated either by clathrin-dependent or clathrin-independent endocytosis [27].

The interdependence of the intracellular machineries required for release and internalization of both exosomes and microvesicles facilitates the development of shared mechanism of deregulation that can be adopted by cancer cells to take over the secretome apparatus [28]. A tumor constantly releases EVs to alter the composition of the microenvironment and deliver the oncogenic signaling to target organs and modify the metastasis organotropism. Cancer EVs are able to modulate the immune response through MHC antigen presentation or trigger T cell apoptosis, influence the cross-talk with the surrounding stroma and induce fibroblast differentiation to eventually favor the migration of tumor cells through a permissive endothelium [29]. For example, patterns of integrin composition are used by cancer cells to target their exosomes to distinct organs and promote the premetastatic niche [30]. Cancer EVs normally have the same set of surface markers to those of healthy cells, being indistinguishable from their normal counterparts, although CD9 and CD81 overall seem to be more abundant, but mutually exclusive, on cancer EVs [25,31].

Apoptotic bodies are a less diffuse type of EVs, with a broad size range of 50 nm to 2 μm [32]. These EVs are released by dying cells, upon fragmentation of their plasma membrane. When derived from cancer cells, apoEVs are able to influence the phenotype of surrounding tumor cells to promote their survival and aggressiveness, through the transfer of spliceosome components that can alter mRNA splicing in recipient cells [33].

3. Content of Extracellular Vesicles and Functional Role in NSCLC

EVs are lipid-bilayer containers of a variety of biologically active signaling molecules, proteins, nucleic acids, functional lipids, and metabolites that are scheduled for secretion in the extracellular space in response to the microenvironment (Figure 1). The cargos associated with EVs can also reflect the pathological state of a cell. Vesiclepedia represents a consistent effort to create a compendium of EV contents, by experimental update from the biomedical community [7]. So far, this manually-curated database has provided constant annotation of the diversity of EV assortments, including the top 100 EV-associated proteins, data obtained from over 38 thousands RNA entries, and more than 600 metabolites identified in EVs.

Cargoes are considered the first regulators of EV formation, when the future fate of a particular vesicle is being determined [10]. Hence, potential cargoes that are enriched in the forming vesicle are targeted at the site of EV production, following a stepwise mechanisms that recruit sorting machineries that will be specific for what type of EVs are going to be produced, whether they are exosomes or microvesicles [27,34]. For example, augmented cellular expression of some small RNA species will probably stimulate the production of exosomes over microvesicles [25]. Size fractionation of small, medium, and large EVs, followed by RNA sequencing, indicated that larger-size or medium-size EVs reflects the protein-coding transcriptome, which is mostly loaded with mRNAs, whereas small EVs, like exosomes, are enriched by small non-coding RNAs [25]. Specifically, two subpopulations of exosomes (namely, small exosomes, Exo-S, and large exosomes, Exo-L) have been identified to be loaded with a different set of miRNAs [35]. These two separate subpopulations, besides having different cargo compositions, are also distinguishable by the superficial distribution of their proteoglycans, suggesting that they might have distinct biological functions and specific types of target cells [36]. Advancement in the isolation techniques have pointed out the fact that functionality of the different types of EVs is indeed mirrored by their composition, in contrast to traditional methods that yielded a mixture of vesicular and non-vesicular particles enriched in exosomes and almost free of larger EVs. Most significantly, contrary to what was generally previously thought, extracellular DNA is not associated with exosomes, but could instead be easily co-purified with the small EV fraction during standard isolation protocols while, revised high-resolution isolation protocols of density gradients followed by direct immunoaffinity capture of EVs, separate DNA together with the non-vesicular fraction [20]. This DNA is wrapped around histones, and it is possibly released by the cell alongside EVs or during apoptosis. Nevertheless, in the past decade, circulating DNA has been defined as the gold standard for clinical diagnosis by liquid biopsy in cancer, with particular relevance in NSCLC. In fact, although cytosolic DNA can also be released by active processes of secretory autophagy [20], it might not directly reflect the ongoing process of functional differentiation of a cell during oncogenesis or metastatization. The need to find dynamic markers of clinical utility has oriented efforts towards the study of miRNA expression and mutations on the mRNA molecules associated to EVs [6]. In particular, extracellular miRNAs can have a primary role in cell-to-cell communication, gene expression regulation, and cell reprogramming of target cells, as well as the exploration of the mechanisms of their release in the body fluids, assuming that an escalating focus on these molecules in the course of the last 2 to 3 years continues [37]. Out of total cell-free biofluid RNA, small non-coding RNA species, including miRNAs, are the most abundant in circulation, and are co-purified with both small EVs and non-EV particles [20,38]. A fraction of the functional miRNAs found into the blood are thought to be passively released by dying cells as complexes with RNA-binding proteins (RBPs) or encapsulated within apoptotic bodies [39,40]. Once outside in the extracellular space, free miRNA are only stable when bound to RBPs, like Argonaute2 (AGO2) [39]. Protein-RNA complexes are necessary to protect circulating miRNAs from plasma RNAases, and AGO2, the same effector of miRNA-mediated silencing [41], can also guide the release in the circulation of a bunch of cellular non-EV associated miRNAs with active biological functions [42]. Cell-released miRNAs can also be transported in the plasma and delivered to recipient cells by other non-vesicular particles, like high-density lipoprotein (HDL) complexes and exomeres [38,43]. However, most of the cell-released miRNAs involved in the direct silencing of cellular mRNAs are selectively delivered to target cells by loading within small EVs [20,38]. The miRNAs destined to be exported possess a conserved recognition motif, called the EXOmotif, which guides the binding to a series of specialized RBPs [44,45]. The selective binding of these RBPs is necessary for miRNA sorting into EVs, and coordinate the intracellular and extracellular miRNA distribution [46]. Hence, only a specific subpopulation of the total miRNAs are actively secreted outside the cell, and the relative abundance of miRNAs to be directed into the EVs is further controlled by dynamic post-transcriptional modifications [47]. Transcriptome profiling of EVs shows that miRNAs are actually a major component of the EV-RNA cargo, along with a minority of other small RNAs and protein-coding and long-noncoding transcripts [48]. Cancer-cell exosomes can even produce miRNAs cell-independently, by performing the entire process of miRNAs maturation from precursor miRNAs (pre-miRNAs), via their own Dicer and AGO2. These exosomes, when isolated from the blood of cancer patients, can alter the target cell transcriptome in a canonical Dicer-AGo2-dependent miRNA-mediated gene silencing, and promote oncogenic transformation of normal epithelial cells [49]. Single-EV analysis reveals that there is a strong heterogeneity of protein and RNA assortments among the EVs [25]. EVs that are in the exosome-range size are less heterogeneous in their protein content, compared to larger vesicles, but may be enriched in distinct miRNA populations [25,36]. Interestingly, cell-free non-vesicular ribonucleoprotein (RNP) particles also can work in the transport of small RNAs that are different from those contained in the exosomes, and these mostly include tRNAs [25].

Isolation of EVs from plasma of early stage NSCLC patients and subsequent miRNA-seq profiling, can define, with increased diagnostic accuracy, patient-specific subsets of miRNAs that are more expressed in lung cancer patients, compared with healthy individuals [50,51]. In fact, some miRNAs, which can work as strong biomarkers to predict disease progression, are selectively loaded into the EVs of NSCLC patients [52]. The composition of EV-incorporated miRNAs distinguish a fraction of miRNAs that is different from the cytoplasm pool, and can have a specific role in cancer development [46]. Classical kinase pathways can also regulate miRNA sorting into EVs to sustain cell proliferation of lung cancer cells [46]. The trafficking of potential oncogenic miRNAs between the cytosol and EVs is controlled by cellular signaling and the protein AGO2 is pivotal in determining which miRNA has to be sorted into EVs [42,53]. For example, during oncogenic activation of the mitogen-activated protein kinase (MAPK) pathway, downstream to mutant RAS GTPase, the interaction of AGO2-binding miRNAs with EVs is suppressed, while it favors AGO2-independent EV sorting of distinct subsets of miRNAs [54]. The majority of miRNAs with a functional role in cancer development are indeed loaded into exosomes, and cancer exosomes are specifically enriched in oncogenic miRNAs [25,49,55]. This property comes along with an empowered diagnostic potentiality when compared to whole-blood recovery miRNA profiles. Possibly, the miRNAs that are transported by EVs of cancer cells can directly prepare the metastatic niche by promoting drug resistance and immune escape. On the other hand, non-EV circulating miRNA might be released to follow different routes of regulation association with RNP for secretion, and provide a different contribution to the metastatic process. In this context, AGO2, which is a major regulator of miRNA activity, could be pivotal in orchestrating the interplay between the cellular localization and the EV-mediated extrusion of miRNA, while protecting those miRNAs that enter the circulation through passive mechanisms [39].

4. Role of Extracellular Vesicles in Drug Resistance

4.1. Immunomodulation, Resistance to Immune Checkpoint Inhibitors, and Premetastatic Niche Formation Caused by Extracellular Vesicles

Besides the functional delivering of their molecular content to recipient cells, EVs can be seen as signalosomes of several cellular processes, from the activation of cell surface receptor pathways or the transfer of membrane-associated molecules, including immunogenic antigens [56]. De facto, EV production originally evolved as a major communication mechanism during the regulation of the immune system. Numerous studies have shown that EVs can exert most of the functions as their parental cell by either working as primary effectors of T cell activation, with a direct role in antigen presentation and transfer of MHC antigen-bound molecules, or by inhibiting the immune cell function [57]. The first examples were EVs released by B cells and dendritic cells that were able to activate antigen presentation on T cells, inducing a specific immune response [58]. Soon after arrived the first demonstration showing that EVs can carry tumor-derived antigens and stimulate specific cytotoxic activity against the tumor [59]. EVs maintain the same topology of the antigen-presenting cell (APC) of origin, exposing the MHC-antigen complex at their surface. Optimal T cell activation occurs when EVs transfer the MHC-antigen complex in conjunction with the action of co-stimulatory molecules that are provided by the APC and are absent on the EV [60]. Classical tetraspanins, including the CD9, CD63, and CD81 are very important in the antigen recognition process [61]. Beyond their role in regulating EV formation, tetraspanins are pivotal in modulating immune-signaling complexes, as cell-surface co-stimulatory molecules, by facilitating the lateral positioning of MHC molecules and receptor-ligand interaction [62]. Co-engagement of CD81 and the antigen-receptor complex is fundamental for the activation of both T cells and B cells. Similarly, CD81 expressed on EVs derived from the same APC is involved in antigen presentation [63].

Induction of immunosuppression is intimately related to oncogenic progression through the development of adaptive mechanisms that bypass inhibitory action of proliferative pathways. In a context of active immune surveillance, there is a selective pressure for cancer cells to either downregulate neoantigen presentation or induce an immunosuppressive microenvironment [64]. In the past 15 years, there has been a drive to improve outcomes of NSCLC patients by shifting from traditional chemotherapy to advanced targeted therapies and immunotherapies. Specifically, most of the efforts have concentrated on the development of antibodies, like nivolumab, ipilimumab, and pembrolizumab, which could block the action of so-called immune checkpoint regulators [65]. Targeted immune checkpoints are the programmed cell death 1 (PD-1) and the cytotoxic T lymphocyte antigen 4 (CTLA-4), both expressed on T cells, while PD-L1 is produced by tumor cells [65].

During therapy, immune escape in NSCLC evolves through a multistep process visible at the level of DNA mutations in MHC genes, epigenome, and changes in RNA expression of immune regulatory genes [64,66]. Acquired resistance to immunotherapy by tumor cells is directly reflected in their EV production [67]. Most importantly, tumors release a continuous amount of immunomodulatory EVs, which can become massive during cytotoxic treatment [68]. Cancer cells can intensify their production of immunosuppressive EVs during the onset of resistance to anti-PDL1 immunotherapy [69]. Tumor-antigen-carrying EVs might be the vehicle by which the immune system spots the presence of cancer cells, and the parallel screening for antigenic EV could bring a great promise for the diagnosis of immune responsive tumors [67]. At the same time, tumor-derived EVs can be the conduit for immune surveillance escaping through cancer cells and for the failure of immunotherapies [70]. Cross-presentation of tumor-specific antigens by tumor-derived EVs can be downregulated, while suppression of T cell activation is often promoted through overexpression of PD-L1, which interacts with PD-1 receptor on T cells to elicit the immune checkpoint response. During treatment with immune checkpoint inhibitors, tumor cells evade immune response by upregulating the surface expression of PD-L1 and releasing EVs that expose PD-L1 to suppress the function of CD8 T cells and facilitate tumor growth in vivo [71]. Presence of interferon-γ (INF-γ) in the microenvironment stimulates the production of these immunosuppressive EVs and, in patients, circulating levels of EV PD-L1 that increase parallel to the levels of INF-γ, and can be used to stratify clinical responders to the PD-1 inhibitor pembrolizumab and non-responders [71]. PD-L1 is present on the surface of EVs isolated from the plasma of NSCLC patients, and the number of PD-L1-positive EVs correlates with PD-L1 expression level in the tumor tissue from the same patient [69]. PD-L1-expressing EVs that are derived from NSCLC cells can induce apoptosis of T lymphocytes, promoting tumor growth in mice. In this context, it seems that these EVs inhibit the secretion of INF-γ from Jurkat T cells to maintain a circuit of immune inactivation [69]. Tumor-derived EVs are also capable of inducing the expression of PD-L1 on monocytes, via transfer of tumor-cell specific noncoding RNAs, to facilitate immune escape accompanied by concurrent release of cytokines that contribute to cancer-related inflammation [72]. EVs that are isolated from the plasma of NSCLC patients can also contain the mRNA of PD-L1, and the number of copies of the vesicular PD-L1 mRNA per ml of blood associates with a response to the anti-PD-1 antibodies nivolumab and pembrolizumab [73]. The expression of PD-L1 mRNA in plasma EVs, collected before and after surgery of NSCLC patients with stage II to stage III tumors and from NSCLC stage IV patients collected before and after anti-PD-1/PD-L1 therapy, shows a high correlation with PD-L1 expression in the corresponding lung tumor tissues, decreasing significantly after surgery and after anti-PD-1/PD-L1 therapy [74].

Every tumor that has grown to a considerable size has also somehow escaped immune surveillance, which is the prerequisite for cancer cells to colonize distant sites. Tumor EVs might trigger adaptive immune responses or suppress inflammation, depending on the status of a particular immune cell, shaping in this way the evolution of metastases under the selective pressure of the immune system. A primary tumor can seed out immunosuppressive EVs to remotely prepare the metastatic niche for the homing of cancer cells. Fusion of EVs with the recipient plasma membrane is regulated by the presence of specific docking proteins, receptors, and integrins. Vesicles released by cancer cells that have evaded the immune system can be targeted to favorite organs and, given the integrin pattern they present on the surface, promote the premetastatic niche of the specific cancer-cell type from which they have originated [30,75]. Organotropic metastasis formation in the lung, for example, is promoted by expression on the EV surface of the integrin $\alpha6\beta4$, while EVs displaying the integrin $\alpha V\beta5$ are directed to the liver [30]. Once seeded at the metastatic niche, tumor EVs may engage the interferon pathway of normal cells, allowing further uptake of malignant EVs [76]. Subsequent internalization and delivery of internal cargos are typically followed by activation of intracellular signaling, which could be pivotal in determining the oncogenic transformation of target cells, and make the organ site hospitable to implantation and growth of cancer cells from the primary tumor [23,77]. Tumor-cell EVs are able to induce oncogenic transformation and anchorage-independent growth of healthy fibroblasts by carrying fibronectin or laminin that alter the intracellular signaling of membrane integrins [23]. The distant microenvironment can be further influenced by the release of immunosuppressive cytokines caused by tumor-derived EV [67].

The metastatic propensity of a primary tumor can also be inferred by the analysis of the vesicular RNA cargo [12]. A multitude of miRNAs that are transferred via EVs have been shown to have a potential role in anti-tumor immunity [78]. These miRNAs can exert key immune-modulatory and pro-oncogenic functions by targeting the mRNAs of T cell receptors [79], or by downregulating the expression of genes involved in the inflammatory response [80]. Moreover, miRNAs transported by tumor-derived EVs, like those secreted by lung cancer cells in hypoxic conditions, can induce infiltration of pro-tumorigenic macrophages via transfer of miR-103a that lead to subsequent downregulation of the phosphatase and tensin homolog (PTEN) mRNA levels [81]. Recently, characteristic EV-miRNA signatures of cancer-related inflammation have been associated with prediction of response to anti PD-1/PD-L1 therapy in NSCLC [82]. Hence, tumor-derived EVs can prepare the organ-specific metastatization of the cancer cell type they have originated from by inducing local and distant suppression of the immune system function, and expression profiling of integrins on the EV surface might reflect metastatization propensity to preferential organs. As well as being a carrier of antigens and immunosuppressive molecules, tissue-targeted EVs also constitute a mechanism of horizontal propagation of RNA-delivered information from APC to immune system cells (Figure 2).

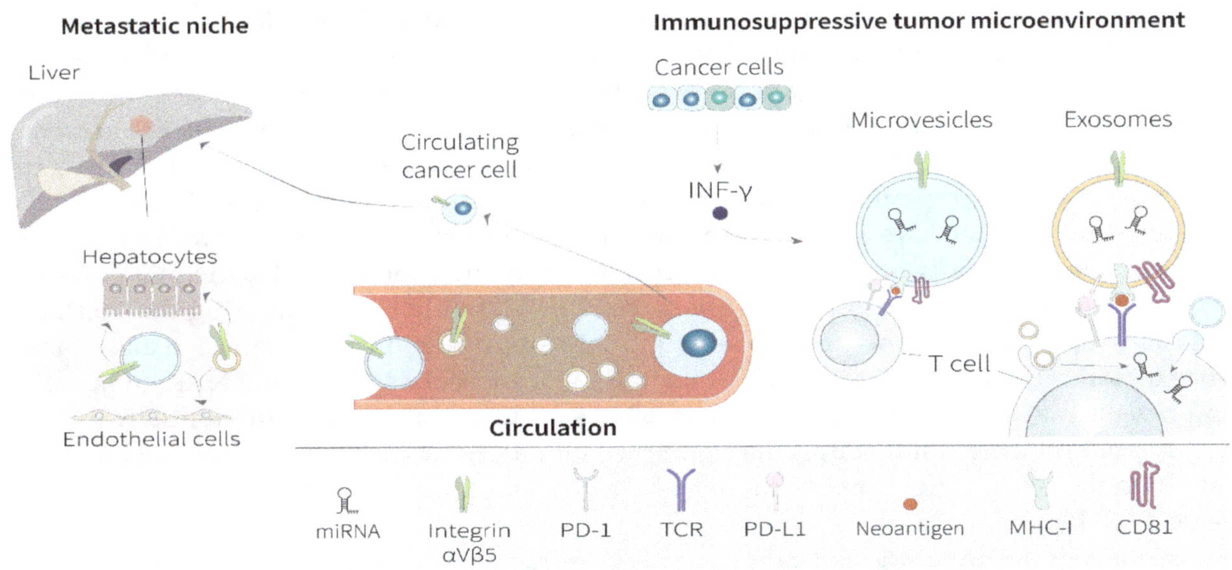

Figure 2. Functional relevance of extracellular vesicles in the oncogenesis of non-small cell lung cancer. In the contexts of non-small cell lung cancer (NSCLC), extracellular vesicles (EV)s, including both microvesicles and exosomes, can be regarded as cell signalosomes. EVs are able to influence the tumor microenvironment and modulate the immune response through downregulation of the T-cell receptor (TCR) activation by the programmed cell death ligand 1 (PD-L1), during neoantigen presentation via the major histocompatibility complex of class I (MHC-I) and co-engagement of the tetraspanin CD81. Production of immunosuppressive EVs in NSCLC patients is favored by circulating interferon-γ (INF-γ) in the tumor microenvironment. Tumor-derived EVs can display specific integrin patterns that target them to favorite organs, where they can influence the organotropism of the premetastatic niche. For example, organotropic metastasis formation of cancer cells in the liver is promoted by EV-expression of αVβ5 integrins. EV cross-talk with the surrounding stroma enables the endothelium to be more permissive for the extravasation of circulating cancer cells to further promote organ-specific invasion. Besides, delivery of the intraluminal microRNA (miRNA) EV cargo into recipient T cells, by membrane fusion or endocytosis, can have a major role in suppressing the anti-tumor inflammatory response.

4.2. Involvement of Extracellular Vesicles in Resistance to Targeted Therapy

With the increasing knowledge of molecular biology and genetics of tumors, the research and clinical application of targeted therapy has become a hot topic, with the aim of improving prognosis and guiding therapeutic decision by understanding the emerging of therapeutic resistance [5]. Each single lung cancer is characterized by an average of 300 DNA mutations, while only a few genes are able to promote tumorigenesis [83]. The two main NSCLC drivers are EGFR and ALK, along with ROS1 and the met proto-oncogene (c-MET), and other genes that are less frequently altered [4,84]. EGFR tyrosine kinase inhibitors (TKIs), including gefitinib, erlotinib, afatinib, and osimertinib, typically improve time to progression, response rates, and overall survival in EGFR-mutated patients, but acquired resistance to anti-EGFR TKIs is inevitable [85]. For this reason, improvement of the most effective diagnostic tools is necessary to predict the risk of developing drug resistance and ameliorate the clinical management of NSCLC patients. The clonogenic selection of the T790M mutation in the EGFR gene is the most frequent and well-known mechanism of resistance. Many studies have demonstrated the possibility to detect the T790M mutation on circulating tumor DNA (ctDNA) liquid biopsy, although the degree of sensitivity of such methodologies is quite low [86,87]. Similarly, after treatment of ALK-mutated patients with the drugs most commonly used at present, like crizotinib, ceritinib, and alectinib, a series of on-target and off-target resistance mutations may occur, which ctDNA analysis is generally insufficient for detecting [88]. Hence, analysis of tumor-derived EVs has emerged as an alternative approach to ctDNA to evaluate therapeutic effect of targeted therapy. Specifically, it has been reported that exosomal RNA can be a useful biomarker for detecting development of the T790M and other activating EGFR mutations, increasing sensitivity and specificity of traditional ctDNA analysis [89,90]. The importance of tumor EVs as clinical biomarkers is supported by the fact that EVs are directly implicated in the propagation of resistance mechanisms between cancer cells. Recently, the EXTRA study was launched with the aim to identify novel predictive biomarkers of resistance to afatinib by analyzing circulating and encapsulated biomarkers in peripheral blood [91]. EVs can be a vehicle to the onset of therapeutic resistance through the transfer of oncogenic miRNAs. For example, gefitinib-resistant cell lines are able to release exosomes enriched in specific miRNAs that are able to confer resistance phenotypes to recipient cells. In particular, the miR-214, the lncRNA H19, and the miR-21 are associated with gefitinib resistance [92–94], while EVs containing the lncRNA RP11 838N2.4 can propagate resistance to erlotinib [95]. ALK-rearranged cancer cells can also transfer the resistance to anti-ALK TKIs by releasing EVs containing a subset of specific miRNAs. In fact, EV-RNA profiling reveals that miR-21-5p, miR-486-3p, lncRNAs MEG3, and XIST are differentially expressed in the EVs secreted by the resistant subclones, and the circulating levels of these EV-associated miRNAs correlate with disease progression of EML4-ALK-translocated lung adenocarcinoma patients treated with ALK-TKIs [96].

EVs, found in the circulation of cancer patients refractory to therapy, can also transport nucleic acids molecules of wild-type EGFR and mutated EGFR that reflect the genetic signature of the original tumor [97]. However, it is know that cancer cell-derived EVs can reprogram quiescent cells towards a pro-tumorigenic phenotype more efficiently by direct transfer of cellular oncogenic protein kinases [98]. Cancer cells use EVs to transfer oncogenic ALK to normal cell in the surrounding tumor microenvironment and confer resistance to neighboring drug-sensitive cancer cells, via activation of the MAPK pathway [99]. Likewise, NSCLC cells can modify the function of adjacent cells by the exchange of EVs that transport constitutively-activated EGFR [100]. Interestingly, transduced EGFR translocate into the nucleus by a mechanism that is dependent on EV fusion with the nuclear envelope, rather than specific nuclear localization signals [101]. In the nucleus, EGFR is capable to activate pathways that are associated with tumor resistance to therapy-induced DNA damage and anti-EGFR treatment, in a mechanism that is independent of its conventional signaling from the plasma membrane [102,103]. Malignant cells can also share EVs that contribute to horizontal transfer of EGFR oncogenic isoforms, such the EGFRvIII, found in most aggressive tumors, and induce oncogenic activity downstream to MAPK and morphological progression towards a more invasive phenotype [97,104]. From the clinical perspective, detection of circulating EGFR protein from NSCLC patients would have important implications in defining reliable biomarkers of clinical prediction of anti-EGFR therapy [105]. Now, information derived from patients' liquid biopsy might be finally complemented with the most advanced techniques of EV purification, along with accurate analysis of EV protein content nucleic acid [106].

5. Detection of Extracellular Vesicles and Their Application as Clinical Biomarkers for NSCLC

Over the past five years, explorative approaches for integrating liquid biopsy into the management of most common cancer types have been developed at a rapid rate [107], with over 360 clinical trials going on, including 32 studies focusing on EVs, of which more than 10% focus on lung cancer (https://clinicaltrials.gov/). Liquid biopsy of cancer patients can provide a comprehensive overview of tumor heterogeneity to help with diagnosis and guide patient care. Regulatory agencies in Europe and the United States have approved liquid biopsy tests in clinical routine for the detection of EGFR mutations on ctDNA in plasma of patients with NSCLC [108], with the aim to expand the diagnostic application of liquid biopsy to other NSCLC oncogenes, including ALK and ROS1 [87]. Although the analysis of ctDNA is currently the gold standard for routine diagnostics of NSCLC, the analysis of circulating EVs has emerged as a potential complementary methodology for implementing the detection sensitivity of drug-resistance mutations and tracking of tumor-specific alterations during disease progression and in the course of therapy [109].

Despite the substantial technical advances, the complexity associated with the enrichment of single or multiple EV subtypes with different composition poses continuous challenges to avoiding method-dependent bias. Methods for isolation from body fluids and analysis of EVs comprise separation on microfluidics, by density gradients, or through size exclusion chromatography, which can be followed by immunoprecipitation for further enrichment of specific subpopulations of EVs [110–112]. Combination of western blot, nanoparticle tracking, and flow cytometry is usually used to verify EV purity, and can be completed by using proteomics, lipidomics, RNA/DNA analysis to study EV composition [110]. The scientific community has agreed on the urgent need to standardize the procedure of EV isolation from patients' liquid biopsies, validate EV-associated biomarkers, and put experimental guidelines into practice [7,113,114]. Moreover, highly stringent and novel isolation methodologies now provide a reappraisal of what classifies the variety of circulating nanoparticles, distinguishing vesicular and non-vesicular components, rather than exosomes or larger microvesicles [20]. The study of the functionality of these revised subpopulations avoided many controversies derived from non-specific isolation techniques and raised further interest in using selected subpopulations of EVs for successful biomarker analysis. However, EV analysis would be most beneficial in diagnostics if the procedure can specifically identify which antigens should be targeted in order to separate cancer-derived EVs from normal host cell-derived EVs.

Clinical application of EV-based diagnostics in NSCLC can be particularly helpful in the detection of oncogene mutations that confer sensitivity to specific drugs or are selected in those patients that develop pharmacological resistance. Tumor-released EVs can also help the early detection of clones that will develop immune escape through the overexpression of PD-L1 or the production of immunosuppressive miRNAs. Moreover, by defining tumor-specific miRNA signatures, EV content analysis can provide direct information on the dynamic transformation of a tumor mass and its clonal outgrowth [38,115]. Composition of small RNA species varies a lot between the human fluids, with a relative abundance of miRNAs in the blood plasma [116]. Mapping the diversity of extracellular RNAs (exRNAs), based on the use of specific RNA isolation methods, has revealed the importance of the related carriers. In fact, the accurate separation of exosomes and microvesicles from non-EV particles can facilitate the use of exRNA, which are present at variable concentrations depending on the body fluid analyzed, as powerful biomarkers [38,40]. Experimental data clearly show that most commercial kits preferentially enrich AGO2-bound miRNAs and only partially or not at all enrich EVs, whereas there is a need to apply methodologies that pre-isolate EVs prior to RNA extraction to specifically identify EV-associated exRNA [40]. Based on the evidence that blood levels of EV miRNAs significantly increases during cancer conditions, EV-associated miRNAs are now considered to be a most promising biomarker for the blood-based early detection of cancer [117]. In fact, when included into EVs, cell released miRNAs are rather stable and easy-to-detect [118], and, although most of the analysis on circulating miRNA is still exploratory, blood-based RNA profiling of cancer patients has provided a valuable source of information to understand tumor's evolution and adaptation to therapeutic pressure in many clinical studies [114].

Compared to exosomes, which clearly show an enrichment of specific miRNA contents that are distinct from the cellular pool, microvesicles more closely reflect the source cell transcriptome, being particularly enriched in mRNA sequences [25]. Larger vesicles can function as shuttles between two cells for the transfer of protein-translating mRNAs [11,119]. These mRNAs could become a complementary genetic material of easy access for molecular diagnostics of oncogene mutations. For example, analysis of vesicles-associated nucleic acids for common B-Raf proto-oncogene (BRAF), KRAS Proto-Oncogene (KRAS), and epidermal growth factor receptor (EGFR) mutations has shown higher sensitivity compared to liquid biopsies of plasma ctDNA in association to clinical outcomes of patients with NSCLC [120]. EVs isolated from plasma of NSCLC patients can be used for EGFR genotyping through sequential liquid biopsies during treatment for the detection drug-resistance mutations, like the pT790M, demonstrating improved concordance with testing of tumor tissue compared to a conventional liquid biopsy of ctDNA [121].

To date, most available protocols to recover EVs from patients' blood resulted in the recovery of a heterogeneous population of vesicles of uncertain origin, but the implementation of more robust technologies have facilitated the isolation of specific subpopulations of as few as ten EVs per microliter of plasma with direct impact on the performance of revealing predictive biomarkers [111]. Novel approaches have for the first time enabled the identification of cancer-derived EVs through a specific immunocapture method that targets membrane proteins of the tumor tissue of origin, separating cancer EVs from the bulk of background blood EVs [122,123]. Selective isolation of tumor-associated EVs can also be coupled with direct PCR-based quantification of oncogenic mRNAs following nanodroplet encapsulation of plasma EVs, and dramatically improve the sensitivity potential of liquid biopsy of BRAF and KRAS mutations [123].

As a vehicle of extracellular RNA, EVs could be a considered as an easily accessible reservoir of biomarkers discovery (Figure 3) and molecular analysis of the different EV contents can provide complementary clinical information (Table 1).

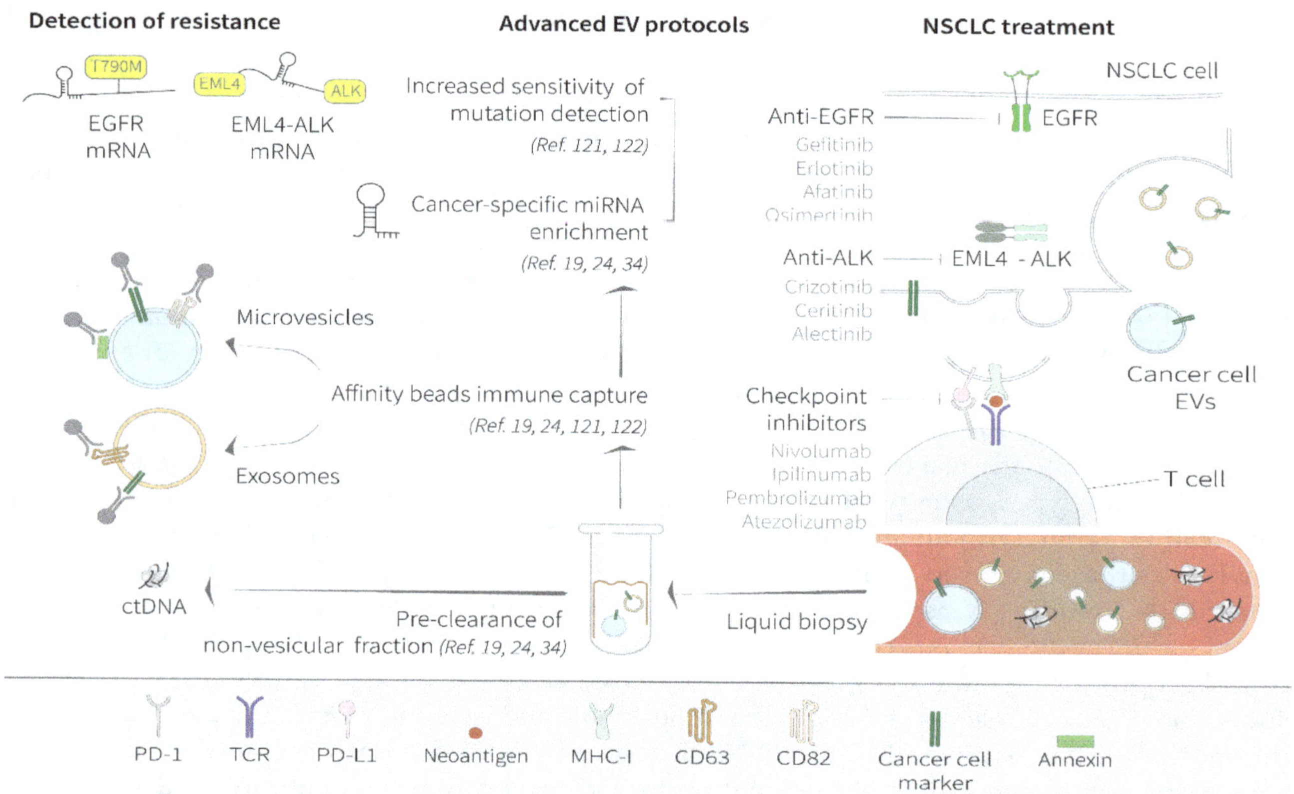

Figure 3. Significance of cancer cell-derived extracellular vesicles as biomarkers of drug resistance in non-small cell lung cancer. Conventional therapies of non-small cell lung cancer (NSCLC), either implying targeted tyrosine kinase inhibitors (TKIs) or anti-checkpoint antibodies, inevitably place the tumor cells under sustained selective pressure that eventually bring to pharmaceutical resistance in patients. Clonal evolution of drug-resistant cells can be captured by liquid biopsy and the data obtained from dynamic monitoring of tumor changes can be used to guide further therapeutic decisions. Clonal selection can be gene specific, by occurring through hotspot mutations in the therapy-targeted genes, for example in the anaplastic lymphoma kinase (ALK), typically fused with the echinoderm microtubule-associated protein-like 4 (EML4), and in the epidermal growth factor receptor (EGFR).

Resistance to treatment in NSCLC can also be mediated by overexpression of the programmed cell death 1 (PD-1) receptor and its ligand, the programmed cell death ligand 1 (PD-L1), to elicit the immune checkpoint response. Currently, the molecular profile of lung cancers in liquid biopsy is obtained by using circulating tumor DNA (ctDNA) as a primary tumor marker. However, substantial improvements of the molecular approaches have been made to assess tumor heterogeneity through the analysis of tumor-released extracellular vesicles (EVs) from liquid biopsy. Revised differential separation protocols are often based on advanced microfluidics, immune capture of specific EV tetraspanins, affinity beads, and EV enrichment by antibody-based selection of surface cancer markers. Specific miRNA cargo can be enriched in subtypes of EVs with distinct biological functions and can be potentially used as biomarkers of oncogenic progression. For example, analysis of EV-associated mRNA might be particularly advantageous for the detection of mutations, like the substitution T790M in resistant NSCLC patients, and gene fusions, like the EML4-ALK in NSCLC, as well as splicing variants and changes in gene-expression profiles. In comparison to ctDNA molecules, which normally present two copies for cancer cell of origin, mRNA likely yields from highly expressed genes, occurring in thousands of copies per cells, and can be shed into the circulation within EVs at massive concentrations. At the moment most of EV-based analysis of circulating miRNA and mRNA remain exploratory and necessitate further clinical validation. miRNA, microRNA; mRNA, messenger RNA; TCR, T-cell receptor; MHC, major histocompatibility complex.

Table 1. Comparison of the main biomarkers derived from extracellular vesicles for clinical application in non-small cell lung cancer.

EV Content	Potential Utility as NSCLC Biomarker
mRNA	gene expression, driving/resistance mutations, splice variants, gene fusions, tumor heterogeneity
miRNA	immune-suppressive signature
lncRNA	drug-resistance response
Protein	PD-L1 expression, neoantigen expression, patient MHC haplotyping

EV, extracellular vesicle; NSCLC, non-small cell lung cancer; mRNA, messenger RNA; miRNA, microRNA; lncRNA, long non-coding RNA; PD-L1, programmed cell death ligand 1; MHC, major histocompatibility complex.

Improvement of efficient and cost-effective methodologies for separation of exosomes and microvesicles population from the background of non-EV particles, based on their surface proteome composition, can foster the ultrasensitive detection of cancer-specific mutations on mRNA that could implement classical ctDNA blood analysis [36,124]. At the same time, construction of reference collections of vesicular RNAs that include predictive biomarkers for immunotherapy and targeted therapy will serve as a shared fundamental resource for future application of exRNAs [114,125].

6. Conclusions

In the wake of assigning an integrative view of the heterogeneity of resistance mechanisms, it is mandatory for to have an implementation of protocols that can assess the ongoing patient mutational status almost real time for the allocation of the specific therapies. Liquid biopsy poses as the best choice for the non-invasive monitoring of emerging resistance mechanisms at different time points, (i.e., at baseline and during treatment), which cannot be captured by conventional tissue biopsy. In most cases, measurement of ctDNA has permitted to recover substantial information about the onset of resistance mutations in NSCLC, even much before clinical progression. To achieve the highest grade of sensitivity, however, it is understandable that the measurement of ctDNA must be incorporated with more deviceful techniques. The discovery of the ubiquitous presence of EVs in the body fluids, coupled with the development of ever more reliable and low-cost technologies for the isolation of EVs and the implementation in the purification of their content promise to provide a substantial improvement in the detection of tumor markers, from routine blood draws. Emission of immunosuppressive organ-specific EVs from the primary tumor can be a precondition to prepare distant sites for metastatic invasion, and sensitive methodologies for tumor EV isolation and analysis can constitute a new frontier of cancer diagnostic to predict the metastatic spread and therapy outcome. Several challenges should be overcome for a possible implementation of EVs in the clinical practice, with the first and foremost challenge being the standardization of the methodologies for EVs isolation and purification from plasma. These methodologies should be easily usable by molecular diagnostic laboratories, with a low turnaround time and a low cost.

References

1. Singh, S.S.; Dahal, A.; Shrestha, L.; Jois, S.D. Genotype driven therapy for non-small cell lung cancer: Resistance, pan inhibitors and immunotherapy. In *Current Medicinal Chemistry*; Bentham Science Publishers: Sharjah, UAE, 2019; p. 26.
2. Halliday, P.R.; Blakely, C.M.; Bivona, T.G. Emerging targeted therapies for the treatment of non-small cell lung cancer. *Curr. Oncol. Rep.* **2019**, *21*, 21. [CrossRef]
3. Shun, L.; Xia, L.; Liu, Y.; Wang, Y. PD-1/PD-L1 Blockade Therapy in Advanced Non-Small-Cell Lung Cancer: Current Status and Future Directions. *Oncologist* **2019**, *24* (Suppl. 1), S31–S41.
4. Abdel Karim, N.; Kelly, K. Role of targeted therapy and immune checkpoint blockers in advanced non-small cell lung cancer: A review. *Oncologist* **2019**, *24*, 1270–1284. [CrossRef] [PubMed]
5. Rotow, J.; Bivona, T.G. Understanding and targeting resistance mechanisms in NSCLC. *Nat. Rev. Cancer* **2017**, *17*, 637–658. [CrossRef] [PubMed]
6. Siravegna, G.; Marsoni, S.; Siena, S.; Bardelli, A. Integrating liquid biopsies into the management of cancer. *Nat. Rev. Clin. Oncol.* **2017**, *14*, 531–548. [CrossRef] [PubMed]
7. Pathan, M.; Fonseka, P.; Chitti, S.V.; Kang, T.; Sanwlani, R.; Van Deun, J.; Hendrix, A.; Mathivanan, S. Vesiclepedia 2019: A compendium of RNA, proteins, lipids and metabolites in extracellular vesicles. *Nucleic Acids Res.* **2019**, *47*, D516–D519. [CrossRef] [PubMed]
8. Théry, C.; Witwer, K.W.; Aikawa, E.; Alcaraz, M.J.; Anderson, J.D.; Andriantsitohaina, R.; Antoniou, A.; Arab, T.; Archer, F.; Atkin-Smith, G.K.; et al. Minimal information for studies of extracellular vesicles 2018 (MISEV2018): A position statement of the International Society for Extracellular Vesicles and update of the MISEV2014 guidelines. *J. Extracell. Vesicles* **2018**, *7*, 1535750. [CrossRef]
9. El Andaloussi, S.; Mäger, I.; Breakefield, X.O.; Wood, M.J.A. Extracellular vesicles: Biology and emerging therapeutic opportunities. *Nat. Rev. Drug Discov.* **2013**, *12*, 347–357. [CrossRef]
10. Van Niel, G.; D'Angelo, G.; Raposo, G. Shedding light on the cell biology of extracellular vesicles. *Nat. Rev. Mol. Cell Biol.* **2018**, *19*, 213–228. [CrossRef]
11. Valadi, H.; Ekström, K.; Bossios, A.; Sjöstrand, M.; Lee, J.J.; Lötvall, J.O. Exosome-mediated transfer of mRNAs and microRNAs is a novel mechanism of genetic exchange between cells. *Nat. Cell Biol.* **2007**, *9*, 654–659. [CrossRef]
12. Syn, N.; Wang, L.; Sethi, G.; Thiery, J.P.; Goh, B.C. Exosome-mediated metastasis: From epithelial-mesenchymal transition to escape from immunosurveillance. *Trends Pharmacol. Sci.* **2016**, *37*, 606–617. [CrossRef] [PubMed]
13. Boriachek, K.; Islam, M.N.; Möller, A.; Salomon, C.; Nguyen, N.T.; Hossain, M.S.A.; Yamauchi, Y.; Shiddiky, M.J.A. Biological functions and current advances in isolation and detection strategies for exosome nanovesicles. *Small* **2018**, *14*, 1702153. [CrossRef] [PubMed]
14. Johnstone, R.M.; Adam, M.; Hammond, J.R.; Orr, L.; Turbide, C. Vesicle formation during reticulocyte maturation. Association of plasma membrane activities with released vesicles (exosomes). *J. Biol. Chem.* **1987**, *262*, 9412–9420. [PubMed]
15. Willms, E.; Johansson, H.J.; Mäger, I.; Lee, Y.; Blomberg, K.E.M.; Sadik, M.; Alaarg, A.; Smith, C.I.E.; Lehtiö, J.; El Andaloussi, S.; et al. Cells release subpopulations of exosomes with distinct molecular and biological properties. *Sci. Rep.* **2016**, *6*, 22519. [CrossRef]
16. Baietti, M.F.; Zhang, Z.; Mortier, E.; Melchior, A.; Degeest, G.; Geeraerts, A.; Ivarsson, Y.; Depoortere, F.; Coomans, C.; Vermeiren, E.; et al. Syndecan-syntenin-ALIX regulates the biogenesis of exosomes. *Nat. Cell Biol.* **2012**, *14*, 677–685. [CrossRef]

17. Colombo, M.; Moita, C.; Van Niel, G.; Kowal, J.; Vigneron, J.; Benaroch, P.; Manel, N.; Moita, L.F.; Théry, C.; Raposo, G. Analysis of ESCRT functions in exosome biogenesis, composition and secretion highlights the heterogeneity of extracellular vesicles. *J. Cell Sci.* **2013**, *126*, 5553–5565. [CrossRef]
18. Théry, C.; Boussac, M.; Véron, P.; Ricciardi-Castagnoli, P.; Raposo, G.; Garin, J.; Amigorena, S. Proteomic analysis of dendritic cell-derived exosomes: A secreted subcellular compartment distinct from apoptotic vesicles. *J. Immunol.* **2001**, *166*, 7309–7318. [CrossRef]
19. Takahashi, A.; Okada, R.; Nagao, K.; Kawamata, Y.; Hanyu, A.; Yoshimoto, S.; Takasugi, M.; Watanabe, S.; Kanemaki, M.T.; Obuse, C.; et al. Exosomes maintain cellular homeostasis by excreting harmful DNA from cells. *Nat. Commun.* **2017**, *8*, 15287. [CrossRef]
20. Jeppesen, D.K.; Fenix, A.M.; Franklin, J.L.; Higginbotham, J.N.; Zhang, Q.; Zimmerman, L.J.; Liebler, D.C.; Ping, J.; Liu, Q.; Evans, R.; et al. Reassessment of exosome composition. *Cell* **2019**, *177*, 428–445. [CrossRef]
21. Villarroya-Beltri, C.; Baixauli, F.; Mittelbrunn, M.; Fernández-Delgado, I.; Torralba, D.; Moreno-Gonzalo, O.; Baldanta, S.; Enrich, C.; Guerra, S.; Sánchez-Madrid, F. ISGylation controls exosome secretion by promoting lysosomal degradation of MVB proteins. *Nat. Commun.* **2016**, *7*, 13588. [CrossRef]
22. Jahn, R.; Scheller, R.H. SNAREs—Engines for membrane fusion. *Nat. Rev. Mol. Cell Biol.* **2006**, *7*, 631–643. [CrossRef] [PubMed]
23. Li, B.; Antonyak, M.A.; Zhang, J.; Cerione, R.A. RhoA triggers a specific signaling pathway that generates transforming microvesicles in cancer cells. *Oncogene* **2012**, *31*, 4740–4749. [CrossRef] [PubMed]
24. McConnell, R.E.; Higginbotham, J.N.; Shifrin, D.A.; Tabb, D.L.; Coffey, R.J.; Tyska, M.J. The enterocyte microvillus is a vesicle-generating organelle. *J. Cell Biol.* **2009**, *185*, 1285–1298. [CrossRef] [PubMed]
25. Gyuris, A.; Navarrete-Perea, J.; Jo, A.; Cristea, S.; Zhou, S.; Fraser, K.; Wei, Z.; Krichevsky, A.M.; Weissleder, R.; Lee, H.; et al. Physical and molecular landscapes of mouse glioma extracellular vesicles define heterogeneity. *Cell Rep.* **2019**, *27*, 3972–3987. [CrossRef] [PubMed]
26. Rana, S.; Yue, S.; Stadel, D.; Zöller, M. Toward tailored exosomes: The exosomal tetraspanin web contributes to target cell selection. *Int. J. Biochem. Cell Biol.* **2012**, *44*, 1574–1584. [CrossRef]
27. Oh, N.; Park, J.H. Endocytosis and exocytosis of nanoparticles in mammalian cells. *Int. J. Nanomed.* **2014**, *9*, 51–63.
28. Shah, R.; Patel, T.; Freedman, J.E. Circulating extracellular vesicles in human disease. *N. Engl. J. Med.* **2018**, *379*, 958–966. [CrossRef]
29. Ruivo, C.F.; Adem, B.; Silva, M.; Melo, S.A. The biology of cancer exosomes: Insights and new perspectives. *Cancer Res.* **2017**, *77*, 6480–6488. [CrossRef]
30. Hoshino, A.; Costa-Silva, B.; Shen, T.L.; Rodrigues, G.; Hashimoto, A.; Tesic Mark, M.; Molina, H.; Kohsaka, S.; Di Giannatale, A.; Ceder, S.; et al. Tumour exosome integrins determine organotropic metastasis. *Nature* **2015**, *527*, 329–335. [CrossRef]
31. Yoshioka, Y.; Kosaka, N.; Konishi, Y.; Ohta, H.; Okamoto, H.; Sonoda, H.; Nonaka, R.; Yamamoto, H.; Ishii, H.; Mori, M.; et al. Ultra-sensitive liquid biopsy of circulating extracellular vesicles using ExoScreen. *Nat. Commun.* **2014**, *5*, 3591. [CrossRef]
32. Caruso, S.; Poon, I.K.H. Apoptotic cell-derived extracellular vesicles: More than just debris. *Front. Immunol.* **2018**, *9*, 1486. [CrossRef] [PubMed]
33. Shin, Y.J.; Minata, M.; Lee, Y.; Yamaguchi, S.; Komarova, S.; Alsheikh, H.A.; Shender, V.O.; Shakhparonov, M.I.; Anufrieva, K.; Shi, J.; et al. Apoptotic cell-derived extracellular vesicles promote malignancy of glioblastoma via intercellular transfer of splicing factors. *Cancer Cell* **2018**, *34*, 119–135.
34. Ostrowski, M.; Carmo, N.B.; Krumeich, S.; Fanget, I.; Raposo, G.; Savina, A.; Moita, C.F.; Schauer, K.; Hume, A.N.; Freitas, R.P.; et al. Rab27a and Rab27b control different steps of the exosome secretion pathway. *Nat. Cell Biol.* **2010**, *12*, 19–30. [CrossRef] [PubMed]
35. Zhang, Q.; Higginbotham, J.N.; Jeppesen, D.K.; Yang, Y.P.; Li, W.; McKinley, E.T.; Graves-Deal, R.; Ping, J.; Britain, C.M.; Dorsett, K.A.; et al. Transfer of functional cargo in exomeres. *Cell Rep.* **2019**, *27*, 940–954. [CrossRef] [PubMed]
36. Zhang, H.; Freitas, D.; Kim, H.S.; Fabijanic, K.; Li, Z.; Chen, H.; Mark, M.T.; Molina, H.; Martin, A.B.; Bojmar, L.; et al. Identification of distinct nanoparticles and subsets of extracellular vesicles by asymmetric flow field-flow fractionation. *Nat. Cell Biol.* **2018**, *20*, 332–343. [CrossRef] [PubMed]

37. Treiber, T.; Treiber, N.; Meister, G. Regulation of microRNA biogenesis and its crosstalk with other cellular pathways. *Nat. Rev. Mol. Cell Biol.* **2019**, *20*, 5–20. [CrossRef]
38. Murillo, O.D.; Thistlethwaite, W.; Rozowsky, J.; Subramanian, S.L.; Lucero, R.; Shah, N.; Jackson, A.R.; Srinivasan, S.; Chung, A.; Laurent, C.D.; et al. exRNA atlas analysis reveals distinct extracellular rna cargo types and their carriers present across human biofluids. *Cell* **2019**, *177*, 463–477. [CrossRef]
39. Schwarzenbach, H.; Nishida, N.; Calin, G.A.; Pantel, K. Clinical relevance of circulating cell-free microRNAs in cancer. *Nat. Rev. Clin. Oncol.* **2014**, *11*, 145–156. [CrossRef]
40. Srinivasan, S.; Yeri, A.; Cheah, P.S.; Chung, A.; Danielson, K.; De Hoff, P.; Filant, J.; Laurent, C.D.; Laurent, L.D.; Magee, R.; et al. Small RNA sequencing across diverse biofluids identifies optimal methods for exRNA Isolation. *Cell* **2019**, *177*, 446–462. [CrossRef]
41. Leung, A.K.L.; Young, A.G.; Bhutkar, A.; Zheng, G.X.; Bosson, A.D.; Nielsen, C.B.; Sharp, P.A. Genome-wide identification of Ago2 binding sites from mouse embryonic stem cells with and without mature microRNAs. *Nat. Struct. Mol. Biol.* **2011**, *18*, 237–245. [CrossRef]
42. Arroyo, J.D.; Chevillet, J.R.; Kroh, E.M.; Ruf, I.K.; Pritchard, C.C.; Gibson, D.F.; Mitchell, P.S.; Bennett, C.F.; Pogosova-Agadjanyan, E.L.; Stirewalt, D.L.; et al. Argonaute2 complexes carry a population of circulating microRNAs independent of vesicles in human plasma. *Proc. Natl. Acad. Sci. USA* **2011**, *108*, 5003–5008. [CrossRef] [PubMed]
43. Vickers, K.C.; Palmisano, B.T.; Shoucri, B.M.; Shamburek, R.D.; Remaley, A.T. MicroRNAs are transported in plasma and delivered to recipient cells by high-density lipoproteins. *Nat. Cell Biol.* **2011**, *13*, 423–435. [CrossRef] [PubMed]
44. Santangelo, L.; Cicchini, C.; Mancone, C.; Battistelli, C.; Tripodi, M.; Giurato, G.; Tarallo, R.; Weisz, A.; Montaldo, C.; Alonzi, T. The RNA-binding protein SYNCRIP is a component of the hepatocyte exosomal machinery controlling microRNA sorting. *Cell Rep.* **2016**, *17*, 799–808. [CrossRef] [PubMed]
45. Villarroya-Beltri, C.; Gutiérrez-Vázquez, C.; Sánchez-Cabo, F.; Pérez-Hernández, D.; Vázquez, J.; Martin-Cofreces, N.; Martinez-Herrera, D.J.; Pascual-Montano, A.; Mittelbrunn, M.; Sánchez-Madrid, F. Sumoylated hnRNPA2B1 controls the sorting of miRNAs into exosomes through binding to specific motifs. *Nat. Commun.* **2013**, *4*, 1–10. [CrossRef]
46. Statello, L.; Maugeri, M.; Garre, E.; Nawaz, M.; Wahlgren, J.; Papadimitriou, A.; Lundqvist, C.; Lindfors, L.; Collén, A.; Sunnerhagen, P.; et al. Identification of RNA-binding proteins in exosomes capable of interacting with different types of RNA: RBP-facilitated transport of RNAs into exosomes. *PLoS ONE* **2018**, *13*, e0195969. [CrossRef]
47. Hackenberg, M.; Würdinger, T.; Middeldorp, J.M.; Sie, D.; Pegtel, D.M.; Sadek, P.; Ylstra, B.; van Eijndhoven, M.A.J.; de Menezes, R.X.; Koppers-Lalic, D.; et al. Nontemplated nucleotide additions distinguish the small rna composition in cells from exosomes. *Cell Rep.* **2014**, *8*, 1649–1658.
48. Amorim, M.G.; Valieris, R.; Drummond, R.D.; Pizzi, M.P.; Freitas, V.M.; Sinigaglia-Coimbra, R.; Calin, G.A.; Pasqualini, R.; Arap, W.; Silva, I.T.; et al. A total transcriptome profiling method for plasma-derived extracellular vesicles: Applications for liquid biopsies. *Sci. Rep.* **2017**, *7*, 1–11. [CrossRef]
49. Melo, S.A.; Sugimoto, H.; O'Connell, J.T.; Kato, N.; Villanueva, A.; Vidal, A.; Qiu, L.; Vitkin, E.; Perelman, L.T.; Melo, C.A.; et al. Cancer exosomes perform cell-independent microrna biogenesis and promote tumorigenesis. *Cancer Cell* **2014**, *26*, 707–721. [CrossRef]
50. Jin, X.; Chen, Y.; Chen, H.; Fei, S.; Chen, D.; Cai, X.; Liu, L.; Lin, B.; Su, H.; Zhao, L.; et al. Evaluation of tumor-derived exosomal miRNA as potential diagnostic biomarkers for early-stage non-small cell lung cancer using next-generation sequencing. *Clin. Cancer Res.* **2017**, *23*, 5311–5319. [CrossRef]
51. Chen, J.; Hu, C.; Pan, P. Extracellular vesicle microRNA transfer in lung diseases. *Front. Physiol.* **2017**, *8*, 1028. [CrossRef]
52. Silva, J.; García, V.; Zaballos, Á.; Provencio, M.; Lombardía, L.; Almonacid, L.; García, J.M.; Domínguez, G.; Peña, C.; Diaz, R.; et al. Vesicle-related microRNAs in plasma of nonsmall cell lung cancer patients and correlation with survival. *Eur. Respir. J.* **2011**, *37*, 617–623. [CrossRef] [PubMed]
53. Li, L.; Zhu, D.; Huang, L.; Zhang, J.; Bian, Z.; Chen, X.; Liu, Y.; Zhang, C.Y.; Zen, K. Argonaute 2 complexes selectively protect the circulating microRNAs in cell-secreted microvesicles. *PLoS ONE* **2012**, *7*, e46957. [CrossRef] [PubMed]

54. McKenzie, A.J.; Hoshino, D.; Hong, N.H.; Cha, D.J.; Franklin, J.L.; Coffey, R.J.; Patton, J.G.; Weaver, A.M. KRAS-MEK signaling controls ago2 sorting into exosomes. *Cell Rep.* **2016**, *15*, 978–987. [CrossRef] [PubMed]
55. Yuan, T.; Huang, X.; Woodcock, M.; Du, M.; Dittmar, R.; Wang, Y.; Tsai, S.; Kohli, M.; Boardman, L.; Patel, T.; et al. Plasma extracellular RNA profiles in healthy and cancer patients. *Sci. Rep.* **2016**, *6*, 1–11. [CrossRef]
56. Robbins, P.D.; Morelli, A.E. Regulation of immune responses by extracellular vesicles. *Nat. Rev. Immunol.* **2014**, *14*, 195–208. [CrossRef]
57. Veerman, R.E.; Güçlüler Akpinar, G.; Eldh, M.; Gabrielsson, S. Immune cell-derived extracellular vesicles —Functions and therapeutic applications. *Trends Mol. Med.* **2019**, *25*, 382–394. [CrossRef]
58. Raposo, G.; Nijman, H.W.; Stoorvogel, W.; Liejendekker, R.; Harding, C.V.; Melief, C.J.; Geuze, H.J. B lymphocytes secrete antigen-presenting vesicles. *J. Exp. Med.* **1996**, *183*, 1161–1172. [CrossRef]
59. Zitvogel, L.; Regnault, A.; Lozier, A.; Wolfers, J.; Flament, C.; Tenza, D.; Ricciardi-Castagnoli, P.; Raposo, G.; Amigorena, S. Eradication of established murine tumors using a novel cell-free vaccine: Dendritic cell-derived exosomes. *Nat. Med.* **1998**, *4*, 594–600. [CrossRef]
60. Théry, C.; Ostrowski, M.; Segura, E. Membrane vesicles as conveyors of immune responses. *Nat. Rev. Immunol.* **2009**, *9*, 581–593. [CrossRef]
61. Levy, S.; Shoham, T. The tetraspanin web modulates immune-signalling complexes. *Nat. Rev. Immunol.* **2005**, *5*, 136–148. [CrossRef]
62. Saiz, M.L.; Rocha-Perugini, V.; Sánchez-Madrid, F. Tetraspanins as organizers of antigen-presenting cell function. *Front. Immunol.* **2018**, *9*, 1074. [CrossRef] [PubMed]
63. Kurywchak, P.; Tavormina, J.; Kalluri, R. The emerging roles of exosomes in the modulation of immune responses in cancer. *Genome Med.* **2018**, *10*, 23. [CrossRef] [PubMed]
64. Jamal-Hanjani, M.; Wilson, G.A.; McGranahan, N.; Birkbak, N.J.; Watkins, T.B.K.; Veeriah, S.; Shafi, S.; Johnson, D.H.; Mitter, R.; Rosenthal, R.; et al. Tracking the evolution of non–small-cell lung cancer. *N. Engl. J. Med.* **2017**, *376*, 2109–2121. [CrossRef] [PubMed]
65. Hoos, A. Development of immuno-oncology drugs-from CTLA4 to PD1 to the next generations. *Nat. Rev. Drug Discov.* **2016**, *15*, 235–247. [CrossRef] [PubMed]
66. Rosenthal, R.; Cadieux, E.L.; Salgado, R.; Al Bakir, M.; Moore, D.A.; Hiley, C.T.; Lund, T.; Tanić, M.; Reading, J.L.; Joshi, K.; et al. Neoantigen-directed immune escape in lung cancer evolution. *Nature* **2019**, *567*, 479–485. [CrossRef]
67. Meehan, K.; Vella, L.J. The contribution of tumour-derived exosomes to the hallmarks of cancer. *Crit. Rev. Clin. Lab. Sci.* **2016**, *53*, 121–131. [CrossRef]
68. Aubertin, K.; Silva, A.K.A.; Luciani, N.; Espinosa, A.; Djemat, A.; Charue, D.; Gallet, F.; Blanc-Brude, O.; Wilhelm, C. Massive release of extracellular vesicles from cancer cells after photodynamic treatment or chemotherapy. *Sci. Rep.* **2016**, *6*, 1–11. [CrossRef]
69. Kim, D.H.; Kim, H.; Choi, Y.J.; Kim, S.Y.; Lee, J.-E.; Sung, K.J.; Sung, Y.H.; Pack, C.-G.; Jung, M.; Han, B.; et al. Exosomal PD-L1 promotes tumor growth through immune escape in non-small cell lung cancer. *Exp. Mol. Med.* **2019**, *51*. [CrossRef]
70. Valenti, R.; Huber, V.; Iero, M.; Filipazzi, P.; Parmiani, G.; Rivoltini, L. Tumor-released microvesicles as vehicles of immunosuppression. *Cancer Res.* **2007**, *67*, 2912–2915. [CrossRef]
71. Chen, G.; Huang, A.C.; Zhang, W.; Zhang, G.; Wu, M.; Xu, W.; Yu, Z.; Yang, J.; Wang, B.; Sun, H.; et al. Exosomal PD-L1 contributes to immunosuppression and is associated with anti-PD-1 response. *Nature* **2018**, *560*, 382–386. [CrossRef]
72. Warnken, U.; Seiffert, M.; Schulz, R.; Nessling, M.; Paggetti, J.; Lichter, P.; Cid, L.L.; Worst, T.; Stilgenbauer, S.; Diederichs, S.; et al. Tumor-derived exosomes modulate PD-L1 expression in monocytes. *Sci. Immunol.* **2017**, *2*, eaah5509.
73. Del Re, M.; Marconcini, R.; Pasquini, G.; Rofi, E.; Vivaldi, C.; Bloise, F.; Restante, G.; Arrigoni, E.; Caparello, C.; Grazia Bianco, M.; et al. PD-L1 mRNA expression in plasma-derived exosomes is associated with response to anti-PD-1 antibodies in melanoma and NSCLC. *Br. J. Cancer* **2018**, *118*, 820–824. [CrossRef] [PubMed]
74. Yu, S.; Shi, M.; Feng, J. Expression of PD-L1 in plasma exosomes of NSCLC patients and its associations with PD-L1 expression of corresponding tumor tissues. In Proceedings of the ESMO 2019 Congress, Barcelona, Spain, 27 September–1 October 2019; Volume 30, pp. v25–v54.

75. Costa-Silva, B.; Aiello, N.M.; Ocean, A.J.; Singh, S.; Zhang, H.; Thakur, B.K.; Becker, A.; Hoshino, A.; Mark, M.T.; Molina, H.; et al. Pancreatic cancer exosomes initiate pre-metastatic niche formation in the liver. *Nat. Cell Biol.* **2015**, *17*, 816–826. [CrossRef] [PubMed]
76. Ortiz, A.; Gui, J.; Zahedi, F.; Yu, P.; Cho, C.; Bhattacharya, S.; Carbone, C.J.; Yu, Q.; Katlinski, K.V.; Katlinskaya, Y.V.; et al. An Interferon-driven oxysterol-based defense against tumor-derived extracellular vesicles. *Cancer Cell* **2019**, *35*, 33–45. [CrossRef] [PubMed]
77. Zhao, H.; Yang, L.; Baddour, J.; Achreja, A.; Bernard, V.; Moss, T.; Marini, J.C.; Tudawe, T.; Seviour, E.G.; San Lucas, F.A.; et al. Tumor microenvironment derived exosomes pleiotropically modulate cancer cell metabolism. *Elife* **2016**, *5*, e10250. [CrossRef] [PubMed]
78. Tsitsiou, E.; Lindsay, M.A. microRNAs and the immune response. *Curr. Opin. Pharmacol.* **2009**, *9*, 514–520. [CrossRef] [PubMed]
79. Jasinski-Bergner, S.; Mandelboim, O.; Seliger, B. The Role of MicroRNAs in the control of innate immune response in cancer. *J. Natl. Cancer Inst.* **2014**, *106*, 1–13. [CrossRef]
80. Muller, L.; Mitsuhashi, M.; Simms, P.; Gooding, W.E.; Whiteside, T.L. Tumor-derived exosomes regulate expression of immune function-related genes in human T cell subsets. *Sci. Rep.* **2016**, *6*, 20254. [CrossRef]
81. Hsu, Y.-L.; Kuo, P.-L.; Hung, J.-Y.; Chang, W.-A.; Jian, S.-F.; Lin, Y.-S.; Pan, Y.-C.; Wu, C.-Y. Hypoxic lung-cancer-derived extracellular vesicle microrna-103a increases the oncogenic effects of macrophages by targeting PTEN. *Mol. Ther.* **2018**, *26*, 568–581. [CrossRef]
82. Shukuya, T.; Amann, J.; Ghai, V.; Wang, K.; Carbone, D.P. Circulating miRNA and extracellular vesicle containing miRNA as response biomarkers of anti PD-1/PD-L1 therapy in non-small-cell lung cancer. *J. Clin. Oncol.* **2018**, *36*, 3058. [CrossRef]
83. Vogelstein, B.; Papadopoulos, N.; Velculescu, V.E.; Zhou, S.; Diaz, L.A.; Kinzler, K.W. Cancer genome landscapes. *Science* **2013**, *340*, 1546–1558. [CrossRef] [PubMed]
84. Ulivi, P.; Zoli, W.; Capelli, L.; Chiadini, E.; Calistri, D.; Amadori, D. Target therapy in NSCLC patients: Relevant clinical agents and tumour molecular characterisation. *Mol. Clin. Oncol.* **2013**, *1*, 575–581. [CrossRef] [PubMed]
85. Wu, S.G.; Shih, J.Y. Management of acquired resistance to EGFR TKI-targeted therapy in advanced non-small cell lung cancer. *Mol. Cancer* **2018**, *17*, 38. [CrossRef] [PubMed]
86. Ulivi, P. Paola non-invasive methods to monitor mechanisms of resistance to tyrosine kinase Inhibitors in non-small-cell lung cancer: Where do we stand? *Int. J. Mol. Sci.* **2016**, *17*, 1186. [CrossRef] [PubMed]
87. Pasini, L.; Ulivi, P. Liquid biopsy for the detection of resistance mechanisms in nsclc: Comparison of different blood biomarkers. *J. Clin. Med.* **2019**, *8*, 998. [CrossRef] [PubMed]
88. Rothenstein, J.M.; Chooback, N. ALK inhibitors, resistance development, clinical trials. *Curr. Oncol.* **2018**, *25*, S59–S67. [CrossRef]
89. Krug, A.K.; Enderle, D.; Karlovich, C.; Priewasser, T.; Bentink, S.; Spiel, A.; Brinkmann, K.; Emenegger, J.; Grimm, D.G.; Castellanos-Rizaldos, E.; et al. Improved EGFR mutation detection using combined exosomal RNA and circulating tumor DNA in NSCLC patient plasma. *Ann. Oncol.* **2018**, *29*, 700–706. [CrossRef]
90. Castellanos-Rizaldos, E.; Grimm, D.G.; Tadigotla, V.; Hurley, J.; Healy, J.; Neal, P.L.; Sher, M.; Venkatesan, R.; Karlovich, C.; Raponi, M.; et al. Exosome-based detection of EGFR T790M in plasma from non–small cell lung cancer patients. *Clin. Cancer Res.* **2018**, *24*, 2944–2950. [CrossRef]
91. Okuma, Y.; Morikawa, K.; Tanaka, H.; Yokoyama, T.; Itani, H.; Horiuchi, K.; Nakagawa, H.; Takahashi, N.; Bessho, A.; Soejima, K.; et al. Prospective exosome-focused translational research for afatinib study of non-small cell lung cancer patients expressing EGFR (EXTRA study). *Thorac. Cancer* **2019**, *10*, 395–400. [CrossRef]
92. Jing, C.; Cao, H.; Qin, X.; Yu, S.; Wu, J.; Wang, Z.; Ma, R.; Feng, J. Exosome-mediated gefitinib resistance in lung cancer HCC827 cells via delivery of miR-21. *Oncol. Lett.* **2018**, *15*, 9811–9817. [CrossRef]
93. Lei, Y.; Guo, W.; Chen, B.; Chen, L.; Gong, J.; Li, W. Tumor-released lncRNA H19 promotes gefitinib resistance via packaging into exosomes in non-small cell lung cancer. *Oncol. Rep.* **2018**, *40*, 3438–3446. [CrossRef] [PubMed]
94. Zhang, Y.; Li, M.; Hu, C. Exosomal transfer of miR-214 mediates gefitinib resistance in non-small cell lung cancer. *Biochem. Biophys. Res. Commun.* **2018**, *507*, 457–464. [CrossRef] [PubMed]

95. Zhang, W.; Cai, X.; Yu, J.; Lu, X.; Qian, Q.; Qian, W. Exosome-mediated transfer of lncRNA RP11-838N2.4 promotes erlotinib resistance in non-small cell lung cancer. *Int. J. Oncol.* **2018**, *53*, 527–538. [CrossRef] [PubMed]
96. Kwok, H.H.; Ning, Z.; Chong, P.W.C.; Wan, T.S.K.; Ng, M.H.L.; Ho, G.Y.F.; Ip, M.S.M.; Lam, D.C.L. Transfer of extracellular vesicle-associated-RNAs induces drug resistance in ALK-translocated lung adenocarcinoma. *Cancers* **2019**, *11*, 104. [CrossRef]
97. Figueroa, J.M.; Skog, J.; Akers, J.; Li, H.; Komotar, R.; Jensen, R.; Ringel, F.; Yang, I.; Kalkanis, S.; Thompson, R.; et al. Detection of wild-type EGFR amplification and EGFRvIII mutation in CSF-derived extracellular vesicles of glioblastoma patients. *Neuro Oncol.* **2017**, *19*, 1494–1502. [CrossRef]
98. Van der Mijn, J.C.; Sol, N.; Mellema, W.; Jimenez, C.R.; Piersma, S.R.; Dekker, H.; Schutte, L.M.; Smit, E.F.; Broxterman, H.J.; Skog, J.; et al. Analysis of AKT and ERK1/2 protein kinases in extracellular vesicles isolated from blood of patients with cancer. *J. Extracell. Vesicles* **2014**, *3*, 1–11. [CrossRef]
99. Cesi, G.; Philippidou, D.; Kozar, I.; Kim, Y.J.; Bernardin, F.; Van Niel, G.; Wienecke-Baldacchino, A.; Felten, P.; Letellier, E.; Dengler, S.; et al. A new ALK isoform transported by extracellular vesicles confers drug resistance to melanoma cells. *Mol. Cancer* **2018**, *17*, 1–14. [CrossRef]
100. Perez-Torres, M.; Valle, B.L.; Maihle, N.J.; Negron-Vega, L.; Nieves-Alicea, R.; Cora, E.M. Shedding of epidermal growth factor receptor is a regulated process that occurs with overexpression in malignant cells. *Exp. Cell Res.* **2008**, *314*, 2907–2918. [CrossRef]
101. Read, J.; Ingram, A.; Al Saleh, H.A.; Platko, K.; Gabriel, K.; Kapoor, A.; Pinthus, J.; Majeed, F.; Qureshi, T.; Al-Nedawi, K. Nuclear transportation of exogenous epidermal growth factor receptor and androgen receptor via extracellular vesicles. *Eur. J. Cancer* **2017**, *70*, 62–74. [CrossRef]
102. Hsu, S.C.; Miller, S.A.; Wang, Y.; Hung, M.C. Nuclear EGFR is required for cisplatin resistance and DNA repair. *Am. J. Transl. Res.* **2009**, *1*, 249–258.
103. Li, C.; Iida, M.; Dunn, E.F.; Ghia, A.J.; Wheeler, D.L. Nuclear EGFR contributes to acquired resistance to cetuximab. *Oncogene* **2009**, *28*, 3801–3813. [CrossRef] [PubMed]
104. Al-Nedawi, K.; Meehan, B.; Micallef, J.; Lhotak, V.; May, L.; Guha, A.; Rak, J. Intercellular transfer of the oncogenic receptor EGFRvIII by microvesicles derived from tumour cells. *Nat. Cell Biol.* **2008**, *10*, 619–624. [CrossRef] [PubMed]
105. Jantus-Lewintre, E.; Sirera, R.; Cabrera, A.; Blasco, A.; Caballero, C.; Iranzo, V.; Rosell, R.; Camps, C. Analysis of the prognostic value of soluble epidermal growth factor receptor plasma concentration in advanced non-small-cell lung cancer patients. *Clin. Lung Cancer* **2011**, *12*, 320–327. [CrossRef] [PubMed]
106. Maramotti, S.; Paci, M.; Manzotti, G.; Rapicetta, C.; Gugnoni, M.; Galeone, C.; Cesario, A.; Lococo, F. Soluble epidermal growth factor receptors (sEGFRs) in cancer: Biological aspects and clinical relevance. *Int. J. Mol. Sci.* **2016**, *17*, 593. [CrossRef] [PubMed]
107. Sheridan, C. Investors keep the faith in cancer liquid biopsies. *Nat. Biotechnol.* **2019**, *37*, 972–974. [CrossRef] [PubMed]
108. Sorensen, B.S.; Wu, L.; Wei, W.; Tsai, J.; Weber, B.; Nexo, E.; Meldgaard, P. Monitoring of epidermal growth factor receptor tyrosine kinase inhibitor-sensitizing and resistance mutations in the plasma DNA of patients with advanced non-small cell lung cancer during treatment with erlotinib. *Cancer* **2014**, *120*, 3896–3901. [CrossRef]
109. Wan, J.C.M.; Massie, C.; Garcia-Corbacho, J.; Mouliere, F.; Brenton, J.D.; Caldas, C.; Pacey, S.; Baird, R.; Rosenfeld, N. Liquid biopsies come of age: Towards implementation of circulating tumour DNA. *Nat. Rev. Cancer* **2017**, *17*, 223–238. [CrossRef]
110. Coumans, F.A.W.; Brisson, A.R.; Buzas, E.I.; Dignat-George, F.; Drees, E.E.E.; El-Andaloussi, S.; Emanueli, C.; Gasecka, A.; Hendrix, A.; Hill, A.F.; et al. Methodological guidelines to study extracellular vesicles. *Circ. Res.* **2017**, *120*, 1632–1648. [CrossRef]
111. Zhang, P.; Zhou, X.; He, M.; Shang, Y.; Tetlow, A.L.; Godwin, A.K.; Zeng, Y. Ultrasensitive detection of circulating exosomes with a 3D-nanopatterned microfluidic chip. *Nat. Biomed. Eng.* **2019**, *3*, 438–451. [CrossRef]
112. Zhang, H.; Lyden, D. Asymmetric-flow field-flow fractionation technology for exomere and small extracellular vesicle separation and characterization. *Nat. Protoc.* **2019**, *14*, 1027–1053. [CrossRef]

113. Van Deun, J.; Mestdagh, P.; Agostinis, P.; Akay, Ö.; Anand, S.; Anckaert, J.; Martinez, Z.A.; Baetens, T.; Beghein, E.; Bertier, L.; et al. EV-TRACK: Transparent reporting and centralizing knowledge in extracellular vesicle research. *Nat. Methods* **2017**, *14*, 228–232. [CrossRef] [PubMed]
114. Das, S.; Abdel-Mageed, A.B.; Adamidi, C.; Adelson, P.D.; Akat, K.M.; Alsop, E.; Ansel, K.M.; Arango, J.; Aronin, N.; Avsaroglu, S.K.; et al. The Extracellular RNA Communication Consortium: Establishing Foundational Knowledge and Technologies for Extracellular RNA Research. *Cell* **2019**, *177*, 231–242. [CrossRef] [PubMed]
115. Yates, L.A.; Norbury, C.J.; Gilbert, R.J.C. Minireview The Long and Short of MicroRNA. *Cell* **2013**, *153*, 516–519. [CrossRef] [PubMed]
116. Godoy, P.M.; Bhakta, N.R.; Barczak, A.J.; Cakmak, H.; Fisher, S.; MacKenzie, T.C.; Patel, T.; Price, R.W.; Smith, J.F.; Woodruff, P.G.; et al. Large differences in small RNA composition between human biofluids. *Cell Rep.* **2018**, *25*, 1346–1358. [CrossRef]
117. Matsumura, T.; Sugimachi, K.; Iinuma, H.; Takahashi, Y.; Kurashige, J.; Sawada, G.; Ueda, M.; Uchi, R.; Ueo, H.; Takano, Y.; et al. Exosomal microRNA in serum is a novel biomarker of recurrence in human colorectal cancer. *Br. J. Cancer* **2015**, *113*, 275–281. [CrossRef]
118. Larrea, E.; Sole, C.; Manterola, L.; Goicoechea, I.; Armesto, M.; Arestin, M.; Caffarel, M.M.; Araujo, A.M.; Araiz, M.; Fernandez-Mercado, M.; et al. New concepts in cancer biomarkers: Circulating miRNAs in liquid biopsies. *Int. J. Mol. Sci.* **2016**, *17*, 627. [CrossRef]
119. Yang, J.; Hagen, J.; Guntur, K.V.; Allette, K.; Schuyler, S.; Ranjan, J.; Petralia, F.; Gesta, S.; Sebra, R.; Mahajan, M.; et al. A next generation sequencing based approach to identify extracellular vesicle mediated mRNA transfers between cells. *BMC Genom.* **2017**, *18*, 987. [CrossRef]
120. Möhrmann, L.; Huang, H.J.; Hong, D.S.; Tsimberidou, A.M.; Fu, S.; Piha-Paul, S.A.; Subbiah, V.; Karp, D.D.; Naing, A.; Krug, A.; et al. Liquid biopsies using plasma exosomal nucleic acids and plasma cell-free DNA compared with clinical outcomes of patients with advanced cancers. *Clin. Cancer Res.* **2018**, *24*, 181–188. [CrossRef]
121. Hur, J.Y.; Kim, H.J.; Lee, J.S.; Choi, C.M.; Lee, J.C.; Jung, M.K.; Pack, C.G.; Lee, K.Y. Extracellular vesicle-derived DNA for performing EGFR genotyping of NSCLC patients. *Mol. Cancer* **2018**, *17*, 15. [CrossRef]
122. Zaborowski, M.P.; Lee, K.; Na, Y.J.; Sammarco, A.; Zhang, X.; Iwanicki, M.; Cheah, P.S.; Lin, H.Y.; Zinter, M.; Chou, C.Y.; et al. Methods for systematic identification of membrane proteins for specific capture of cancer-derived extracellular vesicles. *Cell Rep.* **2019**, *27*, 255–268.e6. [CrossRef]
123. Notarangelo, M.; Zucal, C.; Modelska, A.; Pesce, I.; Scarduelli, G.; Potrich, C.; Lunelli, L.; Pederzolli, C.; Pavan, P.; la Marca, G.; et al. Ultrasensitive detection of cancer biomarkers by nickel-based isolation of polydisperse extracellular vesicles from blood. *EBioMedicine* **2019**, *43*, 114–126. [CrossRef] [PubMed]
124. Ji, H.; Chen, M.; Greening, D.W.; He, W.; Rai, A.; Zhang, W.; Simpson, R.J. Deep sequencing of RNA from three different extracellular vesicle (EV) subtypes released from the human LIM1863 colon cancer cell line uncovers distinct mirna-enrichment signatures. *PLoS ONE* **2014**, *9*, e110314. [CrossRef] [PubMed]
125. Camidge, D.R.; Doebele, R.C.; Kerr, K.M. Comparing and contrasting predictive biomarkers for immunotherapy and targeted therapy of NSCLC. *Nat. Rev. Clin. Oncol.* **2019**, *16*, 341–355. [CrossRef] [PubMed]

2

Extracellular Vesicles-Based Biomarkers Represent a Promising Liquid Biopsy in Endometrial Cancer

Carolina Herrero [1], Alexandre de la Fuente [2], Carlos Casas-Arozamena [1], Victor Sebastian [3,4], Martin Prieto [3,4], Manuel Arruebo [3,4], Alicia Abalo [1], Eva Colás [5,6], Gema Moreno-Bueno [6,7,8], Antonio Gil-Moreno [5,6], Ana Vilar [1], Juan Cueva [1,6], Miguel Abal [1,6] and Laura Muinelo-Romay [1,6,*]

1 Translational Medical Oncology Group (Oncomet), Health Research Institute of Santiago de Compostela (IDIS), University Hospital of Santiago de Compostela (SERGAS), Trav. Choupana s/n, 15706 Santiago de Compostela, Spain; carolina.herrero@rai.usc.es (C.H.); carlos.casas_95@hotmail.es (C.C.-A.); Alicia.Abalo.Pineiro@sergas.es (A.A.); ana.vilar@telefonica.net (A.V.); jfcueva@gmail.com (J.C.); miguel.abal.posada@sergas.es (M.A.)
2 Nasasbiotech, S.L., Canton Grande 3, 15003 A Coruña, Spain; afuegon79@gmail.com
3 Department of Chemical Engineering, Aragon Institute of Nanoscience (INA), University of Zaragoza, 50018 Zaragoza, Spain; victorse@unizar.es (V.S.); mprieto@unizar.es (M.P.); arruebom@unizar.es (M.A.)
4 Networking Research Center on Bioengineering, Biomaterials and Nanomedicine (CIBER-BBN), Monforte de Lemos 3–5, 28029 Madrid, Spain
5 Biomedical Research Group in Gynecology, Vall Hebron Research Institute (VHIR), Universitat Autonoma de Barcelona, 119–129 Pg. Vall d'Hebron, 08035 Barcelona, Spain; eva.colas@vhir.org (E.C.); antonioimma@yahoo.com (A.G.-M.)
6 Centro de Investigación Biomédica en Red de Cáncer (CIBERONC), Monforte de Lemos 3–5, 28029 Madrid, Spain; gmoreno@iib.uam.es
7 Department of Biochemistry, Universidad Autonoma de Madrid (UAM), Instituto de Investigaciones Biomedicas "Alberto Sols" (CSIC-UAM), IdiPaz, Pedro Rico, 6, 28029 Madrid, Spain
8 Fundacion MD Anderson Internacional, C/Gómez Hemans 2, 28033 Madrid, Spain
* Correspondence: Laura.muinelo.romay@sergas.es

Abstract: Tumor-derived extracellular vesicles (EVs) are secreted in large amounts into biological fluids of cancer patients. The analysis of EVs cargoes has been associated with patient´s outcome and response to therapy. However, current technologies for EVs isolation are tedious and low cost-efficient for routine clinical implementation. To explore the clinical value of circulating EVs analysis we attempted a proof-of-concept in endometrial cancer (EC) with ExoGAG, an easy to use and highly efficient new technology to enrich EVs. Technical performance was first evaluated using EVs secreted by Hec1A cells. Then, the clinical value of this strategy was questioned by analyzing the levels of two well-known tissue biomarkers in EC, L1 cell adhesion molecule (L1CAM) and Annexin A2 (ANXA2), in EVs purified from plasma in a cohort of 41 EC patients and 20 healthy controls. The results demonstrated the specific content of ANXA2 in the purified EVs fraction, with an accurate sensitivity and specificity for EC diagnosis. Importantly, high ANXA2 levels in circulating EVs were associated with high risk of recurrence and non-endometrioid histology suggesting a potential value as a prognostic biomarker in EC. These results also confirmed ExoGAG technology as a robust technique for the clinical implementation of circulating EVs analyses.

Keywords: EVs; ExoGAG; endometrial cancer; ANXA2

1. Introduction

Endometrial cancer (EC) is the most common neoplasm of the female genital tract in the developed world, and the incidence has risen over the past years. Despite the progress in early detection and treatment, a significant number of cases of advanced ECs are still diagnosed. Consequently, mortality from this cancer has not improved in recent decades and is primarily driven by high-grade carcinomas that are more likely to present at an advanced stage and, ultimately, are more likely to recur. The prognosis for recurrent EC is poor, especially for the 50% of these advanced ECs that present extra-pelvic disease recurrence. To this regard, effort has been mainly dedicated to reach a consensus on EC risk classification to promote consistency for future clinical trials design, including their molecular characterization and its integration into clinicopathological profiling to develop prognostic and predictive biomarkers [1,2]. Various immunohistochemical markers have been studied for the prediction of recurrences, like loss of the estrogen receptor (ER) and progesterone receptor (PR). In particular, the presence of the L1 cell adhesion molecule (L1CAM) has been described as a strong independent predictor for distant recurrence and overall survival in EC, as well as a marker to select patients who could benefit from more extensive diagnostic and/or therapeutic procedures [3–5]. Likewise, annexin-A2 has been reported as a predictor of recurrent disease in EC and as an independent risk factor for poor prognosis [6–8]. Nevertheless, these biomarkers have not been implemented into clinical practice, in part due to limited sensitivity and/or specificity, absence of robust clinical validation and compromised performance in terms of cost-effectiveness.

In parallel to the discovery and the technical and clinical validation of soluble protein biomarkers, the concept of liquid biopsy refers to the analysis of tumoral material shed from primary tumors and their metastatic sites into peripheral blood like circulating tumor cells (CTCs), circulating tumor DNA (ctDNA), circulating miRNAs, and tumor-derived extracellular vesicles (EVs). Major advantages of liquid biopsies analysis include real-time information about the disease by minimal invasive procedures that can be performed regularly [9]. In particular, small extracellular vesicles (sEVs) secreted by tumor cells are revolutionizing the field of liquid biopsy, as safe carriers of nucleic acids, proteins, and lipids reflective of the disease state. As responsible for myriad roles in physiology and disease, and the main mediators of cell-to-cell communication between the tumor and stromal cells in both local and distant microenvironments, this heterogeneous group of small lipid-enclosed structures might be a relevant source for diagnostic and therapeutic tools [10]. The biological relevance of tumor EVs in disease progression and metastatic potential has been reported [11], with an increased secretion associated with high-grade disease [12]. Limited work has been reported in the field of EVs in EC [13], in particular on their phenotypic characterization and involvement in the metastatic fate of CTCs [14].

Despite its potential, widespread implementation of EV-based biomarkers in clinical practice has not been yet achieved. This is mainly due to the limited performance of current methodologies for the separation and enrichment of EVs for clinical and technical validation with sufficient specificity and sensitivity for routine clinical implementation [15]. Here, we describe a proof-of-concept to evaluate the clinical applicability of EV-based biomarkers for EC management with ExoGAG technology. EVs are decorated on their surface with glycoproteins associated with EV biogenesis and different biological functions [16]. ExoGAG technology is based on the ability of glycosaminoglycans (GAGs), that are negatively charged linear polysaccharides composed of repeating disaccharides with variable sulfation levels, to neutralize the positive charge of cationic agents leading to their aggregation in a complex that can be further purified by centrifugation [17]. We first assessed the ExoGAG technical performance by studying EVs purified from conditioned culture media, resulting in a robust and reproducible methodology with a simple and easy to handle protocol. Moreover, we further validated the ExoGAG clinical performance in plasma samples from EC patients by analyzing L1CAM and ANXA2, known biomarkers for this tumor, in the isolated EVs, resulting in an improved diagnostic and prognostic value.

2. Results

2.1. *Purification of EVs from Conditioned Media from Endometrial Cancer Cells*

First, we applied the EVs purification technology to separate and characterize the fraction of tumor EVs from conditioned cell culture medium from Hec1A endometrial cancer cells. Hec1A culture media, supplemented with fetal bovine serum (FBS) reduced in EVs, was recovered upon conditioning for 48 h and processed with ExoGAG following the manufacturer's instructions. Briefly, the recovered conditioned media was incubated for 5 min with the ExoGAG reactive and centrifuged at 3000× *g* for 30 min. The pellet containing EVs was then resuspended in PBS and directly analyzed by nanoparticle tracking analysis (NTA) Nanosight NS300 as described in Section 4. As shown in Figure 1A, both the representative image of the purified EVs (left panel) and their particle profile (right panel) were consistent with a homogeneous population of particles of approximately 200 nm, similar to the particle profile of EVs purified by ultracentrifugation at 100,000× *g* for 16 h, performed in parallel as a methodological control. Minor differences in particle size and particle subpopulations between EV sample comparison may be related to the limitations of NTA sensibility for the profiling of biological vesicles [18]. In addition, the efficiency of EVs purification by ExoGAG, expressed as the number of EVs per frame in the precipitate, was similar compared to the number of EVs in the pellet obtained by ultracentrifugation (Figure 1B). By contrast, the quantification of protein contaminants from the ExoGAG methodology of EVs separation by bicinchoninic acid assay (BCA) resulted in an improved purification compared to the ultracentrifugation protocol (** $p = 0.0011$ according to paired *t*-test) (Figure 1C). Consistently, transmission electron microscopy (TEM) imaging showed that samples isolated with ExoGAG contained only EVs and crystallized organic remnants of ExoGAG agent (Figure 1D, right panels), while samples precipitated using ultracentrifugation showed a large amount of unwanted organic residues (Figure 1D, left panels). In terms of morphological characterization, EVs obtained using ExoGAG appeared cleaner and better dispersed compared to those isolated by ultracentrifugation presenting large clustering (Figure 1D). Despite the fact that a crystallized ExoGAG layer on the surface of the EVs was observed (Figure 1D, magnification in right panels), it did not prevent the interaction with AuNPs-PEG-CD9 further characterizing the purified EVs. It is likely that when the sample is dispersed in the medium, ExoGAG remains solubilized but as the sample dries for microscopy visualization it begins to crystallize on the EVs' surface. Similar results in terms of NTA Nanosight NS300 profile, purification, efficiency, and characterization were obtained when EVs were purified with ExoGAG from the MDA-MB-231 breast cancer cell line, the endometrial cancer cell line Ishikawa, and the ovarian cancer cell line SKOV3 (Figure S1). From these results, we conclude that ExoGAG technology meets the appropriate criteria to obtain a highly purified population of EVs with an optimal particle profile and recovery rate, and a rapid and simple process, thus making it suitable to translate an EV-based biomarker into the clinical setting.

2.2. *Purification of EVs from Plasma Samples*

In order to provide a useful tool to translate EV-based biomarkers into clinics, we next assessed the performance of ExoGAG technology in plasma, as the most suitable sample for liquid biopsy in EC. Plasma samples were processed with ExoGAG as indicated by the manufacturer. Briefly, 500 µL of plasma were incubated with 1 mL of ExoGAG reactive for 5 min before centrifugation at 16,000× *g* for 15 min, resulting in a precipitate containing EVs. The representative NTA Nanosight NS300 image and particle profile of EVs population purified by ExoGAG from plasma showed a homogeneous population of 200 nm on average (Figure 2A). In this case, the concentration of EVs isolated by ExoGAG was higher compared to ultracentrifugation (Figure S2A, left panel). In addition, cytometry analysis resulted in the specific CD9 labeling of EVs purified by ExoGAG (Figure 2B), also similar to those purified by ultracentrifugation (Figure S2A, right panel). We further compared ExoGAG technology with ExoQuick and ExoSpin, two commercially available technologies for the isolation of EVs also based on precipitation agents. As shown in Figure S2B, ExoGAG showed no significant reduction on

EV purification efficiency, but significantly reduced levels of co-precipitated contaminants. All these results confirmed the performance of ExoGAG to separate the population of EVs also from a complex plasma sample, with optimal quality, purity, and recovery rate.

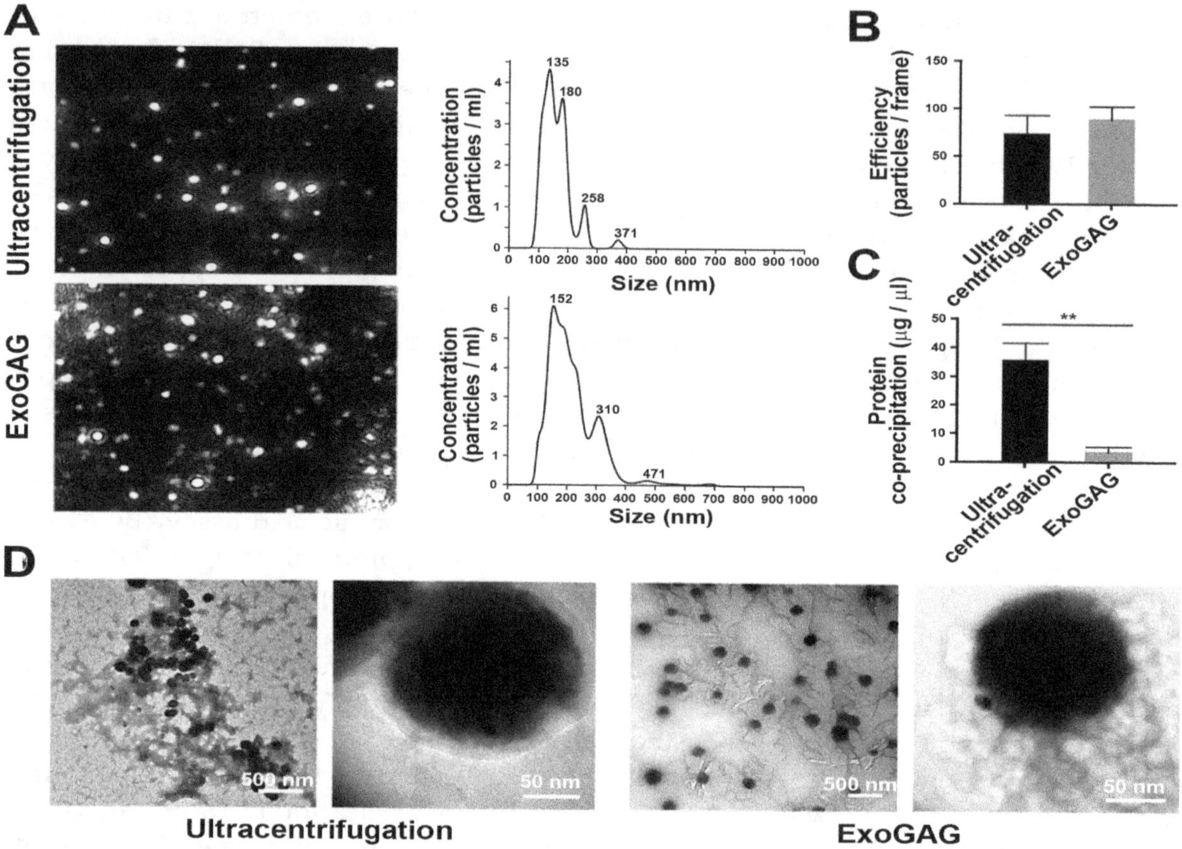

Figure 1. Characterization of Hec1A-derived extracellular vesicles (EVs) obtained by ultracentrifugation and ExoGAG. (**A**) Representative nanoparticle tracking analysis (NTA) Nanosight NS300 image of isolated EVs showing particle size (nm) and concentration (particles/mL). A similar particle profile was obtained in both methodologies (n = 4). (**B**) Efficiency of purified EVs, expressed as the number of EVs per frame, was similar in ExoGAG and ultracentrifugation (n = 4). (**C**) Protein co-precipitation levels quantified by bicinchoninic acid assay (BCA) assay showed a low co-precipitated protein in EVs isolated by ExoGAG compared to those isolated by ultracentrifugation (** p = 0.0011 according to paired t-test; n = 4). (**D**) Representative transmission electron microscopy (TEM) images of purified EVs harvested by ultracentrifugation (left panels) and ExoGAG (right panels), further coupled to AuNP-PEG-CD9 as shown in the magnifications.

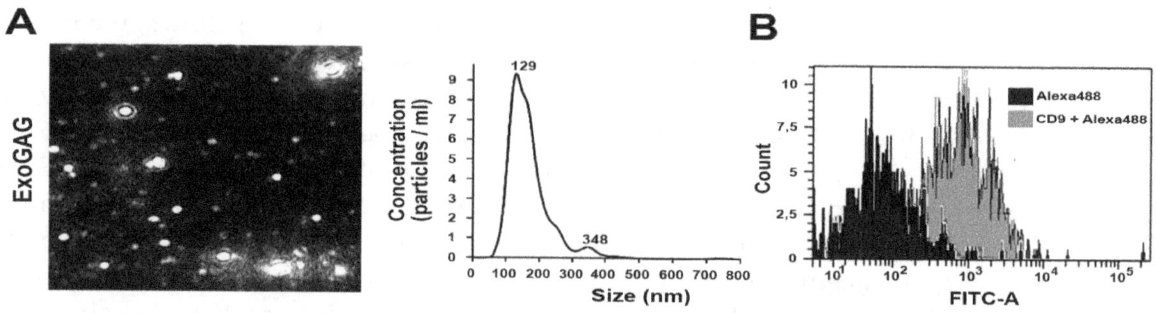

Figure 2. Characterization of plasma EVs isolated by ExoGAG. (**A**) NTA Nanosight NS300 particle tracking profile of harvested EVs expressed in size (nm) and concentration (particles/mL) showed a population of 200 nm in average (n = 4). (**B**) Cytometry analysis of EVs resulted in a specific CD9 labeling of the EVs purified (n = 2). FITCA: Fluorescein isothiocyanate A.

2.3. ANXA2 Levels Are Increased in Circulating EVs of EC Patients

We then applied the ExoGAG technology to a cohort of plasma samples from EC patients and healthy controls to analyze its clinical value. Preliminary data indicated that the number of EVs in patients was increased compared to healthy controls (* $p = 0.0148$ according to Mann–Whitney U test) (Figure S3), in accordance with previous data [19]. We then analyzed the protein levels of L1CAM and ANXA2 by ELISA in the purified EVs, known biomarkers with prognostic value in endometrial tumor tissue samples. We first assessed the distribution of the levels of ANXA2 and L1CAM in complete plasma before ExoGAG purification and in the purified EVs after ExoGAG purification in healthy controls ($n = 10$) and patients ($n = 9$). We observed that ANXA2 expression mainly corresponded to the purified EVs, with a residual component of the free soluble protein in plasma (Figure 3A, left panel). By contrast, the distribution of L1CAM levels indicated that the majority of L1CAM content did not correspond to the purified EVs and resulted from the soluble fraction of plasma samples (Figure 3A, right panel). When analyzed for its diagnostic value in this cohort of 9 patients and 10 controls, the levels of ANXA2 biomarker were found increased in complete plasma samples from patients compared to healthy controls, although no statistical significance was observed ($p = 0.090$, according to Mann–Whitney U test); meanwhile L1CAM showed significantly higher levels in healthy individuals (* $p = 0.027$, according to Mann–Whitney U test) (Figure 3B). Interestingly, when we focused the analysis on ExoGAG EVs, significantly higher levels of ANXA2 were found in patients in comparison with the controls (* $p = 0.027$, according to Mann–Whitney U test), while L1CAM did not show differences between groups ($p = 0.887$, according to Mann–Whitney U test) (Figure 3C). These results evidenced the potential of ANXA2 monitoring in EVs isolated from plasma of EC patients using ExoGAG technology.

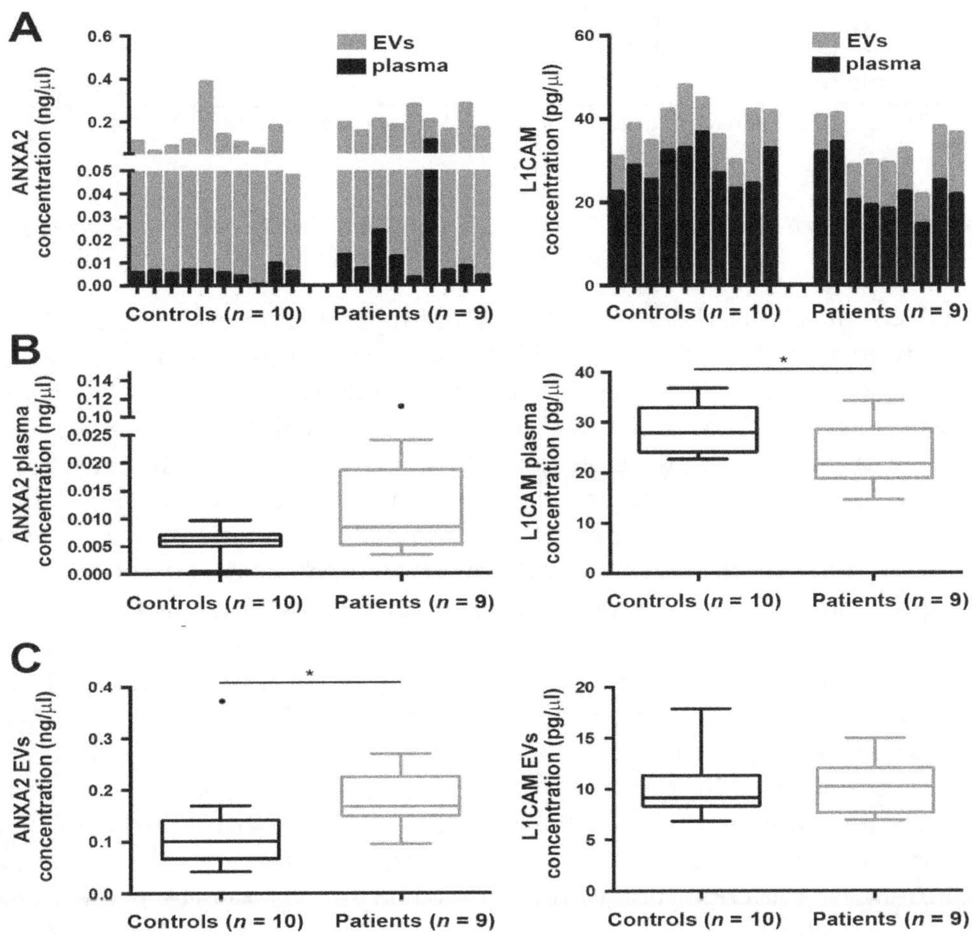

Figure 3. ANXA2 and L1CAM levels in endometrial cancer (EC) plasma samples and circulating EVs

from patients and healthy controls. (**A**) Analysis of ANXA2 and L1CAM levels in plasma showed that ANXA2 levels mainly corresponded to purified EVs while L1CAM levels were mainly present in the soluble fraction of plasma. (**B**) Analysis of both markers in complete plasma samples from healthy controls and patients showed higher levels of L1CAM in controls and no significant differences for ANXA2 levels (* $p = 0.027$ and $p = 0.090$, respectively, according to Mann–Whitney U test). (**C**) Analysis of ANXA2 levels in circulating EVs showed significantly increased levels in patients vs. healthy controls (* $p = 0.027$, according to Mann–Whitney U test) while L1CAM did not show differences ($p = 0.887$, according to Mann–Whitney U test). These comparative analyses were conducted with 9 patients and 10 healthy controls included in the global cohort of patients and healthy controls (see Table 1).

Table 1. Clinical characteristics of the EC cohort.

Feature	*n* (%)	Feature	*n* (%)
Age (mean, range)	68 (46–88)	**Myometrial infiltration**	
FIGO stage		<50%	17 (41.5%)
I	26 (63.4%)	≥50%	23 (56.1%)
II	6 (14.6%)	Unknown	1 (2.4%)
III	8 (19.5%)	**Risk of recurrence**	
IV	1 (2.5%)	Low	11 (26.8%)
Histologic type		Intermediate	6 (14.6%)
Endometrioid	26 (63.4%)	High	24 (58.7%)
Non-endometrioid	12 (29.3%)		
Other	3 (7.3%)		
Histological grade			
Grade 1	15 (36.6%)		
Grade 2	11 (26.8%)		
Grade 3	14 (34.1%)		
Unknown	1 (2.5%)		

FIGO, Federation of Gynecology and Obstetrics.

2.4. ANXA2 Levels in Purified EVs from EC Patients Showed Diagnosis Value and Correlated with Tumor Aggressiveness

Taking into account the results obtained addressing the EV-associated ANXA2 and L1CAM levels, we decided to further study the value of ANXA2 as a diagnostic marker. For that, we isolated the ExoGAG EVs in a cohort of 41 patients with EC and 20 healthy controls. After this analysis, the increased levels of ANXA2 in the cohort of patients was validated (*** $p = 0.001$ according to Mann–Whitney U test) (Figure 4A, upper panel). We also assessed the clinical value of this marker by performing a receiver operating characteristic (ROC) curve analysis to determine its diagnosis accuracy, obtaining an area under receiver operating curve (AUROC) of 0.74 (*** $p = 0.002$) (Figure 4A, lower panel), which evidenced its accuracy and sensitivity to discriminate the patients' population. We continued to question whether plasmatic EV-associated ANXA2 levels were linked with a more aggressive disease, as already described for primary tumors. For this, we analyzed the correlation of ANXA2 in EVs purified by ExoGAG from plasma samples with the clinical characteristics of our complete cohort of EC patients. A positive correlation was found between the levels of ANXA2 and the risk of recurrence and the tumor histology, showing those patients with non-endometrioid tumors and high risk of recurrence had higher ANXA2 levels in purified EVs (* $p = 0.019$ and * $p = 0.012$, respectively according to Mann–Whitney U test) (Table 2, Figure 4B). In addition, higher levels of ANXA2 in EVs purified from plasma of Grade 3 and FIGO (International Federation of Gynecology and Obstetrics) stage III–IV EC patients were found (* $p = 0.038$ and * $p = 0.043$, respectively according to Mann–Whitney U test) (Table 2). These results evidenced the potential of EV-based ANXA2 expression as a prognosis liquid biopsy biomarker in EC.

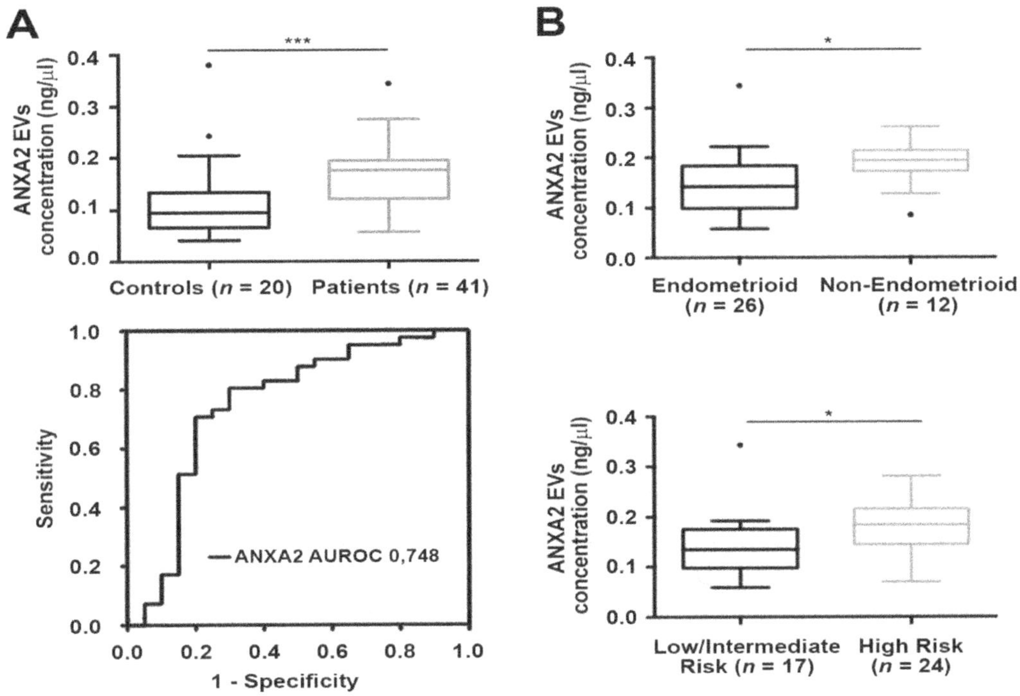

Figure 4. ANXA2 levels in EVs purified by ExoGAG as a diagnosis and prognosis marker in EC. (**A**) Analysis of ANXA2 levels in circulating EVs showed significantly increased levels in patients vs. healthy controls (*** $p = 0.001$, according to Mann–Whitney U test) (upper panel). Receiver operating characteristic (ROC) curves confirmed the power of this biomarker to discriminate EC patients and healthy controls (lower panel). (**B**) ANXA2 levels according to the histology and the risk of recurrence of the EC cohort. Patients with non-endometrioid ($n = 12$) and high risk of recurrence tumors ($n = 24$) showed higher levels of ANXA2 (* $p = 0.019$ and * $p = 0.012$, respectively, according to Mann–Whitney U test). All analyses were conducted on $n = 41$ patients and $n = 20$ healthy controls.

Table 2. Circulating EV-associated ANXA2 levels according to the clinical characteristics of EC patients.

Fiture	*n*	ANXA2 (ng/uL) Mean (SD)	*p*
Age			
<68	19	0.163 (0.059)	
≥68	22	0.179 (0.066)	0.23
FIGO			
I/II	32	0.157 (0.059)	
III/IV	9	0.199 (0.058)	**0.043**
Histologic type			
Endometrioid	26	0.150 (0.061)	
Non-endmetrioid	12	0.189 (0.051)	**0.019**
Histologic grade			
Grade 1–2	26	0.168 (0.063)	
Grade 3	14	0.191 (0.054)	**0.038**
Myometrial infiltration			
<50	17	0.158 (0.066)	
≥50	23	0.175 (0.059)	0.385
Risk of recurrence			
Low/intermediate	17	0.147 (0.070)	
high	24	0.188 (0.049)	**0.012**

SD, standard deviation; p was calculated according to Mann–Whitney U test. $p < 0.05$ are marked in bold.

3. Discussion

The deficit in appropriate technologies for efficient purification of EVs from complex plasma samples remains the main impediment for the implementation of liquid biopsy-based biomarkers with high specificity and sensitivity in clinical practice [20]. To overcome this limitation, in this work we have combined an efficient EVs purification technology with known biomarkers with clinical value. The ExoGAG technology demonstrates optimal abilities to separate tumor EVs from conditioned culture media of representative EC cells and from the more complex plasma samples from EC patients. This includes adequate particle profile and recovery rates, high purity EVs with minimal protein contaminants, and an easily implementable processing into clinical routine. All these advantages, compared to ultracentrifugation as the gold standard technology, as well as to other commercial precipitation strategies, make this technology adequate for the requirements of specificity and sensibility in EVs purification. Of note, regarding the other precipitation-based strategies analyzed in our study, the main advantage of ExoGAG technology is the purity of the EVs obtained, which is a key factor for their posterior characterization. Linked to the extensive literature showing that EVs derived from tumor cell lines contribute to tumor invasion, chemoresistance, angiogenesis, tumor innervation, metastasis or immune escape [21,22], and to those evidences demonstrating that tumor-derived EVs are associated with advanced disease in different indications [23–28] and the response to therapy [29], the analysis of EV-based biomarkers should be expedited for the translation of this form of liquid biopsy into the clinics.

We aimed to validate this relevant technological strategy by selecting L1CAM and ANXA2 as known independent biomarkers with prognostic and predictive value in EC [3,4,6]. The L1CAM cell adhesion molecule is an axonal glycoprotein belonging to the immunoglobulin supergene family. The ectodomain, consisting of several immunoglobulin-like domains and fibronectin-like repeats (type III), is linked via a single transmembrane sequence to a conserved cytoplasmic domain. Importantly, several studies have described the association between L1CAM expression and a poor evolution of EC patients [5,30]. In particular, in early-stage EC with endometrioid histology, expression of L1CAM was related to unfavorable pathological findings, distant recurrence, and poor survival, while in high-grade EC and non-endometrioid EC, L1CAM expression was correlated to an advanced stage and lymphovascular space invasion and distant recurrence [3,31,32]. In this work, we observed that L1CAM expression in plasma samples was mainly present as free soluble and not associated with EVs, resulting in no differences between EC patients and healthy controls.

On the other hand, ANXA2 is a member of the annexin family, a calcium-dependent phospholipid-binding protein family which plays an important role in the regulation of cellular growth and in signal transduction pathways. Furthermore, this membrane protein has been involved in endocytosis and exocytosis processes [33]. More specifically, ANXA2 expression in tissue samples has been associated with a higher risk of recurrence in EC [6]. Importantly, our study evidenced EVs specific expression of ANXA2 in plasma samples from EC patients, with almost no levels of protein expressed as a free soluble protein, demonstrating that EV-based biomarker strategy can improve specificity and sensibility of this biomarker in the prediction of EC outcome. To this regard, the specific expression of ANXA2 has been described to play a role in angiogenesis and metastasis in breast cancer, with promise as a potential biomarker as well as a therapeutic target [34]. Moreover, our results also demonstrate that the levels of ANXA2 in purified EVs from the plasma of EC patients correlate with the tumor histology, grade, stage, and risk of recurrence, reinforcing their interest as a marker linked to more aggressive tumors. This translates the clinical utility of ANXA2 from tissue samples to the liquid biopsy frame and guarantees further studies for early detection of advanced endometrial cancer patients resulting in recurrent disease and for monitoring the response to therapy and early assessment of progressive disease.

4. Materials and Methods

4.1. Patients

The present study was approved by the Autonomic Galician Ethical Committee (code 2017/430). Patients participating in the study were recruited between January 2018–April 2019 at the Gynaecologic Department of Vall d'Hebron University Hospital (Barcelona, Spain), the MDA Anderson Cancer Centre of Madrid, and the University Hospital of Santiago de Compostela (Santiago de Compostela, Spain). Our EC cohort included low- to high-risk, grade 1–3, and stage I–IV, and recently diagnosed carcinomas or recurrences (Table 1). The control group included a set of 20 healthy women with the absence of a previous cancer episode and with an age range similar to patients (mean: 65.5; range: 47–77 years). In all cases, peripheral blood samples were collected at the time of surgery of the primary tumor. Importantly, the study was carried out following the rules of the Declaration of Helsinki of 1975, revised in 2013. Informed consent approved by the pertinent ethical committees was signed by all patients (Autonomic Galician Ethical Committee code 2017/430).

4.2. EV Production and Isolation from Conditioned Medium

The human endometrial cancer cell lines Hec1A and Ishikawa, the ovarian cancer cell line SKOV3, and the breast cancer cell line MDA-MB-231, as representative gynecological aggressive cell lines, were used to test the efficacy of ExoGAG technology. Cancer cell lines Hec1A and SKOV3 were cultured in McCoy's 5A media (Gibco, Grand Island, NY, USA) and Ishikawa and MDA-MB-231 were cultured in Dulbecco's Modified Eagle Medium (DMEM) (Lonza, Walkersville, MD, USA), supplemented with 1% penicillin-streptomycin (Gibco, Grand Island, NY, USA) and 10% FBS (Gibco, Thermo Fisher Scientific) reduced in EVs by ultracentrifugation during 16 h at 4 °C, as described in [14] (see Figure S4). Cells were maintained in a humidified atmosphere at 37 °C and 5% CO_2. After 48 h, the culture medium was recovered and small EVs were harvested by differential centrifugation or ExoGAG.

4.2.1. Differential Centrifugation

Conditioned medium (40 mL) was initially cleared of cell debris (10,000× *g*, 20 min, 4 °C) in a SW32Ti rotor (Beckman Coulter, Brea, CA, USA). EVs were then collected at high-speed ultracentrifugation (100,000× *g*, 16 h, 4 °C) using the same rotor and resuspended in PBS.

4.2.2. ExoGAG

Conditioned medium (40 mL) was incubated with 20 mL of ExoGAG reagent (Nasasbiotech, A Coruña, Spain) for 5 min. EVs were collected by centrifugation (3000× *g*, 30 min, 4 °C; Sorvall ST 8R centrifuge, Thermo Scientific, Osterode am Harz, Germany) and resuspended in PBS.

4.3. Isolation of EVs from Plasma

Peripheral blood from EC patients was collected in CellSave tubes (Menarini, Silicon Biosystem, Huntingdon Valley, PA, USA). Plasma was then extracted after a two-step centrifugation at 1600× *g* and 6000× *g* for 10 min. After the second centrifugation plasma was stored at −80 °C until use. EVs were isolated from 500 µL of human plasma samples by ExoGAG. In this case, plasma samples were incubated with ExoGAG reagent (1 mL) for 5 min. EVs were collected by centrifugation at 16,000× *g*, 15 min, 4 °C (Eppendorf, Hamburg, Germany) and resuspended in PBS. Alternatively, EVs were isolated by differential ultracentrifugation as described above (Figure S2A), and with ExoQuick (System Biosciences, Mountain View, CA, USA) and ExoSpin (Cell Guidance Systems, Cambridge, UK) technologies, also based on precipitation agents, following the manufacturer's instructions (see Figure S2B).

4.4. Characterization of EVs

Size and concentration EVs assessed by nanoparticle tracking analysis (NTA) Nanosight NS300 (Malvern, UK) and isolation efficiency was measured in particles/frame. Total EV protein was extracted using lysis buffer (150 mM NaCl, 50 mM Tris-HCl (pH 8.5), 5 mM Ethylenediaminetetraacetic acid (EDTA), 10% Triton ×100, 2 mM Na_3OV_4, 4 mM NaF, 1× Phenylmethylsulfonyl fluoride (PMSF), 1× Proteasa Inhibitor Cocktail (PIC)) and protein extracts were quantified by bicinchoninic acid assay (BCA assay).

In flow cytometry analysis (BD FACSAria^{TM} IIu, BD biosciences, San José, CA, USA) we analyzed CD9, CD63, and CD81 protein levels as the most abundant membrane proteins on EVs and usually employed as markers for their characterization [35,36]. EVs isolated from MDA-MB-231 cells were incubated for 1 h with CD81 (1:50, sc7637, Santa Cruz Biotechnology, Santa Cruz, CA, USA) and CD63 (1:50, sc5275, Santa Cruz Biotechnology, Santa Cruz, CA, USA), specifically expressed in this breast cancer cell line [37], followed by anti-mouse Alexa488 (1:1000, ab150113, Abcam, Cambridge, UK) as the secondary antibody. This last antibody was used as a negative control for background fluorescence. EVs purified from plasma were incubated with anti-mouse CD9 antibody (1:50, sc13118, Santa Cruz Biotechnology, Santa Cruz, CA, USA)

EVs were visualized by transmission electron microscopy (TEM) (FEI TECNAI T20, Thermo Fisher Scientific, Waltham, MA, USA). For this purpose, anti-mouse CD9 antibody (sc13118, Santa Cruz Biotechnology, Santa Cruz, CA, USA) was functionalized with gold nanoparticles. Gold (III) chloride hydrate (50% Au basis), sodium citrate tribasic dihydrate (99%), O-(3-Carboxypropyl)-O′-(2-(3-mercaptopropionylamino) ethyl)-polyethylene glycol (Mw. 5000 Da) (SH-PEG-COOH), N-(3-Dimethylaminopropyl)-N′-ethylcarbodiimide hydrochloride (98%) (EDC), N-Hydroxysuccinimide (97%) (NHS), and phosphotungstic acid hydrate were purchased from Merck KGaA (Darmstadt, Germany) and used as received. Citrate-capped gold nanoparticles (AuNPs) were synthesized following a slightly modified version of the Turkevich–Frens method [38] as reported by Lafuente et al. [39]. Afterward, the AuNPs were functionalized with the anti-mouse CD9 antibody via carbodiimide/hydroxysuccinimide chemistry [40] using a heterobifunctional linker. In detail, 1 mL of O-(3-Carboxypropyl)-O′-(2-(3-mercaptopropionylamino) ethyl)-polyethylene glycol aqueous solution (2 mg/mL) was added dropwise into a vial containing 1 mL of the initially prepared AuNP aqueous dispersion (2 mg/mL) and left to react for 12 h under stirring. Then SH-PEG-COOH functionalized AuNPs were washed three times in Milli-Q water and three times in borate buffered saline (pH 9) (BBS). Subsequently, 250 µL of AuNPs-PEG-COOH dispersion in BBS (0.2 mg/mL) were mixed with 25 µL of an antibody solution in BBS (0.2 mg/mL) for 1 h under stirring. Then, 96 µg of EDC and 115 µg of NHS were added to the mixture and let to react for 24 h. Obtained AuNPs-PEG-CD9 were washed three times in tris-borate-EDTA quenching solution reacting for 12 h to block unreacted groups. Finally, the resulting NP was washed three times in phosphate buffered saline (pH 7.4) and stored at 4 °C for further use.

AuNPs-PEG-CD9 were incubated with EVs in PBS for 24 h at 37 °C. Then, samples were separated by centrifugation at 3000 rpm for 5 min and at 13,000 rpm for 15 min. First washing at 3000 rpm was performed to remove large aggregates that could interfere with TEM visualization. A second centrifugation at 13,000 rpm was performed to separate free EVs from free AuNP-PEG-Ab and those that interacted with EVs. All samples were contrasted with a phosphotungstic acid aqueous solution (3 wt. %) and analyzed by imaging using a TEM operated at 200 kV.

4.5. ANXA2 and L1CAM Levels in Plasma Samples and Circulating EVs

ANXA2 and L1CAM levels in plasma and EVs were determined by ELISA. For that, a total volume of 500 µL of plasma was used to isolated EVs by ExoGAG as previously described. Total EVs protein was extracted using lysis buffer. ANXA2 and L1CAM protein levels in total plasma and EVs were determined by ELISA (Cloud-Clone Corp, Katy, TX, USA) according to the manufacturer's instructions.

The absorbance was read at 450 nm using a plate reader (Epoch 2 microplate reader, BioTek, Winooski, VT, USA) to quantify the protein content of each marker.

4.6. Statistical Analysis

Statistical analyses were conducted using SPSS (version 19.00 for Windows, Chicago, IL, USA) and GraphPad Prism 6.00 software (GraphPad Software Inc., San Diego, CA, USA). EV yield between conditions was analyzed by paired t-test. Comparisons between controls and patients and correlations between clinical data and ANXA2 and L1CAM levels were analyzed by Mann–Whitney U test. Outliers were established based on Tukey's method. The diagnostic power of ANXA2 levels was determined by ROC curve statistics. For all the analysis a p-value <0.05 was considered statistically significant.

5. Conclusions

The results from this study demonstrated that ExoGAG is a robust, easy to use, and highly efficient technology for the isolation of EVs from different biological fluids such as conditioned culture medium and plasma. Furthermore, ExoGAG technology demonstrated an optimal performance for the clinical implementation of EVs purified from plasma of EC patients. The specific expression of the EC biomarkers ANXA2 in the isolated EVs improved their specificity and sensibility as liquid biopsy. Moreover, we could observe high ANXA2 levels in plasmatic EVs associated with non-endometrioid tumors and high risk of recurrence tumors, evidencing the potential use of EV-based ANXA2 expression as a diagnostic and prognostic liquid biopsy biomarker in EC.

Supplementary Materials
Figure S1: Characterization of MDA-MB-231, Ishikawa, and SKOV3 EVs isolated by ExoGAG; Figure S2: Characterization of plasma EVs isolated by ultracentrifugation, ExoQuick, and ExoSpin; Figure S3: Efficiency (particles/frame) of EVs purified by ExoGAG in a representative cohort of endometrial cancer patients (n = 8) and healthy controls (n = 8); Figure S4: Efficiency of ultracentrifugation (16 h) to reduce the EVs contain from culture media analyzed by NTA Nanosight NS300 and expressed as particles/frame.

Author Contributions: Conceptualization, M.A. (Miguel Abal) and L.M.-R.; methodology and analysis, A.d.l.F., V.S., M.P., M.A. (Manuel Arruebo), and C.H.; clinical samples, processing, and data, C.H., C.C.-A., A.A., E.C., G.M.-B., A.G.-M., A.V., and J.C.; writing—original draft preparation, C.H., M.A. (Miguel Abal), and L.M.-R.; writing—review and editing, all authors; supervision, L.M.-R.; funding acquisition, G.M.-B., A.G.-M., M.A. (Miguel Abal), and L.M.-R.

Acknowledgments: We gratefully thank all the patients for their willingness to participate in the study.

Conflicts of Interest: A.d.l.F. and M.Arruebo have equity ownership in Nasasbiotech. The funders had no role in the design of the study; in the collection, analyses, or interpretation of data; in the writing of the manuscript, or in the decision to publish the results.

References

1. Stelloo, E.; Nout, R.A.; Osse, E.M.; Jürgenliemk-Schulz, I.J.; Jobsen, J.J.; Lutgens, L.C.; van der Steen-Banasik, E.M.; Nijman, H.W.; Putter, H.; Bosse, T.; et al. Improved Risk Assessment by Integrating Molecular and Clinicopathological Factors in Early-stage Endometrial Cancer-Combined Analysis of the PORTEC Cohorts. *Clin. Cancer Res.* **2016**, *22*, 4215–4224. [CrossRef] [PubMed]
2. Soslow, R.A.; Tornos, C.; Park, K.J.; Malpica, A.; Matias-Guiu, X.; Oliva, E.; Parkash, V.; Carlson, J.; McCluggage, W.G.; Gilks, C.B. Endometrial Carcinoma Diagnosis: Use of FIGO Grading and Genomic Subcategories in Clinical Practice: Recommendations of the International Society of Gynecological Pathologists. *Int. J. Gynecol. Pathol.* **2019**, *38* (Suppl. 1), S64–S74. [CrossRef]
3. Zeimet, A.G.; Reimer, D.; Huszar, M.; Winterhoff, B.; Puistola, U.; Azim, S.A.; Müller-Holzner, E.; Ben-Arie, A.; van Kempen, L.C.; Petru, E.; et al. L1CAM in early-stage type I endometrial cancer: results of a large multicenter evaluation. *J. Natl. Cancer Inst.* **2013**, *105*, 1142–1150. [CrossRef]
4. Van der Putten, L.J.; Visser, N.C.; van de Vijver, K.; Santacana, M.; Bronsert, P.; Bulten, J.; Hirschfeld, M.; Colas, E.; Gil-Moreno, A.; Garcia, A.; et al. L1CAM expression in endometrial carcinomas: an ENITEC collaboration study. *Br. J. Cancer* **2016**, *115*, 716–724. [CrossRef]

5. Van der Putten, L.J.M.; Visser, N.C.M.; van de Vijver, K.; Santacana, M.; Bronsert, P.; Bulten, J.; Hirschfeld, M.; Colas, E.; Gil-Moreno, A.; Garcia, A.; et al. Added Value of Estrogen Receptor, Progesterone Receptor, and L1 Cell Adhesion Molecule Expression to Histology-Based Endometrial Carcinoma Recurrence Prediction Models: An ENITEC Collaboration Study. *Int. J. Gynecol. Cancer* **2018**, *28*, 514–523. [CrossRef]
6. Alonso-Alconada, L.; Santacana, M.; Garcia-Sanz, P.; Muinelo-Romay, L.; Colas, E.; Mirantes, C.; Monge, M.; Cueva, J.; Oliva, E.; Soslow, R.A.; et al. Annexin-A2 as predictor biomarker of recurrent disease in endometrial cancer. *Int. J. Cancer* **2015**, *136*, 1863–1873. [CrossRef]
7. Deng, L.; Gao, Y.; Li, X.; Cai, M.; Wang, H.; Zhuang, H.; Tan, M.; Liu, S.; Hao, Y.; Lin, B. Expression and clinical significance of annexin A2 and human epididymis protein 4 in endometrial carcinoma. *J. Exp. Clin. Cancer Res.* **2015**, *34*, 96. [CrossRef]
8. Christensen, M.V.; Høgdall, C.K.; Jochumsen, K.M.; Høgdall, E.V.S. Annexin A2 and cancer: A systematic review. *Int. J. Oncol.* **2018**, *52*, 5–18. [CrossRef]
9. Lianidou, E.; Pantel, K. Liquid biopsies. *Genes. Chromosomes Cancer* **2019**, *58*, 219–232. [CrossRef]
10. Clemmens, H.; Lambert, D.W. Extracellular vesicles: translational challenges and opportunities. *Biochem. Soc. Trans.* **2018**, *46*, 1073–1082. [CrossRef]
11. Steinbichler, T.B.; Dudás, J.; Riechelmann, H.; Skvortsova, I.-I. The role of exosomes in cancer metastasis. *Semin. Cancer Biol.* **2017**, *44*, 170–181. [CrossRef] [PubMed]
12. Peinado, H.; Alečković, M.; Lavotshkin, S.; Matei, I.; Costa-Silva, B.; Moreno-Bueno, G.; Hergueta-Redondo, M.; Williams, C.; García-Santos, G.; Ghajar, C.; et al. Melanoma exosomes educate bone marrow progenitor cells toward a pro-metastatic phenotype through MET. *Nat. Med.* **2012**, *18*, 883–891. [CrossRef] [PubMed]
13. Muinelo-Romay, L.; Casas-Arozamena, C.; Abal, M. Liquid Biopsy in Endometrial Cancer: New Opportunities for Personalized Oncology. *Int. J. Mol. Sci.* **2018**, *19*, 2311. [CrossRef] [PubMed]
14. Mariscal, J.; Fernandez-Puente, P.; Calamia, V.; Abalo, A.; Santacana, M.; Matias-Guiu, X.; Lopez-Lopez, R.; Gil-Moreno, A.; Alonso-Alconada, L.; Abal, M. Proteomic Characterization of Epithelial-Like Extracellular Vesicles in Advanced Endometrial Cancer. *J. Proteome Res.* **2019**, *18*, 1043–1053. [CrossRef] [PubMed]
15. Mathai, R.A.; Vidya, R.V.S.; Reddy, B.S.; Thomas, L.; Udupa, K.; Kolesar, J.; Rao, M. Potential Utility of Liquid Biopsy as a Diagnostic and Prognostic Tool for the Assessment of Solid Tumours: Implications in the Precision Oncology. *J. Clin. Med.* **2019**, *8*, 373. [CrossRef] [PubMed]
16. Williams, C.; Royo, F.; Aizpurua-Olaizola, O.; Pazos, R.; Boons, G.-J.; Reichardt, N.-C.; Falcon-Perez, J.M. Glycosylation of extracellular vesicles: current knowledge, tools and clinical perspectives. *J. Extracell. Vesicles* **2018**, *7*, 1442985. [CrossRef]
17. Kubaski, F.; Osago, H.; Mason, R.W.; Yamaguchi, S.; Kobayashi, H.; Tsuchiya, M.; Orii, T.; Tomatsu, S. Glycosaminoglycans detection methods: Applications of mass spectrometry. *Mol. Genet. Metab.* **2017**, *120*, 67–77. [CrossRef]
18. Tang, Y.-T.; Huang, Y.-Y.; Zheng, L.; Qin, S.-H.; Xu, X.-P.; An, T.-X.; Xu, Y.; Wu, Y.-S.; Hu, X.-M.; Ping, B.-H.; et al. Comparison of isolation methods of exosomes and exosomal RNA from cell culture medium and serum. *Int. J. Mol. Med.* **2017**, *40*, 834–844. [CrossRef]
19. Ratajczak, J.; Wysoczynski, M.; Hayek, F.; Janowska-Wieczorek, A.; Ratajczak, M.Z. Membrane-derived microvesicles: important and underappreciated mediators of cell-to-cell communication. *Leukemia* **2006**, *20*, 1487–1495. [CrossRef]
20. Jablonska, J.; Pietrowska, M.; Ludwig, S.; Lang, S.; Thakur, B.K. Challenges in the Isolation and Proteomic Analysis of Cancer Exosomes-Implications for Translational Research. *Proteomes* **2019**, *7*, 22. [CrossRef]
21. Madeo, M.; Colbert, P.L.; Vermeer, D.W.; Lucido, C.T.; Cain, J.T.; Vichaya, E.G.; Grossberg, A.J.; Muirhead, D.; Rickel, A.P.; Hong, Z.; et al. Cancer exosomes induce tumour innervation. *Nat. Commun.* **2018**, *9*, 4284. [CrossRef] [PubMed]
22. Guo, D.; Lui, G.Y.L.; Lai, S.L.; Wilmott, J.S.; Tikoo, S.; Jackett, L.A.; Quek, C.; Brown, D.L.; Sharp, D.M.; Kwan, R.Y.Q.; et al. RAB27A promotes melanoma cell invasion and metastasis via regulation of pro-invasive exosomes. *Int. J. Cancer* **2019**, *144*, 3070–3085. [CrossRef]
23. Peinado, H.; Lavotshkin, S.; Lyden, D. The secreted factors responsible for pre-metastatic niche formation: Old sayings and new thoughts. *Semin. Cancer Biol.* **2011**, *21*, 139–146. [CrossRef]
24. Nogués, L.; Benito-Martin, A.; Hergueta-Redondo, M.; Peinado, H. The influence of tumour-derived extracellular vesicles on local and distal metastatic dissemination. *Mol. Asp. Med.* **2018**, *60*, 15–26. [CrossRef]

25. Melo, S.A.; Luecke, L.B.; Kahlert, C.; Fernandez, A.F.; Gammon, S.T.; Kaye, J.; LeBleu, V.S.; Mittendorf, E.A.; Weitz, J.; Rahbari, N.; et al. Glypican-1 identifies cancer exosomes and detects early pancreatic cancer. *Nature* **2015**, *523*, 177–182. [CrossRef]
26. An, T.; Qin, S.; Sun, D.; Huang, Y.; Hu, Y.; Li, S.; Zhang, H.; Li, B.; Situ, B.; Lie, L.; et al. Unique Protein Profiles of Extracellular Vesicles as Diagnostic Biomarkers for Early and Advanced Non-Small Cell Lung Cancer. *Proteomics* **2019**, *19*, e1800160. [CrossRef]
27. Zhang, L.; Yu, D. Exosomes in cancer development, metastasis, and immunity. *Biochim. Biophys. Acta. Rev. Cancer* **2019**, *1871*, 455–468. [CrossRef]
28. García-Silva, S.; Benito-Martín, A.; Sánchez-Redondo, S.; Hernández-Barranco, A.; Ximénez-Embún, P.; Nogués, L.; Mazariegos, M.S.; Brinkmann, K.; Amor López, A.; Meyer, L.; et al. Use of extracellular vesicles from lymphatic drainage as surrogate markers of melanoma progression and BRAF V600E mutation. *J. Exp. Med.* **2019**, *216*, 1061–1070. [CrossRef]
29. König, L.; Kasimir-Bauer, S.; Bittner, A.-K.; Hoffmann, O.; Wagner, B.; Santos Manvailer, L.F.; Kimmig, R.; Horn, P.A.; Rebmann, V. Elevated levels of extracellular vesicles are associated with therapy failure and disease progression in breast cancer patients undergoing neoadjuvant chemotherapy. *Oncoimmunology* **2017**, *7*, e1376153. [CrossRef]
30. Dellinger, T.H.; Smith, D.D.; Ouyang, C.; Warden, C.D.; Williams, J.C.; Han, E.S. L1CAM is an independent predictor of poor survival in endometrial cancer - An analysis of The Cancer Genome Atlas (TCGA). *Gynecol. Oncol.* **2016**, *141*, 336–340. [CrossRef]
31. Bosse, T.; Nout, R.A.; Stelloo, E.; Dreef, E.; Nijman, H.W.; Jürgenliemk-Schulz, I.M.; Jobsen, J.J.; Creutzberg, C.L.; Smit, V.T.H.B.M. L1 cell adhesion molecule is a strong predictor for distant recurrence and overall survival in early stage endometrial cancer: pooled PORTEC trial results. *Eur. J. Cancer* **2014**, *50*, 2602–2610. [CrossRef] [PubMed]
32. Fadare, O.; Roma, A.A.; Desouki, M.M.; Gwin, K.; Hanley, K.Z.; Jarboe, E.A.; Liang, S.X.; Quick, C.M.; Zheng, W.; Hecht, J.L.; et al. The significance of L1CAM expression in clear cell carcinoma of the endometrium. *Histopathology* **2018**, *72*, 532–538. [CrossRef] [PubMed]
33. Valapala, M.; Vishwanatha, J.K. Lipid raft endocytosis and exosomal transport facilitate extracellular trafficking of annexin A2. *J. Biol. Chem.* **2011**, *286*, 30911–30925. [CrossRef] [PubMed]
34. Maji, S.; Chaudhary, P.; Akopova, I.; Nguyen, P.M.; Hare, R.J.; Gryczynski, I.; Vishwanatha, J.K. Exosomal Annexin II Promotes Angiogenesis and Breast Cancer Metastasis. *Mol. Cancer Res.* **2017**, *15*, 93–105. [CrossRef] [PubMed]
35. Théry, C.; Witwer, K.W.; Aikawa, E.; Alcaraz, M.J.; Anderson, J.D.; Andriantsitohaina, R.; Antoniou, A.; Arab, T.; Archer, F.; Atkin-Smith, G.K.; et al. Minimal information for studies of extracellular vesicles 2018 (MISEV2018): A position statement of the International Society for Extracellular Vesicles and update of the MISEV2014 guidelines. *J. Extracell. Vesicles* **2018**, *7*, 1535750. [CrossRef]
36. Andreu, Z.; Yáñez-Mó, M. Tetraspanins in extracellular vesicle formation and function. *Front. Immunol.* **2014**, *5*, 442. [CrossRef]
37. Green, T.M.; Alpaugh, M.L.; Barsky, S.H.; Rappa, G.; Lorico, A. Breast Cancer-Derived Extracellular Vesicles: Characterization and Contribution to the Metastatic Phenotype. *Biomed. Res. Int.* **2015**, *2015*, 634865. [CrossRef]
38. Kimling, J.; Maier, M.; Okenve, B.; Kotaidis, V.; Ballot, H.; Plech, A. Turkevich Method for Gold Nanoparticle Synthesis Revisited. *J. Phys. Chem. B* **2006**, *110*, 15700–15707. [CrossRef]
39. Lafuente, M.; Pellejero, I.; Sebastián, V.; Urbiztondo, M.A.; Mallada, R.; Pina, M.P.; Santamaría, J. Highly sensitive SERS quantification of organophosphorous chemical warfare agents: A major step towards the real time sensing in the gas phase. *Sens. Actuators B Chem.* **2018**, *267*, 457–466. [CrossRef]
40. Bartczak, D.; Kanaras, A.G. Preparation of Peptide-Functionalized Gold Nanoparticles Using One Pot EDC/Sulfo-NHS Coupling. *Langmuir* **2011**, *27*, 10119–10123. [CrossRef]

3

Glycan-Modified Apoptotic Melanoma-Derived Extracellular Vesicles as Antigen Source for Anti-Tumor Vaccination

Sophie K. Horrevorts [1], Dorian A. Stolk [1], Rieneke van de Ven [2,3], Myrthe Hulst [1], Bert van Het Hof [1], Sanne Duinkerken [1], Marieke H. Heineke [1], Wenbin Ma [4,5,6], Sophie A. Dusoswa [1], Rienk Nieuwland [7], Juan J. Garcia-Vallejo [1], Arjan A. van de Loosdrecht [8], Tanja D. de Gruijl [2], Sandra J. van Vliet [1] and Yvette van Kooyk [1,*]

1 Amsterdam UMC, Department of Molecular Cell Biology and Immunology, Amsterdam Infection and Immunity Institute, Cancer Center Amsterdam, Vrije Universiteit Amsterdam, 1081 HV Amsterdam, The Netherlands
2 Amsterdam UMC, Department of Medical Oncology, Cancer Center Amsterdam, Vrije Universiteit Amsterdam, 1081 HV Amsterdam, The Netherlands
3 Amsterdam UMC, Department of Otolaryngology/Head and Neck Surgery, Cancer Center Amsterdam, Vrije Universiteit Amsterdam, 1081 HV Amsterdam, The Netherlands
4 Ludwig Institute for Cancer Research, B-1200 Brussels, Belgium
5 Walloon Excellence in Life Sciences and Biotechnology, B-1200 Brussels, Belgium
6 De Duve Institute, Université Catholique de Louvain, B-1200 Brussels, Belgium
7 Amsterdam UMC, Laboratory of Experimental Clinical Chemistry, and Vesicle Observation Centre, Academic Medical Centre, 1081 HV Amsterdam, The Netherlands
8 Amsterdam UMC, Department of Hematology, Cancer Center Amsterdam, Vrije Universiteit Amsterdam, 1081 HV Amsterdam, The Netherlands
* Correspondence: y.vankooyk@amsterdamumc.nl

Abstract: Tumors that lack T cell infiltration are less likely to respond to immune checkpoint inhibition and could benefit from cancer vaccination for the initiation of anti-tumor T cell responses. An attractive vaccine strategy is in vivo targeting of dendritic cells (DCs), key initiators of antigen-specific T cell responses. In this study we generated apoptotic tumor cell-derived extracellular vesicles (ApoEVs), which are potentially an abundant source of tumor-specific neo-antigens and other tumor-associated antigens (TAAs), and which can be manipulated to express DC-targeting ligands for efficient antigen delivery. Our data demonstrates that by specifically modifying the glycocalyx of tumor cells, high-mannose glycans can be expressed on their cell surface and on extracellular vesicles derived after the induction of apoptosis. High-mannose glycans are the natural ligands of dendritic cell-specific intercellular adhesion molecule-3-grabbing non-integrin (DC-SIGN), a dendritic cell associated C-type lectin receptor (CLR), which has the ability to efficiently internalize its cargo and direct it to both major histocompatibility complex (MHC)-I and MHC-II pathways for the induction of $CD8^+$ and $CD4^+$ T cell responses, respectively. Compared to unmodified ApoEVs, ApoEVs carrying DC-SIGN ligands are internalized to a higher extent, resulting in enhanced priming of tumor-specific $CD8^+$ T cells. This approach thus presents a promising vaccination strategy in support of T cell-based immunotherapy of cancer.

Keywords: dendritic cell; apoptotic extracellular vesicles; T cell priming; cancer vaccine; glycan modification; C-type lectin

1. Introduction

Immune checkpoint inhibitors, like anti-PD1 and anti-CTLA-4 blocking antibodies, have improved the long-term survival of patients with metastatic melanoma [1,2]. However, not all patients experience clinical benefit from checkpoint inhibitor therapy [1–4]. Patients who do not have T cell infiltration in the tumor microenvironment, those who carry so–called "cold tumors", are less likely to respond and could benefit from tumor vaccination to initiate anti-cancer immunity [5]. The induction of anti-tumor T cell responses relies on the presentation of antigens by dendritic cells (DCs). After antigen capture in the periphery, DCs mature and migrate via the lymphatic system to the lymph nodes. There, DCs present captured antigens via the conventional antigen presentation route on major histocompatibility complex (MHC) class II and via cross-presentation on MHC-I molecules, and combined with appropriate co-stimulation, subsequently prime and activate $CD4^+$ and cytotoxic $CD8^+$ T cells (CTLs), respectively [6,7]. Efficient delivery of antigens to DCs is key for the induction of tumor-specific T cell responses, and DC-specific C-type-lectin receptors (CLRs) are widely studied as targets for specific antigen delivery [8–10]. CLRs endocytose glycosylated pathogens via their carbohydrate recognition domain, allowing for efficient processing of pathogens and antigen loading onto MHC-I and MHC-II molecules [11]. DEC205 was one of the first CLRs used for DC targeting by conjugating antigens to a DEC205-specific antibody (Ab) and, combined with the administration of anti-CD40 for DC activation, resulted in increased $CD8^+$ T cell activation and growth inhibition of established B16-OVA tumors in vivo [12,13]. Since CLRs are abundantly expressed on multiple DCs subsets and glycans are their natural ligands, the use of glycan motifs for the targeting of vaccines to DCs is an attractive immunization strategy [10,14]. Dendritic cell-specific intercellular adhesion molecule-3-grabbing non-integrin (DC-SIGN), which is expressed on monocyte-derived DCs (moDCs), and on skin and lymph node resident antigen presenting cell (APCs) subsets [10,15–17], is a CLR specific for Lewis-type and high-mannose glycans [18]. After internalization, it efficiently directs its cargo to MHC-I and MHC-II, resulting in enhanced $CD8^+$ and $CD4^+$ T cell responses when targeted with $Lewis^Y$ or high-mannose containing liposomes or proteins [10,19].

Besides efficient DC loading, the choice of antigen is pivotal in anti-cancer vaccine development. There is compelling evidence that the presence of neo-antigens is important in the initiation of tumor-specific T cell responses. Tumors with a high mutational load are associated with increased T cell responses and improved clinical outcome [20,21]. Moreover, vaccination with neo-antigens derived from murine melanoma and colon carcinoma models resulted in tumor control in mice [22,23]. The first neo-antigen-based vaccines clinically tested in melanoma patients also showed promising results with relatively high response rates [24,25]. Neo-antigens are therefore regarded as an attractive vaccine target for the induction of superior anti-tumor T cell responses, but require laborious and costly production pipelines involving genome-wide sequencing, non-synonymous mutation identification, epitope prediction, and RNA or peptide production [24,25]. An abundant source of both patient-specific neo-antigens and shared tumor-associated antigens (TAAs) for vaccination purposes, without the need for prior knowledge of the personalized epitope repertoire, is the autologous tumor itself. We previously showed that apoptotic extracellular vesicles (ApoEVs) derived from tumor cells can induce antigen-specific T cell responses both in vitro and ex vivo [26,27]. ApoEVs are produced when cells undergo apoptosis, resulting in membrane blebbing, membrane protrusion, and apoptotic body (ApoBD) formation [28]. ApoBDs have a diameter of 1–5 μM, express phosphatidylserine (PtdSer) on their membrane surface, and contain, among others, proteins and nuclear fragments [28–33]. Besides ApoBDs, apoptotic cells induce formation of smaller vesicles (<1 μM), called apoptotic microvesicles [28,30,31]. In the literature, no consensus is reached regarding the nomenclature of the smaller apoptotic vesicles; therefore, we decided to use the general term ApoEVs to describe all sizes of extracellular vesicles derived from apoptotic cells.

In this study, we designed a DC-targeted tumor vaccine through glycan modification of melanoma-derived ApoEVs. The induction of DC-SIGN-binding high-mannose glycans on tumor cells prior to the induction of apoptosis with bortezomib resulted in high-mannose expressing ApoEVs, which facilitate enhanced uptake by moDCs via DC-SIGN and enabled antigen-specific CD8$^+$ T cell priming. These data support the use of glycan-modified ApoEVs as vehicles for the selective in vivo delivery of tumor-derived (neo-)antigens to DCs in order to kick-start tumor-specific CD8$^+$ T cell responses.

2. Results

2.1. Kifunensine-Treated Melanoma Cells Express High-Mannose Carbohydrate Structures on Their Cell Surface

To target tumor-specific antigens to moDCs for the induction of superior CTL-responses, we aimed to modify ApoEVs with DC-SIGN binding ligands, such as high-mannose glycan structures (Figure 1A). For the generation of high-mannose expressing ApoEVs (ApoEVs-HM), the melanoma cell line Mel-JuSo was cultured 72 h prior to and during the induction of apoptosis, in the presence of the mannosidase I inhibitor kifunensine, an alkaloid and potent inhibitor of the enzyme mannosidase I [34]. By blocking the trimming of mannose residues on precursor glycoproteins, tumor cells express high-mannose containing glycoproteins [19].

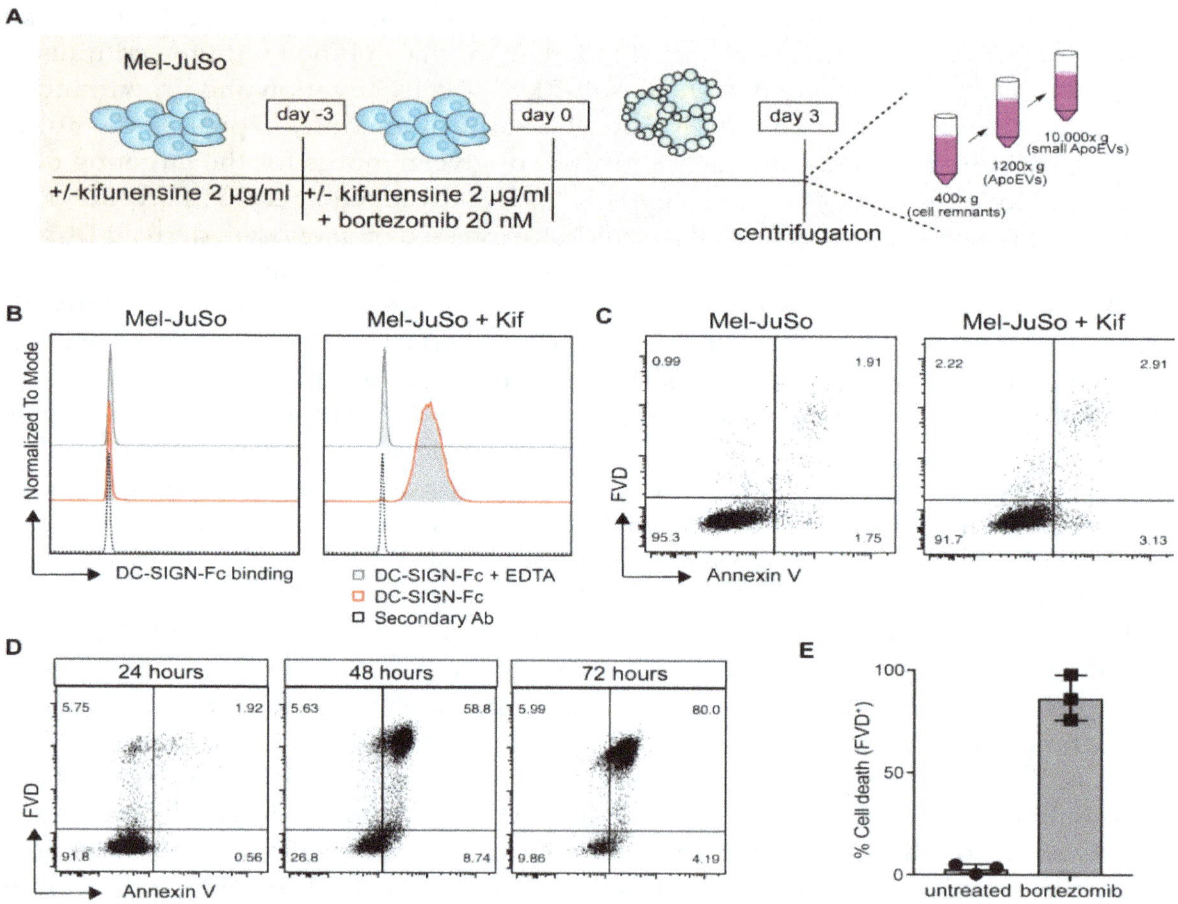

Figure 1. Characterization of kifunensine-treated Mel-JuSo cells. (**A**) Protocol of ApoEV and ApoEV-HM production and isolation procedure. (**B**) Mel-JuSo cells were cultured with kifunensine for 72 h. Dendritic cell-specific intercellular adhesion molecule-3-grabbing non-integrin (DC-SIGN) binding was assessed using DC-SIGN-Fc and measured by flow cytometry. (**C**) Cell viability was measured by flow cytometry after staining with Annexin V and fixable viability dye (FVD). Representative of $n = 3$. (**D**) Mel-JuSo cells were treated with bortezomib to induce apoptosis. After 24, 48, and 72 h Annexin V (as a measure of apoptosis) and the cell viability (FVD) was measured by flow cytometry. Representative plots of $n = 3$. (**E**) After 72 h of apoptosis induction, cell viability ranged between approximately 5–25%. Data shown as mean ± SD of three individual experiments.

Kifunensine treatment induced expression of DC-SIGN binding ligands, as shown by an increased DC-SIGN-Fc binding to Mel-JuSo cells. This is in concordance with previous work where we showed the expression of DC-SIGN binding ligands on a variety of melanoma cell lines after kifunensine treatment [19]. The enhanced DC-SIGN binding was completely abrogated in the presence of EDTA, thereby confirming the specific binding of the DC-SIGN-Fc molecules, as DC-SIGN binding is Ca^{2+} dependent [35] (Figure 1B). The kifunensine treatment did not affect the viability of the cells (Figure 1C). Mel-JuSo cells were treated 72 h with 20 nM bortezomib to induce the formation of early and late ApoEVs. We selected bortezomib for the generation of ApoEVs, as this compound is already used in the clinic for the treatment of multiple myeloma and B cell lymphoma, and potently induces immunogenic cell death [36,37]. Apoptosis induction was monitored every 24 h by membrane staining of PtdSer (Annexin V) in combination with a viability dye (Figure 1D). We observed, 48 h after the induction of apoptosis, an increase in Annexin V staining and decrease in cell viability, with a cell viability below 25% after 72 h (Figure 1E). The ApoEVs were finally isolated using differential centrifugation steps (400× *g* and 1200× *g*) and collected after centrifugation at 10,000× *g* [32,33].

2.2. Glycan Modification Results in ApoEVs with DC-SIGN Binding Properties

We next proceeded to analyze the binding of the different ApoEV and ApoEV-HM batches by DCs. No differences in DC binding could be detected between the unmodified ApoEVs isolated at 1200× *g* or at 10,000× *g*. However, the binding of ApoEVs-HM isolated at 10,000× *g* was significantly increased, compared to the larger vesicles isolated at 1200× *g* (Figure S1). Therefore, we decided to further investigate the immune stimulatory properties of the ApoEVs and ApoEVs-HM isolated at 10,000× *g*. The size distribution of these vesicles was heterogeneous, ranging from approximately 200 nm up to >1000 nm (Figure S2). We checked whether the DC-SIGN binding ligands induced by the kifunensine treatment on the Mel-JuSo cells were preserved on the membranes of the ApoEVs. The vesicles were coated on an ELISA plate and stained with DC-SIGN-Fc (Figure 2A). ApoEVs expressing high-mannose (ApoEVs-HM) bound DC-SIGN-Fc in a dose-dependent manner, in contrast to the unmodified ApoEVs that did not possess DC-SIGN binding properties. The enhanced DC-SIGN binding was completely abrogated in the presence of EDTA, confirming the specificity for DC-SIGN-Fc. Binding of the lectin Narcissus pseudonarcissus agglutinin (NPA) indicated the presence of α1–6 polymannose structures on the high-mannose vesicles (Figure 2B). Next, we generated ApoEVs and ApoEVs-HM from the melanoma cell line SK-Mel28 according to the protocol described in Figure 1A. The SK-Mel28-derived ApoEVs-HM also contained DC-SIGN-binding ligands as detected with DC-SIGN-Fc ELISA (Figure S3), demonstrating that this method can be applied to different tumor cell lines for the generation of DC-SIGN-binding ApoEVs.

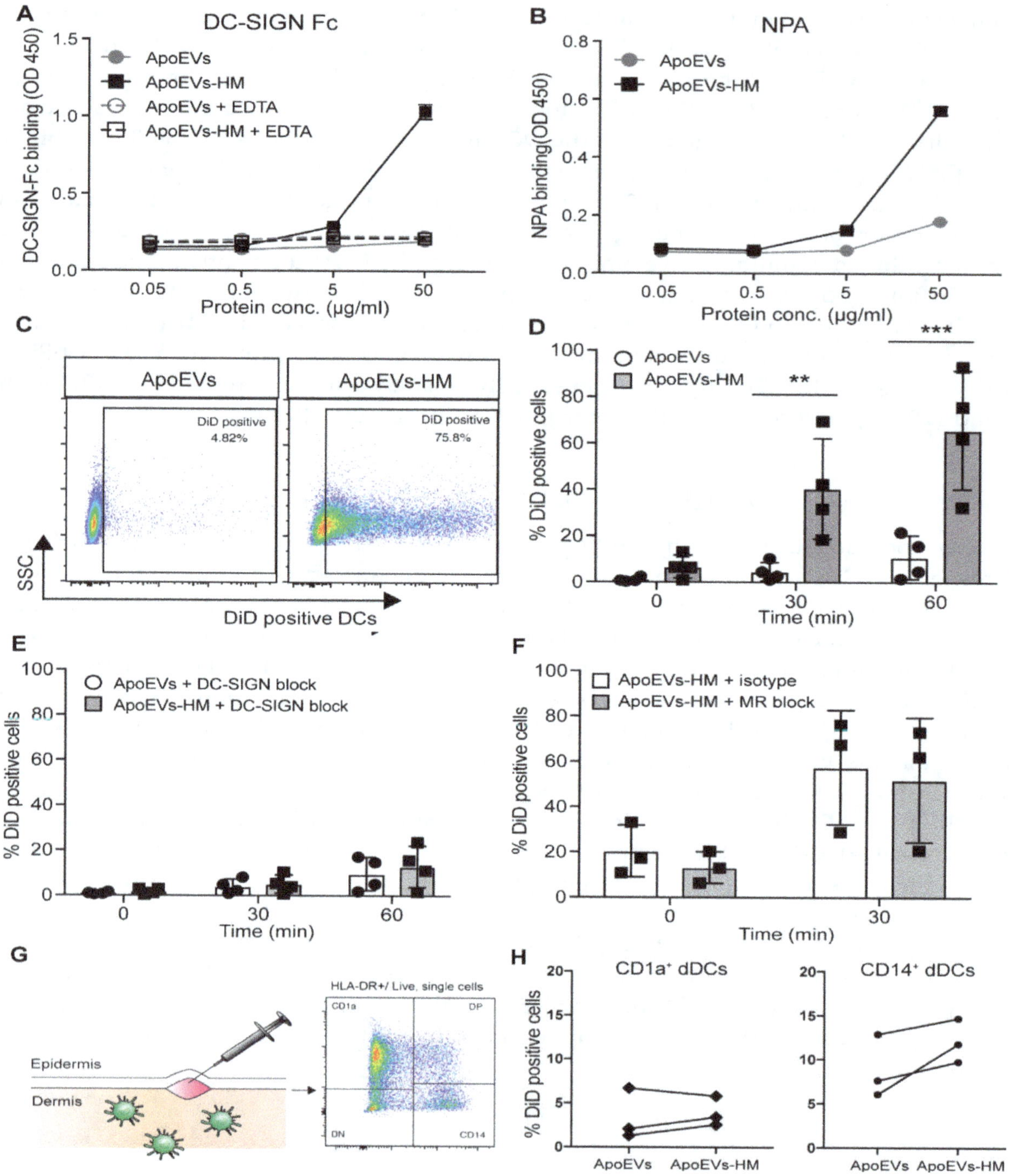

Figure 2. The expression of high-mannose glycans on ApoEVs mediates DC-SIGN binding. (**A**,**B**) Mel-JuSo-derived ApoEVs and ApoEVs-HM were coated, and DC-SIGN binding (**A**) and NPA binding (**B**) were detected by DC-SIGN-Fc or the lectin Narcissus pseudonarcissus agglutinin (NPA), respectively. Data shown as mean ± SD; representative of three individual experiments. (**C**–**F**) MoDCs were pulsed with DiD-labeled ApoEVs and ApoEVs-HM for 45 min at 4 °C, before cells were transferred to 37 °C. Uptake was measured by flow cytometry. (**C**) Representative flow cytometry data at time point 60 min. (**D**) Uptake of ApoEVs and ApoEVs-HM by moDCs after 0, 30, and 60 min. Data shown as mean ± SD of four donors. Statistics performed; two-way repeated measures ANOVA with Sidak post-hoc test. ** $p < 0.01$, *** $p < 0.001$. (**E**) MoDCs were blocked with a DC-SIGN blocking Ab (AZN-D1) or (**F**) MR blocking Ab 30 min prior to the loading with the ApoEVs or ApoEVs-HM. Data shown as mean ± SD of four donors (**E**) or three (**F**) donors. Statistics performed; two-way repeated measures ANOVA with Sidak post-hoc test. (**G**) Gating strategy for the CD1a^{+} and CD14^{+} dermal DCs (dDCs). (**H**) Uptake of DiD-labeled ApoEVs and ApoEVs-HM by migrated dDCs following in situ injection.

2.3. *High-Mannose Expressing ApoEVs Are Internalized via DC-SIGN by moDCs*

To evaluate the DC-SIGN-binding properties of our ApoEVs-HM, we pulsed moDCs with DiD-labeled vesicles for 45 min on ice, before transferring them to 37 °C for an additional 30- or 60-min incubation. The percentage of DiD-positive moDCs was determined as a measure of vesicle binding/uptake. After 60 min at 37 °C, up to 93% of the ApoEV-HM pulsed moDCs were DiD-positive compared to approximately 20% of the moDCs pulsed with the control ApoEVs (Figure 2C,D). Pre-treatment with AZN-D1, a DC-SIGN blocking Ab, completely abrogated uptake of ApoEVs-HM (Figure 2E), showing that the enhanced uptake was completely DC-SIGN-dependent. Because the mannose receptor (MR) on moDCs might also bind mannose structures, we tested whether moDCs could bind ApoEV-HM via MR (Figure 2F). The uptake of ApoEVs-HM was not affected by blocking the MR and was comparable to the uptake of ApoEVs-HM by isotype control treated moDCs. To further investigate the DC-SIGN-targeting properties of ApoEVs-HM, we used a human skin explant model [17], where we injected the vesicles to confirm binding of the ApoEVs-HM to human dermal DCs (dDCs) that naturally express DC-SIGN [16,17]. After two days, the migrated dDCs were analyzed by flow cytometry to identify vesicle uptake (DiD-labeled) by $CD1a^+$ ($HLA\text{-}DR^+/CD1a^+$) and $CD14^+$ ($HLA\text{-}DR^+/CD14^+$) dDCs and Langerhans cells ($HLA\text{-}DR^+/CD1a^{high}/EpCAM^+$) [17] (Figure 2G). A trend of increased ApoEV-HM uptake could be observed in the $CD14^+$ dDCs population compared to the $CD1a^+$ dDCs subset, which is in line with the higher expression of DC-SIGN on the $CD14^+$ subset [16] (Figure 2H). The number of migrated Langerhans cells was, unfortunately, too low to be analyzed.

Since adjuvants can affect the internalization capacity of moDCs [38], we investigated the DC-SIGN-mediated uptake of ApoEVs-HM in the presence of a TLR4-stimulus (Figure S4). MoDCs that received a TLR-4 stimulus (MPLA) showed similar internalization capacity compared to unstimulated moDCs, demonstrating that DC-SIGN is a suitable receptor for vaccine targeting strategies of both immature and maturing moDCs. To investigate if the ApoEVs were internalized by moDCs, we assessed vesicle uptake by imaging flow cytometry. After 30 min incubation at 37 °C, around 50–65% of the moDCs loaded (at 4 °C) with ApoEVs-HM were positive for DiD (Figure 3A). The amount of ApoEV-positive moDCs that could be used for image analysis was low, therefore we only determined the uptake in ApoEV-HM-loaded moDCs. Using imaging flow cytometry, we could discriminate between membrane-bound and internalized DiD-positive vesicles. We created a mask that we applied to all DiD-positive moDCs, which allowed for counting DiD-positive spots, representing the ApoEVs-HM localized at the membrane surface or intracellular (Figure 3B). After 45 min incubation at 4 °C (time point: 0 min), the DiD-labeled ApoEVs-HM were solely localized at the membrane surface of moDCs (Figure 3C). The internalization score was calculated, as the intensity of the DiD signal of the intracellular space versus the intensity of the entire cell. MoDCs that had internalized ApoEVs-HM typically had positive scores, whereas moDCs displaying membrane-bound ApoEVs-HM had negative scores. (NB: the internalization score was 0 when similar amounts of vesicles were located at the membrane, as in the intracellular space). After 30 and 60 min incubation at 37 °C, the ApoEVs-HM were internalized as indicated by the increase in internalization score (Figure 3D,E representative pictures). Collectively, these data show that DC-SIGN facilitates increased binding and internalization of ApoEVs-HM by moDCs.

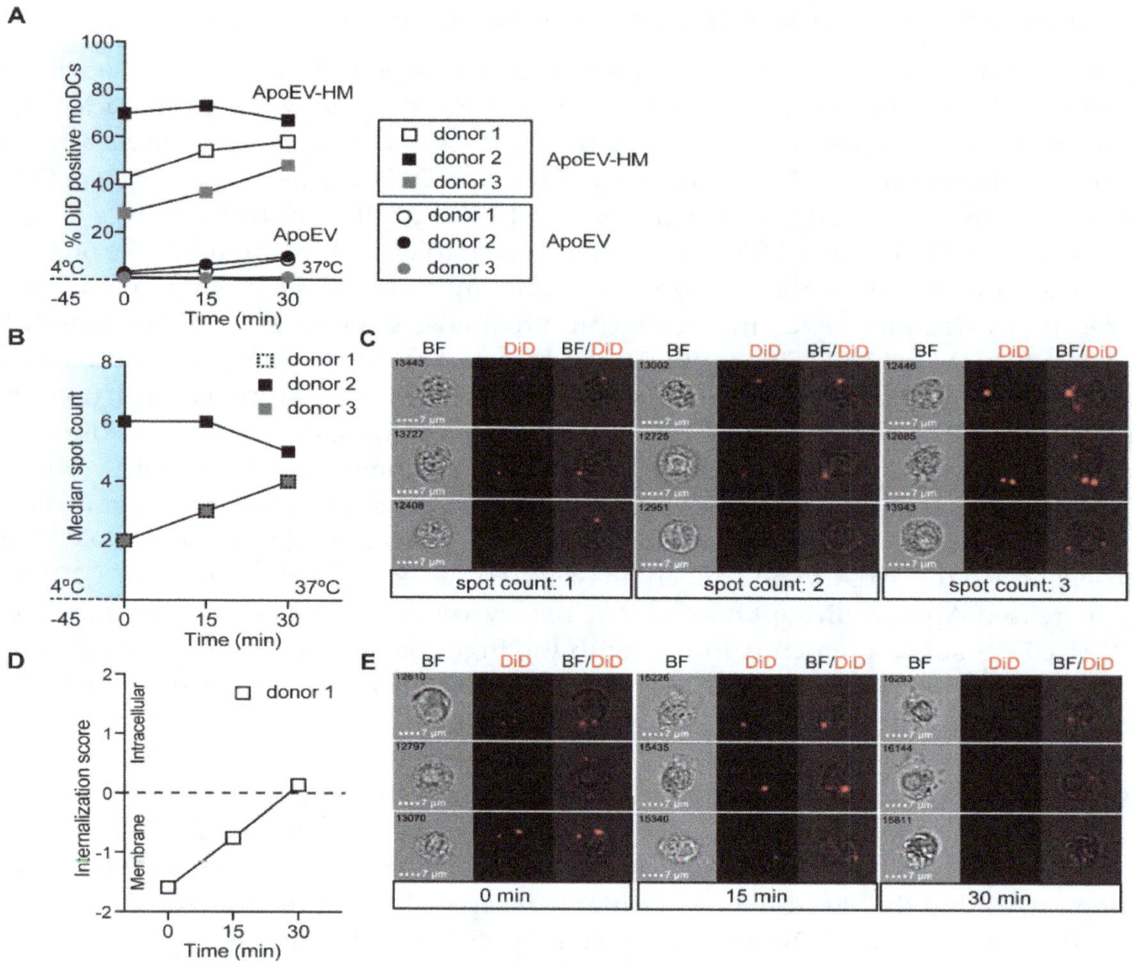

Figure 3. Glycan modified ApoEVs (ApoEVs-HM) are internalized by moDCs. (**A**) MoDCs (white = donor 1, black = donor 2, grey = donor 3) were loaded with DiD-labeled ApoEVs and ApoEVs-HM for 45 min at 4 °C, washed, and incubated at 37 °C for the indicated times. MoDCs were analyzed using imaging flow cytometry. (**B**) Time course of the median spot count in the DiD-positive ApoEV-HM-loaded moDC population using imaging flow cytometry. (**C**) Representative pictures of ApoEV-HM-loaded moDCs at time point 0 min with 1, 2, or 3 positive spots. (**D**) Time course of the internalization score of ApoEVs-HM of donor 1. (**E**) Representative imaging flow cytometry pictures after 0, 15, and 30 min incubation at 37 °C of moDCs (donor 1) that are positive for 2 spots.

2.4. Mel-JuSo-Derived ApoEVs Do Not Mature moDCs

Stress-induced apoptosis can result in immunogenic cell death, a form of cell death that facilitates T cell-mediated immune responses and is characterized by the secretion and cell surface expression of damage-associated molecular patterns (DAMPs), like high mobility group protein B1 (HMGB1), heat-shock proteins, and calreticulin [36]. These molecules have the ability to induce DC maturation by binding pattern recognition receptors (PRRs) [37,39]. The presence of DAMPs on the vesicle membrane would allow for self-adjuvating properties of the vesicle vaccine. We investigated the capacity of our isolated vesicles to mature moDCs by quantifying DC maturation markers by flow cytometric analysis. While TLR4 stimulation with lipopolysaccharide (LPS) induced DC maturation as expected, expression levels of CD80, CD86, and HLA-DR did not differ between untreated and vesicle-loaded moDCs (Figure 4A). However, both LPS-stimulated and ApoEV-HM-loaded moDCs appeared to have slightly lower DC-SIGN expression (Figure 4A). These results can be explained by the fact that DC-SIGN is a non-recycling receptor, and expression decreases after ligand internalization or maturation of DCs [38]. We also checked whether moDC maturation upon TLR stimulation was affected by ApoEV or ApoEV-HM loading, which would not be favorable in a therapeutic setting.

The maturation capacity of LPS was not inhibited in ApoEV-loaded moDCs, as maturation marker expression was comparable between the different conditions (Figure 4B). This was in line with the results observed in a mixed-leukocyte reaction (MLR), where DC-induced $CD4^+$ T cell and $CD8^+$ T cell proliferation did not differ between the two conditions (Figure S5). Together, our results indicate that bortezomib-induced ApoEVs derived from Mel-JuSo cells do not have moDCs maturing properties, nor did they negatively impact moDC maturation upon TLR4 stimulation.

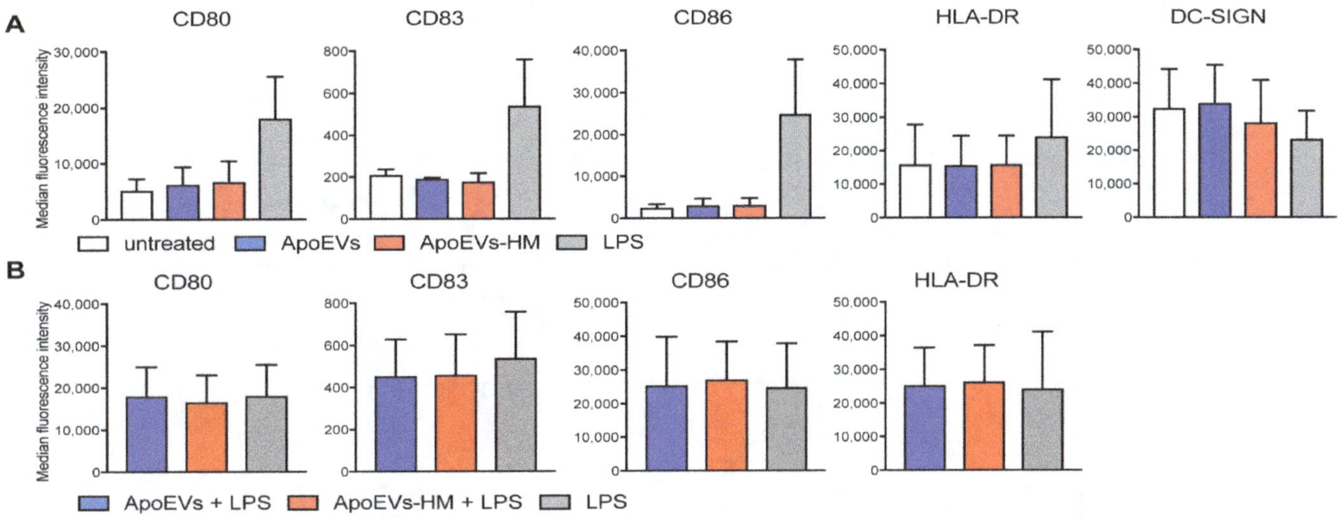

Figure 4. Glycan-modified ApoEVs (ApoEVs-HM) do not mature moDCs. (**A**) MoDCs were loaded with ApoEVs, ApoEVs-HM (100 μg/mL (two donors) or 200 μg/mL (two donors)), or LPS for 3 h, washed, and cultured o/n. The next day, the expression of maturation markers CD80, CD83, CD86, HLA-DR, and DC-SIGN was analyzed by flow cytometry. Graph is combined data of the four donors, showing mean + SD of the median fluorescence intensity. (**B**) MoDCs were loaded with ApoEVs, ApoEVs-HM (100 μg/mL (two donors) or 200 μg/mL (two donors)) in the presence of LPS (10 ng/mL) for 3 h. The next day, the expression of CD80, CD83, CD86, and HLA-DR was analyzed by flow cytometry. Graph is combined data of the four donors, showing mean + SD of the median fluorescence intensity.

2.5. ApoEV-HM-Loaded moDCs Induce MART-1-Specific $CD8^+$ T Cells

Finally, we assessed if the increased uptake of ApoEVs-HM also resulted in the cross-priming of tumor-associated antigen-specific HLA-A2-restricted CTLs. The ApoEVs were isolated from $HLA\text{-}A2^-$ Mel-JuSo cells in order to prevent direct HLA-A2-restricted antigen presentation of the ApoEVs to T cells. To determine the potency of ApoEV-HM-loaded moDC to prime naïve T cells, we used the MART-1 (/Melan-A) as a melanoma-specific antigen. Since endogenous MART-1 was not detected in Mel-JuSo cells, we transduced the cells with MART-1. ApoEVs and ApoEVs-HM derived from the MART-1-expressing Mel-JuSo cell lines contained MART-1 protein, which we confirmed by Western blot (Figure S6). $CD8\beta^+$ T cells were co-cultured with autologous irradiated $CD8\beta^+$ depleted PBMCs (feeders) and allogeneic, HLA-A2-matched vesicle-loaded moDCs for 10 days (Figure 5A). As a read-out, we used dextramers (Dx), recognizing the high-affinity altered-peptide ligand for the immune-dominant MART-1 epitope MART-$1_{26\text{–}35}$, i.e., MART-$1_{26\text{–}35L}$ [40]. MART-1-specific $CD8^+$ T cells could be detected at higher frequencies in cultures stimulated with moDCs loaded with ApoEVs-HM compared to cultures containing the wild-type ApoEV-loaded moDCs (Figure 5B,D). As a positive control, moDCs were pulsed with a short synthetic peptide of the immune-dominant MART-1 epitope MART-$1_{26\text{–}35L}$ (Figure 5C,D), and unpulsed moDCs (no ApoEVs) were taken along as negative control. ApoEVs-HM induced higher percentages of MART-1-specific $CD8^+$ T cells compared to the wild-type ApoEVs, thereby demonstrating that increased internalization of apoptotic vesicles by moDCs results in enhanced $CD8^+$ T cell priming capacity.

In conclusion, we provide evidence that glycan modification of ApoEVs promotes vesicle internalization and CD8$^+$ T cell priming.

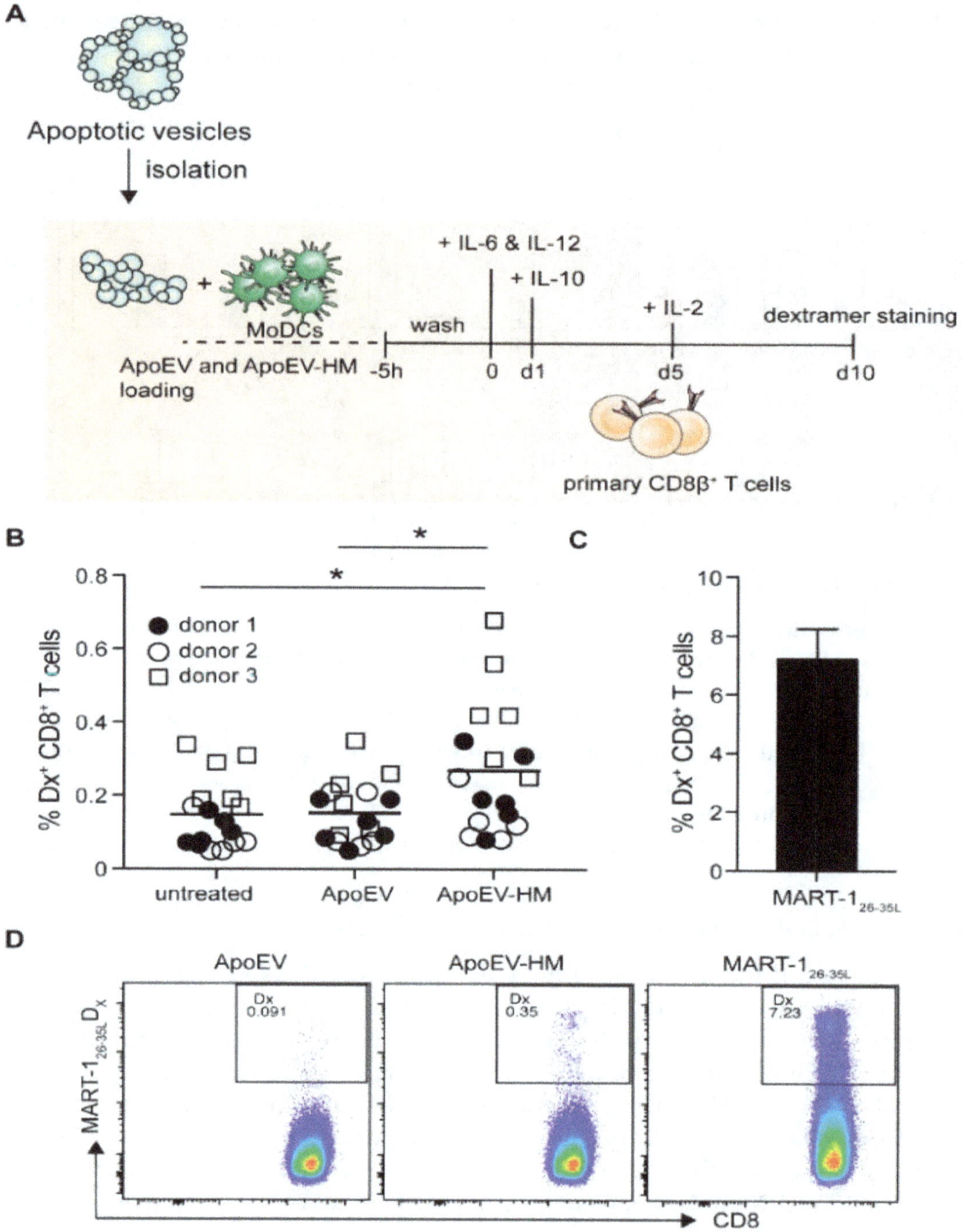

Figure 5. Glycan modification of ApoEVs (ApoEVs-HM) results in priming and expansion of MART-1-specific CD8$^+$ T cells. (**A**) Protocol of priming and expansion of MART-1-specific CD8$^+$ T cells. (**B**) MoDCs were loaded with the different vesicles in the presence of LPS. For the first and third donor six, and for the second donor five separate cultures were evaluated for each condition. Frequencies of MART-$1_{26\text{–}35L}$-dextramer$^+$-CD8$^+$ T cells were analyzed 10 days after induction. Combined data of three donors is shown as percentage of MART-$1_{26\text{–}35L}$-dextramer$^+$-CD8$^+$ T cells. Data were analyzed by one-way ANOVA with Tukey post-hoc test, * $p < 0.05$. (**C**) Frequencies of MART-$1_{26\text{–}35L}$ Dx + CD8$^+$ T cells present in the control cultures using the short MART-$1_{26\text{–}35L}$ peptide. Data is shown as mean + SD of donor 1. (**D**) Representative flow cytometry plots of MART-$1_{26\text{–}35L}$ -Dx$^+$ CD8$^+$ T induced by ApoEV-, ApoEV-HM-, or short peptide (control)-loaded moDCs.

3. Discussion

Targeting tumor-specific antigens to DCs has been shown to increase tumor vaccination efficacy in murine in vivo models [9,13]. However, most of these studies used single antigen formulations, whereas multi-antigenic vaccines hold the power to elicit more (poly)clonal T cell responses directed against multiple tumor-associated (neo-)antigens, thereby broadening anti-tumor immunity and avoiding immune escape. In this study, we investigated multi-antigenic glycan modified ApoEVs as a DC-targeting vaccine to enhance anti-tumor immunity. Our data show that the applied glycan modification of ApoEVs (high mannosylation) resulted in robust targeting to the CLR DC-SIGN. Moreover, it enhanced vesicle uptake and facilitated efficient tumor-specific $CD8^+$ T cell priming, without altering the moDC phenotype or interfering with TLR-mediated DC maturation.

The induction of high-mannose glycans on ApoEVs facilitates efficient DC-SIGN-mediated antigen loading of DCs. The increase in uptake could be fully abrogated by blocking DC-SIGN, thereby excluding a role for the MR. A similar effect was previously observed regarding the uptake of high-mannose gp100 glycoproteins, which were also solely internalized via DC-SIGN [19]. The binding and internalization of ApoEVs did not result in the expression of maturation markers, which was in line with our previous findings showing that heat shock-induced leukemia-derived ApoEVs had no maturing effect on moDCs [26]. However, contradictory data have been published, where uptake of ApoEVs induced upregulation of DC maturation markers (e.g., CD40 and CD86) and secretion of pro-inflammatory cytokines (e.g., IL-6 and TNF-α) by murine bone marrow-derived dendritic cells (BMDCs), and the expression of CD80, CD83 and CD86, and secretion of IL-8 and TNF-α by human moDCs [41,42]. As ApoEVs are not well characterized and the isolation methods differ greatly between studies, these variations in DC-activating effects may be explained by the differences in vesicle origin, the timing of administration, the methods used for apoptosis induction, the isolation of the vesicles, or the size of the vesicles. The effect of vesicle size on immune activation was emphasized by the observation that ApoEVs (size 1–3 μM) derived from murine endothelial cells contained the pro-inflammatory cytokine IL-1α, while the smaller vesicles (<1 μM) did not [43]. When injected in the peritoneal cavity of mice, the larger ApoEVs (1–3 μM) induced a sterile inflammation with neutrophil infiltration [43]. Nevertheless, apoptotic cells can also have an immune inhibitory effect. Apoptotic cell uptake by TLR-stimulated DCs has been reported to result in the inhibition of NF-κB activation, reduced secretion of the pro-inflammatory cytokine IL-12, hampered CD86 expression and the secretion of the regulatory cytokine IL-10 [44–46]. In contrast, our isolated ApoEVs did not inhibit the upregulation of DC maturation markers (CD80, CD83, CD86, and HLA-DR) upon TLR-stimulation, and no differences were observed in the capacity to induce T cell proliferation between maturing ApoEV-loaded and non-ApoEV-loaded DCs in vitro. This is not surprising, as we have seen before that matured moDCs retain their capacity to express CCR7 and migrate towards CCL19 after loading with ApoEVs [26]. Fransen et al. showed an increase in T cell-mediated production of IL-2, IFN-gamma, and, in particular, IL-17 induced by ApoEV-loaded BMDCs. However, their mouse-derived (32Dcl3 cells) ApoEVs had BMDC maturing properties and during their in vitro experiments, no additional adjuvant was administered [41].

Multiple studies have already reported on the ability of ApoEV-loaded DCs to induce antigen-specific $CD8^+$ T cell responses [26,27,47]. Our unmodified ApoEVs were not able to induce robust MART-1_{26-35}-specific $CD8^+$ cell responses. The amount of antigen expressed by the tumor cells, the immunogenicity of the model antigen, and the experimental setup are potential factors influencing T cell priming by ApoEVs. The work of Muhsin-Sharafaldine et al. demonstrated that ApoEVs from B16 tumors endogenously expressing ovalbumin did not result in anti-tumor protection, while after antigen enrichment, ApoEVs were superior in the induction of anti-tumor immunity compared to antigen enriched exosomes and micro-vesicles [47]. We did observe a significant increase in the MART-specific $CD8^+$ T cells frequency in the primed cultures, where moDCs were loaded with glycan-modified ApoEVs, indicating that the enhanced DC-SIGN-mediated uptake augments

T cell responses. These results verify the concept that DC-SIGN targeting, via Abs, glycan-modified liposomes, and modified ApoEVs, can boost $CD8^+$ T cell responses [9,10,38].

Future research should focus on the minimum antigen content of ApoEVs necessary for the induction of strong anti-tumor T cell responses, preferably in an in vivo setting, using neo-antigens or non-self-antigens as model antigens. Additionally, inclusion of multiple neo-antigens in the vaccine is essential, given that the expression of neo-antigens can be selectively lost in tumors and may lead to tumor resistance [48]. Using patient-derived glycan-modified vesicle vaccines could therefore be an abundant and easily accessible source for tumor-specific antigens that circumvents the need for neo-antigen characterization and HLA matching. The content of these vesicles, and the effect of different levels of various (neo-)antigens present in these vesicle preparations, should be assessed to appreciate their impact on the induced anti-tumor immune response and the ultimate vaccination and anti-tumor efficacy achieved.

4. Materials and Methods

4.1. Reagents and Antibodies

The following reagents were used: Saponin (Sigma-Aldrich, Zwijndrecht, The Netherlands), monophosphoryl lipid A (MPLA) from *Salmonella enterica* (InvivoGen, Toulouse, France), lipopolysaccharide (LPS) from *Escherichia coli* (Sigma-Aldrich) Staphylococcal enterotoxin B from Staphylococcus aureus (SEB) (Sigma-Aldrich), Phytohemagglutinin-L (PHA-L) (Vector laboratories, Burlingame, CA, USA), paraformaldehyde (PFA) aqueous solution (Electron Microscopy Sciences, Hatfield, PA, USA), citric acid (Merck Millipore, Amsterdam, The Netherlands), sodium acetate (J.T Baker), TMB (Sigma-Aldrich), 2-mercaptoethanol (Sigma-Aldrich), Laemmli buffer (Bio-Rad, Veenendaal, The Netherlands), bovine serum albumin (BSA) (Roche, Nederland BV, Woerden, The Netherlands), trypsin (ThermoFisher Scientific, Breda, The Netherlands), bortezomib (Selleck, Munich, Germany), human serum (Sigma-Aldrich), human Interleukin-2 (IL-2) (Proleukin (Novartis), Arnhem, The Netherlands), IL-4 (ImmunoTools, Friesoythe, Germany), human IL-6 (R&D Systems, Minneapolis, MN, USA), human IL-10 (R&D Systems), human IL-12 (Peprotech, London, UK), human granulocyte-macrophage colony-stimulating factor (GM-CSF) (ImmunoTools), Kifunensine (Tocris Bioscience, Abingdon, UK) and Bradford protein assay kit (ThermoFisher). The following fluorescent labeled probes and Abs were used: PO-labeled Goat anti-human IgG Fc (Jackson, Ely, UK), anti-CD8β (clone 2ST8.5H7, Beckman Coulter, Brea, CA, USA), anti-MART-1 (clone M2-7C10, ThermoFisher), rabbit anti-GAPDH (14C10, Cell Signaling, Leiden, The Netherlands), anti-mouse-800 (LI-COR Biosciences, Lincoln, NE, USA), anti-rabbit-680 antibodies (LI-COR Biosciences), FITC-labeled anti-human IgG-Fc (Jackson), anti-CD80-FITC (clone 2D10, Biolegend, San Diego, CA, USA), anti-CD83-PE-Cy7 (clone HB15e, eBioscience, Santa Clara, CA, USA), anti-CD86-PE (clone 2331 (FUN-1), BD Biosciences, Temse, Belgium), anti-HLA-DR-BV510 (clone G46-6, BD Biosciences), anti-CD1a-FITC (clone HI149, BD Biosciences), anti-CD14-PE (clone M5E2, Biolegend), anti-EpCAM-BV421 (clone EBA-1, BD Biosciences), unconjugated anti-CD206 (clone 19.2, BD Biosciences), anti-CD8-BV421 (clone RPA-T8, BD Biosciences), anti-CD4-APC (clone RPA-T4, BD Biosciences), streptavidin-PO (Biosource), AF-488-streptavidin (Invitrogen), Violet tracer proliferation kit (ThermoFisher), FITC-Annexin V (BD Biosciences), the monoclonal IgG1 Ab AZN-D1 and AF488 labeled AZN-D1, against the carbohydrate-recognition domain of DC-SIGN (in house production) [35], PE-labeled MART-$1_{26\text{–}35L}$ dextramer (MART-$11_{26\text{–}35L}$ Dx, Immundex, Copenhagen, Denmark), the lipophilic tracer DiD, and the fixable viability dye (FVD) eFluor 780 (both from Invitrogen). The goat-anti-mouse (GaM) magnetic beads and human CD14 beads came from Miltenyi Biotec (Bergisch Gladbach, Germany).

4.2. Cell Lines, Lentiviral Transduction, Glycan Modification, and Apoptosis Induction

The human HLA-A2 negative (HLA-2^-) melanoma cell lines Mel-JuSo and SK-MEL28 (a gift from prof. T.D. de Gruijl, Amsterdam UMC, Amsterdam) were cultured in IMDM medium (Invitrogen) supplemented with 10% FCS, 100 U/mL penicillin/streptomycin, and 2 mM L-glutamine (all from Lonza, Verviers, Belgium) (complete RPMI). For the generation of the MART-1-expressing Mel-JuSo cells, the MART-1 gene (a gift from prof. B.J. van den Eynde, Ludwig Institute, Brussels) was cloned using the NheI and XbaI sites of the PMEL human lentiviral vector (ABM, cat. no. LV801443, accession number NM_006928). An empty control vector was generated by removal of the PMEL human gene using EcoRV5 and subsequent re-ligation. Recombinant lentiviruses were produced by co-transfecting sub-confluent human embryonic kidney (HEK) 293T cells with the lentiviral expression plasmid and packaging plasmids (pMDLg/pRRE and pRSV/Rev) using calcium phosphate as transfection agent. HEK 293T cells were cultured in complete DMEM, in a 37 °C incubator with 5% CO_2. Infectious lentivirus was collected 48 h after transfection and the supernatant was centrifuged to remove cell debris and stored at −80 °C. Mel-JuSo cells were transduced with the lentivirus preparations and after 48 h, cells expressing MART-1/GFP were selected by puromycin (0.2×10^{-7} g/mL) containing selection medium. Three days before induction of apoptosis (day 3), the Mel-JuSo culture medium was supplemented with 2 μg/mL kifunensine. At day 0, 4×10^6 Mel-JuSo cells were seeded in T75 flask (10 mL), supplemented with 2 μg/mL kifunensine. The cells were allowed to adhere for at least 2 h and apoptosis was induced by adding 20 nM bortezomib to the culture medium for 72 h. Cell viability and apoptotic rate was assessed via staining with FITC-labeled Annexin V and fixable viability dye. In short, the culture medium and adherent cells, after trypsinization, were pooled and stained with FVD for 30 min. Hereafter, cells were washed twice and stained with FITC-labeled Annexin V in Annexin V staining buffer for 15 min at RT. Cells were washed and resuspended in Annexin V staining buffer and analyzed by flow cytometry on a ×20 Fortessa SORP flow cytometer (BD Biosciences).

4.3. ApoEV Isolation Protocol

The vesicles were collected 72 h after the induction of bortezomib-induced apoptosis (4.2.) via differential centrifugation steps: 10 min at 400× *g*, 20 min at 1200× *g*, and 30 min 10,000× *g* (ApoEVs) in order to remove cell debris (400× *g* fraction) and larger ApoEVs (1200× *g* fraction) [32,33]. All centrifugation steps were performed at 4 °C. The pellet containing ApoEVs (10,000× *g* fraction) was used for further experiments. Protein concentration was determined by Bradford protein assay kit according to manufacturer's instructions. Protein concentrations in the ApoEV and ApoEV-HM lysed isolated fractions were generally comparable. Size distribution of the ApoEVs and ApoEVs-HM were analyzed by nanoparticle tracking analysis (NTA, Nanosight,).

4.4. Monocyte-Derived DCs

Immature moDCs were generated from monocytes obtained from human peripheral blood mononuclear cells (PBMCs), isolated from buffy coats of healthy donors (Sanquin, Amsterdam, The Netherlands) by a sequential lymphoprep (Axis-Shield, Dundee, UK) and percoll (Amersham, GE Health care, Eindhoven, The Netherlands) gradient. The monocytes were then cultured for 5–6 days in complete RPMI in the presence of recombinant human GM-CSF (800 U/mL) and IL-4 (500 U/mL).

4.5. Lectin Binding Assays

Cells were cultured in T75 flasks and supplemented with kifunensine (2 μg/mL) for 3 days. At day 3, the cells were stained with DC-SIGN-Fc and the plant lectin *Narcissus pseudonarcissus* agglutinin (NPA). DC-SIGN-Fc was produced from transfectants as previously described [49]. After trypsinization, cells were washed twice in TSM buffer (20 mM Tris, pH 7.4, 150 mM NaCL, 1 mM $CaCL_2$, 2 mM $MgCL_2$) supplemented with 1% BSA and stained with biotinylated *Narcissus pseudonarcissus* agglutinin (NPA) (specific for α1–6 polymannose) and purified DC-SIGN-Fc (10 μg/mL) for 30 min at RT. Lectin binding was detected with AF488-conjugated Streptavidin and FITC-labeled anti-human IgG-Fc, viability of the cells was determined with FVD and cells were analyzed by flow cytometry. For lectin binding analysis on ApoEVs, ELISA plates (Nunc maxisorp, ThermoFisher) were coated o/n at 4 °C with 0.05–50 μg/mL ApoEVs in coating buffer (50 mM Na_2CO_3, pH 9.7). Plates were washed in TSM buffer containing 0.05% Tween and blocked with 1% fatty acid free BSA in TSM for 30 min at 37 °C. Plates were incubated with 50 µL DC-SIGN-Fc (2 µg/mL) or biotinylated NPA (1 µg/mL) for 1 h at RT. Binding was detected using PO-labeled Goat anti-human IgG Fc or streptavidin-PO (30 min on RT). Binding was quantified by the addition of substrate buffer (0.1 M citric acid, 0.1 mM sodium acetate, pH 4, 100 µg/mL 3,3′-5,5′-tetramethylbenzidine (TMB) and 0.006% H_2O_2. Absorbance was measured at 450 nm using a spectrophotometer (Bio-Rad).

4.6. Binding and Uptake of ApoEVs

ApoEVs were labeled with 0.5–1 µM DiD for 15 min at 37 °C. After labeling, vesicles were washed twice in PBS and centrifuged at 10,000× *g* for 30 min. Blocking DC-SIGN and MR was done by pre-treating the DCs with AZN-D1 (20 µg/mL), anti-MR (CD206) (40 µg/mL), or isotype control (40 µg/mL) for 20–30 min at 37 °C. Immature or maturing (10 µg/mL MPLA) moDCs (10×10^4 per time point) were loaded with 50 µg/mL DiD labeled ApoEVs or ApoEVs-HM for 45 min on ice to allow binding and thereafter transferred to 37 °C. At different time points (0, 30 and 60 min) cells were washed, and subsequently fixed with 1% PFA. The cells were analyzed by flow cytometry (×20 Fortessa SORP flow cytometer, BD Biosciences).

4.7. In Situ Dermal Dendritic Cell Targeting

Human abdominal skin explants from healthy donors (Bergman Clinics, Bilthoven, The Netherlands) were injected intradermal with 20 µL serum free IMDM (supplemented with penicillin streptomycin, glutamine and gentamycin), containing 50 µg/mL ApoEVs or ApoEVs-HM. At the site of injection, a blister would appear, of which a punch biopsy (8 mm; Microtec) surrounding the blister was taken. Around 8 biopsies per condition were obtained and cultured in a 48-well plate, with the epidermis facing upwards. After 2 days, the crawl-out cells were harvested and stained for analyses by flow cytometry. To distinguish the different migrated dermal DC subsets, we stained with the following antibodies: HLA-DR, CD1a, CD14, EpCAM, and FVD.

4.8. MoDC Maturation Assay

Immature moDCs were seeded in a 96-well U-bottom plate (Greiner) at a concentration of 50×10^3 cells per well and loaded with the different vesicles (2 donors 200 µg/mL and 2 donors 100 µg/mL or LPS (10 ng/mL)). After 3 h, moDCs were washed in PBS and cultured o/n. MoDCs were stained for the expression of DC maturation markers CD80, CD86, and HLA-DR for 30 min. Marker expression was measured by flow cytometry (×20 Fortessa SORP flow cytometer, BD Biosciences).

4.9. Mixed Leukocyte Reaction (MLR) and Cytokine Secretion

$CD14^+$ monocytes were isolated from PBMCs by positive selection using anti-CD14 conjugated magnetics beads according to the manufacturer's instructions. Monocytes were then cultured for 5 days in the presence of 500 U/mL IL-4 and 800 U/mL GM-CSF for the generation of immature moDCs. Peripheral blood lymphocytes (PBLs) from healthy donors were labeled with 5 µM violet tracer, according to the manufacturer's instructions and co-cultured with allogeneic ApoEVs and ApoEVs-HM (200 µg/mL) loaded and LPS-stimulated (10 ng/mL) (24 h) moDCs in complete IMDM medium for 7 days. Control moDCs were stimulated with 5 ng/mL SEB or 5 µg/mL PHA-L. At day 7, cells were harvested and labeled with anti-CD4 and anti-CD8. Proliferation was determined using flow cytometry (×20 Fortessa SORP flow cytometer, BD Biosciences) and analyzed with FlowJo V10 software (Ashland, OR, USA).

4.10. Induction of MART-1-Specific $CD8^+$ Effector T Cells

$CD14^+$ monocytes and $CD8\beta^+$ T cells were isolated from PBMCs of HLA-$A2^+$ healthy donors by positive labeling with magnetic bead-labeled anti-CD14 or anti-CD8β antibody and goat-anti-mouse

(GaM) magnetic beads (Miltenyi Biotec) using a magnetic cell-sorting device (Miltenyi Biotec), according to the manufacturer's instructions. Next, monocytes were cultured for 5 days in in the presence of IL-4 and GM-CSF for the generation of immature moDCs. For the induction of MART-1-specific CTLs, 1×10^6 $CD8\beta^+$ T cells and 1×10^6 irradiated (50 Gy) $CD8\beta^-$autologous PBMCs were co-cultured with 1×10^5 allogeneic HLA-$A2^+$ ApoEV (200 μg/mL, 5 h), ApoEV-HM (200 μg/mL, 5 h), or $MART_{26\text{–}36L}$ peptide-pulsed [40] and LPS-stimulated (10 ng/mL) moDCs for 10 days in Yssel's medium [50] containing 1% human serum. The culture medium was supplemented with IL-6 (10 ng/mL) and IL-12 (10 ng/mL). Six (first en third donor) or five (second donor) equal cultures were initiated per condition. On day 1, 10 ng/mL IL-10 and at day 5, 25 international units (IU) of IL-2 were added to the culture medium. At day 10 the cells were collected and the $CD8^+$ T cells were stained with PE-labeled $MART_{26\text{–}35L}$-specific dextramers (Dx) for 20 min at 37 °C, after which the cells were stained for 30 min at 4 °C for CD8 expression and viability. The frequency of MART-1 specific CTLs was assessed by flow cytometry (×20 Fortessa SORP flow cytometer, BD Biosciences) and analyzed with FlowJo V10 software.

4.11. Imaging Flow Cytometry

MoDCs were pulsed for 45 min on ice with 100 μg/mL ApoEVs or glycan modified ApoEVs (ApoEVs-HM). Thereafter cells were washed and incubated at 37 °C for 0, 15, or 30 min, before fixation with 4% PFA. Images were acquired on the ImageStream ×100 (Amnis Corp., Seattle, WA, USA) and analyzed using Ideas v6.0 software (Amnis-Merck Millipore) as previously described [38]. In short, analysis was done on single cells after compensation on at least 300 cells/ time point for donor 1, 10,000 cells/time point for donor 2 and 2900 cells/time point for donor 3. The spot count was calculated on DiD^+ cells by creating a mask based on the surface of cells in the bright field image. To calculate the internalization score, another mask was created and pixels were eroded from the mask to exclude the membrane. The resulting mask was applied to the fluorescence channel containing the probe of interest. The internalization score was calculated using features provided in the Ideas v6.0 software. DCs that have internalized the bound vesicles have positive scores, while DCs that have most vesicles located at the membrane have negative scores. If the internalization score is 0, equal amounts of vesicles are located at the cell surface as well as intra-cellular.

4.12. Western Blot

To prepare crude cell lysates, ApoEVs and ApoEVs-HM were washed and lysed with RIPA buffer (containing phosphatase and protease inhibitors) on ice. To prepare Western blot samples the lysate was boiled for 5 min at 95 °C in pre-heated 4× Laemmli buffer containing 5% 2-mercaptoethanol. Samples containing equal amounts of protein were loaded on a 15% polyacrylamide gel according to standard procedures. Nitrocellulose membranes (0.45 μm, Bio-Rad) were incubated o/n with mouse anti-MART-1 and rabbit anti-GAPDH. Anti-mouse-800 and anti-rabbit-680 antibodies were used as secondary antibodies. Membranes were analyzed using Odyssey Classic Imager (LI-COR) and Image Studio Lite (LI-COR).

4.13. Statistical Analysis

Data were analyzed with GraphPad Prism 7.0 software (La Jolla, CA, USA) using one-way ANOVA or two-way ANOVA with a post-hoc analysis as indicated in the figure legend. * $p < 0.05$, ** $p < 0.01$, *** $p < 0.001$.

5. Conclusions

In conclusion, mannosylated glycan-enriched melanoma ApoEVs target DC-SIGN on moDCs, thereby enhancing vesicle uptake, cross-presentation and the priming and activation of tumor-specific $CD8^+$ T cells. DC-SIGN-binding patient-derived ApoEVs could therefore be a promising multi-antigenic source for DC-targeted anti-tumor immunotherapy.

Author Contributions: S.K.H., A.A.v.d.L., R.v.d.V., S.J.v.V., J.J.G-V., T.D.d.G. and Y.v.K. were responsible for design of the experiments. Data collection was performed by S.K.H., D.A.S., S.D., M.H.H., B.v.H.H., R.v.d.V., S.A.D. and J.J.G.-V., S.K.H., D.A.S., M.H.H., R.v.d.V., S.A.D. and R.N. analyzed the data. W.M. and R.N. contributed reagents. S.K.H., S.J.v.V. and Y.v.K. drafted the article. All authors critically revised the article and gave their approval for publication.

Acknowledgments: The authors thank T. Konijn and T. O'Toole for technical support at the O2 Flow Facility, Amsterdam UMC, and H. Kalay for providing the MART-1_{26-35L} peptide. We would like to thank the Bergman clinic in Bilthoven, The Netherlands for the supply of healthy donor skin.

References

1. Hodi, F.S.; O'Day, S.J.; McDermott, D.F.; Weber, R.W.; Sosman, J.A.; Haanen, J.B.; Gonzalez, R.; Robert, C.; Schadendorf, D.; Hassel, J.C.; et al. Improved Survival with Ipilimumab in Patients with Metastatic Melanoma. *N. Engl. J. Med.* **2010**, *363*, 711–723. [CrossRef] [PubMed]
2. Topalian, S.L.; Sznol, M.; McDermott, D.F.; Kluger, H.M.; Carvajal, R.D.; Sharfman, W.H.; Brahmer, J.R.; Lawrence, D.P.; Atkins, M.B.; Powderly, J.D.; et al. Survival, durable tumor remission, and long-term safety in patients with advanced melanoma receiving nivolumab. *J. Clin. Oncol.* **2014**, *32*, 1020–1030. [CrossRef] [PubMed]
3. Larkin, J.; Chiarion-Sileni, V.; Gonzalez, R.; Grob, J.J.; Cowey, C.L.; Lao, C.D.; Schadendorf, D.; Dummer, R.; Smylie, M.; Rutkowski, P.; et al. Combined Nivolumab and Ipilimumab or Monotherapy in Untreated Melanoma. *N. Engl. J. Med.* **2015**, *373*, 23–34. [CrossRef] [PubMed]
4. Wolchock, J.D.; Kluger, H.; Callahan, M.K.; Postow, M.A.; Rizvi, N.A.; Lesokhin, A.M.; Segal, N.H.; Ariyan, C.E.; Gordon, R.-A.; Reed, K.; et al. Nivolumab plus Ipilimumab in Advanced Melanoma. *N. Engl. J. Med.* **2013**, *369*, 122–133. [CrossRef] [PubMed]
5. Chen, D.S.; Mellman, I. Oncology meets immunology: The cancer-immunity cycle. *Immunity* **2013**, *39*, 1–10. [CrossRef]
6. Banchereau, J.; Steinman, R.M. Dendritic cells and the control of immunity. *Nature* **1998**, *392*, 245–252. [CrossRef] [PubMed]
7. Steinman, R.M.; Banchereau, J. Taking dendritic cells into medicine. *Nature* **2007**, *449*, 419–426. [CrossRef] [PubMed]
8. Van Dinther, D.; Stolk, D.A.; Van De Ven, R.; Van Kooyk, Y.; De Gruijl, T.D.; Den Haan, J.M.M. Targeting C-type lectin receptors: A high-carbohydrate diet for dendritic cells to improve cancer vaccines. *J. Leukoc. Biol.* **2017**, *102*, 1017–1034. [CrossRef]
9. Unger, W.W.J.; Mayer, C.T.; Engels, S.; Hesse, C.; Perdicchio, M.; Puttur, F.; Streng-ouwehand, I.; Litjens, M.; Kalay, H.; Berod, L.; et al. Antigen targeting to dendritic cells combined with transient regulatory T cell inhibition results in long-term tumor regression. *Oncoimmunology* **2015**, *4*, e970462. [CrossRef]
10. Fehres, C.M.; Kalay, H.; Bruijns, S.C.M.; Musaafir, S.A.M.; Ambrosini, M.; van Bloois, L.; van Vliet, S.J.; Storm, G.; Garcia-Vallejo, J.J.; van Kooyk, Y. Cross-presentation through langerin and DC-SIGN targeting requires different formulations of glycan-modified antigens. *J. Control. Release* **2015**, *203*, 67–76. [CrossRef]
11. Van Kooyk, Y.; Rabinovich, G.A. Protein-glycan interactions in the control of innate and adaptive immune responses. *Nat. Immunol.* **2008**, *9*, 593–601. [CrossRef] [PubMed]
12. Bonifaz, L.; Bonnyay, D.; Mahnke, K.; Rivera, M.; Nussenzweig, M.C.; Steinman, R.M. Efficient Targeting of Protein Antigen to the Dendritic Cell Receptor DEC-205 in the Steady State Leads to Antigen Presentation on Major Histocompatibility Complex Class I Products and Peripheral $CD8^+$ T Cell Tolerance. *J. Exp. Med.* **2002**, *196*, 1627–1638. [CrossRef] [PubMed]
13. Bonifaz, L.C.; Bonnyay, D.P.; Charalambous, A.; Darguste, D.I.; Fujii, S.; Soares, H.; Brimnes, M.K.; Moltedo, B.; Moran, T.M.; Steinman, R.M. In Vivo Targeting of Antigens to Maturing Dendritic Cells via the DEC-205 Receptor Improves T Cell Vaccination. *J. Exp. Med.* **2004**, *199*, 815–824. [CrossRef] [PubMed]
14. García-Vallejo, J.J.; Unger, W.W.J.; Kalay, H.; van Kooyk, Y. Glycan-based DC-SIGN targeting vaccines to enhance antigen cross-presentation. *Oncoimmunology* **2013**, *2*, e23040. [CrossRef] [PubMed]
15. Engering, A.; Van Vliet, S.J.; Hebeda, K.; Jackson, D.G.; Prevo, R.; Singh, S.K.; Geijtenbeek, T.B.H.; Van Krieken, H.; Van Kooyk, Y. Dynamic Populations of Dendritic Cell-Specific ICAM-3 Grabbing

Nonintegrin-Positive Immature Dendritic Cells and Liver/Lymph Node-Specific ICAM-3 Grabbing Nonintegrin-Positive Endothelial Cells in the Outer Zones of the Paracortex of Human Lymph Nodes. *Am. J. Pathol.* **2004**, *164*, 1587–1595. [CrossRef]

16. Fehres, C.M.; Garcia-Vallejo, J.J.; Unger, W.W.J.; van Kooyk, Y. Skin-resident antigen-presenting cells: Instruction manual for vaccine development. *Front. Immunol.* **2013**, *4*, 157. [CrossRef]
17. Fehres, C.M.; Van Beelen, A.J.; Bruijns, S.C.M.; Ambrosini, M.; Kalay, H.; Van Bloois, L.; Unger, W.W.J.; Garcia-vallejo, J.J.; Storm, G.; De Gruijl, T.D.; et al. In situ Delivery of Antigen to DC-SIGN + CD14 + Dermal Dendritic Cells Results in Enhanced CD8 + T-Cell Responses. *J. Investig. Dermatol.* **2015**, *135*, 2228–2236. [CrossRef]
18. Appelmelk, B.J.; van Die, I.; van Vliet, S.J.; Vandenbroucke-Grauls, C.M.J.E.; Geijtenbeek, T.B.H.; van Kooyk, Y. Cutting Edge: Carbohydrate Profiling Identifies New Pathogens That Interact with Dendritic Cell-Specific ICAM-3-Grabbing Nonintegrin on Dendritic Cells. *J. Immunol.* **2003**, *170*, 1635–1639. [CrossRef]
19. Aarnoudse, C.A.; Bax, M.; Sánchez-Hernández, M.; García-Vallejo, J.J.; Van Kooyk, Y. Glycan modification of the tumor antigen gp100 targets DC-SIGN to enhance dendritic cell induced antigen presentation to T cells. *Int. J. Cancer* **2008**, *122*, 839–846. [CrossRef]
20. Rooney, M.S.; Shukla, S.A.; Wu, C.J.; Getz, G.; Hacohen, N. Molecular and Genetic Properties of Tumors Associated with Local Immune Cytolytic Activity. *Cell* **2014**, *160*, 48–61. [CrossRef]
21. Giannakis, M.; Mu, X.J.; Shukla, S.A.; Qian, Z.R.; Cohen, O.; Nishihara, R.; Bahl, S.; Cao, Y.; Amin-Mansour, A.; Yamauchi, M.; et al. Genomic Correlates of Immune-Cell Infiltrates in Colorectal Carcinoma. *Cell Rep.* **2016**, *15*, 857–865. [CrossRef] [PubMed]
22. Yadav, M.; Jhunjhunwala, S.; Phung, Q.T.; Lupardus, P.; Tanguay, J.; Bumbaca, S.; Franci, C.; Cheung, T.K.; Fritsche, J.; Weinschenk, T.; et al. Predicting immunogenic tumour mutations by combining mass spectrometry and exome sequencing. *Nature* **2014**, *515*, 572–576. [CrossRef] [PubMed]
23. Castle, J.C.; Kreiter, S.; Diekmann, J.; Löwer, M.; Van De Roemer, N.; de Graaf, J.; Selmi, A.; Diken, M.; Boegel, S.; Paret, C.; et al. Exploiting the Mutanome for Tumor Vaccination. *Cancer Res.* **2012**, *72*, 1081–1092. [CrossRef] [PubMed]
24. Sahin, U.; Derhovanessian, E.; Miller, M.; Kloke, B.; Simon, P.; Löwer, M.; Bukur, V.; Tadmor, A.D.; Luxemburger, U.; Schrörs, B.; et al. Personalized RNA mutanome vaccines mobilize poly-specific therapeutic immunity against cancer. *Nat. Publ. Gr.* **2017**, *547*, 222–226. [CrossRef] [PubMed]
25. Ott, P.A.; Hu, Z.; Keskin, D.B.; Shukla, S.A.; Sun, J.; Bozym, D.J.; Zhang, W.; Luoma, A.; Giobbie-Hurder, A.; Peter, L.; et al. An immunogenic personal neoantigen vaccine for patients with melanoma. *Nature* **2017**, *547*, 217–221. [CrossRef] [PubMed]
26. Ruben, J.M.; Van Den Ancker, W.; Bontkes, H.J.; Westers, T.M.; Hooijberg, E.; Ossenkoppele, G.J.; De Gruijl, T.D.; Van De Loosdrecht, A.A. Apoptotic blebs from leukemic cells as a preferred source of tumor-associated antigen for dendritic cell-based vaccines. *Cancer Immunol. Immunother.* **2014**, *63*, 335–345. [CrossRef]
27. Ruben, J.M.; Bontkes, H.J.; Westers, T.M.; Hooijberg, E.; Ossenkoppele, G.J.; van de Loosdrecht, A.A.; de Gruijl, T.D. In situ loading of skin dendritic cells with apoptotic bleb-derived antigens for the induction of tumor-directed immunity. *Oncoimmunology* **2014**, *4*, 1–10.
28. Atkin-Smith, G.K.; Poon, I.K.H. Disassembly of the Dying: Mechanisms and Functions. *Trends Cell Biol.* **2017**, *27*, 151–162. [CrossRef]
29. Chulpanova, D.S.; Kitaeva, K.V.; James, V.; Rizvanov, A.A.; Herrmann, M. Therapeutic Prospects of extracellular vesicles in Cancer Treatment. *Front. Immunol.* **2018**, *9*, 1534. [CrossRef]
30. Caruso, S.; Poon, I.K.H. Apoptotic Cell-Derived Extracellular Vesicles: More Than Just Debris. *Front. Immunol.* **2018**, *9*, 1486.
31. Muhsin-Sharafaldine, M.-R.; Mclellan, A.D. Tumor-Derived Apoptotic vesicles: With Death They Do Part. *Front. Immunol.* **2018**, *9*, 957. [CrossRef] [PubMed]
32. Théry, C.; Ostrowski, M.; Segura, E. Membrane vesicles as conveyors of immune responses. *Nat. Rev. Immunol.* **2009**, *9*, 581–593. [CrossRef] [PubMed]
33. Théry, C.; Boussac, M.; Véron, P.; Ricciardi-Castagnoli, P.; Raposo, G.; Garin, J.; Amigorena, S. Proteomic analysis of dendritic cell-derived exosomes: A secreted subcellular compartment distinct from apoptotic vesicles. *J. Immunol.* **2001** *166*, 7309–7318. [CrossRef] [PubMed]

34. Elbein, A.D.; Tropea, J.E.; Mitchell, M.; Kaushal, G.P. Kifunensine, a Potent Inhibitor Mannosidase I * of the Glycoprotein Processing. *J. Biol. Chem.* **1990**, *265*, 15599–15605. [PubMed]
35. Geijtenbeek, T.B.H.; Torensma, R.; Van Vliet, S.J.; Van Duijnhoven, G.C.F.; Adema, G.J.; Van Kooyk, Y.; Figdor, C.G. Identification of DC-SIGN, a novel dendritic cell-specific ICAM-3 receptor that supports primary immune responses. *Cell* **2000**, *100*, 575–585. [CrossRef]
36. Garg, A.D.; More, S.; Rufo, N.; Mece, O.; Livia, M.; Agostinis, P.; Zitvogel, L.; Kroemer, G.; Garg, A.D.; More, S.; et al. Trial watch: Immunogenic cell death induction by anticancer chemotherapeutics. *Oncoimmunology* **2017**, *6*, e1386829. [CrossRef] [PubMed]
37. Galluzzi, L.; Buqué, A.; Kepp, O.; Zitvogel, L.; Kroemer, G. Immunogenic cell death in cancer and infectious disease. *Nat. Rev. Immunol.* **2017**, *17*, 97–111. [CrossRef]
38. Horrevorts, S.K.; Duinkerken, S.; Bloem, K.; Secades, P.; Kalay, H.; Musters, R.J.; van Vliet, S.J.; García-Vallejo, J.J.; van Kooyk, Y. Toll-like receptor 4 triggering promotes cytosolic routing of DC-SIGN-targeted antigens for presentation on MHC class I. *Front. Immunol.* **2018**, *9*, 1231. [CrossRef]
39. Krysko, D.V.; Garg, A.D.; Kaczmarek, A.; Krysko, O.; Agostinis, P.; Vandenabeele, P. Immunogenic cell death and DAMPs in cancer therapy. *Nat. Rev. Cancer* **2012**, *12*, 860–875. [CrossRef]
40. Hangalapura, B.N.; Oosterhoff, D.; Aggarwal, S.; Wijnands, P.G.J.T.B.; van de Ven, R.; Santegoets, S.J.A.M.; van den Tol, M.P.; Hooijberg, E.; Pereboev, A.; van de Eertwegh, A.J.M.; et al. Selective transduction of dendritic cells in human lymph nodes and superior induction of high-avidity melanoma-reactive cytotoxic t cells by a cd40-targeted adenovirus. *J. Immunother.* **2010**, *33*, 706–715. [CrossRef]
41. Fransen, J.H.; Hilbrands, L.B.; Ruben, J.; Stoffels, M.; Adema, G.J.; Van Der Vlag, J.; Berden, J.H. Mouse dendritic cells matured by ingestion of apoptotic blebs induce T cells to produce interleukin-17. *Arthritis Rheum.* **2009**, *60*, 2304–2313. [CrossRef] [PubMed]
42. Fehr, E.M.; Spoerl, S.; Heyder, P.; Herrmann, M.; Bekeredjian-Ding, I.; Blank, N.; Lorenz, H.M.; Schiller, M. Apoptotic-cell-derived membrane vesicles induce an alternative maturation of human dendritic cells which is disturbed in SLE. *J. Autoimmun.* **2013**, *40*, 86–95. [CrossRef] [PubMed]
43. Berda-Haddad, Y.; Robert, S.; Salers, P.; Zekraoui, L.; Farnarier, C.; Dinarello, C.A.; Dignat-George, F.; Kaplanski, G. Sterile inflammation of endothelial cell-derived apoptotic bodies is mediated by interleukin-1. *Proc. Natl. Acad. Sci. USA* **2011**, *108*, 20684–20689. [CrossRef] [PubMed]
44. Municio, C.; Hugo, E.; Alvarez, Y.; Alonso, S.; Blanco, L.; Fernández, N.; Sánchez Crespo, M. Apoptotic cells enhance IL-10 and reduce IL-23 production in human dendritic cells treated with zymosan. *Mol. Immunol.* **2011**, *49*, 97–106. [CrossRef] [PubMed]
45. Sen, P.; Wallet, M.A.; Yi, Z.; Huang, Y.; Henderson, M.; Mathews, C.E.; Earp, H.S.; Matsushima, G.; Baldwin, A.S.; Tisch, R.M. Apoptotic cells induce Mer tyrosine kinase-dependent blockade of NF-κB activation in dendritic cells. *Blood* **2007**, *109*, 653–660. [CrossRef] [PubMed]
46. Stuart, L.M.; Lucas, M.; Simpson, C.; Lamb, J.; Savill, J.; Lacy-Hulbert, A. Inhibitory Effects of Apoptotic Cell Ingestion upon Endotoxin-Driven Myeloid Dendritic Cell Maturation. *J. Immunol.* **2002**, *168*, 1627–1635. [CrossRef] [PubMed]
47. Muhsin-Sharafaldine, M.-R.; Saunderson, S.C.; Dunn, A.C.; Faed, J.M.; Kleffmann, T.; McLellan, A.D. Procoagulant and immunogenic properties of melanoma exosomes, microvesicles and apoptotic vesicles. *Oncotarget* **2016**, *7*, 56279–56294. [CrossRef]
48. Verdegaal, E.M.E.; De Miranda, N.F.C.C.; Visser, M.; Harryvan, T.; Van Buuren, M.M.; Andersen, R.S.; Hadrup, S.R.; Van Der Minne, C.E.; Schotte, R.; Spits, H.; et al. Neoantigen landscape dynamics during human melanoma-T cell interactions. *Nature* **2016**, *536*, 91–95. [CrossRef]
49. Geijtenbeek, T.B.H.; Van Duijnhoven, G.C.F.; Van Vliet, S.J.; Krieger, E.; Vriend, G.; Figdor, C.G.; Van Kooyk, Y. Identification of different binding sites in the dendritic cell-specific receptor DC-SIGN for intercellular adhesion molecule 3 and HIV-1. *J. Biol. Chem.* **2002**, *277*, 11314–11320. [CrossRef]
50. Yssel, H.; De Vries, J.E.; Koken, M.; Van Blitterswijk, W.; Spits, H. Serum-Free Medium for Generation and Propagation of Functional Human Cytotoxic and Helper T cell clones. *J. Immunol. Methods* **1984**, *72*, 219–227. [CrossRef]

4

Three Method-Combination Protocol for Improving Purity of Extracellular Vesicles

Thomas Simon *, Anish Kumaran, Diana-Florentina Veselu and Georgios Giamas *

Department of Biochemistry and Biomedicine, School of Life Sciences, University of Sussex, Brighton BN1 9QG, UK; anish.kumaran26@gmail.com (A.K.); dv80@sussex.ac.uk (D.-F.V.)
* Correspondence: t.simon@sussex.ac.uk (T.S.); g.giamas@sussex.ac.uk (G.G.)

Abstract: Extracellular vesicles (EVs) are nanosized structures able to carry proteins, lipids and genetic material from one cell to another with critical implications in intercellular communication mechanisms. Even though the rapidly growing EVs research field has sparked great interest in the last 20 years, many biological and technical aspects still remain challenging. One of the main issues that the field is facing is the absence of consensus regarding methods for EVs concentration from biofluids and tissue culture medium. Yet, not only can classic methods be time consuming, commercialized kits are also often quite expensive, especially when research requires analyzing numerous samples or concentrating EVs from large sample volumes. In addition, EV concentration often results in either low final yield or significant contamination of the vesicle sample with proteins and protein complexes of similar densities and sizes. Eventually, low vesicle yields highly limit any further application and data reproducibility while contamination greatly impacts extensive functional studies. Hence, there is a need for accessible and sustainable methods for improved vesicle concentration as this is a critical step in any EVs-related research study. In this brief report, we describe a novel combination of three well-known methods in order to obtain moderate-to-high yields of EVs with reduced protein contamination. We believe that such methods could be of high benefits for in vitro and in vivo functional studies.

Keywords: extracellular vesicles; size exclusion chromatography; differential ultracentrifugation

---

1. Introduction

Even though extracellular vesicles (EVs) have been described as 'useless cell debris' for decades, they have been recognized lately as key constituents of inter-cellular communication pathways [1,2]. EVs are lipid bilayer membrane-enclosed particles that are naturally released from cells [3]. Such cargo vessels transfer lipids, proteins, various fragments of nucleic acids and metabolic components to adjacent cells or to distant sites in the body, mainly through the circulatory system. For these reasons, EVs have been reported to play central roles in both normal and pathological conditions, such as pregnancy and cancer [1,4,5]. Similarly, EVs also play a unique role in spreading various pathogens like viruses and prions from one cell to another [6].

EVs can be classified into two clearly defined subtypes based on their sizes, namely "small EVs" (sEVs) with a size between 50–200 nm, or "medium/large EVs" (m/l EVs) with a size range between 200 nm–1 μm in diameter. This nomenclature is now preferred to the classic, yet quite vague, terms "exosomes", "microvesicles" or "oncosomes", as high heterogeneity in terms of size, marker expression and origins have been reported for each subpopulation, leading to overlaps between them [2].

Interest in the EVs has significantly grown in the medical research community over the past decade. Indeed, a thorough and comprehensive description of such EVs-dependent pathways may provide new inputs to develop effective treatments [5,7]. Mostly focusing on the sEVs subtype, the field has failed so far to establish essential technical standards, such as an optimal sEVs concentration/isolation method [1,2]. Thus, there is no consensus regarding that important matter whatsoever, raising important concerns regarding data reliance and reproducibility. As a matter of fact, currently available methods for sEVs concentration can hardly provide both high yield and high purity at once [8,9]. Consequently, such lack of effective techniques directly affects biomarker discovery and functional studies for which description of exclusively sEVs-related mechanisms and cargo is needed.

EVs are most commonly separated and concentrated from cell culture conditioned medium or human biofluids by differential ultracentrifugation (UC). This method allows for the separation of small particles, such as m/l and sEVs, from other larger ones, such as cell debris, based on their respective density and size, through successive increases of centrifugation forces and time. Differential UC is easy to perform, moderately time-consuming and does not require much technical expertise. Nonetheless, even when the parameters are optimized, the process results in a mixture of EVs concentrated along with particles of the same buoyant density and size range. In other words, large proteins and protein aggregates contaminate the EVs preparation. The co-isolated non-EVs structures are most often lipoproteins such as APOA1/2 or APOB, and Albumin [10]. Alternatively, researchers use various different methods such as size exclusion chromatography (SEC), filtration, precipitation, density gradients or immuno-isolation [2]. Yet, such protocols are not perfect as the final yield is often low and the purity is not optimal. In addition, these methods are usually commercialized in the form of expensive kits, altogether making them hardly applicable to extensive in vitro functional studies. For all these reasons, combining some of these methods seems to be the only sensitive strategy to substantially improve both the purity and the concentration of the final EVs preparation [2,10]. In theory, SEC makes it possible to separate the EVs from other particles, mainly proteins complexes and lipoproteins based on their size, through running a sample on a column made of resin with a define pore size. Consequently, such methods should help purify EVs samples obtained through UC [11,12].

For all these reasons, the present study has been undertaken in an attempt to improve the purity of sEVs preparation with the extra goal to maintain an important final concentration so that extensive functional studies are feasible. To do so, we have: (1) concentrated putative sEVs through UC, followed by (2) SEC in order to exclude protein contaminants from the assumed sEVs preparation. Finally, (3) an extra step was performed post-SEC using a centrifugal filter device in order to improve the concentration of the final sEVs samples. The final concentrated sEVs are therefore less contaminated as compare to sEVs separated using UC. Moreover, although there was a marginal loss during the process, we observed that the structure and size of sEVs were intact following all these steps.

2. Results and Discussion

Despite a constant and rapid evolution of the techniques and methods in the EVs field, researchers are yet to reach a consensus regarding the particle concentration step that is critical for any EVs-focused study [13]. However, they largely agree on the higher performance of combinational protocols over single-method approaches, even though proper EVs isolation/purification still seems unrealistic. Indeed, obtaining high concentrations of EVs coupled to acceptable sample purity is still hardly feasible [1,2]. Yet, reaching such a goal would be of high value for current functional studies, as it would make it possible to perform reliable and reproducible experiments for deciphering EVs-specific mechanisms. Indeed, as mentioned in the latest update to the MISEV2018, highly purified EVs should be used when one wants to associate a function or marker expression to vesicles as compared with other potentially present particles [2].

In the present study, conditioned medium was collected from confluent glioblastoma (GBM) cells. Differential UC was then performed so that the original putative sEVs samples could be obtained. Following, nanoparticles analysis (NTA) was employed to determine the concentration of particles in the original samples (1.63×10^{11} particles/mL, Figure 1A).

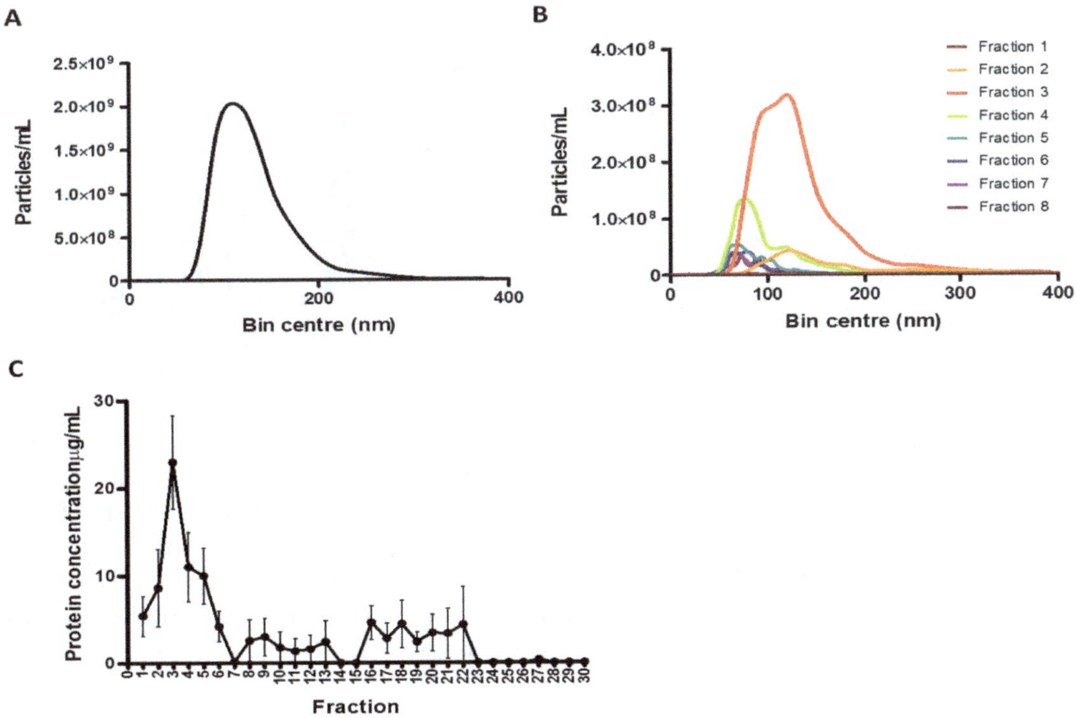

Figure 1. Nanoparticle tracking analysis and protein concentration measurement in initial ultracentrifugation sample and fractions from size exclusion chromatography (Step 1 and 2). Particle samples obtained following step 1 and 2 of the 3 method-combination protocol were processed to nanoparticle analysis (NTA) and protein concentration measurement. (**A**) NTA of the initial ultracentrifugation (UC) sample. Sample was diluted (1/50) in filtered sterile phosphate buffer solution (PBS) and measured using a Nanosight NS300. (**B**) NTA of the fractions from size exclusion chromatography (SEC). The initial UC sample was processed through SEC and fractions were measured by NTA. Fractions were diluted (1/20) in filtered PBS and measured using a Nanosight NS300. (**C**) Protein concentration measurement in the SEC fractions. Protein concentration was measured using a Nanodrop 200. The mean ± SEM of $n = 5$ independent experiments is shown.

SEC was performed following this initial step. As shown in Figure 1B, NTA measurements revealed the presence of particles of the sEVs sizes mainly in SEC fractions 2, 3 and 4 (0.43, 2.70 and 0.74 × 10^{10} particles/mL, respectively). In addition, further Nanodrop analysis showed that SEC fractions 2, 3 and 4 (8.6, 23 and 11 µg/mL, respectively) showed the highest protein content, confirming the detection of putative EVs in the earliest fractions (detection of sEVs-associated proteins) and suggesting the presence of protein contaminants in the latest (Figure 1C).

Accordingly, fraction 3 was pooled with either fraction 2 or 4 and concentrated in a 100 µL of sterile phosphate buffer solution (PBSs). NTA of this final sample showed a particle concentration of 3.84 × 10^{10} particles/mL, which was 4.2× lower as compared to the original concentration obtained by UC (Figure 2A,B).

Western blotting for sEVs markers, namely CD9 and HSP70, fibronectin (FN1) and described EVs sample contaminant albumin (BSA), was then performed. Data revealed exclusive expression of CD9 and HSP70 in both the original UC and final putative sEVs samples, validating the EVs concentration by both the UC method and the three method-combination protocol. The expression of EVs markers was lower in the final sEVs as compared to the UC sample (Figure 3A). BSA expression was mostly observed in the initial UC sample, fraction 5, fraction 6 and fraction 10. In addition, as shown in Figure 3A, a decrease of the FN1 expression was observed in the final sEVs sample as compared to the original UC sample. Yet, FN1 expression was also detected in all the SEC fractions with a slight decrease in fractions 7 and 8, and a slight increase in fractions 9 and 10.

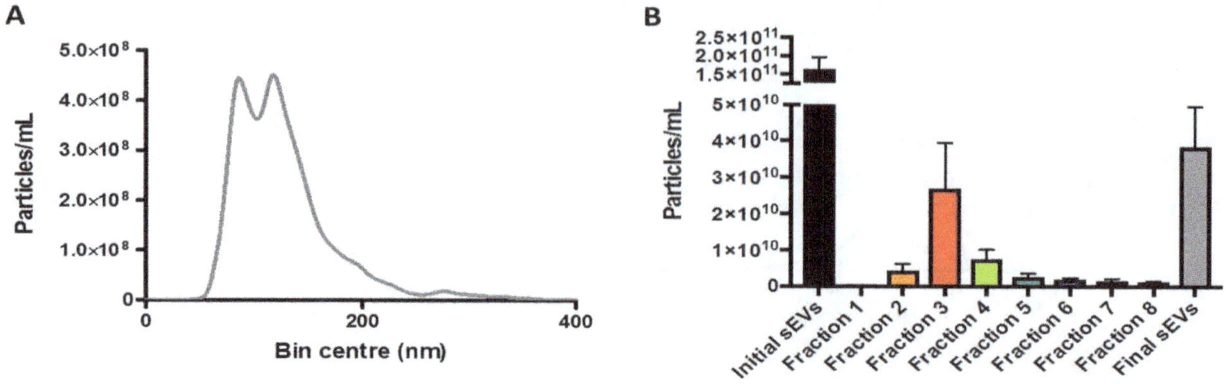

Figure 2. Comparison of particle concentration in final putative sEVs sample versus ultracentrifugation sample and fractions from size exclusion chromatography (Step 3). Final putative sEV sample was obtained following concentration of selected size exclusion chromatography (SEC) fractions using an Amicon Ultra 0.5 device – 30k. (**A**) Nanoparticle tracking analysis (NTA) of the final putative sEV sample. Sample was diluted (1/50) in filtered PBS and analyzed using a Nanosight NS300. (**B**) Particle concentration of the initial UC sample, fractions from SEC and final putative sEVs sample. The mean ± SEM of $n = 5$ independent experiments is shown.

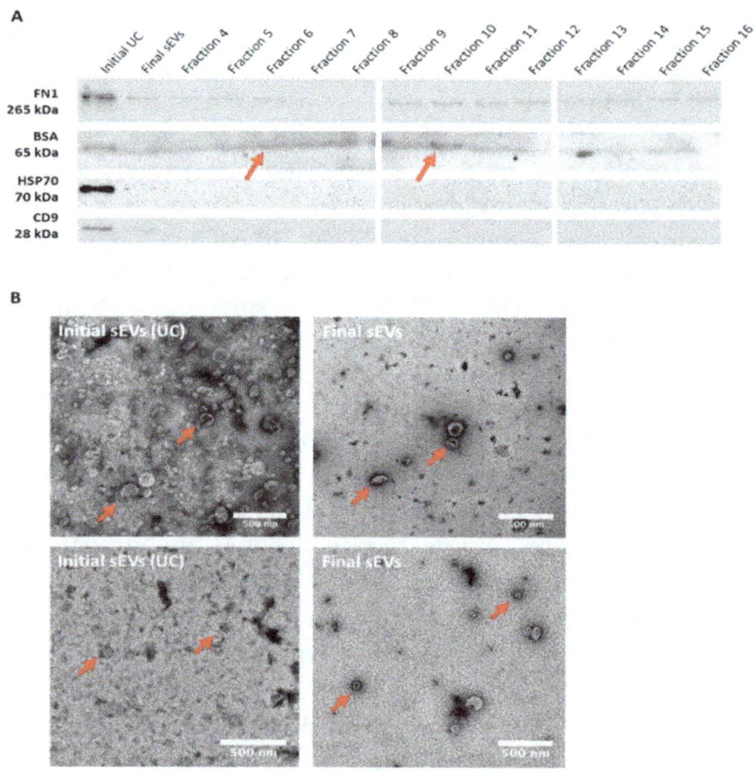

Figure 3. Validation of sEV concentration and decreased protein contamination. (**A**) Western blotting detection of fibronectin (FN1), bovine serum albumin (BSA), HSP70 and CD9 in initial (UC), final sEVs and SEC fractions. (**B**) TEM detection of sEVs (×20k magnification and zoom). Red arrows show sEVs. Representative pictures are shown. Scale bar = 500 µm.

Finally, transmission electron microscopy (TEM) was performed in order to observe the structure/membrane integrity of the final sEVs sample as compare to the original UC one (Figure 3B). As seen in Figure 3B, final particles appear very similar structure wise as compare to original ones from the UC sample, displaying an apparently intact lipid bilayer membrane. Particle concentration appeared much lower in the final sample as compare to the original UC sample, confirming the NTA observations. Moreover, fewer debris and sEVs aggregates could be observed in TEM pictures of the final sEVs sample, as compared to the original UC one.

Using our combination of methods, we observed that our final sEVs samples presented with fewer debris and particle aggregates as compared to our original samples obtained by UC. Altogether, it appears that our three method-combination protocol produced concentrated sEVs samples with enhanced purity as compared to the commonly used UC protocol. We can therefore confirm that combining UC and then SEC, in this order, allows: 1) to use large amounts of cell culture conditioned medium/biofluids for high sEVs concentrations and 2) to separate particles from protein contaminants found in the UC concentrated preparations. Final centrifugal concentration then allows for reducing sample dilution due to SEC.

Our present method might be especially valuable for in vitro functional studies, as one of the strengths here is to make possible using very large volumes (>100 mL) of starting material. The final concentration of sEVs in this way is high enough to perform multiple validation and further functional/phenotypic experiments with the same sample, thus increasing data impact. In addition, even though SEC columns that allow for EVs separation from large volumes are finally emerging, they are still very costly and a few of them would be required in case of repeated usage. Our present UC/SEC combination takes advantage of the SEC impact on sample quality without the requirement of multiple columns in order to process such starting material. Alternatively, here we propose a rather cheap and sustainable method that consecutively has more potential for a wide use and would allow for a better standardization of techniques among teams. As the EVs community is in need of a general improvement of data specificity, we believe our easily accessible alternative could be of great help, especially to small research teams.

For the same reasons, our method combination could also benefit the biomarker discovery in EVs [5]. For instance, better separation of EVs from freely circulating material, such as apoliproteins, in blood would allow for improved identification of EVs-specific biomarkers. As, for example, cancer-derived EVs are believed to travel very long distances to set up metastatic sites, improved plasma-derived EVs concentration could have highly sensitive clinical applications [5]. Nevertheless, as mentioned in the MISEV2018, definitive association of a biomarker with EVs might not be essential to such application. According to the authors, even if it just co-isolates with EVs, such biomarker is valuable as long as it can be associated to any clinical benefit (for diagnosis or prognosis for example) [2]. Yet, one could argue that better EVs separation leading to higher sample purity might provide more specific and thus more effective and stringent EVs-associated biomarkers.

Despite that such a novel method can represent progress towards standardization, there is still room for improvement. For instance, while the decrease of particle concentration we could observe at the end of the protocol should be mostly due, in theory, to the actual purification process, it could also be due to material loss during the many handling, transfer and filtration steps. Furthermore, as we used only one GBM cell line for the present study, we have to acknowledge that our three method-protocol might present variable efficacy when performed with CM derived from cell lines or primary cells of different origins. Indeed, as we also observed in previous studies, EVs production and cargo are highly affected by their cell origin [5]. Nevertheless, even though an extended work would confirm such assumption, we believe that our present three method-protocol to be applicable to any sorts of biofluids, including CM from immortalized and primary cell lines. Furthermore, extended comparison of our present protocol to other available method combinations will be needed in the near future in order to fully assess its efficacy [14]. Finally, while the present protocol is optimized for the specific recovery of sEVs, which received most of the field attention over the last 10 years, interest in m/lEVs and larger particles is slowly growing [15]. As these EVs sub-populations might also be involved in key mechanisms in both normal and pathological conditions, there is a growing need for innovative methods for precise and reliable separation of these different EVs sub-populations. For instance, an additional SEC step could be added to the present protocol, following the 10,000× *g*

UC step in order to separate m/lEVs from other membrane debris and contaminants. Such work would then be a highly valuable follow-up study to the present report.

3. Materials and Methods

3.1. Cells and Reagents

U118 glioblastoma (GBM) cells (ATCC) were maintained in Dulbecco's Modified Eagle Medium (DMEM, Sigma-Aldrich, Gillingham, UK). Cell line culture medium was supplemented with 100 Units mL^{-1} penicillin, 100 μg mL^{-1} streptomycin, 2 mM L-glutamine (PSG, Sigma-Aldrich, Gillingham, UK) and 10% heat inactivated fetal bovine serum (FBS, Sigma-Aldrich, Gillingham, UK).

3.2. Differential Ultracentrifugation for Extracellular Vesicle Concentration

In order to collect sEVs derived from GBM cells, cells were seeded in 4 to 5 × 175 cm^2 flasks and grown in 10% FCS medium until they reach confluence. Then, cells were washed with sterile PBS and 15 mL of corresponding serum free medium was added to each flask for 24 h. Following this incubation, conditioned medium (CM) was collected from each flask, pooled together in 2 × 50 mL falcon tubes and kept at either 4 °C for a very short time (up to 24 h) or at −20 °C for longer periods (up to 6 months) before sEV concentration. In accordance with the latest Minimal Information for Studies of Extracellular Vesicles (MISEV2018), cell count at time of collection was recorded and used to normalize the final sEV concentration (particles mL^{-1} per cell). Cell number and viability were measured using a Countess ™ cell counter (Thermo Fisher Scientific, Life Technologies, Paisley, UK) following mixing of the cell suspension with 0.4% Trypan blue. Only CMs harvested from cell culture with >90% viability were stored.

Concentration of sEVs was performed using an UC-based protocol [13]. Every step of the concentration protocol was performed at 4 °C. In total, 20 mL of stored CM was pipetted into each UC tube. An initial 300× *g* centrifugation was performed for 10 min to discard any floating cells from the CM, followed by a 10 min centrifugation step at 2000× *g* to remove any floating cell debris and dead cells (Hettich Universal 320R centrifuge, DJB Labcare Ltd., Newport Pagnell, UK). A 10,000× *g* UC step (Beckman optima LE 80-k ultracentrifuge, Beckman Type 70 Ti rotor, Beckman polypropylene centrifuge 14 × 89 mm tubes, full dynamic braking, $\mathbf{k_{adj}}$ = 15,638, Beckman Coulter Ltd., High Wycombe, UK) was then performed for 30 min to remove any further cell debris and potential large vesicles (m/lEVs) from the CM. Finally, a first 100,000× *g* UC run was performed for 1 h 30 min to pellet the putative sEVs from the CM (Beckman optima LE 80-k ultracentrifuge, Beckman Type 70 Ti rotor, Beckman polypropylene centrifuge 14 × 89 mm tubes, full dynamic braking, k_{adj} = 494). Supernatant was stored at −20 °C. The UC pellet was then washed in filtered sterile PBS and centrifuged again for 1 h 30 min at 100,000× *g* in order to discard contaminants. The final pellet was re-suspended in 100 μL filtered sterile PBS and immediately characterized through nanoparticle tracking analysis (NTA). Protein concentration (μg/mL) of the final UC preparation was determined using a Nanodrop 200 spectrophotometer (Thermofisher Scientific, Life Technologies, Paisley, UK) (Figure 4).

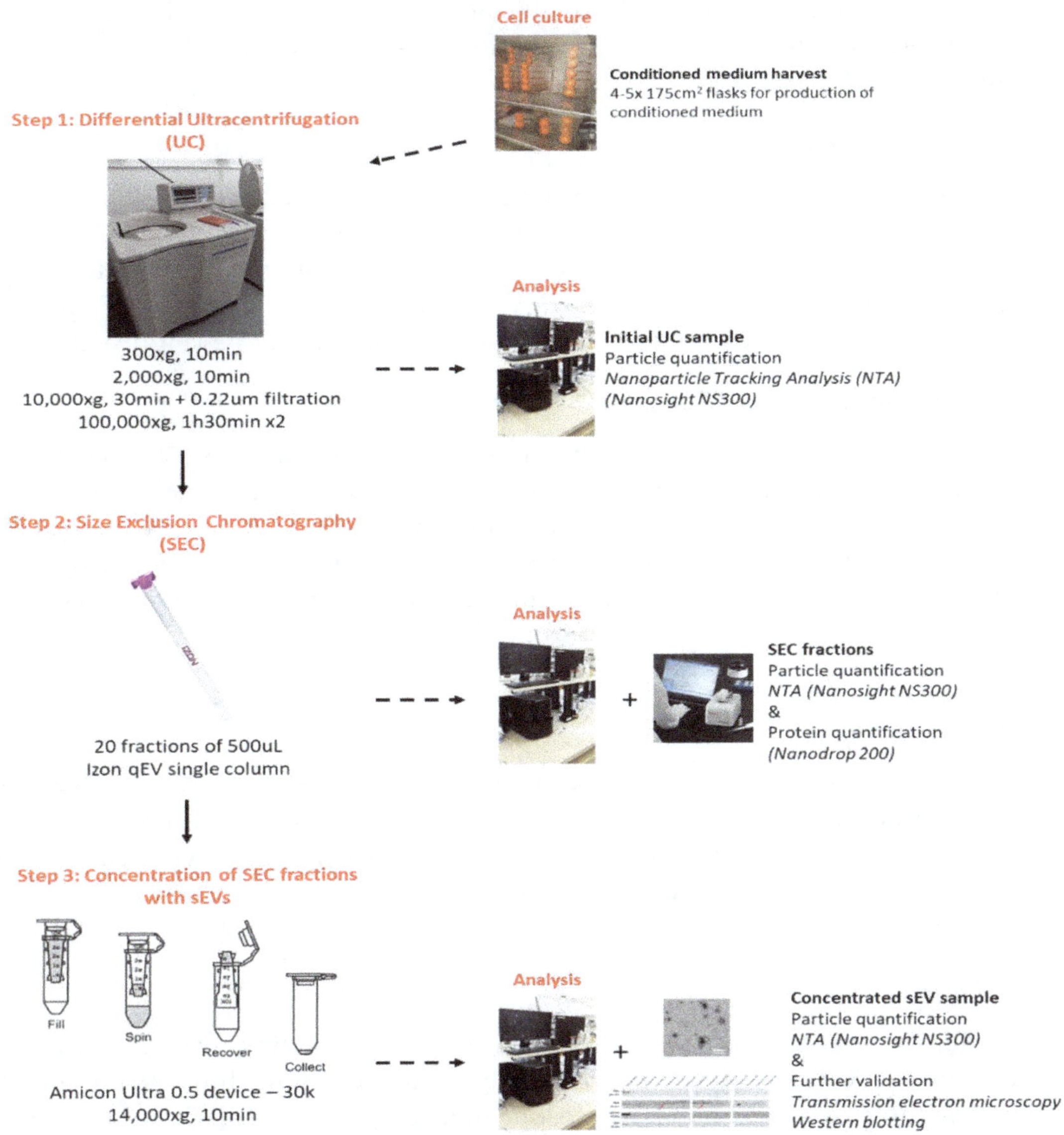

Figure 4. Three method-combination for concentrating EVs derived from cell culture medium. Cells are grown to confluence in 4 × 175 cm^2 flasks to produce conditioned medium. Conditioned medium is then processed through the differential ultracentrifugation (UC) protocol in order to obtain an initial UC sample (step 1). The initial UC sample is then processed through a size exclusion chromatography column (SEC - Izon qEV single column) in the aim to separate putative EVs from protein contaminants (step 2). Following measurement of the particle and protein concentration, SEC fractions of interest are then pooled together and concentrated using Amicon ultra 0.5 devices (step 3). Final validation experiments confirm the sEVs concentration and the decreased protein contamination of the sample.

3.3. Size Exclusion Chromatography for Extracellular Vesicle Separation from Protein Contaminants

Following the initial UC step, 20 μL of the original preparation (out of 100 μL) was kept at −20 °C for further analysis. The rest of the preparation (~80 μL) was diluted in filtered sterile PBS in order to reach a final volume of 150 μL. SEC was performed using qEV single size exclusion columns (separation size = 70 nm, iZON science, Oxford, UK). According to the manufacturer's recommendations, the SEC column was first equilibrated using sterile PBS before the sample (150 μL) was loaded. As stated by the manufacturer, loading a 150 μL sample at the top of the column results in a 1 mL void volume and fractions of 500 μL. Following loading of the sample at the top of the column, fractions (20 in total) of 500 μL were immediately collected and kept on ice. First 7 fractions were then characterized through NTA. Protein concentration (μg/mL) of all fractions was determined using a Nanodrop 200 spectrophotometer (Thermofisher Scientific, Life Technologies, Paisley, UK). Fractions were then stored at −20 °C (Figure 4).

3.4. Concentration of SEC Fractions

SEC fractions with the highest concentrations of particles (as stated, based on NTA data) were concentrated in an Amicon Ultra 0.5 device – 30k (Merck milipore, Watford, UK). The centrifugal filter device was pre-rinsed with filtered PBS. Samples (500 μL at once) were loaded to the filter device and centrifuged at 14,000× *g* for 5–10 min at 4 °C. The putative concentrated sEVs preparation was characterized through NTA and was further processed or stored at −20 °C (Figure 4).

3.5. Nanoparticles Tracking Analysis (NTA)

Vesicle concentration and size were determined using a Nanosight© NS300 and the Nanosight© NTA 3.2 software (Malvern Instruments, Malvern, UK). The following conditions were applied for the NTA analysis at the Nanosight instrument: temperature was 20–25 °C; viscosity was ~0.98cP; camera type was sCMOS; laser type was Blue488; camera levels were either 14 or 15; syringe Pump Speed was set to 70 AU; 5 measurements of 60 s each were recorded. Graphs show an average of at least 4 experiments.

3.6. Coomassie Blue Staining

Samples were loaded, as stated, on 10% tris-glycine gels and run at 180 V and 40 mA for 100 min. The gels were then stained with Quick Coomassie Stain (Generon, Slough, UK) at room temperature overnight. Excess stain was removed through deionized water washes. Gels were viewed and captured by Criterion Stain Free Imager (Biorad, Watford, UK).

3.7. Western Blotting

Characterization of the sEVs was performed through western blotting by measuring the expression of the EV membrane associated marker CD9 (mainly associated with light sEVs) and Fibronectin (mainly associated with dense sEVs), and EV cytosolic marker HSP70 [2,15]. Standard western blotting protocol was performed as described before [16]. For the EV marker analysis, comparable amount of sEVs (as stated) was loaded on the SDS gel. Primary antibodies: anti-BSA (Merck Millipore 07–248, 1/500 dilution, Merck-Millipore, Watford, UK), anti-CD-9 (System Biosciences EXOAB-CD9A-1, 1/10000 dilution), anti-Fibronectin (Abcam ab2413, 1/1000 dilution), anti-HSP-70 (System Biosciences EXOAB-HSP70A-1,1/10000 dilution, Cambridge Bioscience, Cambridge, UK). Secondary antibodies used: Polyclonal Goat Anti-Rabbit/Mouse Immunoglobulins/HRP (Dako P0447/8, 1/3000 dilution, Agilent, CA, USA) antibodies and Anti-Rabbit Immunoglobulins/HRP (ExoAb antibody Kit, System Biosciences EXO-AB-HRP, 1/3000 dilution, Cambridge Bioscience, Cambridge, UK). Chemiluminescence was observed using a UVP Chemstudio instrument (Analytik Jena, London, UK) and the Vision Works software. All experiments have been repeated at least 3 times.

3.8. Transmission Electron Microscopy

Transmission electron microscopy (TEM) has been performed on putative sEVs preparation in order to visualize and assess/confirm the size range of the vesicles, as described before [13]. Samples were visualized using a JEOL JEM1400-Plus (120 kV, LaB6) microscope (JEOL Ltd., Welwyn Garden City, UK) equipped with a Gatan OneView 4K camera at × 20 k magnification. In total, 10–15 pictures per grid were taken.

4. Conclusions

Overall, the present study establishes an easy and affordable method for sEVs separation that provides both improved sample purity and particles' concentration. We believe that the present 3 method-combination protocol will have a great potential for future in vitro functional studies.

Author Contributions: T.S. and G.G. conceived the idea. A.K., D.-F.V. and T.S. performed the experiments and the analysis of the data. T.S. and G.G. wrote the manuscript. All authors have read and agreed to the published version of the manuscript.

Acknowledgments: We would like to thank Pascale Schellenberger for helping with transmission electron microscopy at the University of Sussex's Electron microscopy imaging center, funded by the School of Life Sciences, the Wellcome Trust (095605/Z/11/A, 208348/Z/17/Z) and the RM Philips Trust.

References

1. Margolis, L.; Sadovsky, Y. The biology of extracellular vesicles: The known unknowns. *PLoS Boil.* **2019**, *17*, e3000363. [CrossRef] [PubMed]
2. Théry, C.; Witwer, K.W.; Aikawa, E.; Alcaraz, M.J.; Anderson, J.D.; Andriantsitohaina, R.; Antoniou, A.; Arab, T.; Archer, F.; Atkin-Smith, G.K.; et al. Minimal information for studies of extracellular vesicles 2018 (MISEV2018): A position statement of the International Society for Extracellular Vesicles and update of the MISEV2014 guidelines. *J. Extracell. Vesicles* **2018**, *7*, 1535750. [CrossRef] [PubMed]
3. Wendler, F.; Favicchio, R.; Simon, T.; Alifrangis, C.; Stebbing, J.; Giamas, G. Extracellular vesicles swarm the cancer microenvironment: From tumor–stroma communication to drug intervention. *Oncogene* **2016**, *36*, 877–884. [CrossRef] [PubMed]
4. Wendler, F.; Stamp, G.W.; Giamas, G. Tumor–Stromal Cell Communication: Small Vesicles Signal Big Changes. *Trends Cancer* **2016**, *2*, 326–329. [CrossRef] [PubMed]
5. Lane, R.; Simon, T.; Vintu, M.; Solkin, B.; Koch, B.; Stewart, N.; Benstead-Hume, G.; Pearl, F.M.; Critchley, G.; Stebbing, J.; et al. Cell-derived extracellular vesicles can be used as a biomarker reservoir for glioblastoma tumor subtyping. *Commun. Boil.* **2019**, *2*, 1–12. [CrossRef] [PubMed]
6. Qin, J.; Xu, Q. Functions and application of exosomes. *Acta Pol. Pharm. - Drug Res.* **2014**, *71*, 537–543.
7. Simon, T.; Pinioti, S.; Schellenberger, P.; Rajeeve, V.; Wendler, F.; Cutillas, P.R.; King, A.; Stebbing, J.; Giamas, G. Shedding of bevacizumab in tumour cells-derived extracellular vesicles as a new therapeutic escape mechanism in glioblastoma. *Mol. Cancer* **2018**, *17*, 132. [CrossRef] [PubMed]
8. Gardiner, C.; Di Vizio, L.; Sahoo, S.; Théry, C.; Witwer, K.W.; Wauben, M.; Hill, A.F. Techniques used for the isolation and characterization of extracellular vesicles: Results of a worldwide survey. *J. Extracell. Vesicles* **2016**, *5*, 27066. [CrossRef] [PubMed]
9. Helwa, I.; Cai, J.; Drewry, M.D.; Zimmerman, A.; Dinkins, M.B.; Khaled, M.L.; Seremwe, M.; Dismuke, W.M.; Bieberich, E.; Stamer, W.D.; et al. A Comparative Study of Serum Exosome Isolation Using Differential Ultracentrifugation and Three Commercial Reagents. *PLoS ONE* **2017**, *12*, e0170628. [CrossRef] [PubMed]
10. Takov, K.; Yellon, D.M.; Davidson, S.M. Comparison of small extracellular vesicles isolated from plasma by ultracentrifugation or size-exclusion chromatography: Yield, purity and functional potential. *J. Extracell. Vesicles* **2018**, *8*, 1560809. [CrossRef] [PubMed]
11. Lozano-Ramos, I.; Bancu, I.; Oliveira-Tercero, A.; Armengol, M.P.; Menezes-Neto, A.; Del Portillo, H.A.; Lauzurica-Valdemoros, R.; Borràs, F.E. Size-exclusion chromatography-based enrichment of extracellular vesicles from urine samples. *J. Extracell. Vesicles* **2015**, *4*, 27369. [CrossRef] [PubMed]
12. Guerreiro, E.M.; Vestad, B.; Steffensen, L.A.; Aass, H.C.D.; Saeed, M.; Øvstebø, R.; Costea, D.E.; Galtung, H.K.; Søland, T.M. Efficient extracellular vesicle isolation by combining cell media modifications, ultrafiltration, and size-exclusion chromatography. *PLoS ONE* **2018**, *13*, e0204276. [CrossRef] [PubMed]
13. Théry, C.; Amigorena, S.; Raposo, G.; Clayton, A. Isolation and Characterization of Exosomes from Cell Culture Supernatants and Biological Fluids. *Curr. Protoc. Cell Boil.* **2006**, *30*, 3.22.1–3.22.29. [CrossRef] [PubMed]
14. Kowal, J.; Arras, G.; Colombo, M.; Jouve, M.; Morath, J.P.; Primdal-Bengtson, B.; Dingli, F.; Loew, D.; Tkach, M.; Théry, C. Proteomic comparison defines novel markers to characterize heterogeneous populations of extracellular vesicle subtypes. *Proc. Natl. Acad. Sci. USA* **2016**, *113*, E968–E977. [CrossRef] [PubMed]

15. Giamas, G.; Hirner, H.; Shoshiashvili, L.; Grothey, A.; Gessert, S.; Kuhl, M.; Henne-Bruns, D.; Vorgias, C.E.; Knippschild, U. Phosphorylation of CK1δ: Identification of Ser370 as the major phosphorylation site targeted by PKA in vitro and in vivo. *Biochem. J.* **2007**, *406*, 389–398. [CrossRef] [PubMed]
16. Corso, G.; Mäger, I.; Lee, Y.; Görgens, A.; Bultema, J.; Giebel, B.; Wood, M.; Nordin, J.; El Andaloussi, S. Reproducible and scalable purification of extracellular vesicles using combined bind-elute and size exclusion chromatography. *Sci. Rep.* **2017**, *7*, 11561. [CrossRef] [PubMed]

5

Signaling of Tumor-Derived sEV Impacts Melanoma Progression

Aneta Zebrowska [1], Piotr Widlak [1], Theresa Whiteside [2,3] and Monika Pietrowska [1,*]

1 Maria Sklodowska-Curie National Research Institute of Oncology, Gliwice Branch, 44-100 Gliwice, Poland; aneta7zebrowska@gmail.com (A.Z.); piotr.widlak@io.gliwice.pl (P.W.)
2 UPMC Hillman Cancer Center, University of Pittsburgh, Pittsburgh, PA 15213, USA; whitesidetl@upmc.edu
3 Department of Pathology, University of Pittsburgh School of Medicine Pittsburgh, Pittsburgh, PA 15261, USA
* Correspondence: monika.pietrowska@io.gliwice.pl;

Abstract: Small extracellular vesicles (sEV or exosomes) are nanovesicles (30–150 nm) released both in vivo and in vitro by most cell types. Tumor cells produce sEV called TEX and disperse them throughout all body fluids. TEX contain a cargo of proteins, lipids, and RNA that is similar but not identical to that of the "parent" producer cell (i.e., the cargo of exosomes released by melanoma cells is similar but not identical to exosomes released by melanocytes), possibly due to selective endosomal packaging. TEX and their role in cancer biology have been intensively investigated largely due to the possibility that TEX might serve as key component of a "liquid tumor biopsy." TEX are also involved in the crosstalk between cancer and immune cells and play a key role in the suppression of anti-tumor immune responses, thus contributing to the tumor progression. Most of the available information about the TEX molecular composition and functions has been gained using sEV isolated from supernatants of cancer cell lines. However, newer data linking plasma levels of TEX with cancer progression have focused attention on TEX in the patients' peripheral circulation as potential biomarkers of cancer diagnosis, development, activity, and response to therapy. Here, we consider the molecular cargo and functions of TEX as potential biomarkers of one of the most fatal malignancies—melanoma. Studies of TEX in plasma of patients with melanoma offer the possibility of an in-depth understanding of the melanoma biology and response to immune therapies. This review features melanoma cell-derived exosomes (MTEX) with special emphasis on exosome-mediated signaling between melanoma cells and the host immune system.

Keywords: small extracellular vesicles (sEV); tumor-derived exosomes (TEX); melanoma cell-derived exosomes (MTEX); proteomics; tumor microenvironment; biomarkers

---

1. Introduction

Small extracellular vesicles (sEV), also known as exosomes (EX), are virus-size (30–150 nm) membrane-bound vesicles released by different cell types under both normal and pathological conditions. They represent a subset of the heterogeneous group of extracellular vesicles (EV) that in addition to sEV include larger (250–1000 nm) microvesicles (MV, ectosomes) and the largest (>1000 nm) apoptotic bodies (AB). EV vary in size, biogenesis, release mechanisms, and biochemical properties. sEV or exosomes are formed in the endosomal network as intraluminal vesicles (ILV) within the multivesicular bodies (MVB) and are released to the extracellular space when MVBs fuse with the cellular plasma membrane. In contrast, MV are produced by outward budding ("blebbing") of the plasma membrane, while apoptotic bodies are released when cells undergo the programmed cell death [1–7]. At present, inconsistency in the EV nomenclature exists causing much confusion in the field, which extends to the methodology for sEV isolation and characterization. Currently, the most common vesicle isolation methods, including ultracentrifugation, do not adequately discriminate between various EV subsets. To ease confusion, a simplified nomenclature has been recently adopted in the literature that distinguishes small EV (i.e., <200 nm) and medium/large EV (>200 nm). The class of small EV consists mostly of exosomes, yet other types of EV, e.g., small MV, could also copurify with this fraction [8–11]. In this review, the terms "exosomes" (EX) and (small) extracellular vesicles (sEV) are used interchangeably for simplicity and to stay in line with the recent guidelines from International Society for Extracellular Vesicles (ISEV) [8].

Among various subsets of EV in body fluids of cancer patients, tumor-derived exosomes, called TEX, have attracted much attention as major mediators of intercellular communication in the tumor microenvironment (TME) and as potentially promising diagnostic, prognostic, and predictive biomarkers in cancer and other diseases. The knowledge of the molecular profiles and biology of TEX offers the possibility of a deeper understanding of pathological processes involved in cancer development and may provide important clinical information about disease activity and response to treatment. Today, TEX are considered as prime candidates for a liquid tumor biopsy, and much effort is being invested in validation of this concept. In this review, we summarize recent insights into the biology and composition of melanoma cell-derived exosomes (MTEX) and provide an up-to-date account of their pleiotropic role in melanoma progression and response to anti-melanoma therapies.

2. General Characteristics of sEV

sEV are produced and released by various cell types, including hematopoietic cells and a broad variety of normal or malignant tissue cells [5,7,12,13]. sEV can be isolated from supernatants of cultured cells or diverse body fluids, including blood, urine, saliva, breast milk, ascites effusions, bile, tears, nasal secretions, amniotic, synovial, cerebrospinal, lymphatic, and seminal fluids [5,7,10,14–23]. The molecular content of sEV is of special interest, as it reflects the nature of parental cells. sEV originating from different cell types share their general features, such as the structure of the bilayer lipid–protein membrane and key molecular components. Their molecular cargo consists of proteins (including cytoskeletal proteins, transmembrane proteins, tetraspanins, heat shock proteins, adhesion proteins, enzymes, immunocompetent proteins e.g., death receptor ligands: tumor necrosis factor ligand (FasL, CD95L or CD178) or TNF-related apoptosis-inducing ligand (TRAIL), check-point receptor ligands such as: programmed death-ligand 1 (PD-L1), inhibitory cytokines such as: interleukin 10 (IL-10), IL-6, TNF-α, IL-1β, and TGF-β1, prostaglandin E2, major histocompatibility molecules MHC-I and II, and tumor-associated antigens), nucleic acids (including DNA, RNA, miRNA, non-coding RNA), lipids, and low-molecular-weight metabolites (including alcohols, amides, amino acids, carboxylic acids, sugars). Proteomic analysis of exosome cargos revealed that some proteins are typical for most of these vesicles (including proteins such as Rab2, Rab7, flotillin, and annexin; cytoskeletal proteins, including actin, myosin, tubulin; or heat shock proteins, such as Hsc70 and Hsc90). Tetraspanins, such as CD9, CD63, CD81, CD83, along with housekeeping proteins, ALIX (programmed cell death 6-interacting protein) and TSG101 (tumor susceptibility gene 101 protein), are widely considered as exosome markers. However, in addition to the "common" set of components, sEV of different cellular origins may also carry proteins that are cell-type specific [8–12,24]. It has been shown in many studies that the molecular profile of TEX is distinct from that of sEV derived from non-malignant cells such as dendritic cells (DC), T cells, and others [6,7,11–13,25]. However, it should be noted that the discrimination of TEX from other types of sEV in patients' plasma using, e.g., tetraspanins as sEV-specific markers has been limited and currently, separation of TEX from total sEV in plasma has not been readily available or reliably performed.

sEV circulating in body fluids represent a complex mixture of vesicles released by many different cell types. The majority of studies of TEX present in body fluids of cancer patients are based on analyzes performed with a mixture of sEV derived from different normal or pathological cells. Separation of TEX from this mixture remains a challenge due to the lack of universal cancer-specific antigens that could be targeted for TEX isolation. Nevertheless, a few studies that used specific membrane markers for isolation of TEX from body fluids have been reported. Chondroitin sulfate proteoglycan 4 (CSPG4) was used for separation of melanoma-derived TEX, referred to as MTEX, from vesicles released by non-malignant cells [26–28]. Glypican 1 (GPC1) was used to isolate TEX from plasma of patients with pancreatic cancer [29], and prostate-specific membrane antigen (PSMA) was used to isolate TEX from plasma of prostate cancer patients [30]. CD34 antigen, a unique marker of AML blasts, was used to isolate TEX from the plasma of patients with acute myeloid leukemia [31]. Moreover, MAGE3/6 antigen was used to identify TEX present in sera of patients with melanoma or head and neck squamous cell carcinoma (HNSCC) [25]. In contrast to plasma-derived sEV, TEX isolated from supernatants of cancer cell lines are putatively homogenous. These TEX derived from tumor cell lines are an excellent in vitro model for investigations of interactions of TEX with other cells. In fact, much of what is known about TEX signaling, uptake by responder cells, and reprogramming in TME is derived from in vitro and in vivo studies of TEX isolated from supernatants of tumor cell lines.

sEV have gained interest due to their essential role in "normal" intercellular signaling and communication, which impact the physiological balance and homeostasis as well as disease progression. [1,6,32–36]. Importantly, sEV can modulate the phenotype/functions of recipient cells, even those located in distant organs [2,3,35–38]. Moreover, the role of TEX in cancer progression has been reported for many cancer types, including ovarian, prostate, breast, lung, colorectal and gastric cancers, melanoma, and acute myeloid leukemia [26,37–44]. TEX are being intensively investigated because they play a key role in the reorganization of the TME, remodeling functions of the cells residing in the TME, and enhancing their contribution to tumorigenesis, metastasis, cancer immune escape, as well as resistance to cancer treatment [34,44–54]. The potential role of TEX in cancer biology is schematically illustrated in Figure 1. A better understanding of the mechanisms underlying TEX-mediated reprogramming of normal cells in TME is expected to be clinically significant, leading to improved cancer diagnosis/prognosis and treatment. In addition, TEX are considered to be an attractive source of cancer biomarkers.

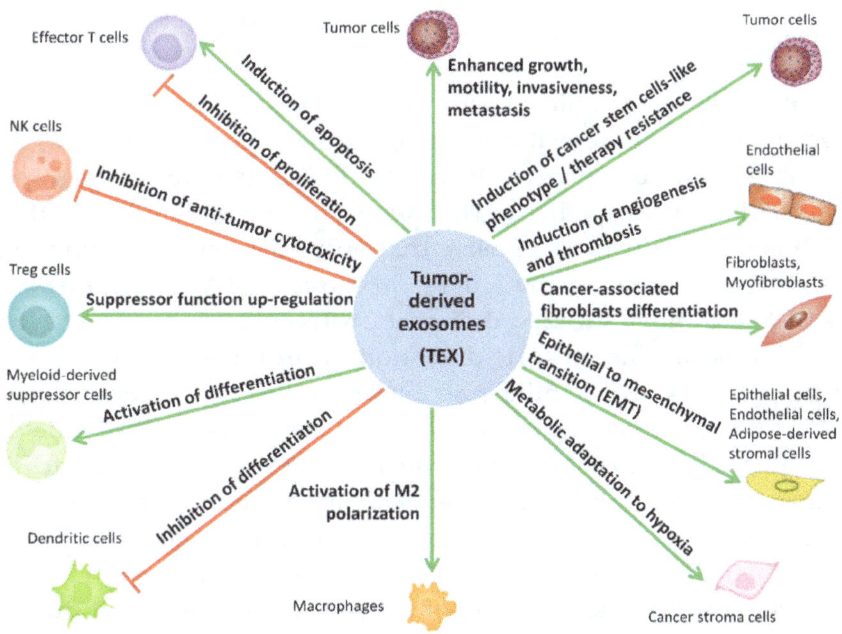

Figure 1. Role of tumor-derived small extracellular vesicles (sEV) (TEX) in cancer biology. Tumor-derived exosomes (TEX) are involved in intercellular signaling and communication between tumor cells and non-malignant cells residing in the tumor microenvironment. TEX reprogram these cells to acquire functions favoring tumor growth and metastasis. TEX-induced changes include enhancing cancer immune escape, remodeling of the tumor stroma, molecular and metabolic reprogramming, and promotion of angiogenesis. Green arrows indicate the processes stimulated by TEX. Red lines with blunt ends indicate processes inhibited by TEX.

3. General Features of Melanoma

Melanoma is the most aggressive skin cancer whose incidence has been increasing worldwide. Melanoma prognosis is generally poor, as it has a high potential for vascular invasion, metastasis, and recurrence [55–57]. Primary melanomas detected in an early stage and completely removed surgically show favorable outcome, with 5-year disease-specific survival rates of 99% [58]. However, melanoma cells tend to metastasize to distant sites, most often to lungs and brain, while evading the host immune system. Hence, the survival rate dramatically decreases when the cancer metastasizes, and 1-year survival rate drops to 35%–62% [57,59], while 5-year survival rate drops to 25% [58]. Currently, surgery is the treatment of choice for patients with cutaneous melanoma, and the adjuvant treatment scheme is usually tailored individually. Melanoma is sensitive to immune checkpoint inhibitors, such as anti-CTLA4 (anti-cytotoxic T-lymphocyte-associated protein 4 or anti-CD 152) and anti-PD1 (anti-programmed cell death protein 1 or anti-CD279) monoclonal antibodies (mAb), and to small-molecule targeted drugs, such as serine/threonine-protein kinase B-Raf (BRAF) or mitogen-activated protein kinase (MEK) inhibitors. Hence, different treatment schemes, including radiotherapy and/or adjuvant treatments with anti-BRAF/MEK inhibitors and anti-PD-1 mAb are currently used, depending on the patient's clinical situation [60]. Despite these novel combinatorial therapies, tumor escape from the immune control and development of primary or acquired therapy resistance that occurs in about half of melanoma patients remain the major therapeutic barrier [61–63]. Melanoma cells communicate with other cells present in the TME, including components of the immune system, via melanoma cell-derived exosomes (MTEX). Hence, in-depth knowledge of MTEX composition and function is expected to bring better understanding of the mechanisms determining the response of melanoma to treatments.

4. The Molecular Cargo of MTEX

4.1. The Proteome of MTEX

Recently, studies of proteomic profiles of cancer-derived sEV have been much intensified [64–67]. However, only limited data are available about the proteome of melanoma-derived TEX. A great part of the available literature focuses on in vitro studies with TEX isolated from supernatants of various melanoma cell lines [68–73]. Ex vivo studies performed with TEX isolated from the blood of melanoma patients are rare [74]. It is important to emphasize that only sEV derived from melanoma cell lines are "pure" MTEX, as those present in the plasma will be mixtures of sEV derived from many different cells. The available studies of MTEX have utilized different proteomic approaches. Most of them are based on shotgun LC-MS/MS strategies (i.e., tryptic digestion of proteins, followed by nano HPLC-MS/MS analysis of the resulting peptides), while others are based on LC-MS/MS analysis of proteins separated by 1D or 2D SDS-PAGE [75]. Currently available proteomic studies demonstrate differences in protein profiles of MTEX in comparison to melanocyte-derived sEV [68,69]. In addition, MTEX derived from melanoma cell lines with a different tumorigenic potential appeared to have distinct proteomic profiles [70]. Moreover, proteomic analysis of sEV present in the plasma of melanoma patients and healthy donors showed clear differences and revealed increased levels of TYRP2, VLA-4, and HSP70 in patients' samples [74]. However, the knowledge of the proteome in TEX produced by melanoma cells remains rather limited, and the available data are difficult to compare because they represent distinct experimental models. A summary of data on proteomics profiling of MTEX released by melanoma cell lines is presented in Table 1.

Table 1. Proteomics Profiling of Melanoma Cell-Derived Exosomes (MTEX) Released in Vitro by Melanoma Cell Lines.

Cell Line	Method of MTEX Purification and Characterization	MS Approach	Major Findings	Ref.
MeWo, SK-MEL-28 (human)	UC/TEM, WB, 1D/2D SDS-PAGE	MALDI-TOF MS/MS	A few proteins identified in MTEX for the first time: prostaglandin regulatory-like protein (PGRL), p120 catenin, syntaxin-binding proteins 1 and 2, septin 2 (Nedd5), ezrin, radixin, tryptophan/aspartic acid (WD) repeat-containing protein 1	[68]
A375 (human)	UC/TEM, NTA, WB	LC-MS/MS	Different sets of proteins present in MTEX and melanocyte-derived EV, including annexin A1, HAPLN1, GRP78, endoplasmin precursor (gp 96), TUBA1B, PYGB), ferritin, heavy polypeptide 1 (MTEX-upregulated), annexin A2, syntenin-1, MFGE8, OXCT (MTEX-downregulated)	[69]
MNT-1, G1, 501 mel, SKMEL28, Daju, A375M, 1205Lu (human)	UC+SEC/WB, TEM, NTA	nanoLC-MS/MS	Different sets of proteins present in MTEX from nontumorigenic, tumorigenic, and metastatic cell lines, including EGFR, PTK2/FAK1, EPHB2, SRC, LGALS1/LEG1, LGALS3/LEG3, NT5E/5NTD-CD73, NRAS, KIT, MCAM/MUC18, MET specific for metastatic cell lines	[70]
B16-F1 (murine)	UC+SEC/CEM, DLS, IA-FCM	uHPLC-MS	10 most abundant proteins: CD81, CD9, histones (H2A, H2B, H3.1, H4), heat shock proteins (HSPA5/GRP78, HSC71), syntetin-1	[71]
B16-F10 (murine)	UC+SEC, UF+SEC/TEM, NTA, WB	nanoLC-MS/MS	Different sets of proteins identified in low- and high-density MTEX, including ACTN4 and CCNY enriched in LD-MTEX and EPHA2 enriched in HD-MTEX	[72]
Mel501 (human)	UC+SEC/WB, CM	RPLC-MS/MS	Different sets of proteins identified in MTEX released in neutral and acidic environment (pH 6.7 and 6.0, respectively), including HRAS, NRAS, TIMP3, HSP90AB1, HSP90B1, HSPAIL, HSPA5, GANAB, gelsolin, and cofilin upregulated in acidic conditions.	[73]

Abbreviations: methods of EX purification: UC—ultracentrifugation, UF—ultrafiltration, SEC—size-exclusion chromatography; methods of EX characterization: NTA—nanoparticle tracking analysis, DLS—dynamic light scattering, TEM—transmission electron microscopy, CEM—cryo-electron microscopy, CM—confocal microscopy, WB—Western blotting.

4.2. Micro RNA Component of MTEX

Micro RNAs are small (19–25 nucleotides) non-coding RNAs that play an important role as regulators of cell differentiation, metabolism, proliferation, or innate and adaptive immunity [76–78]. Micro RNA profiles of TEX differ from miRNA profiles of their donor cancer cells as well as from profiles of sEV released by normal cells [78–81]. The majority of available works report miRNA signatures of pure MTEX released in vitro by melanoma cell lines [69,82–90]. Moreover, a few studies addressed miRNA composition of sEV derived from serum/plasma of melanoma patients [82,91–94]. Similar to proteomics data, few consistencies were observed among these studies due to different models applied. Nevertheless, there were 6 MTEX-upregulated miRNA species reported in more than one study: miR-494 [82,83], let-7c [69,84], miR-690 [84,85], miR-17 [84,91], and miR-494 [82,83], while miR-125b was reported to be downregulated in MTEX or sEV from plasma of melanoma patients [83,92]. Noteworthy, all the above mentioned miRs are known to be involved in cancer cell invasion, migration, and proliferation as well as in inflammatory processes linked to tumorigenesis and cancer progression [82–86,91,92].

Xiao et al. showed significant differences in miRNA content of exosomes isolated from normal melanocytes and malignant cell lines (HEMa-LP and A375), respectively [69]. In this study, 130 miRNAs were upregulated and 98 miRNAs downregulated in MTEX versus melanocyte-derived EX. The majority of differently expressed miRNAs were associated with tumor aggressiveness, including fifteen miRNAs known to be associated with melanoma metastasis: miR-138, miR-125b, miR-130a, miR-34a, miR-196a, miR-199a-3p, miR-25, miR-27a, miR-200b, miR-23b, miR-146a, miR-613, miR-205, miR-149, let-7c [69]. Another study reported enrichment of miRNA-494, which is known for its high metastatic potential, in MTEX released by A375 cells. A series of functional experiments performed by Li et al. demonstrated that intercellular transport of miR-494 in MTEX was responsible for melanoma metastasis [82]. Blocking of exosomal transfer of miR-494 by a knockdown (KO) of Rab27a induced cellular apoptosis and inhibited tumor growth and metastasis in vitro and in human xenografts [82].

Another analysis of the miRs upregulated in sEV of patients with metastatic melanoma (miR-17-5p, miR-19a-3p, miR-149-5p, miR-21, and miR-126-3p) focused on discovery of putative targets of these miRNAs [91]. Among their targets were genes associated with skin response to UV irradiation, genes coding the tumor protein p53 (TP53)/retinoblastoma protein (RB1) and genes related to the TGF-β/SMAD pathway. Upregulation of miRNAs controlling TP53/RB1 activation and the TGF-β/SMAD signaling pathway might play an important role in melanoma progression, as the TGFβ/SMAD pathway regulates the G1/S checkpoint in normal melanocytes [91]. Moreover, miR-17 was identified as a potential oncomiR not only in melanoma but also in other malignancies [93,94]. Association of miR-19a upregulation with increased melanoma invasiveness was confirmed by Levy et al. [95]. Upregulation of miR-21 and miR-19a is associated with increased proliferation, low apoptosis, invasiveness, and high metastatic potential, as reported for various human tumor cells [96,97], while KO of miR-21 in B16 melanoma cells reduced their metastatic potential [98,99]. The oncogenic properties of miR-21 may be a result of down-regulation of the tumor suppressors: PTEN, PDCD4 and the antiproliferative protein BTG2. In addition, miR-21 induced the IFN pathway with protumorigenic effects [98,99]. High abundance of another oncomiR—miR-1246 was detected in MTEX isolated from patient-derived melanoma cell lines, namely, DMBC9, -10, -11, and -12 [83,100]. Many other studies have confirmed its high concentration in sEV from the plasma of patients with various cancers, including melanoma [101,102]. The overexpression of miR-222 in MTEX and cells is also associated with tumor initiation, differentiation, increased cell motility, and invasion, as well as cancer progression [87]. MiR-222 inhibits anti-neoplastic functions of p27, CDKN1B, and c-Fos by down-modulation of their gene expression, reduces apoptosis, and allows proliferation by induction of the PI3K/AKT pathway [89]. Müller et al. showed the importance of let-7a in melanoma development [103]. Let-7a regulates the expression of integrin β-3, the promotor of melanoma progression. The loss of let-7a expression in MTEX derived from 8 different melanoma cell lines resulted in higher integrin β-3 levels in melanoma cells, enhancing their migratory and invasive potential [103]. Finally, let-7a was detected in serum EX as a factor differentiating stage I melanoma patients from non-melanoma subjects [104]. Altogether, the literature supports the key role played by miRs transferred by melanoma-associated TEX in oncogenesis and melanoma metastasis.

In addition to microRNA, MTEX contain mRNA transcripts of genes expressed in melanoma. Sets of mRNAs with different abundance in MTEX and in melanosome-derived exosomes were identified, including 945 transcripts associated with cancer and 364 associated with dermatological diseases [69]. Among upregulated transcripts there was DNA topoisomerase I (TOP1), which is known to be associated with aggressive, advanced tumors and poor prognosis in melanoma [69,105]. Among downregulated transcripts there were ATP-binding cassette, sub-family B, member 5 (ABCB5), which activates the NF-κB pathway enhancing p65 protein stability [106] and is also known to be closely co-regulated with melanoma tumor antigen p97 (tumor growth regulator—melanotransferrin, MTf) [107], and TYRP1 encoding tyrosinase-related protein 1, which is considered as an inhibitor of TYRP1-dependent miR-16 mediating tumor suppression [108,109].

5. Biological Activity of MTEX

The multi-level contribution of MTEX to tumorigenesis accounts for activation of biological processes enabling cancer immune evasion, as well as molecular and metabolic remodeling of tumor micro- and macro-environment, favoring cancer growth and metastasis. The in-depth knowledge of the pleiotropic role of MTEX in the natural history of melanoma has a great potential clinical application in the disease diagnosis, treatment design, and prognosis of patient's outcomes. MTEX are involved in a plethora of functions involved in initiation, progression, and metastasis of tumors, which is schematically depicted in Figure 2 (according to [26,37,73,74,83–150]). The most essential functions of MTEX are addressed more specifically in the following sub-chapters.

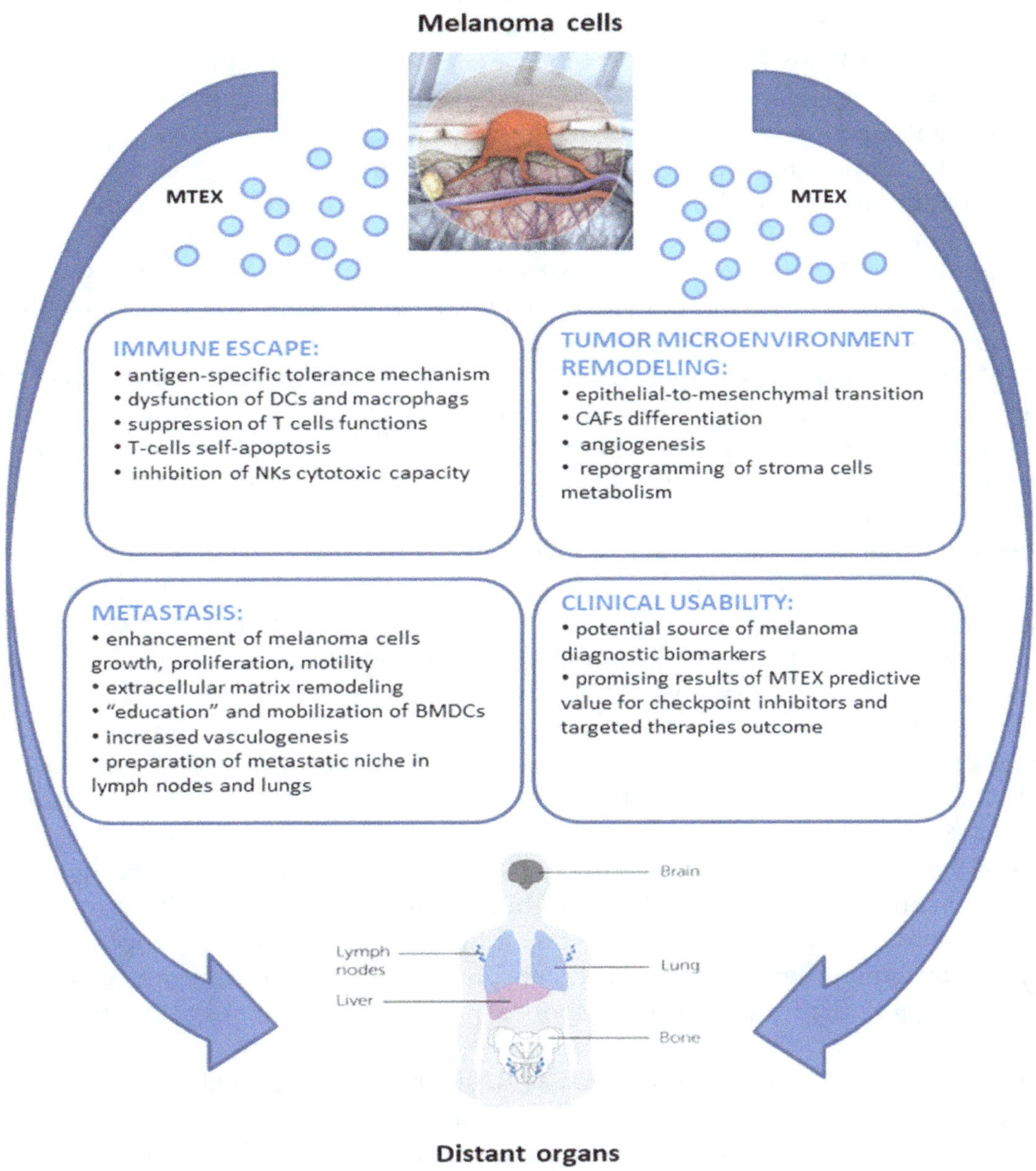

Figure 2. Pleiotropic effect of MTEX on melanoma biology.

5.1. *MTEX Participate in the Reprogramming of Immune Cells*

Growth and progression of cancer involve the escape from the immune surveillance as the sine qua non condition. Emerging evidence supports the idea that MTEX are involved in facilitating tumor escape from the host immune system [33,44,47,49,110–117]. However, in most of the studies reported in this context, melanoma cell lines were used as a source of MTEX. Düchler et al. showed that cancer-induced immunosuppression was mediated by MTEX, and involved an antigen-specific mechanism [118]. The authors provided evidence that MTEX transferred MHC class I receptor proteins from cancer cells to the surface of antigen-presenting cells (APC). At the same time, CD86 and CD40 (co-stimulatory molecules required for differentiation and proliferation of T cells) were down-regulated, while the production of immunosuppressive cytokine IL-6 was induced. Collaboration of TGF-β transported by

MTEX was also demonstrated in this process. The authors hypothesized that MTEX-mediated transfer of the combination of TAA-derived peptide-MHC complexes with immunosuppressive cytokines was a part of antigen-specific tolerance induction enabling melanoma immune escape [118]. The mechanism of melanoma immune escape is also related to the suppression of T cell functions. This can be attributed to an increased level of PD-L1 in MTEX. This immunosuppression is driven by the interaction between PD-L1 carried by MTEX and PD-1 receptors on $CD8^+$ T cells, leading to inhibition of T-cell functions [111,119]. MTEX are also enriched in Fas ligand (FasL) and APO2 ligand (APO2L)/TRAIL, both known as inducing factors of T cell-apoptosis [120]. Another possible mechanism for the suppression of T cell function by MTEX is through the upregulation of PTPN11 protein, which was found to negatively regulate interferon, IL-2, and T cell receptor signaling pathways [121]. Wu et al. confirmed that B16F0-derived MTEX are enriched with Ptpn11 mRNA and can increase PTPN11 dose-dependently in recipient cells. In addition to upregulating PTPN11 in lymphocytes, MTEX derived from B16F0 locally suppressed responses of cells to IL-12 (anti-tumor immunity enhancer) via inhibition of IL12RB2 expression in primary $CD8^+$ T cells. These inhibitory mechanisms of the immune cell response to IL-12 are complemented by B16F0 release of the Wnt-inducible signaling protein 1 (WISP1) that blocks T cell response to IL-12 [121,122]. Furthermore, the cargo of MTEX might alter mitochondrial respiration of cytotoxic T cells and up-regulate genes associated with the Notch signaling pathway [84]. Immunosuppressive activity of MTEX depends on their ligands that engage the T cell receptor (TCR) and IL-2 receptor (IL-2R). Recent studies showed that MTEX inhibited signaling and proliferation of activated primary $CD8^+$ T cells, inducing their apoptosis [25,32,90]. Furthermore, MTEX significantly promoted conversion of CD4(+) T cells to CD4(+)CD25(+)FOXP3(+) T regulatory cells (Treg) enhancing their suppressor functions [25]. Vignard et al. additionally confirmed decreased TCR signaling in T cells as a result of the enrichment in miRNAs regulating TCR signaling and TNF-α secretion (miR-3187-3p, miR-498, miR-122, miR149, miR-181a/b) in MTEX [90].

Accumulating evidence reveals that TNF is negatively regulated by miR-21. This may explain the effects of miR-21 on cell proliferation, migration, invasion, and transformation associated with excessive miR-21 levels in MTEX. Moreover, some results suggest that TNF can promote miR-21 biogenesis [123] as well as the turnover of PDCD4 in macrophages [124]. Yang et al. also showed that increased levels of miR-21 associated with a decreased level of TNF were consistent with elevated IL-10 protein expression and increased Arg1 macrophage expression, which could explain poor immune responses against cancer cells [98]. On the other hand, Fabri et al. reported that miR-21 which was found to be enriched in MTEX might also act as a ligand by binding to receptors of the Toll-like receptor (TLR) family members, murine TLR7, and human TLR8, in immune cells. Triggering the TLR-mediated prometastatic inflammatory response in responder cells might promote tumor growth and metastasis [125].

Stimulation of $TLR2^+$ DC by tumor-derived TLR2 ligands was reported to drive inhibitory signals leading to dysfunctional activity of DC in murine melanoma [126]. Modulation of immune response by MTEX was confirmed by Zhou et al. [85]. They observed that B16-derived MTEX induced apoptosis of $CD4^+$ T cells in vitro and promoted the growth of tumor cells implanted in mice. The opposite results were reported by blocking MTEX release (disrupting the expression of Rab27a), thus confirming the proposed mechanism. Further, they showed that B16-derived MTEX induced activation of caspase-3, caspase-7, and caspase-9, reducing the level of anti-apoptotic proteins, such as BCL-2, BCL-xL, and MCL-1 in $CD4^+$ T cells [85]. MTEX can also alter the functions of natural killer (NK) cells. They were found to modulate the tumor immune responses by inhibiting the cytotoxic activity of NKs and downregulating the expression of NKG2D, NKp30, NKP46, and NKG2C proteins on the surface of NK cells [26,42,127].

5.2. MTEX Participate in the Reprogramming of TME

TME plays a major role in cancer growth and evolution. Diverse cells such as fibroblasts, endothelial, epithelial, and mesenchymal cells or immune cells present in the TME might be reprogrammed by MTEX to favor tumor growth [45,56]. Accumulating data provide evidence that MTEX promote epithelial-to-mesenchymal transition (EMT), which promotes metastasis [46,48,55,57,128]. The mitogen-activated protein kinase (MAPK) signaling pathway is activated during the MTEX-mediated EMT, with the involvement of Let-7i, a miRNA modulator of EMT [104]. Furthermore, acquisition of the EMT-like phenotype is enforced by expression of other key regulators of EMT induction, including ZEB2 and Snail 2 [119,129]. Upregulation of ZEB2 and Snail 2 in primary melanocytes after co-culture with MTEX was confirmed by Xiao et al. This process was accompanied by decreased expression of E-cadherin and increased expression of vimentin [104]. The interplay between MTEX and myeloid stem cells (MSCs) induce the emergence of a tumor-like phenotype with PD-1 and mTOR overexpression in naïve MSCs in vitro and fast tumor progression in vivo [119]. Interaction networks build basing on genes overexpressed in recipient cells upon co-incubation with MTEX identified a variety of other exosomal molecules, apart from PD-1 and mTOR, which might affect tumor progressions, such as MET, Ras, RAF1, Mek, ERK1/2, MITF, BCL2, PI3K, Akt, KIT, JAK STAT3, or ETS1 [119].

MTEX transform fibroblasts into proangiogenic cancer-associated fibroblasts (CAF) in vitro and in vivo. CAF are known to support development of pre-cancerous micro- and macro-environments [86,130,131]. Zhao et al. discovered that incubation of MTEX with fibroblasts resulted in a significant increase of VCAM-1 expression, and this enhancement was even stronger when EX were derived from highly metastatic melanoma cells [131]. Overexpression of miR-155 in MTEX was found to be the trigger factor for the proangiogenic switch of fibroblasts into CAF [86]. MTEX-mediated delivery of miR-155 to fibroblasts suppressed expression of cytokine signaling 1 (SOCS1), that activates the JAK2/STAT3 signaling pathway which, in turn, regulates the expression of proangiogenic factors. Elevated expression of vascular endothelial growth factor A (VEGFa), fibroblast growth factor 2 (FGF2), and matrix metalloproteinase 9 (MMP9) in fibroblasts after incubation with MTEX was confirmed in this study [86]. Shu et al. also reported the presence of exosomal miR-155 and miR-210 across six melanoma cell lines [89] and showed that miRNA cargo of MTEX was capable of reprogramming the metabolism of human adult dermal fibroblasts (HADF). In this study, miR-155 upregulated glucose metabolism (i.e., increased glycolysis), while miR-210 decreased oxidative phosphorylation under non-hypoxic conditions. Exposure of HADF to MTEX resulted in upregulated aerobic glycolysis and downregulated oxidative phosphorylation in stromal fibroblasts, with consequently increasing extracellular acidification [89]. Furthermore, the acidic environment led to upregulation of over 50% of exosomal proteins involved in cancer progression in MTEX derived from the primary non-tumorigenic MEL501 cell line [73]. The upregulated proteins were associated with focal adhesion, actin cytoskeleton regulation, leukocyte trans-endothelial migration, regulation and modification of cell morphology, HSP family proteins, proteoglycans related to cancer, small GTPase mediated signal transduction, and epidermal growth factor receptor (EGFR) signaling pathways [73]. This shows that MTEX are important contributors to changes in the TME that are responsible for creating favorable conditions for the pre-metastatic niche. On the one hand accelerated aerobic glycolysis ensures more effective energy production, but on the other hand, the acidic microenvironment drives immune suppression and creates a pro-metastatic environment [73,89,132].

The pro-angiogenic effects of MTEX are well-documented. MTEX cargos are enriched in pro-angiogenic cytokines, including IL-1α, FGF, GCS-F, TNFα, leptin, TGFα, and VEGF [107]. MTEX also mediate the transfer of miR-9 from melanoma to endothelial cells (EC), which triggers the JAK-STAT pathway and enhances the migratory propensity of vascular cells as well as the formation of a tumor-supporting vascular net [133]. Additionally, it was reported that increased WNT5A signaling, which is known to promote melanoma metastasis, induced a Ca^{2+}-dependent release of exosomes containing the pro-angiogenic VEGF and MMP2 factors in melanoma cells [134].

5.3. MTEX Can Modulate Tumor Progression and Invasiveness

In general, TEX may induce tumor growth in vitro and in vivo [135,136]. It was reported that B16BL6-derived MTEX induced proliferation and inhibited apoptosis of murine melanoma B16BL6 cells, while inhibition of MTEX release by the N-Smase inhibitor suppressed melanoma growth. Noteworthy, the uptake of MTEX resulted in increased levels of cyclin D1, p-Akt (cell proliferation-related proteins), Bcl-2 (survival-related protein), and decreased level of Bax (apoptosis-related proteins) [137]. Peinado et al. reported that the oncoprotein MET selectively enriched in MTEX released by metastatic melanoma cells promoted the tumorigenic potential of melanoma [74]. Pre-conditioning of bone marrow (BM) with MTEX obtained from a highly metastatic melanoma B16-F10 cell line promoted mobilization of bone marrow-derived cells (BMDC), increasing tumor vasculogenesis, invasion, and metastasis. Comparative analysis of the protein content in MTEX from highly metastatic and poorly metastatic melanoma cells confirmed MET signaling as the principal mediator of BM progenitor cell "education". Pre-treatment of BM cells with B16-F10 MTEX resulted in HGF-induced S6 and ERK phosphorylation compared to non-treated controls. Effectors of MET-mediated BM progenitor cell mobilization, i.e., S6-kinase (mTOR pathway) and ERK (MAPK pathway), are known mediators of HGF/MET signaling. Further, the metastatic spread and organotropism of highly metastatic B16-F10 primary tumors were reduced by the BM of mice "educated" with the low-metastatic B16-F1 MTEX that lacked the MET receptor. These data suggested that non-metastatic MTEX might educate the BM and prevent metastatic disease, a finding that is worth further exploration. Finally, it was confirmed that MET expression was elevated in sEV circulating in the plasma of patients with metastatic melanoma [74]. Additionally, influence of metabotropic glutamate receptor 1 (GRM1) expressed on melanoma cells was tested for cell migration and invasiveness [138]. This neuronal receptor induces in vitro melanocytic transformation and spontaneous malignant melanoma development in vivo. Moreover, modulation (decrease) of GRM1 expression results in a decrease in both cell proliferation in vitro and tumor progression in vivo. Isola et al. verified a hypothesis that exosomes released by a GRM1-positive (metastatic) cell line made GRM1-negative (non-metastatic) cells acquire features characteristic for GRM1-positive cells, i.e., to migrate, invade, form colonies, and exhibit anchorage-independent cell growth. They found that acquiring these features was not connected with expression of this receptor on GRM1-negative cells. Another aspect of the potential role of MTEX in tumorigenesis is analysis of specific RAB genes involved in sEV secretion (RAB1A, RAB5B, RAB7, RAB27A) [74]. Rab27a is a regulator of protein trafficking and melanoma proliferation [139]. Reduced expression of Rab27a resulted in decreased sEV production, and in decreased release of pro-angiogenic factors (PlGF-2, osteopontin, and PDGF-AA) from tumor cells, interfering with BMDC mobilization and tumor invasiveness [74]. These results are in line with the latest findings of Guo and colleagues [140], who reported that the GTPase RAB27A was overexpressed in melanoma patients and correlated with poor patient survival. A loss of RAB27A expression in melanoma cell lines blocked invasion and cell motility in vitro, and spontaneous metastasis in vivo. Furthermore, RAB27A-expressing MTEX promoted the invasion phenotype of melanoma cells in contrast to MTEX without RAB27A [140]. All in all, these results suggest that RAB27A promotes the biogenesis of a distinct pro-invasive MTEX subpopulation [74,140].

MTEX are also involved in preparation of metastatic niche for melanoma in lymph nodes and lungs and in reprogramming of innate osteotropism of melanoma cells [74,141,142]. MTEX from a highly-metastatic B16-F10 cell line promoted lymph nodes (LN) metastasis in mice [142] and were detected after 24h in the interstitium of the lung, BM, liver, and spleen (organotropic sites for B16-F10 metastasis), but not in the circulatory system [74]. Several genes responsible for the recruitment of melanoma cells (stabilin 1, ephrin receptor β4, and αv integrin), extracellular matrix remodeling (Mapk14, uPA, laminin 5, Col 18α1, G-α13, p38), vascular growth (TNF-α, TNF-αip2, VEGF-B, HIF-1α, Thbs1) [142], and effectors of pre-metastatic niche formation such as S100A8, S100A9 [74] were upregulated by B16-F10 MTEX. The osteotropism of melanoma cells is related to the activation of the SDF-1/CXCR4/CXCR7 axis. MTEX were found to promote osteotropism of not-osteotropic melanoma cells (SK-Mel28, WM266) in vitro through membrane CXCR7 up-regulation. Thus, MTEX were found to contribute to bone metastasis in melanoma [141].

6. MTEX as Potential Clinical Biomarkers

MTEX present in body fluids of melanoma patients are a promising source of prognostic biomarkers as a new type of so-called liquid biopsy. Alegre et al. performed an analysis of the established melanoma biomarkers such as: MIA, S100B, and tyrosinase-related protein 2 (TYRP2) in sEV isolated from sera of stage IV melanoma patients, patients with no evidence of disease (NED), and healthy donors (HD) [37]. The levels of MIA and S100B were significantly higher in melanoma patients in comparison to HD and NED patients. Furthermore, patients with high EV concentration of MIA had shorter median survival compared to those with lower MIA levels (4 versus 11 months; $p < 0.05$). Hence, the data suggest the potential diagnostic and prognostic utility of MIA in plasma sEV [37]. Levels of MIA, along with growth/differentiation factor 15 precursor protein (GDF15) showed a significant increase in the whole secretome of uveal melanoma versus non-malignant cells [143], which was in line with the results of Alegre et al. [37]. Tenga et al. showed that miR-532-5p and miR-106b present in serum sEV could be used for classification of melanoma patients, including differentiation of patients with metastatic and non-metastatic disease and stage I-II patients from stage III-IV patients [144]. In addition, miR-17, miR-19a, miR-21, miR-126, and miR-149 were found to be expressed at significantly higher levels in patients with metastatic sporadic melanoma compared to familial melanoma patients or healthy controls [91]. On the other hand, levels of miR-125b in sEV were significantly lower in patients with advanced melanoma compared with disease-free patients with melanoma and healthy controls, while there was no statistical difference in the miR-125b levels between patients and controls when analyzing serum samples [92].

Melanoma is sensitive to immune checkpoint inhibitors (such as anti-CTLA4 and anti-PD1 monoclonal antibodies) and small-molecule targeted drugs (such as BRAF inhibitors and MEK inhibitors). However, many patients with melanoma fail to respond to these therapies, and the mechanisms of resistance to a therapy are not understood [61–63,145,146]. The accumulating data suggest the importance of MTEX in understanding these mechanisms and the role of MTEX as predictive biomarkers of response to immune therapies and outcome [55–57,147]. Higher levels of miR-497-5p in circulating sEV during MAPKi-based therapy of cutaneous metastatic melanoma patients (with BRAFV600 mutations) were significantly correlated with progression-free survival (hazard ratio of 0.27) [147]. Increased level of miR-497-5p was also associated with prolonged post-recurrence survival in resected metastases from patients with metastatic III (lymph nodes) and metastatic IV cutaneous malignant melanoma (CMM) [148]. Treatment with vemurafenib and dabrafenib induced miR-211-5p up-regulation in melanoma-derived EV, both in vitro and in vivo, thus promoting survival in parent melanoma cells despite a down-regulation of pERK1/2 by BRAF inhibitors [146]. What is more, transfection of miR-211 in low-expressing miR-211–5p melanoma cells resulted in enhanced proliferation of melanoma cells. What is more, 100-fold increase in miR-211–5p expression in vemurafenib-treated miR-211-5p-transfected cells was found with no reduction of cells proliferation upon BRAF inhibitor treatment. These findings suggest that miR-211-5p up-regulation upon vemurafenib treatment allows these cells to survive and grow into a population of cells that have reduced sensitivity to vemurafenib. Going further, inhibition of miR-211-5p in a vemurafenib resistant cell line decreased cell proliferation. The outcome of the study of Lunavat et al. leads to better understanding of possible mechanisms of acquiring by patients' resistance to the BRAF inhibitors treatment by showing that miR-211-5p can reduce the sensitivity to vemurafenib treatment in melanoma cells by regulating cellular proliferation. [146]. Another group of "new drugs" used in the treatment of melanoma are immune checkpoint inhibitors. Anti-PD-1 antibodies are frequently used in melanoma treatment to rejuvenate anti-tumor immunity, and in the majority of patients the response is durable, yet not all melanoma patients respond to this therapy [60,149]. Chen et al. reported positive correlation between exosomal-PD-L1 (Exo-PD-L1) level and IFN-γ, both in vitro using melanoma cell lines and in vivo in patients with metastatic melanoma [111]. Upregulation of PD-L1 by IFN-γ in metastatic melanoma leads to functional suppression of CD8+ T effector cells enabling melanoma growth and metastasis. In part, this explains low response rate to anti-PD-1 therapy (pembrolizumab). The level of circulating Exo-PD-L1 distinguished clinical responders from non-responders to pembrolizumab treatment. Since the level of exosomal PD-L1 was altered early during the anti-PD-1 therapy, the authors suggest that it might be an indicator of response to treatment [111]. A recent paper by Cordonnier et al. describes monitoring of circulating Exo-PD-L1 in melanoma patients treated with immune checkpoint inhibitors and BRAF/MEK inhibitors. This prospective clinical study confirmed a significantly higher level of Exo-PD-L1 in plasma of melanoma patients compared to soluble PD-L1 and demonstrated that the level of Exo-PD-L1 inversely correlated with patients' response to therapy [150]. The results of this clinical study provide a rationale for monitoring Exo-PD-L1 level as a potential predictor of the melanoma patients' response to treatment and outcome [150].

Clinical relevance of MTEX-based biomarkers is currently limited by the necessity of separation of MTEX from other fractions of sEV circulating in body fluids. Recently, however, the anti-CSPG4 mAb was used for the separation of MTEX and sEV produced by normal tissue from the plasma of melanoma patients [26–28]. CSPG4$^+$ MTEX captured from the plasma of melanoma patients are highly enriched in melanoma-associated antigens (MAA) in comparison to CSPG4(-) non-MTEX, including CSPG4, TYRP2, MelanA, Gp100, VLA4. Moreover, several immunostimulatory (CD40, CD40L, CD80, OX40, OX40L) and immunosuppressive (PDL-1, CD39, CD73, FasL, LAP-TGFβ, TRAIL, CTLA-4) proteins were enriched in MTEX compared to sEV purified from plasma of healthy donors [26,28]. Noteworthy, looking at individual differences among proteins in the cargo of MTEX and non-MTEX, significant correlations with disease activity were observed for both fractions of vesicles. For example, non-MTEX ability to induce apoptosis of T cells positively correlated with the disease stage [28]. The obtained data suggest that features of both MTEX and non-MTEX, as well as individual MTEX/total sEV ratios, might be useful for monitoring melanoma progression [26,28]. In addition to CSPG4, other melanoma-specific or enhanced proteins might also be considered as potential markers of MTEX. This includes several melanoma-associated antigens (MAA-4, MAA-B2, and melanoma antigen recognized by T-cells) found in MTEX released by 7 different melanoma cell lines with various phenotypic features (non-tumorigenic, tumorigenic, metastatic) [70]. Moreover, several other cancer-related proteins (NRAS, Src, c-Met, c-Kit, EGFR, MCAM, annexin A1, HAPLN1, LGALS1, GALS3, NT5E, and PMEL) were detected in MTEX originating from various melanoma cell lines [69,70]. Therefore, several candidates for MTEX-markers are known that could be used for the immune capture of MTEX circulating in the body fluids of melanoma patients. Hence, the emerging concept of MTEX-based biomarkers of melanoma will meet the necessary methodological support in the nearest future.

7. Future Directions

Although the number of publications reporting on sEV in melanoma is growing exponentially, the resulting knowledge remains limited. Most likely, this is due to several factors that impede research of sEV. First, no uniformly accepted nomenclature for EV has been established, creating havoc in the definition of investigated EV. Further, no standardized procedures for the isolation of different EV types exist, leading to differences in contamination levels and co-isolation of various vesicles. The criteria and methods of EV characterization are also not clear and seem to change as we progress in the understanding of the EV heterogeneity. Despite the recommendations updated yearly by the International Society of Extracellular Vesicles (ISEV), published papers often provide incomplete data creating further confusion. The emerging view of the complex biology of EV requires strict criteria for the definition of phenotypes, genotypes, and functions of participating EV. Specifically, in a large body of available data on melanoma-associated sEV in body fluids, their origin is often unclear. Until recently, melanoma cell lines had been the only reliable source of MTEX. However, research performed with vesicles produced by cell lines does not adequately reflect interactions taking place in body fluids or tissues. Separation of MTEX from plasma by immune capture allowed for a more relevant evaluation of their characteristics and functions in disease and comparisons of data between individual patients. While this represents considerable progress, ex vivo analysis of MTEX also provides only a limited view of their biological agenda in the TME and the periphery. In vivo studies of MTEX in murine models of melanoma are critical for the translation of signaling mediated by MTEX in vitro to cells, tissues, and organs in animals. Correlative studies of MTEX and clinical endpoints in melanoma progression, resistance, or response to therapies are growing in numbers and the concept of MTEX as a liquid tumor biopsy is slowly crystallizing. Understanding of multicellular MTEX-mediated signaling and their reprogramming activities in the TME opens a way for the use of MTEX-induced changes as yet another biomarker of disease activity. The next step is to develop and implement reliable means for the isolation and molecular characterization of MTEX from body fluids and tissues of patients with melanoma. At present, these methods are in the discovery stage, and the emerging results are promising not only due to successful subsetting of sEV into MTEX and non-MTEX, but also because of evidence that mechanistic and functional studies of MTEX can yield new and previously unsuspected information. For example, the ability of MTEX to simultaneously deliver to recipient cells multiple and often contradictory, i.e., suppressive vs. stimulatory signals have alerted us to the possibility of regulatory functions MTEX might exercise in vivo. Similarly, the realization that MTEX utilize surface proteins as well as miRs to transmit signals to recipient cells alerts us to ask why these two signaling pathways co-exist and how they impact the biology. As melanoma biomarkers, MTEX might provide a more reliable diagnostic, prognostic, or outcome data than total sEV isolated from body fluids. Future validation studies encompassing all aspects of MTEX isolation, characterization, and signaling will be necessary to move the field forward and translate current knowledge to clinically applicable strategies and methods. In this respect, antibody-based microarrays, multiparameter quantitative flow cytometry, and targeted proteomics are emerging as the tools applicable to serial monitoring of MTEX in body fluids of patients with melanoma. The future will likely see numerous such studies performed as part of clinical trials designed to validate the roles of MTEX in the biology of melanoma.

Author Contributions: All authors have read and agree to the published version of the manuscript. A.Z.—conception and drafting the article; P.W. and T.W.—critical revision of the article; M.P.—design and revision of the manuscript, funding and final approval of the version to be published.

Abbreviations

AB	apoptotic bodies
APC	antigen presenting cells
BM	bone marrow
BMDC	bone marrow derived cells
CAF	cancer-associated fibroblasts
CEM	cryo-electron microscopy
CM	confocal microscopy
DC	differential centrifugation
DC	dendritic cells
DLS	dynamic light scattering
EC	endothelial cells
EMT	epithelial-to-mesenchymal transition
EV	extracellular vesicles
EX	exosomes
HADF	human adult dermal fibroblasts
IA-FCM	immune-affinity flow cytometry
KO	knockdown
MAA	melanoma associated antigens
mAb	monoclonal antibodies

MSC	myeloid stem cells
MTEX	melanoma cell-derived exosomes
MV	microvesicles
MVB	multivesicular bodies
NTA	nanoparticle tracking analysis
SEC	size-exclusion chromatography
SEM	scanning electron microscopy
sEV	small extracellular vesicles
TEM	transmission electron microscopy
TEX	tumor-derived exosomes
TME	tumor microenvironment
UC	ultracentrifugation
UF	ultrafiltration
WB	western blotting

References

1. Kalluri, R.; LeBleu, V.S. The biology, function, and biomedical applications of exosomes. *Science* **2020**, *367*, eaau6977. [CrossRef] [PubMed]
2. Willms, E.; Cabañas, C.; Mäger, I.; Wood, M.J.A.; Vader, P. Extracellular Vesicle Heterogeneity: Subpopulations, Isolation Techniques, and Diverse Functions in Cancer Progression. *Front. Immunol.* **2018**, *9*, 738. [CrossRef] [PubMed]
3. van Niel, G.; D'Angelo, G.; Raposo, G. Shedding light on the cell biology of extracellular vesicles. *Nat. Rev. Mol. Cell Biol.* **2018**, *19*, 213–228. [CrossRef]
4. Zaborowski, M.P.; Balaj, L.; Breakefield, X.O.; Lai, C.P. Extracellular Vesicles: Composition, Biological Relevance, and Methods of Study. *Bioscience* **2015**, *65*, 783–797. [CrossRef] [PubMed]
5. Yáñez-Mó, M.; Siljander, P.R.; Andreu, Z.; Zavec, A.B.; Borràs, F.E.; Buzas, E.I.; Buzas, K.; Casal, E.; Cappello, F.; Carvalho, J.; et al. Biological properties of extracellular vesicles and their physiological functions. *J. Extracell. Vesicles* **2015**, *4*, 27066. [CrossRef]
6. Minciacchi, V.R.; Freeman, M.R.; Di Vizio, D. Extracellular Vesicles in Cancer: Exosomes, Microvesicles and the Emerging Role of Large Oncosomes. *Semin. Cell Dev. Biol.* **2015**, *40*, 41–51. [CrossRef]
7. György, B.; Szabó, T.G.; Pásztói, M.; Pál, Z.; Misják, P.; Aradi, B.; László, V.; Pállinger, E.; Pap, E.; Kittel, A.; et al. Membrane vesicles, current state-of-the-art: Emerging role of extracellular vesicles. *Cell. Mol. Life Sci.* **2011**, *68*, 2667–2688. [CrossRef]
8. Théry, C.; Witwer, K.W.; Aikawa, E.; Alcaraz, M.J.; Anderson, J.D.; Andriansitohaina, R.; Antoniou, A.; Arab, T.; Archer, F.; Atkin-Smith, G.K.; et al. Minimal information for studies of extracellular vesicles 2018 (MISEV2018): A position statement of the International Society for Extracellular Vesicles and update of the MISEV2014 guidelines. *J. Extracell. Vesicles* **2019**, *8*, 1535750. [CrossRef]
9. Kowal, J.; Arras, G.; Colombo, M.; Jouve, M.; Morath, P.J.; Primdal-Bengtson, B.; Dingli, F.; Leow, D.; Tkach, M.; Théry, C. Proteomic comparison defines novel markers to characterize heterogeneous populations of extracellular vesicle subtypes. *Proc. Natl. Acad. Sci. USA* **2016**, *113*, 968–977. [CrossRef]
10. Abramowicz, A.; Widlak, P.; Pietrowska, M. Proteomic analysis of exosomal cargo: The challenge of high purity vesicle isolation. *Mol. Biosyst.* **2016**, *12*, 1407–1419. [CrossRef]
11. Gurunathan, S.; Kang, M.H.; Jeyaraj, M.; Qasim, M.; Kim, J.H. Review of the Isolation, Characterization, Biological Function, and Multifarious Therapeutic Approaches of Exosomes. *Cells* **2019**, *8*, 307. [CrossRef]
12. Zebrowska, A.; Skowronek, A.; Wojakowska, A.; Widlak, P.; Pietrowska, M. Metabolome of Exosomes: Focus on Vesicles Released by Cancer Cells and Present in Human Body Fluids. *Int. J. Mol. Sci.* **2019**, *20*, 3461. [CrossRef] [PubMed]
13. Théry, C.; Ostrowski, M.; Segura, E. Membrane vesicles as conveyors of immune responses. *Nat. Rev. Immunol.* **2009**, *9*, 581–593. [CrossRef]

14. Dixon, C.L.; Sheller-Miller, S.; Saade, G.R.; Fortunato, S.J.; Lai, A.; Palma, C.; Guanzon, D.; Salomon, C.; Menon, R. Amniotic Fluid Exosome Proteomic Profile Exhibits Unique Pathways of Term and Preterm Labor. *Endocrinology* **2018**, *159*, 2229–2240. [CrossRef]
15. Yu, J.; Lin, Y.; Xiong, X.; Li, K.; Yao, Z.; Dong, H.; Jiang, Z.; Yu, D.; Yeung, S.J.; Zhang, H. Detection of Exosomal PD-L1 RNA in Saliva of Patients With Periodontitis. *Front. Genet.* **2019**, *10*, 202. [CrossRef]
16. Grigor'eva, A.E.; Tamkovich, S.N.; Eremina, A.V.; Tupikin, A.E.; Kabilov, M.R.; Chernykh, V.V.; Vlassov, V.V.; Laktionov, P.P.; Ryabchikova, E.I. Exosomes in tears of healthy individuals: Isolation, identification, and characterization. *Biochem. Moscow Suppl. Ser. B* **2016**, *10*, 165. [CrossRef]
17. Wu, G.; Yang, G.; Zhang, R. Altered microRNA Expression Profiles of Extracellular Vesicles in Nasal Mucus From Patients With Allergic Rhinitis. *Allergy Asthma Immunol. Res.* **2015**, *7*, 449–457. [CrossRef]
18. Saunderson, S.C.; Dunn, A.C.; Crocker, P.R.; McLellan, A.D. CD169 mediates the capture of exosomes in spleen and lymph node. *Blood* **2014**, *123*, 208–216. [CrossRef] [PubMed]
19. Chiasserini, D.; van Weering, J.R.; Piersma, S.R.; Pham, T.V.; Malekzadeh, A.; Teunissen, C.E.; de Wit, H.; Jiménez, C.R. Proteomic analysis of cerebrospinal fluid extracellular vesicles: A comprehensive dataset. *J. Proteom.* **2014**, *106*, 191–204. [CrossRef]
20. Romancino, D.P.; Paterniti, G.; Campos, Y.; De Luca, A.; Di Felice, V.; d'Azzo, A.; Bongiovanni, A. Identification and characterization of the nano-sized vesicles released by muscle cells. *FEBS Lett.* **2013**, *587*, 1379–1384. [CrossRef]
21. Masyuk, A.I.; Huang, B.Q.; Ward, C.J.; Gradilone, S.A.; Banales, J.M.; Masyuk, T.V.; Radtke, B.; Splinter, P.L.; LaRusso, N.F. Biliary exosomes influence cholangiocyte regulatory mechanisms and proliferation through interaction with primary cilia. *Am. J. Physiol. Gastrointest. Liver Physiol.* **2010**, *299*, G990–G999. [CrossRef] [PubMed]
22. Runz, S.; Keller, S.; Rupp, C.; Stoeck, A.; Issa, Y.; Koensgen, D.; Mustea, A.; Sehouli, J.; Kristiansen, G.; Altevogt, P. Malignant ascites-derived exosomes of ovarian carcinoma patients contain CD24 and EpCAM. *Gynecol. Oncol.* **2007**, *107*, 563–571. [CrossRef] [PubMed]
23. Admyre, C.; Johansson, S.M.; Qazi, K.R.; Filén, J.J.; Lahesmaa, R.; Norman, M.; Neve, E.P.; Scheynius, A.; Gabrielsson, S. Exosomes with immune modulatory features are present in human breast milk. *J. Immunol.* **2007**, *179*, 1969–1978. [CrossRef]
24. Andreau, Z.; Yanez-Mo, M. Tetraspanins in extracellular vesicle formation and function. *Front. Immunol.* **2014**, *5*, 442. [CrossRef]
25. Wieckowski, E.U.; Visus, C.; Szajnik, M.; Szczepanski, M.J.; Storkus, W.J.; Whiteside, T.L. Tumor-Derived Microvesicles Promote Regulatory T Cell Expansion and Induce Apoptosis in Tumor-Reactive Activated CD8+ T Lymphocytes. *J. Immunol.* **2009**, *183*, 3720–3730. [CrossRef]
26. Sharma, P.; Diergaarde, B.; Ferrone, S.; Kirkwood, J.M.; Whietside, T.L. Melanoma cell-derived exosomes in plasma of melanoma patients suppress functions of immune effector cells. *Sci. Rep.* **2020**, *10*, 92. [CrossRef]
27. Ferrone, S.; Whiteside, T.L. Targeting CSPG4 for isolation of melanoma cell-derived exosomes from body fluids. *HNO* **2020**, *68*, 100–105. [CrossRef]
28. Sharma, P.; Ludwig, S.; Muller, L.; Hong, C.S.; Kirkwood, J.M.; Ferrone, S.; Whiteside, T.L. Immunoaffinity-based isolation of melanoma cell-derived exosomes from plasma of patients with melanoma. *J. Extracell. Vesicles* **2018**, *7*, 1435138. [CrossRef]
29. Melo, S.A.; Luecke, L.B.; Kahlert, C.; Fernandez, A.F.; Gammon, S.T.; Kaye, J.; LeBleu, V.S.; Mittendorf, E.A.; Weitz, J.; Rahbari, N.; et al. Glypican-1 identifies cancer exosomes and detects early pancreatic cancer. *Nature* **2015**, *523*, 177–182. [CrossRef]
30. Mizutani, K.; Terazawa, R.; Kameyama, K.; Kato, T.; Horie, K.; Tsuchiya, T.; Seike, K.; Ehara, H.; Fujita, Y.; Kawakami, K.; et al. Isolation of prostate cancer-related exosomes. *Anticancer Res.* **2014**, *34*, 3419–3423.
31. Hong, C.S.; Muller, L.; Boyiadzis, M.; Whiteside, T.L. Isolation and characterization of CD34+ blast-derived exosomes in acute myeloid leukemia. *PLoS ONE* **2014**, *9*, e103310. [CrossRef]
32. Raimondo, S.; Pucci, M.; Alessandro, R.; Fontana, S. Extracellular Vesicles and Tumor-Immune Escape: Biological Functions and Clinical Perspectives. *Int. J. Mol. Sci.* **2020**, *21*, 2286. [CrossRef] [PubMed]
33. Whiteside, T.L. Exosomes in Cancer: Another Mechanism of Tumor-Induced Immune Suppression. In *Tumor Immune Microenvironment in Cancer Progression and Cancer Therapy*; Kalinski, P., Ed.; Springer: Cham, Switzerland, 2017; Volume 1036, pp. 81–89. [CrossRef]

34. Whiteside, T.L. Tumor-Derived Exosomes and Their Role in Cancer Progression. *Adv. Clin. Chem.* **2016**, *74*, 103–141. [CrossRef] [PubMed]
35. Meldolesi, J. Exosomes and Ectosomes in Intercellular Communication. *Curr. Biol.* **2018**, *28*, R435–R444. [CrossRef]
36. Ludwig, A.K.; Giebel, B. Exosomes: Small vesicles participating in intercellular communication. *Int. J. Biochem. Cell Biol.* **2012**, *44*, 11–15. [CrossRef] [PubMed]
37. Alegre, E.; Zubiri, L.; Perez-Gracia, J.L.; González-Cao, M.; Soria, L.; Martín-Algarra, S.; González, A. Circulating melanoma exosomes as diagnostic and prognosis biomarkers. *Clin. Chim. Acta* **2016**, *454*, 28–32. [CrossRef] [PubMed]
38. Whiteside, T.L. The potential of tumor-derived exosomes for noninvasive cancer monitoring. *Expert Rev. Mol. Diagn.* **2015**, *15*, 1293–1310. [CrossRef] [PubMed]
39. Duijvesz, D.; Versluis, C.Y.; van der Fels, C.A.; Vredenbregt-van den Berg, M.S.; Leivo, J.; Peltola, M.T.; Bangma, C.H.; Pettersson, K.S.; Jenster, G. Immuno-based detection of extracellular vesicles in urine as diagnostic marker for prostate cancer. *Int. J. Cancer* **2015**, *137*, 2869–2878. [CrossRef]
40. Szajnik, M.; Derbis, M.; Lach, M.; Patalas, P.; Michalak, M.; Drzewiecka, H.; Szpurek, D.; Nowakowski, A.; Spaczynski, M.; Baranowski, W.; et al. Exosomes in Plasma of Patients with Ovarian Carcinoma: Potential Biomarkers of Tumor Progression and Response to Therapy. *Gynecol. Obstet.* **2013**, *4*, 003. [CrossRef]
41. Silva, J.; Garcia, V.; Rodriguez, M.; Compte, M.; Cisneros, E.; Veguillas, P.; Garcia, J.M.; Dominguez, G.; Campos-Martin, Y.; Cuevas, J.; et al. Analysis of exosome release and its prognostic value in human colorectal cancer. *Genes Chromosomes Cancer* **2012**, *51*, 409–418. [CrossRef]
42. Szczepanski, M.J.; Szajnik, M.; Welsh, A.; Whiteside, T.L.; Boyiadzis, M. Blast-derived microvesicles in sera from patients with acute myeloid leukemia suppress natural killer cell function via membrane-associated transforming growth factor-beta1. *Haematologica* **2011**, *96*, 1302–1309. [CrossRef] [PubMed]
43. Baran, J.; Baj-Krzyworzeka, M.; Weglarczyk, K.; Szatanek, R.; Zembala, M.; Barbasz, J.; Czupryna, A.; Szczepanik, A.; Zembala, M. Circulating tumour-derived microvesicles in plasma of gastric cancer patients. *Cancer Immunol. Immunother.* **2010**, *59*, 841–850. [CrossRef] [PubMed]
44. Zhang, L.; Yu, D. Exosomes in cancer development, metastasis, and immunity. *Biochim. Biophys. Acta Rev. Cancer* **2019**, *1871*, 455–468. [CrossRef] [PubMed]
45. Mashouri, L.; Yousefi, H.; Aref, A.R.; Ahadi, A.M.; Molaei, F.; Alahari, S.K. Exosomes: Composition, biogenesis, and mechanisms in cancer metastasis and drug resistance. *Mol. Cancer* **2019**, *18*, 75. [CrossRef]
46. Tung, K.H.; Ernstoff, M.S.; Allen, C.; Shu, S. A Review of Exosomes and their Role in The Tumor Microenvironment and Host-Tumor "Macroenvironment". *J. Immunol. Sci.* **2019**, *3*, 4–8. [CrossRef]
47. Wang, T.; Nasser, M.I.; Shen, J.; Qu, S.; He, Q.; Zhao, M. Functions of Exosomes in the Triangular Relationship between the Tumor, Inflammation, and Immunity in the Tumor Microenvironment. *J. Immunol. Res.* **2019**, *2019*, 4197829. [CrossRef]
48. Syn, N.; Wang, L.; Sethi, G.; Thiery, J.P.; Goh, B.C. Exosome-Mediated Metastasis: From Epithelial-Mesenchymal Transition to Escape from Immunosurveillance. *Trends Pharmacol. Sci.* **2016**, *37*, 606–617. [CrossRef]
49. Whiteside, T.L. Exosomes and tumor-mediated immune suppression. *J. Clin. Investig.* **2016**, *126*, 1216–1223. [CrossRef]
50. Whiteside, T.L. The role of tumor-derived exosomes in epithelial mesenchymal transition (EMT). *Transl. Cancer Res.* **2017**, *6*, S90–S92. [CrossRef]
51. Ludwig, N.; Yerneni, S.S.; Razzo, B.M.; Whiteside, T.L. Exosomes from HNSCC Promote Angiogenesis through Reprogramming of Endothelial Cells. *Mol. Cancer Res.* **2018**, *16*, 1798–1808. [CrossRef] [PubMed]
52. Barros, F.M.; Carneiro, F.; Machado, J.C.; Melo, S.A. Exosomes and Immune Response in Cancer: Friends or Foes? *Front. Immunol.* **2018**, *9*, 730. [CrossRef] [PubMed]
53. Czernek, L.; Düchler, M. Functions of Cancer-Derived Extracellular Vesicles in Immunosuppression. *Arch. Immunol. Ther. Exp.* **2017**, *65*, 311–323. [CrossRef] [PubMed]
54. Whiteside, T.L. Immune modulation of T-cell and NK (natural killer) cell activities by TEX (tumour-derived exosomes). *Biochem. Soc. Trans.* **2013**, *41*, 245–251. [CrossRef] [PubMed]
55. Tucci, M.; Mannavola, F.; Passarelli, A.; Stucci, L.S.; Cives, M.; Silvestris, F. Exosomes in melanoma: A role in tumor progression, metastasis and impaired immune system activity. *Oncotarget* **2018**, *9*, 20826–20837. [CrossRef]

56. Tucci, M.; Passarelli, A.; Mannavola, F.; Felici, C.; Stucci, L.S.; Cives, M.; Silvestris, F. Immune System Evasion as Hallmark of Melanoma Progression: The Role of Dendritic Cells. *Front. Oncol.* **2019**, *9*, 1148. [CrossRef]
57. Isola, A.L.; Eddy, K.; Chen, S. Biology, Therapy and Implications of Tumor Exosomes in the Progression of Melanoma. *Cancers* **2016**, *8*, 110. [CrossRef]
58. Siegel, R.L.; Miller, K.D.; Jemal, A. Cancer statistics, 2020. *CA Cancer J. Clin.* **2020**, *70*, 7–30. [CrossRef]
59. Osella-Abate, S.; Ribero, S.; Sanlorenzo, M.; Maule, M.M.; Richiardi, L.; Merletti, F.; Tomasini, C.; Marra, E.; Macripò, G.; Fierro, M.T.; et al. Risk factors related to late metastases in 1,372 melanoma patients disease free more than 10 years. *Int. J. Cancer* **2015**, *136*, 2453–2457. [CrossRef]
60. Rutkowski, P.; Wysocki, P.J.; Nasierowska-Guttmejer, A.; Jeziorski, A.; Wysocki, W.M.; Kalinka-Warzocha, E.; Świtaj, T.; Kozak, K.; Kamińska-Winciorek, G.; Czarnecka, A.M.; et al. Cutaneous melanomas. *Oncol. Clin. Pract.* **2019**, *15*, 1–19. [CrossRef]
61. Weiss, S.A.; Wolchok, J.D.; Sznol, M. Immunotherapy of Melanoma: Facts and Hopes. *Clin. Cancer Res.* **2019**, *25*, 5191–5201. [CrossRef]
62. Fujimura, T.; Fujisawa, Y.; Kambayashi, Y.; Aiba, S. Significance of BRAF Kinase Inhibitors for Melanoma Treatment: From Bench to Bedside. *Cancers* **2019**, *11*, 1342. [CrossRef] [PubMed]
63. Pasquali, S.; Hadjinicolaou, A.V.; Chiarion Sileni, V.; Rossi, C.R.; Mocellin, S. Systemic treatments for metastatic cutaneous melanoma. *Cochrane Database Syst. Rev.* **2018**, *2*, CD011123. [CrossRef] [PubMed]
64. Sun, Y.; Huo, C.; Qiao, Z.; Shang, Z.; Uzzaman, A.; Liu, S.; Jiang, X.; Fan, L.Y.; Ji, L.; Guan, X.; et al. Comparative Proteomic Analysis of Exosomes and Microvesicles in Human Saliva for Lung Cancer. *J. Proteome Res.* **2018**, *17*, 1101–1107. [CrossRef] [PubMed]
65. Luo, D.; Zhan, S.; Xia, W.; Huang, L.; Ge, W.; Wang, T. Proteomics study of serum exosomes from papillary thyroid cancer patients. *Endocr. Relat. Cancer* **2018**, *25*, 879–891. [CrossRef]
66. Gangoda, L.; Liem, M.; Ang, C.S.; Keerthikumar, S.; Adda, C.G.; Parker, B.S.; Mathivanan, S. Proteomic Profiling of Exosomes Secreted by Breast Cancer Cells with Varying Metastatic Potential. *Proteomics* **2017**, *17*, 23–24. [CrossRef]
67. Ludwig, S.; Marczak, L.; Sharma, P.; Abramowicz, A.; Gawin, M.; Widlak, P.; Whiteside, T.L.; Pietrowska, M. Proteomes of exosomes from HPV(+) or HPV(−) head and neck cancer cells: Differential enrichment in immunoregulatory proteins. *Oncoimmunology* **2019**, *8*, 1593808. [CrossRef]
68. Mears, R.; Craven, R.A.; Hanrahan, S.; Totty, N.; Upton, C.; Young, S.L.; Patel, P.; Selby, P.J.; Banks, R.E. Proteomic analysis of melanoma-derived exosomes by two-dimensional polyacrylamide gel electrophoresis and mass spectrometry. *Proteomics* **2004**, *4*, 4019–4031. [CrossRef] [PubMed]
69. Xiao, D.; Ohlendorf, J.; Chen, Y.; Taylor, D.D.; Rai, S.N.; Waigel, S.; Zacharias, W.; Hao, H.; McMasters, K.M. Identifying mRNA, microRNA and protein profiles of melanoma exosomes. *PLoS ONE* **2012**, *7*, e46874. [CrossRef]
70. Lazar, I.; Clement, E.; Ducoux-Petit, M.; Denat, L.; Soldan, V.; Dauvillier, S.; Balor, S.; Burlet-Schiltz, O.; Larue, L.; Muller, C.; et al. Proteome characterization of melanoma exosomes reveals a specific signature for metastatic cell lines. *Pigment Cell Melanoma Res.* **2015**, *28*, 464–475. [CrossRef]
71. Muhsin-Sharafaldine, M.R.; Saunderson, S.C.; Dunn, A.C.; Faed, J.M.; Kleffmann, T.; McLellan, A.D. Procoagulant and immunogenic properties of melanoma exosomes, microvesicles and apoptotic vesicles. *Oncotarget* **2016**, *7*, 56279–56294. [CrossRef]
72. Willms, E.; Johansson, H.J.; Mäger, I.; Lee, Y.; Blomberg, K.E.; Sadik, M.; Alaarg, A.; Smith, C.I.; Lehtiö, J.; El Andaloussi, S.; et al. Cells release subpopulations of exosomes with distinct molecular and biological properties. *Sci. Rep.* **2016**, *6*, 22519. [CrossRef] [PubMed]
73. Boussadia, Z.; Lamberti, J.; Mattei, F.; Pizzi, E.; Puglisi, R.; Zanetti, C.; Pasquini, L.; Fratini, F.; Fantozzi, L.; Felicetti, F.; et al. Acidic microenvironment plays a key role in human melanoma progression through a sustained exosome mediated transfer of clinically relevant metastatic molecules. *J. Exp. Clin. Cancer Res.* **2018**, *37*, 245. [CrossRef] [PubMed]
74. Peinado, H.; Alečković, M.; Lavotshkin, S.; Matei, I.; Costa-Silva, B.; Moreno-Bueno, G.; Hergueta-Redondo, M.; Williams, C.; García-Santos, G.; Ghajar, C.; et al. Melanoma exosomes educate bone marrow progenitor cells toward a pro-metastatic phenotype through MET. *Nat. Med.* **2012**, *18*, 883–891. [CrossRef] [PubMed]
75. Surman, M.; Stępień, E.; Przybyło, M. Melanoma-Derived Extracellular Vesicles: Focus on Their Proteome. *Proteomes* **2019**, *7*, 21. [CrossRef]
76. Kalluri, R. The biology and function of exosomes in cancer. *J. Clin. Investig.* **2016**, *126*, 1208–1215. [CrossRef]

77. Tsitsiou, E.; Lindsay, M.A. microRNAs and the immune response. *Trends Immunol.* **2008**, *29*, 343–351. [CrossRef]
78. Salehi, M.; Sharifi, M. Exosomal miRNAs as novel cancer biomarkers: Challenges and opportunities. *J. Cell. Physiol.* **2018**, *233*, 6370–6380. [CrossRef]
79. Sun, Z.; Shi, K.; Yang, S.; Liu, J.; Zhou, Q.; Wang, G.; Song, J.; Li, Z.; Zhang, Z.; Yuan, W. Effect of exosomal miRNA on cancer biology and clinical applications. *Mol. Cancer* **2018**, *17*, 147. [CrossRef]
80. Hessvik, N.P.; Phuyal, S.; Brech, A.; Sandvig, K.; Llorente, A. Profiling of microRNAs in exosomes released from PC-3 prostate cancer cells. *Biochim. Biophys. Acta* **2012**, *1819*, 1154–1163. [CrossRef]
81. Gajos-Michniewicz, A.; Duechler, M.; Czyz, M. MiRNA in melanoma-derived exosomes. *Cancer Lett.* **2014**, *347*, 29–37. [CrossRef]
82. Li, J.; Chen, J.; Wang, S.; Li, P.; Zheng, C.; Zhou, X.; Tao, Y.; Chen, X.; Sun, L.; Wang, A.; et al. Blockage of transferred exosome-shuttled miR-494 inhibits melanoma growth and metastasis. *J. Cell. Physiol.* **2019**, *234*, 15763–15774. [CrossRef] [PubMed]
83. Wozniak, M.; Peczek, L.; Czernek, L.; Düchler, M. Analysis of the miRNA Profiles of Melanoma Exosomes Derived Under Normoxic and Hypoxic Culture Conditions. *Anticancer Res.* **2017**, *37*, 6779–6789. [CrossRef] [PubMed]
84. Bland, C.L.; Byrne-Hoffman, C.N.; Fernandez, A.; Rellick, S.L.; Deng, W.; Klinke, D.J., 2nd. Exosomes derived from B16F0 melanoma cells alter the transcriptome of cytotoxic T cells that impacts mitochondrial respiration. *FEBS J.* **2018**, *285*, 1033–1050. [CrossRef] [PubMed]
85. Zhou, J.; Yang, Y.; Wang, W.; Zhang, Y.; Chen, Z.; Hao, C.; Zhang, J. Melanoma-released exosomes directly activate the mitochondrial apoptotic pathway of CD4+ T cells through their microRNA cargo. *Exp. Cell Res.* **2018**, *371*, 364–371. [CrossRef]
86. Zhou, X.; Yan, T.; Huang, C.; Xu, Z.; Wang, L.; Jiang, E.; Wang, H.; Chen, Y.; Liu, K.; Shao, Z.; et al. Melanoma cell-secreted exosomal miR-155-5p induce proangiogenic switch of cancer-associated fibroblasts via SOCS1/JAK2/STAT3 signaling pathway. *J. Exp. Clin. Cancer Res.* **2018**, *37*, 242. [CrossRef]
87. Felicetti, F.; De Feo, A.; Coscia, C.; Puglisi, R.; Pedini, F.; Pasquini, L.; Bellenghi, M.; Errico, M.C.; Pagani, E.; Carè, A. Exosome-mediated transfer of miR-222 is sufficient to increase tumor malignancy in melanoma. *J. Transl. Med.* **2016**, *1*, 56. [CrossRef]
88. Chen, L.; Karisma, V.W.; Liu, H.; Zhong, L. MicroRNA-300: A Transcellular Mediator in Exosome Regulates Melanoma Progression. *Front. Oncol.* **2019**, *9*, 1005. [CrossRef]
89. Shu, S.L.; Yang, Y.; Allen, C.L.; Maguire, O.; Minderman, H.; Sen, A.; Ciesielski, M.J.; Collins, K.A.; Bush, P.J.; Singh, P.; et al. Metabolic reprogramming of stromal fibroblasts by melanoma exosome microRNA favours a pre-metastatic microenvironment. *Sci. Rep.* **2018**, *8*, 12905. [CrossRef]
90. Vignard, V.; Labbé, M.; Marec, N.; André-Grégoire, G.; Jouand, N.; Fonteneau, J.F.; Labarrière, N.; Fradin, D. MicroRNAs in Tumor Exosomes Drive Immune Escape in Melanoma. *Cancer Immunol. Res.* **2020**, *8*, 255–267. [CrossRef]
91. Pfeffer, S.R.; Grossmann, K.F.; Cassidy, P.B.; Yang, C.H.; Fan, M.; Kopelovich, L.; Leachman, S.A.; Pfeffer, L.M. Detection of Exosomal miRNAs in the Plasma of Melanoma Patients. *J. Clin. Med.* **2015**, *4*, 2012–2027. [CrossRef]
92. Alegre, E.; Sanmamed, M.F.; Rodriguez, C.; Carranza, O.; Martín-Algarra, S.; González, A. Study of circulating microRNA-125b levels in serum exosomes in advanced melanoma. *Arch. Pathol. Lab. Med.* **2014**, *138*, 828–832. [CrossRef] [PubMed]
93. Greenberg, E.; Hershkovitz, L.; Itzhaki, O.; Hajdu, S.; Nemlich, Y.; Ortenberg, R.; Gefen, N.; Edry, L.; Modai, S.; Keisari, Y.; et al. Regulation of cancer aggressive features in melanoma cells by microRNAs. *PLoS ONE* **2011**, *6*, e18936. [CrossRef] [PubMed]
94. Rao, E.; Jiang, C.; Ji, M.; Huang, X.; Iqbal, J.; Lenz, G.; Wright, G.; Staudt, L.M.; Zhao, Y.; McKeithan, T.W.; et al. The miRNA-17 approximately 92 cluster mediates chemoresistance and enhances tumor growth in mantle cell lymphoma via pi3k/akt pathway activation. *Leukemia* **2012**, *26*, 1064–1072. [CrossRef] [PubMed]
95. Levy, C.; Khaled, M.; Iliopoulos, D.; Janas, M.M.; Schubert, S.; Pinner, S.; Chen, P.H.; Li, S.; Fletcher, A.L.; Yokoyama, S.; et al. Intronic miR-211 assumes the tumor suppressive function of its host gene in melanoma. *Mol. Cell* **2010**, *40*, 841–849. [CrossRef]
96. Si, M.L.; Zhu, S.; Wu, H.; Lu, Z.; Wu, F.; Mo, Y.Y. MiR-21-mediated tumor growth. *Oncogene* **2007**, *26*, 2799–2803. [CrossRef]

97. Zhu, S.; Wu, H.; Wu, F.; Nie, D.; Sheng, S.; Mo, Y.Y. MicroRNA-21 targets tumor suppressor genes in invasion and metastasis. *Cell Res.* **2008**, *18*, 350–359. [CrossRef]
98. Yang, C.H.; Yue, J.; Pfeffer, S.R.; Handorf, C.R.; Pfeffer, L.M. MicroRNA miR-21 regulates the metastatic behavior of b16 melanoma cells. *J. Biol. Chem.* **2011**, *286*, 39172–39178. [CrossRef] [PubMed]
99. Yang, C.H.; Yue, J.; Fan, M.; Pfeffer, L.M. IFN induces miR-21 through a signal transducer and activator of transcription 3-dependent pathway as a suppressive negative feedback on IFN-induced apoptosis. *Cancer Res.* **2010**, *70*, 8108–8116. [CrossRef] [PubMed]
100. Xu, Y.F.; Hannafon, B.N.; Khatri, U.; Gin, A.; Ding, W.Q. The origin of exosomal miR-1246 in human cancer cells. *RNA Biol.* **2019**, *16*, 770–784. [CrossRef]
101. Bhagirath, D.; Yang, T.L.; Bucay, N.; Sekhon, K.; Majid, S.; Shahryari, V.; Dahiya, R.; Tanaka, Y.; Saini, S. microRNA-1246 Is an Exosomal Biomarker for Aggressive Prostate Cancer. *Cancer Res.* **2018**, *78*, 1833–1844. [CrossRef]
102. Felicetti, F.; Errico, M.C.; Bottero, L.; Segnalini, P.; Stoppacciaro, A.; Biffoni, M.; Felli, N.; Mattia, G.; Petrini, M.; Colombo, M.P.; et al. The promyelocytic leukemia zinc finger-microRNA-221/-222 pathway controls melanoma progression through multiple oncogenic mechanisms. *Cancer Res.* **2008**, *68*, 2745–2754. [CrossRef]
103. Müller, D.; Bosserhoff, A. Integrin β3 expression is regulated by let-7a miRNA in malignant melanoma. *Oncogene* **2008**, *27*, 6698–6706. [CrossRef]
104. Xiao, D.; Barry, S.; Kmetz, D.; Egger, M.; Pan, J.; Rai, S.N.; Qu, J.; McMasters, K.M.; Hao, H. Melanoma cell-derived exosomes promote epithelial-mesenchymal transition in primary melanocytes through paracrine/autocrine signaling in the tumor microenvironment. *Cancer Lett.* **2016**, *376*, 318–327. [CrossRef] [PubMed]
105. Ryan, D.; Rafferty, M.; Hegarty, S.; O'Leary, P.; Faller, W.; Gremel, G.; Bergqvist, M.; Agnarsdottir, M.; Strömberg, S.; Kampf, C.; et al. Topoisomerase I amplification in melanoma is associated with more advanced tumours and poor prognosis. *Pigment Cell Melanoma Res.* **2010**, *23*, 542–553. [CrossRef] [PubMed]
106. Wang, S.; Tang, L.; Lin, J.; Shen, Z.; Yao, Y.; Wang, W.; Tao, S.; Gu, C.; Ma, J.; Xie, Y.; et al. ABCB5 promotes melanoma metastasis through enhancing NF-κB p65 protein stability. *Biochem. Biophys. Res. Commun.* **2017**, *492*, 18–26. [CrossRef] [PubMed]
107. Suryo Rahmanto, Y.; Dunn, L.; Richardson, D. The melanoma tumor antigen, melanotransferrin (p97): A 25-year hallmark—From iron metabolism to tumorigenesis. *Oncogene* **2007**, *26*, 6113–6124. [CrossRef] [PubMed]
108. Soengas, M.; Hernando, E. TYRP1 mRNA goes fishing for miRNAs in melanoma. *Nat. Cell Biol.* **2017**, *19*, 1311–1312. [CrossRef]
109. Gilot, D.; Migault, M.; Bachelot, L.; Journé, F.; Rogiers, A.; Donnou-Fournet, E.; Mogha, A.; Mouchet, N.; Pinel-Marie, M.L.; Mari, B.; et al. A non-coding function of TYRP1 mRNA promotes melanoma growth. *Nat. Cell Biol.* **2017**, *19*, 1348–1357. [CrossRef] [PubMed]
110. Li, X.; Wang, Y.; Wang, Q.; Liu, Y.; Bao, W.; Wu, S. Exosomes in cancer: Small transporters with big functions. *Cancer Lett.* **2018**, *435*, 55–65. [CrossRef]
111. Chen, G.; Huang, A.C.; Zhang, W.; Zhang, G.; Wu, M.; Xu, W.; Yu, Z.; Yang, J.; Wang, B.; Sun, H.; et al. Exosomal PD-L1 contributes to immunosuppression and is associated with anti-PD-1 response. *Nature* **2018**, *560*, 382–386. [CrossRef]
112. Kaiser, J. Malignant messengers. *Science* **2016**, *352*, 164–166. [CrossRef] [PubMed]
113. Chen, D.S.; Mellman, I. Elements of cancer immunity and the cancer-immune set point. *Nature* **2017**, *541*, 321–330. [CrossRef]
114. Muller, L.; Mitsuhashi, M.; Simms, P.; Gooding, W.E.; Whiteside, T.L. Tumor-derived exosomes regulate expression of immune function-related genes in human T cell subsets. *Sci. Rep.* **2016**, *6*, 20254. [CrossRef]
115. Bergmann, C.; Strauss, L.; Wang, Y.; Szczepanski, M.J.; Lang, S.; Johnson, J.T.; Whiteside, T.L. T regulatory Type 1 cells (Tr1) in squamous cell carcinoma of the head and neck: Mechanisms of suppression and expansion in advanced disease. *Clin. Cancer Res.* **2008**, *14*, 3706–3715. [CrossRef]
116. Clayton, A.; Al-Taei, S.; Webber, J.; Mason, M.D.; Tabi, Z. Cancer exosomes express CD39 and CD73, which suppress T cells through adenosine production. *J. Immunol.* **2011**, *187*, 676–683. [CrossRef] [PubMed]

117. Szajnik, M.; Czystowska, M.; Szczepanski, M.J.; Mandapathil, M.; Whiteside, T.L. Tumor-derived microvesicles induce, expand and up-regulate biological activities of human regulatory T cells (Treg). *PLoS ONE* **2010**, *5*, e11469. [CrossRef] [PubMed]

118. Düchler, M.; Czernek, L.; Peczek, L.; Cypryk, W.; Sztiller-Sikorska, M.; Czyz, M. Melanoma-Derived Extracellular Vesicles Bear the Potential for the Induction of Antigen-Specific Tolerance. *Cells* **2019**, *8*, 665. [CrossRef]

119. Gyukity-Sebestyén, E.; Harmati, M.; Dobra, G.; Németh, I.B.; Mihály, J.; Zvara, Á.; Hunyadi-Gulyás, É.; Katona, R.; Nagy, I.; Horváth, P.; et al. Melanoma-Derived Exosomes Induce PD-1 Overexpression and Tumor Progression via Mesenchymal Stem Cell Oncogenic Reprogramming. *Front. Immunol.* **2019**, *10*, 2459. [CrossRef]

120. Martínez-Lorenzo, M.J.; Anel, A.; Alava, M.A.; Piñeiro, A.; Naval, J.; Lasierra, P.; Larrad, L. The human melanoma cell line MelJuSo secretes bioactive FasL and APO2L/TRAIL on the surface of microvesicles. Possible contribution to tumor counterattack. *Exp. Cell Res.* **2004**, *295*, 315–329. [CrossRef]

121. Wu, Y.; Deng, W.; McGinley, E.C.; Klinke, D.J., 2nd. Melanoma exosomes deliver a complex biological payload that upregulates PTPN11 to suppress T lymphocyte function. *Pigment Cell Melanoma Res.* **2017**, *30*, 203–218. [CrossRef]

122. Kulkarni, Y.M.; Chambers, E.; McGray, A.J.R.; Ware, J.S.; Bramson, J.L.; Klinke, D.J. A quantitative systems approach to identify paracrine mechanisms that locally suppress immune response to Interleukin-12 in the B16 melanoma model. *Integr. Biol.* **2012**, *4*, 925–936. [CrossRef] [PubMed]

123. Cottonham, C.L.; Kaneko, S.; Xu, L. miR-21 and miR-31 converge on TIAM1 to regulate migration and invasion of colon carcinoma cells. *J. Biol. Chem.* **2010**, *285*, 35293–35302. [CrossRef]

124. Yasuda, M.; Schmid, T.; Rubsamen, D.; Colburn, N.H.; Irie, K.; Murakami, A. Downregulation of programmed cell death 4 by inflammatory conditions contributes to the generation of the tumor promoting microenvironment. *Mol. Carcinog.* **2010**, *49*, 837–848. [CrossRef] [PubMed]

125. Fabbri, M.; Paone, A.; Calore, F.; Galli, R.; Gaudio, E.; Santhanam, R.; Lovat, F.; Fadda, P.; Mao, C.; Nuovo, G.J.; et al. MicroRNAs bind to Toll-like receptors to induce prometastatic inflammatory response. *Proc. Natl. Acad. Sci. USA* **2012**, *109*, E2110–E2116. [CrossRef] [PubMed]

126. Tang, M.; Diao, J.; Gu, H.; Khatri, I.; Zhao, J.; Cattral, M.S. Toll-like Receptor 2 Activation Promotes Tumor Dendritic Cell Dysfunction by Regulating IL-6 and IL-10 Receptor Signaling. *Cell Rep.* **2015**, *13*, 2851–2864. [CrossRef]

127. Ashiru, O.; Boutet, P.; Fernández-Messina, L.; Agüera-González, S.; Skepper, J.N.; Valés-Gómez, M.; Reyburn, H.T. Natural killer cell cytotoxicity is suppressed by exposure to the human NKG2D ligand MICA*008 that is shed by tumor cells in exosomes. *Cancer Res.* **2010**, *70*, 481–489. [CrossRef] [PubMed]

128. Guo, Y.; Ji, X.; Liu, J.; Fan, D.; Zhou, Q.; Chen, C.; Wang, W.; Wang, G.; Wang, H.; Yuan, W.; et al. Effects of exosomes on pre-metastatic niche formation in tumors. *Mol. Cancer* **2019**, *18*, 39. [CrossRef]

129. Caramel, J.; Papadogeorgakis, E.; Hill, L.; Browne, G.J.; Richard, G.G.; Wierinckx, A.; Saldanha, G.; Osborne, J.; Hutchinson, P.; Tse, G.; et al. A switch in the expression of embryonic EMT-inducers drives the development of malignant melanoma. *Cancer Cell.* **2013**, *24*, 466–480. [CrossRef]

130. Hu, T.; Hu, J. Melanoma-derived exosomes induce reprogramming fibroblasts into cancer-associated fibroblasts via Gm26809 delivery. *Cell Cycle* **2019**, *18*, 3085–3094. [CrossRef]

131. Zhao, X.-P.; Wang, M.; Song, Y.; Song, K.; Yan, T.L.; Wang, L.; Li, K.; Shang, Z.J. Membrane microvesicles as mediators for melanoma-fibroblasts communication: Roles of the VCAM-1/VLA-4 axis and the ERK1/2 signal pathway. *Cancer Lett.* **2015**, *360*, 125–133. [CrossRef]

132. Hood, J.; Pan, H.; Lanza, G.; Wickline, S.A. Paracrine induction of endothelium by tumor exosomes. *Lab. Investig.* **2009**, *89*, 1317–1328. [CrossRef] [PubMed]

133. Zhuang, G.; Wu, X.; Jiang, Z.; Kasman, I.; Yao, J.; Guan, Y.; Oeh, J.; Modrusan, Z.; Bais, C.; Sampath, D.; et al. Tumour-secreted miR-9 promotes endothelial cell migration and angiogenesis by activating the JAK-STAT pathway. *EMBO J.* **2012**, *31*, 3513–3523. [CrossRef] [PubMed]

134. Ekström, E.J.; Bergenfelz, C.; von Bülow, V.; Serifler, F.; Carlemalm, E.; Jönsson, G.; Andersson, T.; Leandersson, K. WNT5A induces release of exosomes containing pro-angiogenic and immunosuppressive factors from malignant melanoma cells. *Mol. Cancer* **2014**, *13*, 88. [CrossRef] [PubMed]

135. Sento, S.; Sasabe, E.; Yamamoto, T. Application of a persistent heparin treatment inhibits the malignant potential of oral squamous carcinoma cells induced by tumor cell-derived exosomes. *PLoS ONE* **2016**, *11*, e0148454. [CrossRef]
136. Hazan-Halevy, I.; Rosenblum, D.; Weinstein, S.; Bairey, O.; Raanani, P.; Peer, D. Cell-specific uptake of mantle cell lymphoma-derived exosomes by malignant and non-malignant B-lymphocytes. *Cancer Lett.* **2015**, *364*, 59–69. [CrossRef] [PubMed]
137. Matsumoto, A.; Takahashi, Y.; Nishikawa, M.; Sano, K.; Morishita, M.; Charoenviriyakul, C.; Saji, H.; Takakura, Y. Accelerated growth of B16BL6 tumor in mice through efficient uptake of their own exosomes by B16BL6 cells. *Cancer Sci.* **2017**, *108*, 1803–1810. [CrossRef]
138. Isola, A.L.; Eddy, K.; Zembrzuski, K.; Goydos, J.S.; Chen, S. Exosomes released by metabotropic glutamate receptor 1 (GRM1) expressing melanoma cells increase cell migration and invasiveness. *Oncotarget* **2018**, *9*, 1187–1199. [CrossRef]
139. Ostrowski, M.; Carmo, N.B.; Krumeich, S.; Fanget, I.; Raposo, G.; Savina, A.; Moita, C.F.; Schauer, K.; Hume, A.N.; Freitas, R.P.; et al. Rab27a and Rab27b control different steps of the exosome secretion pathway. *Nat. Cell Biol.* **2009**, *12*, 19–30. [CrossRef] [PubMed]
140. Guo, D.; Lui, G.Y.L.; Lai, S.L.; Wilmott, J.S.; Tikoo, S.; Jackett, L.A.; Quek, C.; Brown, D.L.; Sharp, D.M.; Kwan, R.Y.Q.; et al. RAB27A promotes melanoma cell invasion and metastasis via regulation of pro-invasive exosomes. *Int. J. Cancer* **2019**, *144*, 3070–3085. [CrossRef]
141. Mannavola, F.; Tucci, M.; Felici, C.; Passarelli, A.; D'Oronzo, S.; Silvestris, F. Tumor-derived exosomes promote the in vitro osteotropism of melanoma cells by activating the SDF-1/CXCR4/CXCR7 axis. *J. Transl. Med.* **2019**, *17*, 230. [CrossRef] [PubMed]
142. Hood, J.L.; San, R.S.; Wickline, S.A. Exosomes released by melanoma cells prepare sentinel lymph nodes for tumor metastasis. *Cancer Res.* **2011**, *71*, 3792–3801. [CrossRef] [PubMed]
143. Angi, M.; Kalirai, H.; Prendergast, S.; Simpson, D.; Hammond, D.E.; Madigan, M.C.; Beynon, R.J.; Coupland, S.E. In-depth proteomic profiling of the uveal melanoma secretome. *Oncotarget* **2016**, *7*, 49623–49635. [CrossRef] [PubMed]
144. Tengda, L.; Shuping, L.; Mingli, G.; Jie, G.; Yun, L.; Weiwei, Z.; Anmei, D. Serum exosomal microRNAs as potent circulating biomarkers for melanoma. *Melanoma Res.* **2018**, *28*, 295–303. [CrossRef] [PubMed]
145. Larkin, J.; Ascierto, P.A.; Dréno, B.; Atkinson, V.; Liszkay, G.; Maio, M.; Mandalà, M.; Demidov, L.; Stroyakovskiy, D.; Thomas, L.; et al. Combined vemurafenib and cobimetinib in BRAF-mutated melanoma. *N. Engl. J. Med.* **2014**, *371*, 1867–1876. [CrossRef]
146. Lunavat, T.R.; Cheng, L.; Einarsdottir, B.O.; Olofsson Bagge, R.; Veppil Muralidharan, S.; Sharples, R.A.; Lässer, C.; Gho, Y.S.; Hill, A.F.; Nilsson, J.A.; et al. BRAFV600 inhibition alters the microRNA cargo in the vesicular secretome of malignant melanoma cells. *Proc. Natl. Acad. Sci. USA* **2017**, *114*, E5930–E5939. [CrossRef]
147. Svedman, F.C.; Lohcharoenkal, W.; Bottai, M.; Brage, S.E.; Sonkoly, E.; Hansson, J.; Pivarcsi, A.; Eriksson, H. Extracellular microvesicle microRNAs as predictive biomarkers for targeted therapy in metastastic cutaneous malignant melanoma. *PLoS ONE* **2018**, *13*, e0206942. [CrossRef]
148. Segura, M.F.; Belitskaya-Lévy, I.; Rose, A.E.; Zakrzewski, J.; Gaziel, A.; Hanniford, D.; Darvishian, F.; Berman, R.S.; Shapiro, R.L.; Pavlick, A.C.; et al. Melanoma MicroRNA signature predicts post-recurrence survival. *Clin. Cancer Res.* **2010**, *16*, 1577–1586. [CrossRef]
149. Ribas, A.; Hamid, O.; Daud, A.; Hodi, F.S.; Wolchok, J.D.; Kefford, R.; Joshua, A.M.; Patnaik, A.; Hwu, W.I.; Weber, J.S.; et al. Association of Pembrolizumab With Tumor Response and Survival Among Patients With Advanced Melanoma. *JAMA* **2016**, *315*, 1600–1609. [CrossRef]
150. Cordonnier, M.; Nardin, C.; Chanteloup, G.; Derangere, V.; Algros, M.P.; Arnould, L.; Garrido, C.; Aubin, F.; Gobbo, J. Tracking the evolution of circulating exosomal-PD-L1 to monitor melanoma patients. *J. Extracell. Vesicles* **2020**, *9*, 1. [CrossRef]

6

Extracellular Vesicles in the Development of Cancer Therapeutics

Haoyao Sun [1,2], Stephanie Burrola [1], Jinchang Wu [2,3,*] and Wei-Qun Ding [1,*]

1 Department of Pathology, University of Oklahoma Health Science Center, Oklahoma City, OK 73104, USA; haoyao-sun@ouhsc.edu (H.S.); Stephanie-Burrola@ouhsc.edu (S.B.)

2 Department of Radiation Oncology, The Affiliated Suzhou Hospital of Nanjing Medical University, Suzhou 215001, China

3 Section of Oncology, The Second Affiliated Hospital of Xuzhou Medical University, Xuzhou 221006, China

* Correspondence: dr.wjinchang@gmail.com (J.W.); weiqun-ding@ouhsc.edu (W.-Q.D.);

Abstract: Extracellular vesicles (EVs) are small lipid bilayer-delimited nanoparticles released from all types of cells examined thus far. Several groups of EVs, including exosomes, microvesicles, and apoptotic bodies, have been identified according to their size and biogenesis. With extensive investigations on EVs over the last decade, it is now recognized that EVs play a pleiotropic role in various physiological processes as well as pathological conditions through mediating intercellular communication. Most notably, EVs have been shown to be involved in cancer initiation and progression and EV signaling in cancer are viewed as potential therapeutic targets. Furthermore, as membrane nanoparticles, EVs are natural products with some of them, such as tumor exosomes, possessing tumor homing propensity, thus leading to strategies utilizing EVs as drug carriers to effectively deliver cancer therapeutics. In this review, we summarize recent reports on exploring EVs signaling as potential therapeutic targets in cancer as well as on developing EVs as therapeutic delivery carriers for cancer therapy. Findings from preclinical studies are primarily discussed, with early phase clinical trials reviewed. We hope to provide readers updated information on the development of EVs as cancer therapeutic targets or therapeutic carriers.

Keywords: extracellular vesicle; microvesicle; exosome; cancer therapeutic; drug carrier

---

1. Introduction

Extracellular vesicles (EVs) are a generic term referring to several groups of small lipid bilayer-delimited particles generated through various cellular processes and released from all types of cells investigated thus far. These membrane vesicles, including microvesicles (also known as microparticles or ectosomes), exosomes, and apoptotic bodies, all lack a functional nucleus and are unable to replicate themselves. They are constantly released from cells and are involved in a variety of physiological as well as pathological processes. The initial discovery of EVs can be tracked back to 1946 when ultracentrifugation pellets were found to be associated with the activation of platelets and procoagulant properties in human plasma [1]. In the 1980s, EVs released by reticulocytes were captured by electronic microscopy and were considered "waste disposals" to remove waste materials during red blood cell maturation [2,3]. However, EV-mediated transfer of genetic and cellular materials between different cell types was recognized in the late 2000s by several research groups [4–8], thus establishing EVs as messengers for intercellular communication with biological consequences.

Among all the EVs described, exosomes are defined by their small sizes (40–120 nm) and endocytic origin and are most extensively characterized over the years. In the context of cancer, it has been demonstrated that exosomes play a pivotal role in the tumor microenvironment by mediating intercellular communication among cancer cells and stromal cells, thereby promoting tumor proliferation, metastasis, and chemo-resistance [9]. The contribution of exosomal signaling to tumor progression has led to the development of therapeutic strategies targeting various steps of the exosomal signaling processes (see Section 2). On the other hand, since exosomes are endogenously produced and can be transferred among various types of cells, the potential of using these small vesicles as vehicles for drug delivery has been actively explored (see Section 3). In this review, we will focus on recent work in the development of cancer therapeutics targeting EV-mediated cellular processes or utilizing EVs as vehicles for drug delivery. Furthermore, we will discuss the clinical trials that are ongoing or completed using naturally produced EVs as cancer therapeutic vehicles. A simplified view of general aspects of EVs is provided at the first section of this review.

2. EV Cargos and Functions

2.1. EV Nomenclature

EVs were initially called platelet dust, as they were vesicles derived from platelets. In the 1970s, the term "extracellular vesicles" was used to describe calcifying globules in epiphyseal cartilage that were observed by histochemical staining [10]. Since then, the nomenclature of EVs has significantly evolved and EVs are now named primarily according to their sizes and biogenesis processes or the way of release [11]. It is well accepted that there are three main subgroups of EVs that have been identified thus far: (a) exosomes, (b) microvesicles (MVs, also named microparticles/ectosomes), and (c) apoptotic bodies [12]. The most researched EVs are exosomes, which were firstly termed in the 1980s as a group of vesicles ranging from 40 to 120 nm in diameter, formed by the invagination of the multi-vesicular bodies (MVBs) during the late endosome formation [2,3,13]. Differing from exosomes, MVs are larger membrane vesicles (up to 1000 nm in diameter) which are produced by direct budding from cellular membranes, whereas apoptotic bodies are even larger vesicles with 800–5000 nm in diameter and formed during programmed cell death [14,15]. Recently, a smaller group of non-membranous nanoparticles termed "exomeres" (~35 nm) was also reported, which is likely to be generated through a unique cellular process [16]. The overlap in sizes of different EV groups and the difficulty in separating individual EV groups by current isolation techniques have hindered our understanding of their biogenesis, molecular compositions, biodistributions, and functions. For this reason, the International Society for Extracellular Vesicles (ISEV) provided guidelines on the terminology and minimum requirements for defining EV populations in experimental research in 2014, which was updated in 2018 (MISEV2018) [17]. Most notably, instead of using the terms exosomes or MVs, the guidelines urge authors to name EV subtypes based on their physical characteristics, such as size or density, with ranges being defined, biochemical compositions, and the experimental conditions or cell of origin. In accordance with this recommendation, exosomes are considered small EVs (sEVs), which is the term we used interchangeably with exosomes, wherever appropriate, throughout this review.

2.2. EV Surface Markers and Cargos

EVs carry various biomolecules including proteins, RNA, DNA, and lipids. Each group of biomolecules in EVs is often heterogeneous, primarily relating to different EV types, experimental conditions, and their cellular origins [11]. The most characterized EV components are EV proteins and RNAs, especially small RNAs [18]. EV surface protein markers have been critically examined in order to establish specific markers for validating isolated EVs. The MISEV2018 guidelines provide several groups of protein markers in evaluating isolated EVs as well as minimal requirements in experimental data presentation when it comes to EV isolation and characterization [17].

It has come to a consensus that sEVs stably express specific transmembrane proteins such as tetraspanins (most notably CD63, CD9, CD81), Major Histocompatibility Complex (MHC) class I proteins (such as HLA-A/B/C), transferrin receptor, LAMP1/2, and others. These membrane proteins, especially tetraspanins, are frequently applied to validate isolated sEVs. In addition, cytosolic proteins can also be specific markers for sEVs, including Alix, TSG-101, flotillins-1/2, annexins, and heat shock proteins, among others. Cell- or tissue-specific EV markers have also been reported, such as TSPAN8 and EPCAM (epithelial cell), CD37 and CD53 (leukocytes), PECAM1 (endothelial cells), and ERBB2 (breast cancer). Given the heterogeneity of EVs, it is recommended that at least one membrane protein marker, one cytosolic protein marker, and one non-EV protein marker have to be used to validate the isolated sEVs from large EVs [16,17]. It has been recognized that proteins from the nucleus, mitochondria, endoplasmic reticulum, and the Golgi complex are mostly absent in sEVs, which can serve as negative control markers for these vesicles [19]. Enormous efforts have been placed on profiling proteomes of sEVs and the comprehensive databases of sEV proteins can be found at: Vesiclepedia [15], EVpedia [20], and ExoCarta [21].

sEVs contain various RNA species. However, most studies demonstrated that small non-coding RNAs, such as microRNAs, are the major RNA species contained in sEVs, although the presence of mRNA, rRNA, and tRNA in sEVs was also reported [22,23]. Typically, sEVs may contain hundreds of microRNA species in various quantities that play important roles in intercellular communication [7,23–25]. Both coding and non-coding RNAs seem to be functional through transferring from host cells to the recipient cells [26–28]. Specific RNA profiles of sEVs derived from different biofluids or tissues are categorized by several databases, including: Exobase [29], exRNA Atlas [30], and miRandola [31].

DNA in sEVs has also been described, with DNA fragments originating either from the nucleus or from the mitochondria. It seems that all genome DNA are represented randomly in sEVs, which eliminates the possibility of selective DNA packaging [32–34]. While cancer cell-derived sEVs may contain more genomic DNA than that from non-cancer cells [34], whether and how sEV DNA contributes to intercellular communications in the tumor microenvironment, thereby affecting tumor progression, remains to be determined.

2.3. EV Functions

It has long been known that cell-to-cell communication is a strategy utilized to facilitate physiological and pathological processes in various organisms. However, the EV-mediated intercellular communication was only recognized in recent years [7,23]. The double-layer lipid membrane of EVs protects inside contents, allowing transfer of EV materials to surrounding cells or to distal organs via the circulatory system. Most notably, sEVs have been considered potent vehicles to mediate intercellular communication [11]. By transferring signaling molecules among different cell types, sEVs have been shown to play pleiotropic roles in regulating cellular and physiological processes. This includes participating in hemostasis by enhancing coagulation, regulating both innate and acquired immune responses, involvement in pregnancy and embryonic development, as well as other physiological events [35–44].

In addition to mediating intercellular communication, EVs may function as waste disposals to remove unwanted cellular materials. In fact, sEVs were first observed to facilitate reticulocyte maturation via cargo disposing [2,3]. In supporting this function of sEVs, several recent studies revealed the cross-regulation of the EV pathway and lysosomal degradation pathway [45]. Two established lysosome inhibitors, chloroquine and bafilomycin A1, were shown to enhance sEV release [46–48], suggesting that sEVs may act as an alternate pathway for cell component degradation and clearance. The involvement of sEVs in cellular homeostasis is further supported by the findings showing that ubiquitin and ubiquitinated proteins are present in sEVs [49], along with selective lipids and other soluble cellular components [50,51].

The role of sEVs in pathological processes has been evident, especially in cancer. Cancer progression is a dysregulated and uncontrolled pathological process [52]. It is well described that cancer-derived sEVs promote tumor development [53–55] by acting at different stages of cancer progression [56] through various mechanisms. Evidence is provided to indicate that cancer sEVs are involved in enhancing tumorigenesis of epithelial cells [53,57], sustaining tumor angiogenesis [58,59], promoting tumor growth [60,61], facilitating cancer cell invasion and metastasis [54,55,62,63], and contributing to chemo-resistance [64,65] and immunosuppression [66,67]. These important findings of the tumor-promoting effects of cancer sEVs lead to new cancer therapeutic opportunities that aim at targeting cancer exosomal signaling processes, as discussed below.

3. EVs as Potential Therapeutic Targets in Cancer

Given the growing evidence of sEVs' involvement in cancer progression, several strategies have been tested or envisioned to target various steps of the sEVs signaling in order to block their tumor promoting effect. These include targeting cancer sEV biogenesis and release, blocking sEV uptake by recipient cells, eliminating circulating cancer sEVs, and removing specific components from sEVs that contribute to cancer pathogenesis [68–70].

3.1. Suppressing sEV Biogenesis and Release

At the cellular level, sEVs are derived from the endosomal pathway. The invagination of endosomal membranes generates intraluminal vesicles inside of the endosome, forming MVBs. These vesicles are released by cells upon fusion of the endosome with the cellular plasma membrane and the released vesicles are termed exosomes or sEVs [71,72]. The process of forming sEVs and releasing them from cells requires a coordinated effort by various cytoplasmic proteins. This includes endosomal sorting complexes required for transport (ESCRT) and tetraspanins necessary for intraluminal vesicle formation, sphingomyelinase to generate ceramides vital for intraluminal vesicles' formation and sorting, and Rab27a and Rab27b critical for cellular endosomal trafficking [55,71,73]. In an early effort to suppress sEVs' biogenesis, GW4869, a sphingomyelinase inhibitor, was used, which reduced ceramide generation and inhibited sEV formation [74]. Furthermore, attenuation of neutral sphingomyelinase 2 (nSMase2) in breast cancer cells by a knockdown approach reduced sEV formation and attenuated sEV-associated miR-210 transfer, leading to the suppression of cancer cell metastasis in vitro and in a xenograft mouse model [75]. However, the role of nSMase2 in sEV formation and secretion from other cultured cancer cell lines remains unclear [76,77], compromising this approach of targeting sEV biogenesis. Other potential strategies in targeting sEV biogenesis that have been tested or envisioned include the use of Amiloride, an anti-hypotension drug, which reduced sEV yields by blocking membrane-associated heat shock protein 72 (HSP72) in a STAT3-dependent manner in myeloid-derived suppressor cells [78]; inhibiting the syndecan-syntenin-Alix signaling process, since the syndecan heparan sulphate proteoglycans and their cytoplasmic adaptor syntenin, along with Alix and ESCRT, control the formation of sEVs [79]; and targeting cellular molecules, such as Rab27a/b [70,73,80], Rab11, Rab35 [81,82], TSG101, and TSAP6 [70], which are either related to sEV formation or trafficking and secretion from cancer cells. Using a high-throughput screening approach, a recent study identifies that manumycin-A (MA), a natural microbial metabolite, inhibits sEV biogenesis and secretion via the Ras/Raf/ERK1/2 signaling in castration-resistant prostate cancer cells but not in normal prostate epithelial cells [83], indicating a new compound that may serve as a cancer therapeutic via inhibiting sEV biogenesis and secretion. In another high-throughput screening study, miR-26a was identified as being involved in sEV secretion from prostate cancer cells [84], suggesting a new molecular target for suppressing cancer sEV secretion.

3.2. Preventing EV Uptake

Several sEV uptake mechanisms have been recently proposed (Figure 1), including sEV membrane direct fusion with plasma membrane, thereby releasing sEV contents to recipient cells [85,86], and receptor-mediated endocytosis [87], clathrin- and caveolin-mediated endocytosis [88,89], phagocytosis [90], and macropinocytosis [88,91]. Detailed regulation of each of the pathways and their proportional contributions to sEV uptake remains to be further elucidated. It seems reasonable to assume that the uptake to a large extent depends on sEV surface protein compositions and the type of cells in which the sEVs are internalized. Furthermore, irrespective of the uptake pathways, internalized materials will be processed via the endosomal/lysosomal pathway [92]. While limiting cancer sEV uptake by recipient cells is a potential strategy to block cancer sEV signaling and attenuate cancer sEVs' tumor-promoting effect, few studies have been published to support this strategy. Nevertheless, evidence has been provided to indicate that it is feasible to modulate the sEV uptake process in order to attenuate the sEV-induced effect in the recipient cells. Some examples include the following. Autophagy inhibitors such as chloroquine, bafilomycin A, and monensin, were shown to significantly inhibit sEV internalization into microglia, likely through altering vacuolar acidification [91]. Two potent PI3K inhibitors, Wortmannin and LY294002, concentration-dependently inhibited sEV uptake by macrophages, indicating that PI3K is essential for sEV phagocytosis [90]. Disruption of the actin cytoskeleton using Cytochalasin D or Lantrunculin A inhibited sEV uptake by Human Umbilical Vein Cells (HUVECs), confirming that an intact cytoskeleton facilitates sEV internalization [93]. Chlorpromazine, which blocks clathrin-mediated endocytosis, inhibited sEV uptake by ovarian cancer cells [94] and endothelial cells [95], and heparin dose-dependently inhibited sEV uptake by glioblastoma (GBM) cells [96] and bone marrow stromal cells [97]. These findings reinforce the notion that targeting the uptake of cancer sEVs is a promising strategy in the development of new cancer therapeutics, and future efforts should focus on small molecules capable of inhibiting cancer sEV uptake and suppressing tumor progression.

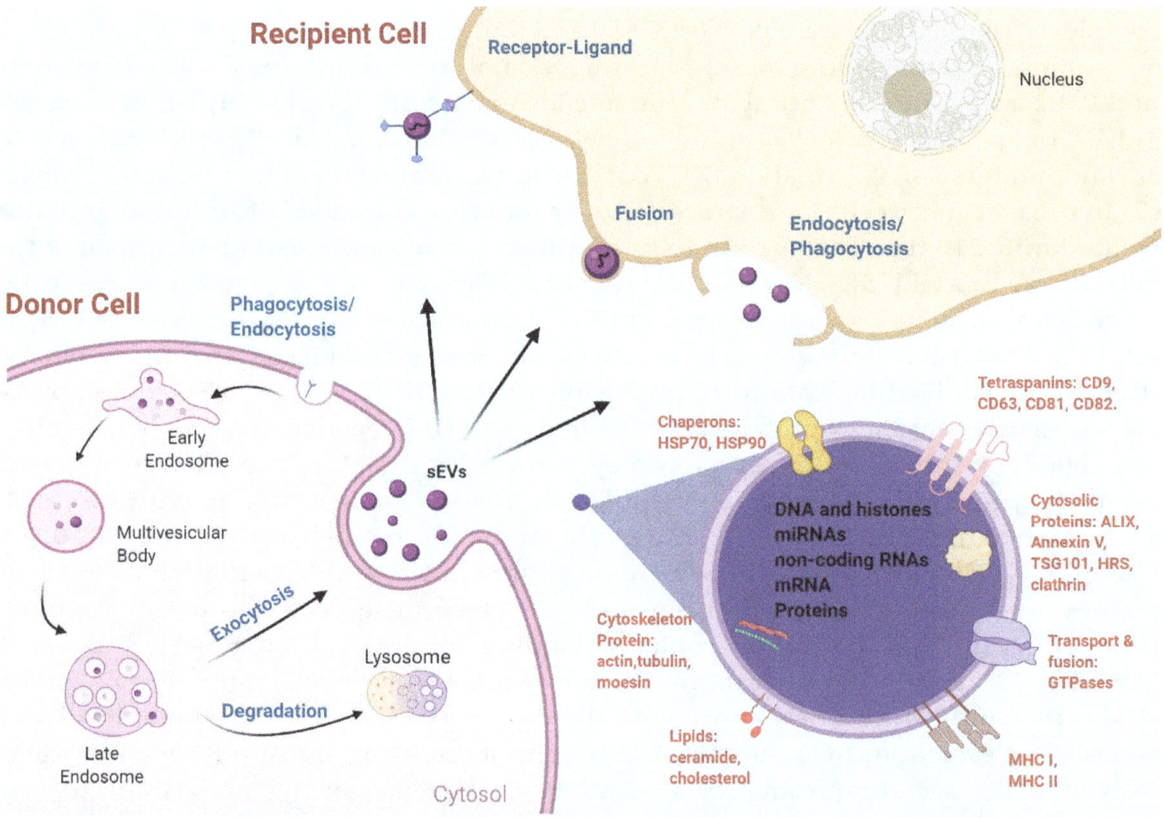

Figure 1. sEV biogenesis, release, uptake, and contents. Created with BioRender.com.

3.3. Eliminating Circulating Cancer sEVs

The transfer of cancer sEVs through the circulatory system to distal organs has been reported to promote tumor metastasis via various mechanisms, most notably by forming pre-metastatic niches in the distal organs [55,62,98]. Considering that most cancer deaths are due to metastatic disease, eliminating circulating cancer sEVs is presumably a great strategy to prevent cancer metastasis, thereby reducing cancer mortality. The idea of "cleaning" the blood to prevent cancer metastasis has been tested many years ago. In the late 1980s, using a continuous whole blood UltraPheresis procedure, plasma fractions with molecular weight less than 150 kDa were removed from patients with metastatic cancer, which reduced tumor size and improved patient immune response and Karnofsky Performance Status [99]. While this technique did not consider removing blood sEVs at the time, it inspired others to develop new devices to remove cancer sEVs from patient plasma [64]. For instance, Hemopurifier®, an affinity-based purifier developed by Aethlon Medical Inc. (San Diego, CA, USA), has been shown to selectively capture viral particles (which have similar size as sEVs) in the plasmas of individuals infected with Hepatitis C and Human Immunodeficiency Virus (HIV) [100,101], and this device is being modified and tested for removal of Her2-positive breast cancer exosomes from patient plasma ([64], https://grantome.com/grant/NIH/R43-CA232977-01). Moreover, a phase I clinical trial using Hemopurifier® in conjunction with pembrolizumab (Keytruda) in patients with advanced head and neck cancer has been recently approved by the Food and Drug Administration (FDA) (NCT04453046).

In line with the strategy of eliminating circulating cancer sEVs, a recent report demonstrated, in a xenograft nude mouse model, that treatment of the mice with human anti-CD9 and anti-CD63 antibodies (intravenous injection) disrupts cancer sEVs in the circulation and suppresses the pulmonary metastasis of implanted human breast cancer cells, yet, has no effect on primary tumor growth of the implants or metastatic ability of the cells in vitro [102]. These findings support the strategy to suppress cancer metastasis via inhibiting the pro-metastatic functions of cancer-derived sEVs using antibodies against their surface proteins. In addition, an innovative design of aptamer-functionalized nanoparticles was shown to eliminate blood oncogenic sEVs into the small intestine, and attenuate oncogenic sEV-induced lung metastasis in mice [103]. This technology utilized positively charged mesoporous silica nanoparticles equipped with Epidermal Growth Factor Receptor (EGFR)-targeting aptamers specifically recognizing and binding the negatively charged oncogenic sEVs and towing them from blood to bile duct for elimination. This interesting study proves that it is feasible to remove oncogenic sEVs selectively from the blood stream, thereby reducing tumor metastatic potential. Further investigations are warranted along this line of research.

3.4. *Targeting Specific sEV Cargo Components*

Specific sEV components that mediate sEVs' tumor-promoting activity are obvious potential cancer therapeutic targets. Some of the targets have been recently explored in order to develop new cancer therapeutics. As discussed above, antibodies against human CD9 and CD63, two well-established sEV surface markers [17], were shown to disrupt oncogenic sEVs and inhibit tumor metastasis in a breast cancer xenograft nude mouse model [102]. However, this experiment strategy of targeting human CD9 and CD63 is only applicable in a xenograft nude mouse model for selectively eliminating human cancer sEVs from the blood, since CD9 and CD63 are expressed in sEVs released from both noncancerous and cancerous cells in humans. Targeting of cancer-specific sEV components will be preferred to achieve a cancer-specific effect. In this context, a recent report demonstrated that cytoskeleton-associated protein 4 (CKAP4), a novel Dickkopf1 (DKK1) receptor, was selectively contained in sEVs from pancreatic ductal adenocarcinoma (PDAC) cells, not in sEVs from normal cells. Various anti-CKAP4 antibodies were then utilized to block the interaction of DKK1 with sEV-associated CKAP4, resulting in an inhibition of the proliferation and migration of PDAC cells and a prolonged survival of PDAC xenograft nude mice [104], supporting further development of this targeting strategy.

In another report, miR-365 in macrophage-derived sEVs was found to significantly decrease the sensitivity of PDAC cells to gemcitabine, and a miR-365 antagonist was able to reverse the gemcitabine resistance of PDAC cells in vitro and in vivo [105], thus suggesting that targeting miR-365 in macrophage-derived sEVs renders PDAC cells more sensitive to gemcitabine. Similarly, miR-155 was found in PDAC cell-derived sEVs that mediates transfer of the gemcitabine resistance traits from resistant PDAC cells to sensitive PDAC cells, conferring gemcitabine resistance of PDAC cells. Targeting miR-155 or the exosome secretion of PDAC cells effectively attenuated the gemcitabine resistance in PDAC cell lines and in xenograft nude mice [106]. Other cancer sEV-associated microRNAs, such as miR-21 and miR-1246, have also been found to be selectively enriched in cancer sEVs and considered as therapeutic targets [107,108]. Since cancer sEVs selectively encapsulate certain microRNA species [25,109–111], targeting cancer sEV-associated microRNAs will continue to be an attractive strategy for the development of new cancer therapeutics.

Immune checkpoint protein inhibitors, such as PD1/PD-L1 inhibitors, are novel cancer therapeutic targets which have revolutionized cancer therapy with great efficacy, even for those cancer patients whom standard therapy has failed [112]. However, only 10%–30% of patients responded to checkpoint inhibitor therapy [113]. The immune escape is partially due to the fact that tumor-derived sEVs contain PD-L1, a PD1 ligand, which binds to PD1 on the surface of T cells and suppresses T cell activation [66,67]. The sEV PD-L1 level was thus suggested to be a prognostic marker for anti-PD1 therapy response [114], and blocking sEV PD-L1 has been proposed to overcome the resistance to anti-PD-1/PD-L1 antibody therapy [115]. Indeed, anti-PD-L1 antibodies were shown to block sEV PD-L1, induce systemic anti-tumor immunity, and suppress tumor growth in a syngeneic colorectal cancer model [67].

New oncogenic components in cancer sEVs are continuously being identified which may contribute to tumor progression or chemo-resistance [116,117]. Efforts on targeting these sEV-associated oncogenic molecules for cancer therapeutic development are expected to expand.

4. EVs as Drug Carriers in Cancer Treatment

Compared to artificial drug vehicles, such as liposomes, EVs are favored drug carriers [118] because of their autologous nature that would prevent undesired immunogenicity and toxicity [119,120]. sEVs also possess high capacity of homing toward tumor cells when compared to liposomes [62,121], implying that sEVs are more efficient in delivering drugs for cancer therapy. Furthermore, studies have shown that sEVs are stable membrane vesicles under different pH values, temperatures, or freeze–thaw cycles [122], and these properties can be further enhanced by surface modification [123], supporting their potential compliance with good manufacturing practices (GMPs) in future clinical use. In addition, as nano-sized particles, sEVs were shown to be able to cross the blood–brain barrier and the tumor vasculature via enhanced permeability and retention (EPR), thereby potentially increasing accumulation of nanoparticles in brain tumors [124–126].

Diverse techniques have been practiced to encapsulate cancer therapeutics by sEVs in order to develop more efficient tumor-targeting vehicles. Here, we review the sEV loading strategies reported in recent literature.

4.1. EV Sources and Loading Efficiency

Based on the heterogeneity of sEVs derived from various biological sources [18], it is safe to assume that the source of the sEVs may relate to their drug loading efficiency and their therapeutic efficacy. Indeed, experimental evidence has been provided to show that drug loading efficiency of sEVs derived from pancreatic stellate cells (PSCs), pancreatic cancer cells (PCCs), and macrophages significantly differ when doxorubicin was simply incubated with the sEVs, with those from PCCs being most efficient. However, the doxorubicin-loaded macrophage sEVs are most effective in killing cancer cells, indicating that higher loading capacity does not equal to high anticancer activity of the drug-loaded sEVs [127]. This implies that both the biological source of the sEVs and the drug loading efficiency need to be evaluated when sEVs are applied as drug carriers for cancer therapy. In line with this concept, sEVs derived from mesenchymal stem cells (MSCs) are considered good carriers for drug delivery because they possess low immunogenicity [9,128] and are well tolerated in mice [129] and humans [130]. Both a miR-9 inhibitor and the chemo drug paclitaxel have been successfully incorporated into sEVs derived from MSCs which inhibited tumor cell growth [131,132]. However, allogeneic MSCs may also be able to transfer immunogenic proteins, such as MHC molecules, via secreted EVs, which might cause immunological responses [133]. Furthermore, the immunogenicity of MSCs-derived EVs varies, depending on experimental conditions by which the

EVs are produced [134]. Future efforts are required to closely monitor immunologic responses post administration of MSCs-derived EVs and develop uniform procedures in preparing MSCs-derived EVs. In addition to MSCs, sEVs from immature dendritic cells or self-derived dendritic cells were also considered, possessing low immunogenicity and used to encapsulate siRNA or doxorubicin for therapeutic applications [135,136]. Interestingly, cancer cell-derived sEVs were shown to have unique targeting abilities homing to tumorous microenvironments [137]. sEVs from HeLa and patient ascites were shown to deliver heterologous siRNAs to HeLa cells and cause cell death [138]. Autologous sEVs were found to be safe and effective in delivering gemcitabine for pancreatic cancer therapy in experimental model systems [139]. These results show that cancer cell-derived sEVs are promising carriers for effective delivery of chemotherapeutic drugs or nucleotides. Given the tumor-promoting activity of cancer-derived sEVs [53–55], the safety and long-term effect of these membrane vesicles as drug-delivery carriers needs to be carefully evaluated.

4.2. Loading Therapeutics into sEVs via Donor Cells

Efficient loading of cancer therapeutics into a given sEV population can be critical when it comes to drug efficacy. In this context, one loading strategy that has been described in packaging cancer therapeutics into sEVs is to load cancer therapeutics into sEVs via donor cells, which is in contrast to directly loading therapeutics into isolated sEVs. In this case, microRNAs have been most often loaded into sEVs via the donor cells. For example, adipose tissue-derived MSCs were transfected with a miR-122 expression plasmid to overexpress this microRNA and the sEVs derived from these cells were highly enriched in miR-122. An intra-tumor injection of miR-122-enriched sEVs significantly increased the efficacy of Sorafenib on inhibiting hepatocellular carcinoma in a xenograft nude mouse model [140]. Functional delivery of miR-21 derived from glioma cells to the surrounding microglia led to downregulation of specific miR-21 mRNA target genes [141]; likewise, sEVs from primary glioma cells, stably expressing miR-302-367, were shown to enrich in miR-302-367 by internalizing neighboring glioblastoma cells, and altering tumor development in vivo [142], and overexpression of miR-146b in marrow stromal cells generated sEVs with high miR-146b content, which significantly reduced glioma xenograft growth in rats [143]. More studies have been reported in testing the strategy of loading microRNA inhibitors or mimics into sEVs via the donor cells for therapeutic applications, as was recently reviewed [144].

An interesting study demonstrated that the chemotherapeutic paclitaxel (PTX) could be added directly to the culture of MSCs to generate sEVs that are highly associated with PTX and significantly suppress cancer cell proliferation [132]. However, this strategy of loading chemotherapeutics into sEVs has been less explored, likely because of the loading efficiency, considering the potential metalizing of PTX in treated cells. Instead, direct loading of chemotherapeutics and microRNA/siRNAs into the isolated sEVs has been widely adapted for testing sEVs as drug carriers for therapeutic delivery.

4.3. Loading Therapeutics into Isolated sEVs

The lipid-bilayer membrane structure of sEVs favors encapsulating hydrophobic compounds and molecules, which may directly integrate into the sEVs without disturbing their membrane barrier. In contrast, hydrophilic compounds and molecules require permeabilization of the bilayer membrane in order to be incorporated into the sEVs [145,146]. Various approaches have been proposed to load hydrophobic and hydrophilic drugs or biological molecules into sEVs. The most common approaches include opening up the pores in lipid membranes by physic forces, such as electroporation, sonication, freeze and thaw cycles, and extrusion, and by chemical means, such as using transfection reagents. The pros and cons of these methods for membrane permeabilization and cargo loading has been reviewed elsewhere [147]. Therefore, we will only briefly discuss these loading approaches in the following.

Direct incubation of therapeutics with sEVs at given temperatures and durations is a simple strategy for loading drugs into sEVs. The loading efficiency mainly relies on the concentration of the drugs or molecules and their hydrophobicity. A proper loading can usually be achieved for hydrophobic compounds without disturbing the integrity of the sEV membrane [132]. Nevertheless, the loading efficiency is often lower compared to other loading approaches.

Electroporation has been a method widely used to introduce DNA or RNA into mammalian cells [148,149], and is often applied for drug or nucleotide loading into sEVs [150,151]. The desired sEVs will be co-incubated with the therapeutics and exposed to certain volts of electric fields to open up the pores of the sEV membrane to allow the therapeutics to enter into the permeabilized sEVs. This method has been preferentially applied when incorporating nucleic acids like siRNA, mRNA, DNA, and microRNA, into sEVs [152]. Its loading efficiency is usually higher than incubation [139]. However, the main drawback of this method is the risk of damaging the EV membranes that may cause aggregation of sEVs and precipitation of nucleic acids.

Sonication uses ultrasound energy transmitted through a sonicating probe that reduces the rigidity of sEV membranes, thus allowing more therapeutic molecules to be incorporated into sEVs. For example, PTX was loaded into sEVs more efficiently by sonication than electroporation and incubation [139]. However, the sonicating probe produces consistent heat during the sonication and the operation has to be done on ice, with intervals between strokes [153]. There is no doubt that sonication may compromise the membrane integrity of sEVs, with the therapeutics occasionally being attached to the outer membrane of the sEV, which affects the drug distribution in vivo [139].

The freeze and thaw approach takes advantage of the formation of ice crystals that temporarily disrupt the sEV membrane, allowing therapeutic compounds to enter into the sEVs prior to membrane reconstitution [154]. This method shows lower cargo loading compared to sonication- and extrusion-based methods [155]. One to three cycles of freeze and thaw were usually performed during drug incorporation, which may accelerate the degradation and aggregation of the sEVs [122,156].

Extrusion utilizes a lipid syringe extruder with pore sizes between 100 and 400 nm, which break the sEV membrane physically and then mix with therapeutics. This method possesses high loading efficiency when compared to freeze and thaw, sonication, and saponin treatment [155,157]. One can imagine that the extrusion approach may cause damage of the sEV membranes as it does by sonication and electroporation.

Saponin treatment and the use of common transfection reagents, such as cationic lipids, have also been applied to load exogenous materials into sEVs. It was demonstrated to be an effective approach for sEV encapsulation of therapeutic drugs when compared to electroporation [155]. While we would expect more studies using the transfection approach for sEV loading, especially for loading of nucleic acids, the chemical transfection reagent itself will need to be removed prior to delivering the sEVs to target cells [157].

Through the above-mentioned approaches, multiple therapeutic agents in the forms of DNA, microRNA, siRNA, porphyrins, proteins (catalase, stress-induced heat shock proteins), and chemotherapeutics (curcumin, paclitaxel, docetaxel, gemcitabine) have been successfully loaded into sEVs and tested for their therapeutic value [158–161]. Nonetheless, it remains to be determined which approaches are most appropriate for loading specific agents into desired sEVs.

5. Clinical Trials Testing sEVs as Cancer Therapeutic Carriers

The potential of sEVs to serve as cancer therapeutic carriers and the promising results from preclinical studies have led to clinical trials aimed to develop sEV-based cancer therapy. We searched ClinicalTrials.gov and Pubmed.gov on 7 July 2020 and found 12 clinical trials testing sEVs as potential cancer therapeutics or therapeutic carriers, with 8 of them being registered in ClinicalTrials.gov (Table 1). Some of the clinical trials have reported their end results and others are still ongoing [162]. These clinical trials can be categorized according to their biological source of sEVs that are used as therapeutic carriers, as discussed below. Note that these clinical trials are mostly in early stages, and the definitive therapeutic value of sEVs for cancer therapy has yet to be determined.

Table 1. Clinical trials of EV-based cancer therapy.

Disease	Drug	EV Source	Phase, *n* of Patients	Status	Reference
Malignant Pleural Effusion	Methotrexate	Autologous Tumor-Derived Microparticles	Phase 2, $n = 90$	Recruiting	NCT02657460 [1] Guo, M. [163]
	Methotrexate	Microparticles	N/A, $n = 248$	Recruiting	NCT04131231 [1]
	Chemotherapeutic Drugs	Tumor Cell- Derived Microparticles	Phase 2, $n = 30$	Unknown	NCT01854866 [1] Tang, K. [164]
	Cisplatin	Tumor Cell- Derived Microparticles	N/A, $n = 6$	Completed	Ma, J. [165]
Metastatic Pancreatic Cancer	KRAS [2] G12D siRNA	MSC [3]-Derived Exosomes	Phase 1, $n = 28$	Recruiting	NCT03608631 [1] Kamerkar, S. [166]
Head and Neck Cancer	Grape Extract	Plant Exosomes	Phase 1, $n = 60$	Active, Not Recruiting	NCT01668849 [1]
	Hemopurifier Pembro-lizumab	Blood-Derived Exosomes	N/A, $n = 12$	Not Yet Recruiting	NCT04453046 [1]
Colorectal Cancer	Curcumin	Plant Exosomes	Phase 1, $n = 7$	Active, Not Recruiting	NCT01294072 [1]
	GM-CSF [4]	AEX [5]	Phase 1, $n = 40$	Completed	Dai, S. [167]
Non-Small Cell Lung Cancer	Antigens	Tumor Dex2 [6]	Phase 2, $n = 41$	Completed	NCT01159288 [1] Besse, B. [168]
	MAGE[7] Tumor Antigens	Autologous DEX [6]	Phase 1, $n = 13$	Completed	Morse, M.A. [169]
Metastatic Melanoma	MAGE[7] 3 Peptides	Autologous DEX [6]	Phase 1, $n = 15$	Completed	Escudier, B. [170]

[1] The NCT# refers to a registered National Clinical Trial (NCT) which can be found at Clinicaltrials.gov, [2] Kirsten Rat Sarcoma (KRAS), [3] Mesenchymal Stem Cells (MSC), [4] Granulocyte- Macrophage Colony-Stimulating Factor (GM-CSF), [5] Ascites- Derived Exosomes (AEX), [6] Dendritic Cell- Derived Exosomes (DEX), [7] Melanoma Antigens (MAGE).

5.1. Clinical Trials Using Dendritic Cell-Derived sEVs (DEX)

In 2005, two phase I clinical trials were reported using autologous DEX as immune stimulants, one for patients with metastatic melanoma, and another for patients with non-small cell lung cancer (NSCLC) [169,170]. Similar procedures were used in isolating sEVs from patients and loading MAGE-3 antigens to the sEVs for these trials. In the metastatic melanoma trial, 15 patients were included and received a 4-week outpatient vaccination course with antigen-loaded DEX given intradermally (1/10th) and subcutaneously (9/10th) per week for 4 weeks. There was no major toxicity being observed and some patients showed partial response and tumor repression. This is the first study to show the feasibility and safety of DEX-based vaccination in melanoma patients. In the NSCLC trial, 13 patients were enrolled, with 9 completing the therapy. The antigen-loaded DEX was given, intradermally (1/10th) and subcutaneously (9/10th), 4 times at weekly intervals. Similar to the melanoma trial, no major toxicity was observed during a 24-month follow up, and immune activation and stability of disease was observed in some patients with advanced NSCLC. The success of this phase I trial led to a phase II clinical trial for NSCLC in France (NCT01159288). In the phase II trial, DEX was upgraded from the first-generation interferon gamma-free DEX (IFN-γ-free DEX) to a second generation (IFN-γ-DEX) in order to enhance DEX-induced T cell responses. Twenty-four patients were recruited, and the results confirmed that DEX boosts antitumor immunity in patients with advanced NSCLC with outstanding safety [168]. Together, these clinical trials indicate a potential safe immunotherapy using DEX in metastatic melanoma and NSCLC, and an enhanced efficacy of DEX when administered in combination with IFN-γ.

5.2. Clinical Trials Using Ascites-Derived sEVs (AEX)

In 2008, a phase I study using autologous AEX combined with granulocyte-macrophage colony-stimulating factor (GM-CSF) for colorectal cancer was completed [167]. Forty patients with advanced colorectal cancer were included in the study and randomly assigned to AEX alone or AEX plus GM-CSF groups. Patients received 4 subcutaneous immunizations at weekly intervals. Results showed that both groups of patients tolerated the treatment well and AEX plus GM-CSF rather than

AEX alone induces beneficial antitumor cytotoxic T lymphocyte (CTL) response. These findings suggest that the immunotherapy of colorectal cancer with AEX in combination with GM-CSF is feasible and safe, and may be applied for immunotherapy of colorectal cancer.

5.3. Clinical Trials Using Tumor Cell-Derived EVs

A preclinical study has confirmed the feasibility of using apoptotic tumor cells induced by chemotherapeutic drugs to produce drug-packaging EVs [164]. Several anti-cancer drugs, including methotrexate, doxorubicin, and cisplatin, were shown to be packaged into EVs released by tumor cells, such as the mouse hepatocarcinoma tumor cell line H22 or the human ovarian cancer A2780. These drug-containing EVs effectively killed tumor cells in murine models without typical side effects, such as hair and/or weight loss or liver and/or kidney function impairment. Inspired by these preclinical results, three clinical trials were consecutively registered to test the effects of chemotherapeutic packed EVs in cancer patients (NCT01854866, NCT02657460, and NCT04131231). Whereas findings from two of the trials remain to be reported, one of the trials published their results in 2019 [163], showing that autologous tumor EVs packed with methotrexate symptomatically improved 10 of 11 lung cancer patients with malignant pleural effusion. The methotrexate-packed EVs activated antitumor effector cells including CTLs and TH1 in the pleural microenvironment and only caused mild (grades 1 to 2) adverse events.

Tumor EVs packed with chemotherapeutics also contributed to reverse drug resistance of malignant cells. Intrathoracic injection of cisplatin-packed tumor EVs in three end-stage lung cancer patients eliminated 95% of tumor cells in the malignant fluids and ameliorated patient symptoms. These therapeutic effects were absent in another three patients with intrathoracic injection of cisplatin alone [165].

5.4. Clinical Trials Using Plant-Derived sEVs

sEVs derived from plants are unquestionably safer than those from tumor cells. Grapefruits were found to yield higher sEVs (2.21 g/kg raw material) than grapes, tomatoes, bovine milk, or ginger [171]. Grapefruit-derived nanovectors (GNVs) were demonstrated to transport chemotherapeutic agents, siRNA, DNA expression vectors, and proteins to different kinds of cells. Co-delivery of folic acid and PTX by GNVs showed a therapeutic benefit in a mouse model of colon cancer [172]. These preclinical results led to a phase I clinical study investigating the efficacy of plant sEVs conjugated with curcumin that was orally delivered to patients with colon cancer (NCT01294072). Another phase I clinical trial was designed to evaluate the ability of plant sEVs to prevent oral mucositis during chemo-radiation of head and neck cancer (NCT01668849), which will shed light on the potential of using plant sEVs to alleviate side effects during cancer therapy.

5.5. Clinical Trials Using Normal Fibroblast-Like Mesenchymal Cell-Derived EVs

A preclinical study has demonstrated that sEVs, derived from fibroblast-like mesenchymal cells and loaded with siRNA or shRNA targeting KRAS mutation (KrasG12D), are significantly more effective than other drug carriers in inhibiting pancreatic ductal adenocarcinoma (PDAC) progression in vitro and in vivo [166]. Following the report, this research group initiated a phase I clinical trial (NCT03608631) aimed at testing this approach in patients with stage IV PDAC bearing the KrasG12D mutation. They will also evaluate median progression-free survival (PFS) and median overall survival (OS) as secondary objectives.

6. Conclusions

Research on EVs in cancer has been intensified over the last decade. The involvement of EVs, especially sEVs, in promoting cancer progression through intercellular communication is well recognized. This leads to efforts focusing on targeting EV signaling or utilizing EVs as drug carriers to develop novel cancer therapeutics. In this review, we have summarized recent progress in the development of EVs as cancer therapeutics, both in preclinical studies and clinical trials. Clearly, most of the studies reported on targeting sEV signaling, such as EV microRNA signaling, are at preclinical stages, and clinical trials are primarily related to developing EVs as therapeutic carriers at relatively early phases. This indicates that, on one hand, significant progress has been made in understanding how to better target EV signaling for the development of cancer therapeutics and the safety of delivering EVs into humans as therapeutic carriers, and on the other hand, clinical efficacy of EVs as therapeutic targets or therapeutic carriers remains to be determined. Compared to targeting EV signaling, utilizing EVs as therapeutic carriers seems to be a more practical strategy in therapeutic development and has advanced from preclinical studies to clinical trials. This is likely due to the fact that targeting cancer-specific EV signaling remains a challenge, as clear distinction of cancer EVs from healthy EVs has not been firmly established, and the heterogeneity of EVs is well recognized, which renders it difficult in specific targeting of EV signaling. In addition, current technology in EV isolation and validation needs to be improved, which also limits the effort in exploring EV signaling in cancer. Ongoing EV research needs to focus on these challenges in order to establish clinically applicable therapeutics targeting EV signaling in cancer. There are also challenges in the development of EVs as therapeutic carriers [173], including production and purification of EVs on an industrial scale, potential EV contamination with virus [100,174], and long-term side effects of tumor-derived EVs when they are applied as therapeutic carriers. However, these challenges are mostly technological, not conceptual, and hopefully can be overcome with concentrated effort. It is expected that EVs as therapeutic targets or delivery carriers may soon open up new avenues in clinical management of malignant diseases.

Author Contributions: H.S. searched the literature and drafted the manuscript. S.B. helped with drafting the manuscript and prepared the figure and table. J.W. conceived of and participated in drafting the manuscript. W.-Q.D. conceived of and finalized the manuscript. All authors have read and approved the final manuscript.

Acknowledgments: We thank the Department of Pathology at the University of Oklahoma Health Sciences Center for administrative support.

Abbreviations

EV	Extracellular Vesicle
MV	Microvesicle
ISEV	International Society for Extracellular Vesicles
sEV	Small Extracellular Vesicle
MHC	Major Histocompatibility Complex
EPCAM	Epithelial Cell Adhesion Molecule
PECAM1	Platelet Endothelial Cell Adhesion Molecule 1
ERBB2	Erb-B2 Receptor Tyrosine Kinase 2
ESCRT	Endosomal Sorting Complexes Required for Transport
nSMase2	Neutral sphingomyelinase 2
HSP72	Heat-Shock Protein 72
STAT-3	Signal Transducer and Activator of Transcription 3
MA	Manumycin-A
PI3K	Phosphatidylinositol 3-Kinase
HUVEC	Human Umbilical Vein Endothelial Cell
GBM	Glioblastoma
CKAP4	Cytoskeleton-Associated Protein 4
DKK1	Dickkopf1
PDAC	Pancreatic Ductal Adenocarcinoma
PD1	Programmed Cell Death Protein 1

PD-L1	Programmed Death-Ligand 1
GMP	Good Manufacturing Practice
EPR	Enhanced Permeability and Retention
PSC	Pancreatic Stellate Cell
MSC	Mesenchymal Stem Cells
HeLa	Henrietta Lacks Cells
PTX	Paclitaxel
KRAS	Kirsten Rat Sarcoma
DEX	Dendrite Cell- Derived Exosomes
NSCLC	Non-Small Cell Lung Cancer
MAGE-3	Melanoma-Associated Antigen 3
IFN-γ-free DEX	Interferon Gamma-free Exosomes
IFN-γ-DEX	Interferon Gamma-containing Exosomes
AEX	Ascites-Derived Exosomes
GM-CSF	Granulocyte-Macrophage Colony-Stimulating Factor
CTL	Cytotoxic T Lymphocyte
TH1	T-cell Helper 1
GNV	Grapefruit-Derived Nanovectors
PFS	Progression-Free Survival
OS	Overall Survival

References

1. Chargaff, E.; West, R. The biological significance of the thromboplastic protein of blood. *J. Biol. Chem.* **1946**, *166*, 189–197. [PubMed]
2. Harding, C.; Heuser, J.; Stahl, P. Receptor-mediated endocytosis of transferrin and recycling of the transferrin receptor in rat reticulocytes. *J. Cell Biol.* **1983**, *97*, 329–339. [CrossRef] [PubMed]
3. Pan, B.T.; Teng, K.; Wu, C.; Adam, M.; Johnstone, R.M. Electron microscopic evidence for externalization of the transferrin receptor in vesicular form in sheep reticulocytes. *J. Cell Biol.* **1985**, *101*, 942–948. [CrossRef] [PubMed]
4. Baj-Krzyworzeka, M.; Szatanek, R.; Weglarczyk, K.; Baran, J.; Urbanowicz, B.; Branski, P.; Ratajczak, M.Z.; Zembala, M. Tumour-derived microvesicles carry several surface determinants and mRNA of tumour cells and transfer some of these determinants to monocytes. *Cancer Immunol. Immun.* **2006**, *55*, 808–818. [CrossRef] [PubMed]
5. Ratajczak, J.; Wysoczynski, M.; Hayek, F.; Janowska-Wieczorek, A.; Ratajczak, M.Z. Membrane-derived microvesicles: Important and underappreciated mediators of cell-to-cell communication. *Leukemia* **2006**, *20*, 1487–1495. [CrossRef]
6. Aliotta, J.M.; Sanchez-Guijo, F.M.; Dooner, G.J.; Johnson, K.W.; Dooner, M.S.; Greer, K.A.; Greer, D.; Pimentel, J.; Kolankiewicz, L.M.; Puente, N.; et al. Alteration of marrow cell gene expression, protein production, and engraftment into lung by lung-derived microvesicles: A novel mechanism for phenotype modulation. *Stem Cells* **2007**, *25*, 2245–2256. [CrossRef]
7. Valadi, H.; Ekstrom, K.; Bossios, A.; Sjostrand, M.; Lee, J.J.; Lotvall, J.O. Exosome-mediated transfer of mRNAs and microRNAs is a novel mechanism of genetic exchange between cells. *Nat. Cell Biol.* **2007**, *9*, 654–659. [CrossRef]
8. Skog, J.; Wurdinger, T.; van Rijn, S.; Meijer, D.H.; Gainche, L.; Sena-Esteves, M.; Curry, W.T., Jr.; Carter, B.S.; Krichevsky, A.M.; Breakefield, X.O. Glioblastoma microvesicles transport RNA and proteins that promote tumour growth and provide diagnostic biomarkers. *Nat. Cell Biol.* **2008**, *10*, 1470–1476. [CrossRef]
9. Kalluri, R.; LeBleu, V.S. The biology, function, and biomedical applications of exosomes. *Science* **2020**, 367. [CrossRef]
10. Bonucci, E. Fine structure and histochemistry of "calcifying globules" in epiphyseal cartilage. *Z. Zellforsch. Mikrosk. Anat.* **1970**, *103*, 192–217. [CrossRef]

11. Yanez-Mo, M.; Siljander, P.R.; Andreu, Z.; Zavec, A.B.; Borras, F.E.; Buzas, E.I.; Buzas, K.; Casal, E.; Cappello, F.; Carvalho, J.; et al. Biological properties of extracellular vesicles and their physiological functions. *J. Extracell. Vesicles* **2015**, *4*, 27066. [CrossRef] [PubMed]
12. Gould, S.J.; Raposo, G. As we wait: Coping with an imperfect nomenclature for extracellular vesicles. *J. Extracell. Vesicles* **2013**, 2. [CrossRef] [PubMed]
13. Hessvik, N.P.; Llorente, A. Current knowledge on exosome biogenesis and release. *Cell Mol. Life Sci.* **2018**, *75*, 193–208. [CrossRef]
14. Crescitelli, R.; Lasser, C.; Szabo, T.G.; Kittel, A.; Eldh, M.; Dianzani, I.; Buzas, E.I.; Lotvall, J. Distinct RNA profiles in subpopulations of extracellular vesicles: Apoptotic bodies, microvesicles and exosomes. *J. Extracell. Vesicles* **2013**, *2*. [CrossRef] [PubMed]
15. Kalra, H.; Simpson, R.J.; Ji, H.; Aikawa, E.; Altevogt, P.; Askenase, P.; Bond, V.C.; Borras, F.E.; Breakefield, X.; Budnik, V.; et al. Vesiclepedia: A compendium for extracellular vesicles with continuous community annotation. *Plos Biol.* **2012**, *10*, e1001450. [CrossRef] [PubMed]
16. Zhang, H.; Freitas, D.; Kim, H.S.; Fabijanic, K.; Li, Z.; Chen, H.; Mark, M.T.; Molina, H.; Martin, A.B.; Bojmar, L.; et al. Identification of distinct nanoparticles and subsets of extracellular vesicles by asymmetric flow field-flow fractionation. *Nat. Cell Biol.* **2018**, *20*, 332–343. [CrossRef]
17. Thery, C.; Witwer, K.W.; Aikawa, E.; Alcaraz, M.J.; Anderson, J.D.; Andriantsitohaina, R.; Antoniou, A.; Arab, T.; Archer, F.; Atkin-Smith, G.K.; et al. Minimal information for studies of extracellular vesicles 2018 (MISEV2018): A position statement of the International Society for Extracellular Vesicles and update of the MISEV2014 guidelines. *J. Extracell. Vesicles* **2018**, *7*, 1535750. [CrossRef]
18. Wortzel, I.; Dror, S.; Kenific, C.M.; Lyden, D. Exosome-Mediated Metastasis: Communication from a Distance. *Dev. Cell* **2019**, *49*, 347–360. [CrossRef]
19. Charoenviriyakul, C.; Takahashi, Y.; Morishita, M.; Matsumoto, A.; Nishikawa, M.; Takakura, Y. Cell type-specific and common characteristics of exosomes derived from mouse cell lines: Yield, physicochemical properties, and pharmacokinetics. *Eur. J. Pharm. Sci.* **2017**, *96*, 316–322. [CrossRef]
20. Kim, D.K.; Kang, B.; Kim, O.Y.; Choi, D.S.; Lee, J.; Kim, S.R.; Go, G.; Yoon, Y.J.; Kim, J.H.; Jang, S.C.; et al. EVpedia: An integrated database of high-throughput data for systemic analyses of extracellular vesicles. *J. Extracell. Vesicles* **2013**, *2*. [CrossRef]
21. Simpson, R.J.; Kalra, H.; Mathivanan, S. ExoCarta as a resource for exosomal research. *J. Extracell. Vesicles* **2012**, *1*. [CrossRef] [PubMed]
22. Wei, Z.; Batagov, A.O.; Schinelli, S.; Wang, J.; Wang, Y.; El Fatimy, R.; Rabinovsky, R.; Balaj, L.; Chen, C.C.; Hochberg, F.; et al. Coding and noncoding landscape of extracellular RNA released by human glioma stem cells. *Nat. Commun.* **2017**, *8*, 1145. [CrossRef] [PubMed]
23. Hannafon, B.N.; Ding, W.Q. Intercellular Communication by Exosome-Derived microRNAs in Cancer. *Int. J. Mol. Sci.* **2013**, *14*, 14240–14269. [CrossRef] [PubMed]
24. Ratajczak, J.; Miekus, K.; Kucia, M.; Zhang, J.; Reca, R.; Dvorak, P.; Ratajczak, M.Z. Embryonic stem cell-derived microvesicles reprogram hematopoietic progenitors: Evidence for horizontal transfer of mRNA and protein delivery. *Leukemia* **2006**, *20*, 847–856. [CrossRef]
25. Hannafon, B.N.; Carpenter, K.J.; Berry, W.L.; Janknecht, R.; Dooley, W.C.; Ding, W.Q. Exosome-mediated microRNA signaling from breast cancer cells is altered by the anti-angiogenesis agent docosahexaenoic acid (DHA). *Mol. Cancer* **2015**, *14*, 133. [CrossRef]
26. Pegtel, D.M.; Cosmopoulos, K.; Thorley-Lawson, D.A.; van Eijndhoven, M.A.; Hopmans, E.S.; Lindenberg, J.L.; de Gruijl, T.D.; Wurdinger, T.; Middeldorp, J.M. Functional delivery of viral miRNAs via exosomes. *Proc. Natl. Acad. Sci. USA* **2010**, *107*, 6328–6333. [CrossRef]
27. Montecalvo, A.; Larregina, A.T.; Shufesky, W.J.; Stolz, D.B.; Sullivan, M.L.; Karlsson, J.M.; Baty, C.J.; Gibson, G.A.; Erdos, G.; Wang, Z.; et al. Mechanism of transfer of functional microRNAs between mouse dendritic cells via exosomes. *Blood* **2012**, *119*, 756–766. [CrossRef]
28. Ismail, N.; Wang, Y.; Dakhlallah, D.; Moldovan, L.; Agarwal, K.; Batte, K.; Shah, P.; Wisler, J.; Eubank, T.D.; Tridandapani, S.; et al. Macrophage microvesicles induce macrophage differentiation and miR-223 transfer. *Blood* **2013**, *121*, 984–995. [CrossRef]
29. Li, S.; Li, Y.; Chen, B.; Zhao, J.; Yu, S.; Tang, Y.; Zheng, Q.; Li, Y.; Wang, P.; He, X.; et al. exoRBase: A database of circRNA, lncRNA and mRNA in human blood exosomes. *Nucleic Acids Res.* **2018**, *46*, D106–D112. [CrossRef]

30. Murillo, O.D.; Thistlethwaite, W.; Rozowsky, J.; Subramanian, S.L.; Lucero, R.; Shah, N.; Jackson, A.R.; Srinivasan, S.; Chung, A.; Laurent, C.D.; et al. exRNA Atlas Analysis Reveals Distinct Extracellular RNA Cargo Types and Their Carriers Present across Human Biofluids. *Cell* **2019**, *117*, 463–477.e415. [CrossRef]
31. Russo, F.; Di Bella, S.; Vannini, F.; Berti, G.; Scoyni, F.; Cook, H.V.; Santos, A.; Nigita, G.; Bonnici, V.; Lagana, A.; et al. miRandola 2017: A curated knowledge base of non-invasive biomarkers. *Nucleic Acids Res.* **2018**, *46*, D354–D359. [CrossRef] [PubMed]
32. Kahlert, C.; Melo, S.A.; Protopopov, A.; Tang, J.; Seth, S.; Koch, M.; Zhang, J.; Weitz, J.; Chin, L.; Futreal, A.; et al. Identification of double-stranded genomic DNA spanning all chromosomes with mutated KRAS and p53 DNA in the serum exosomes of patients with pancreatic cancer. *J. Biol. Chem.* **2014**, *289*, 3869–3875. [CrossRef] [PubMed]
33. Sansone, P.; Savini, C.; Kurelac, I.; Chang, Q.; Amato, L.B.; Strillacci, A.; Stepanova, A.; Iommarini, L.; Mastroleo, C.; Daly, L.; et al. Packaging and transfer of mitochondrial DNA via exosomes regulate escape from dormancy in hormonal therapy-resistant breast cancer. *Proc. Natl. Acad. Sci. USA* **2017**, *114*, E9066–E9075. [CrossRef] [PubMed]
34. Thakur, B.K.; Zhang, H.; Becker, A.; Matei, I.; Huang, Y.; Costa-Silva, B.; Zheng, Y.; Hoshino, A.; Brazier, H.; Xiang, J.; et al. Double-stranded DNA in exosomes: A novel biomarker in cancer detection. *Cell Res.* **2014**, *24*, 766–769. [CrossRef] [PubMed]
35. Magnette, A.; Chatelain, M.; Chatelain, B.; Ten Cate, H.; Mullier, F. Pre-analytical issues in the haemostasis laboratory: Guidance for the clinical laboratories. *Thromb. J.* **2016**, *14*, 49. [CrossRef] [PubMed]
36. Carayon, K.; Chaoui, K.; Ronzier, E.; Lazar, I.; Bertrand-Michel, J.; Roques, V.; Balor, S.; Terce, F.; Lopez, A.; Salome, L.; et al. Proteolipidic composition of exosomes changes during reticulocyte maturation. *J. Biol. Chem.* **2011**, *286*, 34426–34439. [CrossRef] [PubMed]
37. Gasser, O.; Schifferli, J.A. Activated polymorphonuclear neutrophils disseminate anti-inflammatory microparticles by ectocytosis. *Blood* **2004**, *104*, 2543–2548. [CrossRef]
38. Montecalvo, A.; Shufesky, W.J.; Stolz, D.B.; Sullivan, M.G.; Wang, Z.; Divito, S.J.; Papworth, G.D.; Watkins, S.C.; Robbins, P.D.; Larregina, A.T.; et al. Exosomes as a short-range mechanism to spread alloantigen between dendritic cells during T cell allorecognition. *J. Immunol.* **2008**, *180*, 3081–3090. [CrossRef]
39. Mincheva-Nilsson, L.; Baranov, V. Placenta-derived exosomes and syncytiotrophoblast microparticles and their role in human reproduction: Immune modulation for pregnancy success. *Am. J. Reprod. Immunol.* **2014**, *72*, 440–457. [CrossRef]
40. Lakkaraju, A.; Rodriguez-Boulan, E. Itinerant exosomes: Emerging roles in cell and tissue polarity. *Trends Cell Biol.* **2008**, *18*, 199–209. [CrossRef]
41. Quesenberry, P.J.; Aliotta, J.M. The paradoxical dynamism of marrow stem cells: Considerations of stem cells, niches, and microvesicles. *Stem Cell Rev.* **2008**, *4*, 137–147. [CrossRef] [PubMed]
42. Nahar, N.N.; Missana, L.R.; Garimella, R.; Tague, S.E.; Anderson, H.C. Matrix vesicles are carriers of bone morphogenetic proteins (BMPs), vascular endothelial growth factor (VEGF), and noncollagenous matrix proteins. *J. Bone Min. Metab.* **2008**, *26*, 514–519. [CrossRef] [PubMed]
43. Shirakami, Y.; Lee, S.A.; Clugston, R.D.; Blaner, W.S. Hepatic metabolism of retinoids and disease associations. *Biochim. Biophys. Acta* **2012**, *1821*, 124–136. [CrossRef] [PubMed]
44. Lai, C.P.; Breakefield, X.O. Role of exosomes/microvesicles in the nervous system and use in emerging therapies. *Front. Physiol.* **2012**, *3*, 228. [CrossRef] [PubMed]
45. Guo, H.; Chitiprolu, M.; Roncevic, L.; Javalet, C.; Hemming, F.J.; Trung, M.T.; Meng, L.; Latreille, E.; Tanese de Souza, C.; McCulloch, D.; et al. Atg5 Disassociates the V1V0-ATPase to Promote Exosome Production and Tumor Metastasis Independent of Canonical Macroautophagy. *Dev. Cell* **2017**, *43*, 716–730 e717. [CrossRef]
46. Villarroya-Beltri, C.; Baixauli, F.; Mittelbrunn, M.; Fernandez-Delgado, I.; Torralba, D.; Moreno-Gonzalo, O.; Baldanta, S.; Enrich, C.; Guerra, S.; Sanchez-Madrid, F. ISGylation controls exosome secretion by promoting lysosomal degradation of MVB proteins. *Nat. Commun.* **2016**, *7*, 13588. [CrossRef]
47. Ilie, A.; Gao, A.Y.L.; Boucher, A.; Park, J.; Berghuis, A.M.; Hoffer, M.J.V.; Hilhorst-Hofstee, Y.; McKinney, R.A.; Orlowski, J. A potential gain-of-function variant of SLC9A6 leads to endosomal alkalinization and neuronal atrophy associated with Christianson Syndrome. *Neurobiol. Dis.* **2019**, *121*, 187–204. [CrossRef]
48. Edgar, J.R.; Manna, P.T.; Nishimura, S.; Banting, G.; Robinson, M.S. Tetherin is an exosomal tether. *Elife* **2016**, *5*. [CrossRef]

49. Putz, U.; Howitt, J.; Lackovic, J.; Foot, N.; Kumar, S.; Silke, J.; Tan, S.S. Nedd4 family-interacting protein 1 (Ndfip1) is required for the exosomal secretion of Nedd4 family proteins. *J. Biol. Chem.* **2008**, *283*, 32621–32627. [CrossRef]
50. Phillips, M.C. Molecular mechanisms of cellular cholesterol efflux. *J. Biol. Chem.* **2014**, *289*, 24020–24029. [CrossRef]
51. Amzallag, N.; Passer, B.J.; Allanic, D.; Segura, E.; Thery, C.; Goud, B.; Amson, R.; Telerman, A. TSAP6 facilitates the secretion of translationally controlled tumor protein/histamine-releasing factor via a nonclassical pathway. *J. Biol. Chem.* **2004**, *279*, 46104–46112. [CrossRef] [PubMed]
52. Hanahan, D.; Weinberg, R.A. Hallmarks of cancer: The next generation. *Cell* **2011**, *144*, 646–674. [CrossRef] [PubMed]
53. Melo, S.A.; Sugimoto, H.; O'Connell, J.T.; Kato, N.; Villanueva, A.; Vidal, A.; Qiu, L.; Vitkin, E.; Perelman, L.T.; Melo, C.A.; et al. Cancer exosomes perform cell-independent microRNA biogenesis and promote tumorigenesis. *Cancer Cell* **2014**, *26*, 707–721. [CrossRef]
54. Costa-Silva, B.; Aiello, N.M.; Ocean, A.J.; Singh, S.; Zhang, H.; Thakur, B.K.; Becker, A.; Hoshino, A.; Mark, M.T.; Molina, H.; et al. Pancreatic cancer exosomes initiate pre-metastatic niche formation in the liver. *Nat. Cell Biol.* **2015**, *17*, 816–826. [CrossRef] [PubMed]
55. Peinado, H.; Aleckovic, M.; Lavotshkin, S.; Matei, I.; Costa-Silva, B.; Moreno-Bueno, G.; Hergueta-Redondo, M.; Williams, C.; Garcia-Santos, G.; Ghajar, C.; et al. Melanoma exosomes educate bone marrow progenitor cells toward a pro-metastatic phenotype through MET. *Nat. Med.* **2012**, *18*, 883–891. [CrossRef] [PubMed]
56. Xavier, C.P.R.; Caires, H.R.; Barbosa, M.A.G.; Bergantim, R.; Guimaraes, J.E.; Vasconcelos, M.H. The Role of Extracellular Vesicles in the Hallmarks of Cancer and Drug Resistance. *Cells* **2020**, *9*, 1141. [CrossRef]
57. Antonyak, M.A.; Li, B.; Boroughs, L.K.; Johnson, J.L.; Druso, J.E.; Bryant, K.L.; Holowka, D.A.; Cerione, R.A. Cancer cell-derived microvesicles induce transformation by transferring tissue transglutaminase and fibronectin to recipient cells. *Proc. Natl. Acad. Sci. USA* **2011**, *108*, 4852–4857. [CrossRef]
58. Sato, S.; Vasaikar, S.; Eskaros, A.; Kim, Y.; Lewis, J.S.; Zhang, B.; Zijlstra, A.; Weaver, A.M. EPHB2 carried on small extracellular vesicles induces tumor angiogenesis via activation of ephrin reverse signaling. *Jci Insight* **2019**, *4*. [CrossRef]
59. Bao, L.; You, B.; Shi, S.; Shan, Y.; Zhang, Q.; Yue, H.; Zhang, J.; Zhang, W.; Shi, Y.; Liu, Y.; et al. Metastasis-associated miR-23a from nasopharyngeal carcinoma-derived exosomes mediates angiogenesis by repressing a novel target gene TSGA10. *Oncogene* **2018**, *37*, 2873–2889. [CrossRef]
60. Pavlyukov, M.S.; Yu, H.; Bastola, S.; Minata, M.; Shender, V.O.; Lee, Y.; Zhang, S.; Wang, J.; Komarova, S.; Wang, J.; et al. Apoptotic Cell-Derived Extracellular Vesicles Promote Malignancy of Glioblastoma Via Intercellular Transfer of Splicing Factors. *Cancer Cell* **2018**, *34*, 119–135 e110. [CrossRef]
61. Setti, M.; Osti, D.; Richichi, C.; Ortensi, B.; Del Bene, M.; Fornasari, L.; Beznoussenko, G.; Mironov, A.; Rappa, G.; Cuomo, A.; et al. Extracellular vesicle-mediated transfer of CLIC1 protein is a novel mechanism for the regulation of glioblastoma growth. *Oncotarget* **2015**, *6*, 31413–31427. [CrossRef] [PubMed]
62. Hoshino, A.; Costa-Silva, B.; Shen, T.L.; Rodrigues, G.; Hashimoto, A.; Tesic Mark, M.; Molina, H.; Kohsaka, S.; Di Giannatale, A.; Ceder, S.; et al. Tumour exosome integrins determine organotropic metastasis. *Nature* **2015**, *527*, 329–335. [CrossRef] [PubMed]
63. Wang, X.; Luo, G.; Zhang, K.; Cao, J.; Huang, C.; Jiang, T.; Liu, B.; Su, L.; Qiu, Z. Hypoxic Tumor-Derived Exosomal miR-301a Mediates M2 Macrophage Polarization via PTEN/PI3Kgamma to Promote Pancreatic Cancer Metastasis. *Cancer Res.* **2018**, *78*, 4586–4598. [CrossRef] [PubMed]
64. Marleau, A.M.; Chen, C.S.; Joyce, J.A.; Tullis, R.H. Exosome removal as a therapeutic adjuvant in cancer. *J. Transl. Med.* **2012**, *10*, 134. [CrossRef] [PubMed]
65. Shedden, K.; Xie, X.T.; Chandaroy, P.; Chang, Y.T.; Rosania, G.R. Expulsion of small molecules in vesicles shed by cancer cells: Association with gene expression and chemosensitivity profiles. *Cancer Res.* **2003**, *63*, 4331–4337.
66. Chen, G.; Huang, A.C.; Zhang, W.; Zhang, G.; Wu, M.; Xu, W.; Yu, Z.; Yang, J.; Wang, B.; Sun, H.; et al. Exosomal PD-L1 contributes to immunosuppression and is associated with anti-PD-1 response. *Nature* **2018**, *560*, 382–386. [CrossRef]

67. Poggio, M.; Hu, T.; Pai, C.C.; Chu, B.; Belair, C.D.; Chang, A.; Montabana, E.; Lang, U.E.; Fu, Q.; Fong, L.; et al. Suppression of Exosomal PD-L1 Induces Systemic Anti-tumor Immunity and Memory. *Cell* **2019**, *177*, 414–427 e413. [CrossRef]
68. Andaloussi, S.E.L.; Mager, I.; Breakefield, X.O.; Wood, M.J. Extracellular vesicles: Biology and emerging therapeutic opportunities. *Nat. Rev. Drug Discov.* **2013**, *12*, 347–357. [CrossRef]
69. Urabe, F.; Kosaka, N.; Ito, K.; Kimura, T.; Egawa, S.; Ochiya, T. Extracellular vesicles as biomarkers and therapeutic targets for cancer. *Am. J. Physiol. Cell Physiol.* **2020**, *318*, C29–C39. [CrossRef]
70. Yamamoto, T.; Kosaka, N.; Ochiya, T. Latest advances in extracellular vesicles: From bench to bedside. *Sci. Technol. Adv. Mater.* **2019**, *20*, 746–757. [CrossRef]
71. D'Souza-Schorey, C.; Schorey, J.S. Regulation and mechanisms of extracellular vesicle biogenesis and secretion. *Essays Biochem.* **2018**, *62*, 125–133. [CrossRef] [PubMed]
72. Mathieu, M.; Martin-Jaular, L.; Lavieu, G.; Thery, C. Specificities of secretion and uptake of exosomes and other extracellular vesicles for cell-to-cell communication. *Nat. Cell Biol.* **2019**, *21*, 9–17. [CrossRef] [PubMed]
73. Ostrowski, M.; Carmo, N.B.; Krumeich, S.; Fanget, I.; Raposo, G.; Savina, A.; Moita, C.F.; Schauer, K.; Hume, A.N.; Freitas, R.P.; et al. Rab27a and Rab27b control different steps of the exosome secretion pathway. *Nat. Cell Biol.* **2010**, *12*, 19–30, sup pp. 11–13. [CrossRef] [PubMed]
74. Middleton, R.C.; Rogers, R.G.; De Couto, G.; Tseliou, E.; Luther, K.; Holewinski, R.; Soetkamp, D.; Van Eyk, J.E.; Antes, T.J.; Marban, E. Newt cells secrete extracellular vesicles with therapeutic bioactivity in mammalian cardiomyocytes. *J. Extracell. Vesicles* **2018**, *7*, 1456888. [CrossRef] [PubMed]
75. Kosaka, N.; Iguchi, H.; Hagiwara, K.; Yoshioka, Y.; Takeshita, F.; Ochiya, T. Neutral sphingomyelinase 2 (nSMase2)-dependent exosomal transfer of angiogenic microRNAs regulate cancer cell metastasis. *J. Biol. Chem.* **2013**, *288*, 10849–10859. [CrossRef]
76. Phuyal, S.; Hessvik, N.P.; Skotland, T.; Sandvig, K.; Llorente, A. Regulation of exosome release by glycosphingolipids and flotillins. *Febs J.* **2014**, *281*, 2214–2227. [CrossRef]
77. Yuyama, K.; Sun, H.; Mitsutake, S.; Igarashi, Y. Sphingolipid-modulated exosome secretion promotes clearance of amyloid-beta by microglia. *J. Biol. Chem.* **2012**, *287*, 10977–10989. [CrossRef]
78. Chalmin, F.; Ladoire, S.; Mignot, G.; Vincent, J.; Bruchard, M.; Remy-Martin, J.P.; Boireau, W.; Rouleau, A.; Simon, B.; Lanneau, D.; et al. Membrane-associated Hsp72 from tumor-derived exosomes mediates STAT3-dependent immunosuppressive function of mouse and human myeloid-derived suppressor cells. *J. Clin. Investig.* **2010**, *120*, 457–471. [CrossRef]
79. Baietti, M.F.; Zhang, Z.; Mortier, E.; Melchior, A.; Degeest, G.; Geeraerts, A.; Ivarsson, Y.; Depoortere, F.; Coomans, C.; Vermeiren, E.; et al. Syndecan-syntenin-ALIX regulates the biogenesis of exosomes. *Nat. Cell Biol.* **2012**, *14*, 677–685. [CrossRef]
80. Bobrie, A.; Krumeich, S.; Reyal, F.; Recchi, C.; Moita, L.F.; Seabra, M.C.; Ostrowski, M.; Thery, C. Rab27a supports exosome-dependent and -independent mechanisms that modify the tumor microenvironment and can promote tumor progression. *Cancer Res.* **2012**, *72*, 4920–4930. [CrossRef]
81. Savina, A.; Fader, C.M.; Damiani, M.T.; Colombo, M.I. Rab11 promotes docking and fusion of multivesicular bodies in a calcium-dependent manner. *Traffic* **2005**, *6*, 131–143. [CrossRef] [PubMed]
82. Hsu, C.; Morohashi, Y.; Yoshimura, S.; Manrique-Hoyos, N.; Jung, S.; Lauterbach, M.A.; Bakhti, M.; Gronborg, M.; Mobius, W.; Rhee, J.; et al. Regulation of exosome secretion by Rab35 and its GTPase-activating proteins TBC1D10A-C. *J. Cell Biol.* **2010**, *189*, 223–232. [CrossRef] [PubMed]
83. Datta, A.; Kim, H.; Lal, M.; McGee, L.; Johnson, A.; Moustafa, A.A.; Jones, J.C.; Mondal, D.; Ferrer, M.; Abdel-Mageed, A.B. Manumycin A suppresses exosome biogenesis and secretion via targeted inhibition of Ras/Raf/ERK1/2 signaling and hnRNP H1 in castration-resistant prostate cancer cells. *Cancer Lett.* **2017**, *408*, 73–81. [CrossRef] [PubMed]
84. Duffy, M.J.; Sturgeon, C.; Lamerz, R.; Haglund, C.; Holubec, V.L.; Klapdor, R.; Nicolini, A.; Topolcan, O.; Heinemann, V. Tumor markers in pancreatic cancer: A European Group on Tumor Markers (EGTM) status report. *Ann. Oncol.* **2010**, *21*, 441–447. [CrossRef] [PubMed]
85. Parolini, I.; Federici, C.; Raggi, C.; Lugini, L.; Palleschi, S.; De Milito, A.; Coscia, C.; Iessi, E.; Logozzi, M.; Molinari, A.; et al. Microenvironmental pH is a key factor for exosome traffic in tumor cells. *J. Biol. Chem.* **2009**, *284*, 34211–34222. [CrossRef]
86. Jahn, R.; Lang, T.; Sudhof, T.C. Membrane fusion. *Cell* **2003**, *112*, 519–533. [CrossRef]

87. Gonda, A.; Kabagwira, J.; Senthil, G.N.; Wall, N.R. Internalization of Exosomes through Receptor-Mediated Endocytosis. *Mol Cancer Res.* **2019**, *17*, 337–347. [CrossRef]
88. Tian, T.; Zhu, Y.L.; Zhou, Y.Y.; Liang, G.F.; Wang, Y.Y.; Hu, F.H.; Xiao, Z.D. Exosome uptake through clathrin-mediated endocytosis and macropinocytosis and mediating miR-21 delivery. *J. Biol. Chem.* **2014**, *289*, 22258–22267. [CrossRef]
89. Nanbo, A.; Kawanishi, E.; Yoshida, R.; Yoshiyama, H. Exosomes derived from Epstein-Barr virus-infected cells are internalized via caveola-dependent endocytosis and promote phenotypic modulation in target cells. *J. Virol.* **2013**, *87*, 10334–10347. [CrossRef]
90. Feng, D.; Zhao, W.L.; Ye, Y.Y.; Bai, X.C.; Liu, R.Q.; Chang, L.F.; Zhou, Q.; Sui, S.F. Cellular internalization of exosomes occurs through phagocytosis. *Traffic* **2010**, *11*, 675–687. [CrossRef]
91. Fitzner, D.; Schnaars, M.; van Rossum, D.; Krishnamoorthy, G.; Dibaj, P.; Bakhti, M.; Regen, T.; Hanisch, U.K.; Simons, M. Selective transfer of exosomes from oligodendrocytes to microglia by macropinocytosis. *J. Cell Sci.* **2011**, *124*, 447–458. [CrossRef] [PubMed]
92. Mulcahy, L.A.; Pink, R.C.; Carter, D.R. Routes and mechanisms of extracellular vesicle uptake. *J. Extracell. Vesicles* **2014**, *3*. [CrossRef] [PubMed]
93. Svensson, K.J.; Christianson, H.C.; Wittrup, A.; Bourseau-Guilmain, E.; Lindqvist, E.; Svensson, L.M.; Morgelin, M.; Belting, M. Exosome uptake depends on ERK1/2-heat shock protein 27 signaling and lipid Raft-mediated endocytosis negatively regulated by caveolin-1. *J. Biol. Chem.* **2013**, *288*, 17713–17724. [CrossRef] [PubMed]
94. Escrevente, C.; Keller, S.; Altevogt, P.; Costa, J. Interaction and uptake of exosomes by ovarian cancer cells. *BMC Cancer* **2011**, *11*, 108. [CrossRef]
95. Banizs, A.B.; Huang, T.; Nakamoto, R.K.; Shi, W.; He, J. Endocytosis Pathways of Endothelial Cell Derived Exosomes. *Mol. Pharm.* **2018**, *15*, 5585–5590. [CrossRef]
96. Christianson, H.C.; Svensson, K.J.; van Kuppevelt, T.H.; Li, J.P.; Belting, M. Cancer cell exosomes depend on cell-surface heparan sulfate proteoglycans for their internalization and functional activity. *Proc. Natl. Acad. Sci. USA* **2013**, *110*, 17380–17385. [CrossRef]
97. Zheng, Y.; Tu, C.; Zhang, J.; Wang, J. Inhibition of multiple myelomaderived exosomes uptake suppresses the functional response in bone marrow stromal cell. *Int. J. Oncol.* **2019**, *54*, 1061–1070. [CrossRef]
98. Fong, M.Y.; Zhou, W.; Liu, L.; Alontaga, A.Y.; Chandra, M.; Ashby, J.; Chow, A.; O'Connor, S.T.; Li, S.; Chin, A.R.; et al. Breast-cancer-secreted miR-122 reprograms glucose metabolism in premetastatic niche to promote metastasis. *Nat. Cell Biol.* **2015**, *17*, 183–194. [CrossRef]
99. Lentz, M.R. Continuous whole blood UltraPheresis procedure in patients with metastatic cancer. *J. Biol. Response Mod.* **1989**, *8*, 511–527.
100. Tullis, R.H.; Duffin, R.P.; Handley, H.H.; Sodhi, P.; Menon, J.; Joyce, J.A.; Kher, V. Reduction of hepatitis C virus using lectin affinity plasmapheresis in dialysis patients. *Blood Purif.* **2009**, *27*, 64–69. [CrossRef]
101. Tullis, R.H.; Duffin, R.P.; Zech, M.; Ambrus, J.L. Affinity hemodialysis for antiviral therapy. II. Removal of HIV-1 viral proteins from cell culture supernatants and whole blood. *Blood Purif.* **2003**, *21*, 58–63. [CrossRef] [PubMed]
102. Nishida-Aoki, N.; Tominaga, N.; Takeshita, F.; Sonoda, H.; Yoshioka, Y.; Ochiya, T. Disruption of Circulating Extracellular Vesicles as a Novel Therapeutic Strategy against Cancer Metastasis. *Mol. Ther.* **2017**, *25*, 181–191. [CrossRef] [PubMed]
103. Xie, X.; Nie, H.; Zhou, Y.; Lian, S.; Mei, H.; Lu, Y.; Dong, H.; Li, F.; Li, T.; Li, B.; et al. Eliminating blood oncogenic exosomes into the small intestine with aptamer-functionalized nanoparticles. *Nat. Commun.* **2019**, *10*, 5476. [CrossRef] [PubMed]
104. Kimura, H.; Yamamoto, H.; Harada, T.; Fumoto, K.; Osugi, Y.; Sada, R.; Maehara, N.; Hikita, H.; Mori, S.; Eguchi, H.; et al. CKAP4, a DKK1 Receptor, Is a Biomarker in Exosomes Derived from Pancreatic Cancer and a Molecular Target for Therapy. *Clin. Cancer Res.* **2019**, *25*, 1936–1947. [CrossRef]
105. Binenbaum, Y.; Fridman, E.; Yaari, Z.; Milman, N.; Schroeder, A.; Ben David, G.; Shlomi, T.; Gil, Z. Transfer of miRNA in Macrophage-Derived Exosomes Induces Drug Resistance in Pancreatic Adenocarcinoma. *Cancer Res.* **2018**, *78*, 5287–5299. [CrossRef]
106. Mikamori, M.; Yamada, D.; Eguchi, H.; Hasegawa, S.; Kishimoto, T.; Tomimaru, Y.; Asaoka, T.; Noda, T.; Wada, H.; Kawamoto, K.; et al. MicroRNA-155 Controls Exosome Synthesis and Promotes Gemcitabine Resistance in Pancreatic Ductal Adenocarcinoma. *Sci. Rep.* **2017**, *7*, 42339. [CrossRef]

107. Li, X.J.; Ren, Z.J.; Tang, J.H.; Yu, Q. Exosomal MicroRNA MiR-1246 Promotes Cell Proliferation, Invasion and Drug Resistance by Targeting CCNG2 in Breast Cancer. *Cell Physiol. Biochem.* **2017**, *44*, 1741–1748. [CrossRef]
108. Chen, J.H.; Wu, A.T.H.; Bamodu, O.A.; Yadav, V.K.; Chao, T.Y.; Tzeng, Y.M.; Mukhopadhyay, D.; Hsiao, M.; Lee, J.C. Ovatodiolide Suppresses Oral Cancer Malignancy by Down-Regulating Exosomal Mir-21/STAT3/beta-Catenin Cargo and Preventing Oncogenic Transformation of Normal Gingival Fibroblasts. *Cancers* **2019**, *12*, 56. [CrossRef]
109. Hannafon, B.N.; Trigoso, Y.D.; Calloway, C.L.; Zhao, Y.D.; Lum, D.H.; Welm, A.L.; Zhao, Z.J.; Blick, K.E.; Dooley, W.C.; Ding, W.Q. Plasma exosome microRNAs are indicative of breast cancer. *Breast Cancer Res.* **2016**, *18*, 90. [CrossRef]
110. Xu, Y.F.; Hannafon, B.N.; Khatri, U.; Gin, A.; Ding, W.Q. The origin of exosomal miR-1246 in human cancer cells. *Rna Biol.* **2019**, *16*, 770–784. [CrossRef] [PubMed]
111. Xu, Y.F.; Hannafon, B.N.; Zhao, Y.D.; Postier, R.G.; Ding, W.Q. Plasma exosome miR-196a and miR-1246 are potential indicators of localized pancreatic cancer. *Oncotarget* **2017**, *8*, 77028–77040. [CrossRef] [PubMed]
112. Chen, D.S.; Mellman, I. Oncology meets immunology: The cancer-immunity cycle. *Immunity* **2013**, *39*, 1–10. [CrossRef] [PubMed]
113. Page, D.B.; Postow, M.A.; Callahan, M.K.; Allison, J.P.; Wolchok, J.D. Immune modulation in cancer with antibodies. *Annu. Rev. Med.* **2014**, *65*, 185–202. [CrossRef] [PubMed]
114. Zhang, C.; Fan, Y.; Che, X.; Zhang, M.; Li, Z.; Li, C.; Wang, S.; Wen, T.; Hou, K.; Shao, X.; et al. Anti-PD-1 Therapy Response Predicted by the Combination of Exosomal PD-L1 and CD28. *Front. Oncol.* **2020**, *10*, 760. [CrossRef]
115. Xie, F.; Xu, M.; Lu, J.; Mao, L.; Wang, S. The role of exosomal PD-L1 in tumor progression and immunotherapy. *Mol. Cancer* **2019**, *18*, 146. [CrossRef]
116. Dong, H.; Wang, W.; Chen, R.; Zhang, Y.; Zou, K.; Ye, M.; He, X.; Zhang, F.; Han, J. Exosome-mediated transfer of lncRNASNHG14 promotes trastuzumab chemoresistance in breast cancer. *Int. J. Oncol.* **2018**, *53*, 1013–1026. [CrossRef]
117. Milman, N.; Ginini, L.; Gil, Z. Exosomes and their role in tumorigenesis and anticancer drug resistance. *Drug Resist. Updat* **2019**, *45*, 1–12. [CrossRef]
118. Elsharkasy, O.M.; Nordin, J.Z.; Hagey, D.W.; de Jong, O.G.; Schiffelers, R.M.; Andaloussi, S.E.; Vader, P. Extracellular vesicles as drug delivery systems: Why and how? *Adv. Drug Deliv. Rev.* **2020**, in press. [CrossRef]
119. Bastos, N.; Ruivo, C.F.; da Silva, S.; Melo, S.A. Exosomes in cancer: Use them or target them? *Semin Cell Dev. Biol.* **2018**, *78*, 13–21. [CrossRef]
120. Ohno, S.; Drummen, G.P.; Kuroda, M. Focus on Extracellular Vesicles: Development of Extracellular Vesicle-Based Therapeutic Systems. *Int. J. Mol. Sci.* **2016**, *17*, 172. [CrossRef]
121. Smyth, T.J.; Redzic, J.S.; Graner, M.W.; Anchordoquy, T.J. Examination of the specificity of tumor cell derived exosomes with tumor cells in vitro. *Biochim. Biophys. Acta* **2014**, *1838*, 2954–2965. [CrossRef] [PubMed]
122. Cheng, Y.; Zeng, Q.; Han, Q.; Xia, W. Effect of pH, temperature and freezing-thawing on quantity changes and cellular uptake of exosomes. *Protein. Cell* **2019**, *10*, 295–299. [CrossRef]
123. Luan, X.; Sansanaphongpricha, K.; Myers, I.; Chen, H.; Yuan, H.; Sun, D. Engineering exosomes as refined biological nanoplatforms for drug delivery. *Acta Pharm. Sin.* **2017**, *38*, 754–763. [CrossRef] [PubMed]
124. Zhuang, X.; Xiang, X.; Grizzle, W.; Sun, D.; Zhang, S.; Axtell, R.C.; Ju, S.; Mu, J.; Zhang, L.; Steinman, L.; et al. Treatment of brain inflammatory diseases by delivering exosome encapsulated anti-inflammatory drugs from the nasal region to the brain. *Mol. Ther.* **2011**, *19*, 1769–1779. [CrossRef]
125. Wolfram, J.; Ferrari, M. Clinical Cancer Nanomedicine. *Nano Today* **2019**, *25*, 85–98. [CrossRef] [PubMed]
126. Li, X.; Tsibouklis, J.; Weng, T.; Zhang, B.; Yin, G.; Feng, G.; Cui, Y.; Savina, I.N.; Mikhalovska, L.I.; Sandeman, S.R.; et al. Nano carriers for drug transport across the blood-brain barrier. *J. Drug Target.* **2017**, *25*, 17–28. [CrossRef] [PubMed]
127. Kanchanapally, R.; Deshmukh, S.K.; Chavva, S.R.; Tyagi, N.; Srivastava, S.K.; Patel, G.K.; Singh, A.P.; Singh, S. Drug-loaded exosomal preparations from different cell types exhibit distinctive loading capability, yield, and antitumor efficacies: A comparative analysis. *Int. J. Nanomed.* **2019**, *14*, 531–541. [CrossRef] [PubMed]
128. Borrelli, D.A.; Yankson, K.; Shukla, N.; Vilanilam, G.; Ticer, T.; Wolfram, J. Extracellular vesicle therapeutics for liver disease. *J. Control. Release* **2018**, *273*, 86–98. [CrossRef]

129. Mendt, M.; Kamerkar, S.; Sugimoto, H.; McAndrews, K.M.; Wu, C.C.; Gagea, M.; Yang, S.; Blanko, E.V.R.; Peng, Q.; Ma, X.; et al. Generation and testing of clinical-grade exosomes for pancreatic cancer. *JCI Insight* **2018**, *3*. [CrossRef]
130. Kordelas, L.; Rebmann, V.; Ludwig, A.K.; Radtke, S.; Ruesing, J.; Doeppner, T.R.; Epple, M.; Horn, P.A.; Beelen, D.W.; Giebel, B. MSC-derived exosomes: A novel tool to treat therapy-refractory graft-versus-host disease. *Leukemia* **2014**, *28*, 970–973. [CrossRef]
131. Munoz, J.L.; Bliss, S.A.; Greco, S.J.; Ramkissoon, S.H.; Ligon, K.L.; Rameshwar, P. Delivery of Functional Anti-miR-9 by Mesenchymal Stem Cell-derived Exosomes to Glioblastoma Multiforme Cells Conferred Chemosensitivity. *Mol. Nucleic Acids* **2013**, *2*, e126. [CrossRef] [PubMed]
132. Pascucci, L.; Cocce, V.; Bonomi, A.; Ami, D.; Ceccarelli, P.; Ciusani, E.; Vigano, L.; Locatelli, A.; Sisto, F.; Doglia, S.M.; et al. Paclitaxel is incorporated by mesenchymal stromal cells and released in exosomes that inhibit in vitro tumor growth: A new approach for drug delivery. *J. Control. Release* **2014**, *192*, 262–270. [CrossRef] [PubMed]
133. Lohan, P.; Treacy, O.; Griffin, M.D.; Ritter, T.; Ryan, A.E. Anti-Donor Immune Responses Elicited by Allogeneic Mesenchymal Stem Cells and Their Extracellular Vesicles: Are We Still Learning? *Front. Immunol.* **2017**, *8*, 1626. [CrossRef] [PubMed]
134. Kilpinen, L.; Impola, U.; Sankkila, L.; Ritamo, I.; Aatonen, M.; Kilpinen, S.; Tuimala, J.; Valmu, L.; Levijoki, J.; Finckenberg, P.; et al. Extracellular membrane vesicles from umbilical cord blood-derived MSC protect against ischemic acute kidney injury, a feature that is lost after inflammatory conditioning. *J. Extracell. Vesicles* **2013**, *2*. [CrossRef] [PubMed]
135. Alvarez-Erviti, L.; Seow, Y.; Yin, H.; Betts, C.; Lakhal, S.; Wood, M.J. Delivery of siRNA to the mouse brain by systemic injection of targeted exosomes. *Nat. Biotechnol.* **2011**, *29*, 341–345. [CrossRef]
136. Tian, Y.; Li, S.; Song, J.; Ji, T.; Zhu, M.; Anderson, G.J.; Wei, J.; Nie, G. A doxorubicin delivery platform using engineered natural membrane vesicle exosomes for targeted tumor therapy. *Biomaterials* **2014**, *35*, 2383–2390. [CrossRef]
137. Kharaziha, P.; Ceder, S.; Li, Q.; Panaretakis, T. Tumor cell-derived exosomes: A message in a bottle. *Biochim. Biophys. Acta* **2012**, *1826*, 103–111. [CrossRef]
138. Shtam, T.A.; Kovalev, R.A.; Varfolomeeva, E.Y.; Makarov, E.M.; Kil, Y.V.; Filatov, M.V. Exosomes are natural carriers of exogenous siRNA to human cells in vitro. *Cell Commun. Signal.* **2013**, *11*, 88. [CrossRef]
139. Becker, A.; Thakur, B.K.; Weiss, J.M.; Kim, H.S.; Peinado, H.; Lyden, D. Extracellular Vesicles in Cancer: Cell-to-Cell Mediators of Metastasis. *Cancer Cell* **2016**, *30*, 836–848. [CrossRef]
140. Lou, G.; Song, X.; Yang, F.; Wu, S.; Wang, J.; Chen, Z.; Liu, Y. Exosomes derived from miR-122-modified adipose tissue-derived MSCs increase chemosensitivity of hepatocellular carcinoma. *J. Hematol. Oncol.* **2015**, *8*, 122. [CrossRef]
141. Abels, E.R.; Maas, S.L.N.; Nieland, L.; Wei, Z.; Cheah, P.S.; Tai, E.; Kolsteeg, C.J.; Dusoswa, S.A.; Ting, D.T.; Hickman, S.; et al. Glioblastoma-Associated Microglia Reprogramming Is Mediated by Functional Transfer of Extracellular miR-21. *Cell Rep.* **2019**, *28*, 3105–3119 e3107. [CrossRef] [PubMed]
142. Fareh, M.; Almairac, F.; Turchi, L.; Burel-Vandenbos, F.; Paquis, P.; Fontaine, D.; Lacas-Gervais, S.; Junier, M.P.; Chneiweiss, H.; Virolle, T. Cell-based therapy using miR-302-367 expressing cells represses glioblastoma growth. *Cell Death Dis.* **2017**, *8*, e2713. [CrossRef] [PubMed]
143. Katakowski, M.; Buller, B.; Zheng, X.; Lu, Y.; Rogers, T.; Osobamiro, O.; Shu, W.; Jiang, F.; Chopp, M. Exosomes from marrow stromal cells expressing miR-146b inhibit glioma growth. *Cancer Lett.* **2013**, *335*, 201–204. [CrossRef] [PubMed]
144. O'Brien, K.; Breyne, K.; Ughetto, S.; Laurent, L.C.; Breakefield, X.O. RNA delivery by extracellular vesicles in mammalian cells and its applications. *Nat. Rev. Mol. Cell Biol.* **2020**. [CrossRef]
145. Fuhrmann, G.; Serio, A.; Mazo, M.; Nair, R.; Stevens, M.M. Active loading into extracellular vesicles significantly improves the cellular uptake and photodynamic effect of porphyrins. *J. Control. Release* **2015**, *205*, 35–44. [CrossRef]
146. O'Loughlin, A.J.; Mager, I.; de Jong, O.G.; Varela, M.A.; Schiffelers, R.M.; El Andaloussi, S.; Wood, M.J.A.; Vader, P. Functional Delivery of Lipid-Conjugated siRNA by Extracellular Vesicles. *Mol. Ther.* **2017**, *25*, 1580–1587. [CrossRef]

147. Srivastava, A.; Amreddy, N.; Pareek, V.; Chinnappan, M.; Ahmed, R.; Mehta, M.; Razaq, M.; Munshi, A.; Ramesh, R. Progress in extracellular vesicle biology and their application in cancer medicine. *Wiley Interdiscip Rev. Nanomed. Nanobiotechnol.* **2020**, *12*, e1621. [CrossRef]
148. Potter, H. Transfection by electroporation. *Curr. Protoc. Mol. Biol.* **2003**. [CrossRef]
149. Bak, R.O.; Dever, D.P.; Porteus, M.H. CRISPR/Cas9 genome editing in human hematopoietic stem cells. *Nat. Protoc.* **2018**, *13*, 358–376. [CrossRef]
150. Hood, J.L.; Scott, M.J.; Wickline, S.A. Maximizing exosome colloidal stability following electroporation. *Anal. Biochem.* **2014**, *448*, 41–49. [CrossRef]
151. Liu, C.; Su, C. Design strategies and application progress of therapeutic exosomes. *Theranostics* **2019**, *9*, 1015–1028. [CrossRef] [PubMed]
152. Asadirad, A.; Hashemi, S.M.; Baghaei, K.; Ghanbarian, H.; Mortaz, E.; Zali, M.R.; Amani, D. Phenotypical and functional evaluation of dendritic cells after exosomal delivery of miRNA-155. *Life Sci.* **2019**, *219*, 152–162. [CrossRef] [PubMed]
153. Li, Y.J.; Wu, J.Y.; Wang, J.M.; Hu, X.B.; Cai, J.X.; Xiang, D.X. Gemcitabine loaded autologous exosomes for effective and safe chemotherapy of pancreatic cancer. *Acta Biomater.* **2020**, *101*, 519–530. [CrossRef] [PubMed]
154. Sato, Y.T.; Umezaki, K.; Sawada, S.; Mukai, S.A.; Sasaki, Y.; Harada, N.; Shiku, H.; Akiyoshi, K. Engineering hybrid exosomes by membrane fusion with liposomes. *Sci. Rep.* **2016**, *6*, 21933. [CrossRef] [PubMed]
155. Haney, M.J.; Klyachko, N.L.; Zhao, Y.; Gupta, R.; Plotnikova, E.G.; He, Z.; Patel, T.; Piroyan, A.; Sokolsky, M.; Kabanov, A.V.; et al. Exosomes as drug delivery vehicles for Parkinson's disease therapy. *J. Control. Release* **2015**, *207*, 18–30. [CrossRef]
156. Goh, W.J.; Lee, C.K.; Zou, S.; Woon, E.C.; Czarny, B.; Pastorin, G. Doxorubicin-loaded cell-derived nanovesicles: An alternative targeted approach for anti-tumor therapy. *Int. J. Nanomed.* **2017**, *12*, 2759–2767. [CrossRef]
157. Kalimuthu, S.; Gangadaran, P.; Rajendran, R.L.; Zhu, L.; Oh, J.M.; Lee, H.W.; Gopal, A.; Baek, S.H.; Jeong, S.Y.; Lee, S.W.; et al. A New Approach for Loading Anticancer Drugs Into Mesenchymal Stem Cell-Derived Exosome Mimetics for Cancer Therapy. *Front. Pharm.* **2018**, *9*, 1116. [CrossRef]
158. Bunggulawa, E.J.; Wang, W.; Yin, T.; Wang, N.; Durkan, C.; Wang, Y.; Wang, G. Recent advancements in the use of exosomes as drug delivery systems. *J. Nanobiotechnol.* **2018**, *16*, 81. [CrossRef] [PubMed]
159. Rahbarghazi, R.; Jabbari, N.; Sani, N.A.; Asghari, R.; Salimi, L.; Kalashani, S.A.; Feghhi, M.; Etemadi, T.; Akbariazar, E.; Mahmoudi, M.; et al. Tumor-derived extracellular vesicles: Reliable tools for Cancer diagnosis and clinical applications. *Cell Commun. Signal.* **2019**, *17*, 73. [CrossRef] [PubMed]
160. Zhang, Y.F.; Shi, J.B.; Li, C. Small extracellular vesicle loading systems in cancer therapy: Current status and the way forward. *Cytotherapy* **2019**, *21*, 1122–1136. [CrossRef]
161. Walker, S.; Busatto, S.; Pham, A.; Tian, M.; Suh, A.; Carson, K.; Quintero, A.; Lafrence, M.; Malik, H.; Santana, M.X.; et al. Extracellular vesicle-based drug delivery systems for cancer treatment. *Theranostics* **2019**, *9*, 8001–8017. [CrossRef] [PubMed]
162. Chen, Y.S.; Lin, E.Y.; Chiou, T.W.; Harn, H.J. Exosomes in clinical trial and their production in compliance with good manufacturing practice. *Tzu-Chi Med. J.* **2019**, *32*, 113–120. [CrossRef]
163. Guo, M.; Wu, F.; Hu, G.; Chen, L.; Xu, J.; Xu, P.; Wang, X.; Li, Y.; Liu, S.; Zhang, S.; et al. Autologous tumor cell-derived microparticle-based targeted chemotherapy in lung cancer patients with malignant pleural effusion. *Sci. Transl. Med.* **2019**, *11*. [CrossRef]
164. Tang, K.; Zhang, Y.; Zhang, H.; Xu, P.; Liu, J.; Ma, J.; Lv, M.; Li, D.; Katirai, F.; Shen, G.X.; et al. Delivery of chemotherapeutic drugs in tumour cell-derived microparticles. *Nat. Commun.* **2012**, *3*, 1282. [CrossRef]
165. Ma, J.; Zhang, Y.; Tang, K.; Zhang, H.; Yin, X.; Li, Y.; Xu, P.; Sun, Y.; Ma, R.; Ji, T.; et al. Reversing drug resistance of soft tumor-repopulating cells by tumor cell-derived chemotherapeutic microparticles. *Cell Res.* **2016**, *26*, 713–727. [CrossRef]
166. Kamerkar, S.; LeBleu, V.S.; Sugimoto, H.; Yang, S.; Ruivo, C.F.; Melo, S.A.; Lee, J.J.; Kalluri, R. Exosomes facilitate therapeutic targeting of oncogenic KRAS in pancreatic cancer. *Nature* **2017**, *546*, 498–503. [CrossRef]
167. Dai, S.; Wei, D.; Wu, Z.; Zhou, X.; Wei, X.; Huang, H.; Li, G. Phase I clinical trial of autologous ascites-derived exosomes combined with GM-CSF for colorectal cancer. *Mol. Ther.* **2008**, *16*, 782–790. [CrossRef]
168. Besse, B.; Charrier, M.; Lapierre, V.; Dansin, E.; Lantz, O.; Planchard, D.; Le Chevalier, T.; Livartoski, A.; Barlesi, F.; Laplanche, A.; et al. Dendritic cell-derived exosomes as maintenance immunotherapy after first line chemotherapy in NSCLC. *Oncoimmunology* **2016**, *5*, e1071008. [CrossRef]

169. Morse, M.A.; Garst, J.; Osada, T.; Khan, S.; Hobeika, A.; Clay, T.M.; Valente, N.; Shreeniwas, R.; Sutton, M.A.; Delcayre, A.; et al. A phase I study of dexosome immunotherapy in patients with advanced non-small cell lung cancer. *J. Transl. Med.* **2005**, *3*, 9. [CrossRef]
170. Escudier, B.; Dorval, T.; Chaput, N.; Andre, F.; Caby, M.P.; Novault, S.; Flament, C.; Leboulaire, C.; Borg, C.; Amigorena, S.; et al. Vaccination of metastatic melanoma patients with autologous dendritic cell (DC) derived-exosomes: Results of thefirst phase I clinical trial. *J. Transl. Med.* **2005**, *3*, 10. [CrossRef]
171. Somiya, M.; Yoshioka, Y.; Ochiya, T. Drug delivery application of extracellular vesicles; insight into production, drug loading, targeting, and pharmacokinetics. *Aims Bioeng.* **2017**, *4*, 73–92. [CrossRef]
172. Wang, Q.; Zhuang, X.; Mu, J.; Deng, Z.B.; Jiang, H.; Zhang, L.; Xiang, X.; Wang, B.; Yan, J.; Miller, D.; et al. Delivery of therapeutic agents by nanoparticles made of grapefruit-derived lipids. *Nat. Commun.* **2013**, *4*, 1867. [CrossRef] [PubMed]
173. Burnouf, T.; Agrahari, V.; Agrahari, V. Extracellular Vesicles As Nanomedicine: Hopes And Hurdles In Clinical Translation. *Int. J. Nanomed.* **2019**, *14*, 8847–8859. [CrossRef] [PubMed]
174. Agrahari, V.; Agrahari, V.; Burnouf, P.A.; Chew, C.H.; Burnouf, T. Extracellular Microvesicles as New Industrial Therapeutic Frontiers. *Trends Biotechnol.* **2019**, *37*, 707–729. [CrossRef] [PubMed]

7

Phenotypic and Functional Characteristics of Exosomes Derived from Irradiated Mouse Organs and their Role in the Mechanisms Driving Non-Targeted Effects

Seda Tuncay Cagatay [1], Ammar Mayah [1], Mariateresa Mancuso [2], Paola Giardullo [2,3], Simonetta Pazzaglia [2], Anna Saran [2], Amuthachelvi Daniel [4], Damien Traynor [4], Aidan D. Meade [4], Fiona Lyng [4], Soile Tapio [5] and Munira Kadhim [1,*]

1 Department of Biological and Medical Sciences, Oxford Brookes University, Oxford OX3 0BP, UK; stuncay-cagatay@brookes.ac.uk (S.T.C.); amayah@brookes.ac.uk (A.M.)
2 Laboratory of Biomedical Technologies, Italian National Agency for New Technologies, Energy and Sustainable Economic Development (ENEA), 00123 Rome, Italy; mariateresa.mancuso@enea.it (M.M.); paola.giardullo@enea.it (P.G.); simonetta.pazzaglia@enea.it (S.P.); annasaran60@gmail.com (A.S.)
3 Department of Radiation Physics, Guglielmo Marconi University, 00193 Rome, Italy
4 Centre for Radiation and Environmental Science, FOCAS Research Institute, Technological University Dublin, D08 NF82 Dublin, Ireland; amuthachelvi.daniel@dit.ie (A.D.); damien.traynor@tudublin.ie (D.T.); aidan.meade@tudublin.ie (A.D.M.); fiona.lyng@tudublin.ie (F.L.)
5 Institute of Radiation Biology, Helmholtz Zentrum München, 85764 Munich, Germany; soile.tapio@helmholtz-muenchen.de
* Correspondence: mkadhim@brookes.ac.uk

Abstract: Molecular communication between irradiated and unirradiated neighbouring cells initiates radiation-induced bystander effects (RIBE) and out-of-field (abscopal) effects which are both an example of the non-targeted effects (NTE) of ionising radiation (IR). Exosomes are small membrane vesicles of endosomal origin and newly identified mediators of NTE. Although exosome-mediated changes are well documented in radiation therapy and oncology, there is a lack of knowledge regarding the role of exosomes derived from inside and outside the radiation field in the early and delayed induction of NTE following IR. Therefore, here we investigated the changes in exosome profile and the role of exosomes as possible molecular signalling mediators of radiation damage. Exosomes derived from organs of whole body irradiated (WBI) or partial body irradiated (PBI) mice after 24 h and 15 days post-irradiation were transferred to recipient mouse embryonic fibroblast (MEF) cells and changes in cellular viability, DNA damage and calcium, reactive oxygen species and nitric oxide signalling were evaluated compared to that of MEF cells treated with exosomes derived from unirradiated mice. Taken together, our results show that whole and partial-body irradiation increases the number of exosomes, instigating changes in exosome-treated MEF cells, depending on the source organ and time after exposure.

Keywords: exosomes; ionising radiation; non-targeted effects; signalling

---

1. Introduction

Radiation affects not only targeted cells but also non-irradiated neighbouring cells, a response described as radiation-induced bystander effects (RIBE). Molecular communication between irradiated and unirradiated neighbouring cells initiates RIBE and out-of-field (abscopal) effects [1]. Abscopal effects are not always categorised as completely separate from RIBE, and both are examples of non-targeted effects of ionising radiation (NTE) [2]. NTE, particularly RIBE, represents a plethora of other biological effects such as DNA damage, epigenetic changes, changes in proliferation, and apoptosis observed in non-targeted cells and tissues that have received molecular signals produced by irradiated cells via intercellular communication through cell gap junctions or through soluble secreted factors [3–8]. A relatively new mechanism that has been identified for mediating RIBE by soluble secreted factors is intercellular communication via exosomes.

Exosomes, which are one type of cell-derived vesicle, exist in different biological conditions and serve as an important additional pathway for signal exchange between cells. They are small (30–120 nm of diameter) membrane vesicles of endosomal origin that are secreted by normal or pathological cells into the microenvironment [9–12]. Exosomes carry a variety of bioactive molecules including proteins, mRNA, microRNA, DNA, and lipids [13–17]. Exosomes can serve as mediators of cell–cell communication. Upon internalisation, they can release their bioactive cargo molecules, which can change molecular profile, signalling pathways, and gene regulation in the recipient cells [18]. Although functions of exosomes have been extensively studied in many fields, including neurodegenerative diseases [19–21] and cancer [22–27], their roles in radiobiology have not been recognised until recently. Exosomes secreted by irradiated cells are likely to engage in various aspects of the systemic response to ionising radiation (IR), including RIBE, as well as abscopal effects. Recent studies have shown that exosomes derived from irradiated cells can cause ionising radiation-induced effects in unirradiated recipient cells [28–31]. Moreover, studies have also shown that exosomes play a role in RIBE in cancer cells and resistance to radiotherapy [32–36]. Mechanistically, IR causes changes in exosomal secretion patterns and content of exosomal cargo of the target cells, which can cause RIBE in the bystander cells as shown in several in vitro studies. [30,32,37–39].

In vivo partial body irradiation (PBI) constitutes a major problem in radiation protection, with contradictory evidence suggesting that PBI may contribute to and/or protect against detrimental health effects. PBI exposures are the norm in diagnostic radiology, radiation therapy and occupational exposures and may have significant implications for systemic consequences and human health effects at low and intermediate doses of ionising radiation [40–43]. However, to date, only limited mechanistic studies are available in understanding the consequences of PBI effects.

Diagnostic X-rays are the primary human-made source of radiation exposure to the general population, accounting for 14% of the total annual exposure worldwide from all sources. Estimates of risk of cancer from these exposures are ranged from 0.6% to 3.0% based on the annual number of diagnostic X-rays undertaken in developed countries [44]. For a better understanding of the underlying processes, epidemiological evidence mounting this risk should be augmented with radiobiological justifications. Moreover, radiation is the mainstay of cancer therapy, as radiotherapy is used in more than 50% of localised patients and is an indispensable component of comprehensive cancer treatment and care [45,46]. Therefore, it is vital to elucidate key mechanisms that could increase the risk of secondary malignancies, which could be developed as a result of non-targeted effects.

The risks posed by partial body as well as whole body exposures after both low, and therapeutic doses for cancer and non-cancer endpoints can be evaluated for radiation protection purposes if there is a plausible and consistent mechanism for detriment, a dose–response relationship that allows risk assessment, and biomarkers of response available for molecular epidemiological analysis. Exosomes can be ideal candidates as both prognostic and predictive biomarkers to monitor the radiation response and risk assessment given that radiation affects not only the production of exosomes but also their composition.

Growing evidence has suggested that radiation therapy can result in an increase in the release of exosomes and increase in oncogenic materials within the exosomes. Studies with glioblastoma multiforme cell lines have shown that radiation can elevate exosome release with a molecular profile containing an abundance of molecules essential for cell motility such as connective tissue growth factor (CTGF) mRNA and insulin-like growth factor binding protein 2 (IGFBP2) protein [32]. Evidence also showed that head and neck squamous cell carcinoma (HNSCC) FaDu cells release exosomes with a different proteome profile compared to the unirradiated control cells [47]. Other studies carried out with irradiated HNSCC cells demonstrated that exosomes carry pro-survival signals following ionising radiation [34]. Exosome transfer from stromal to breast cancer cells can regulate therapy resistance pathways including the STAT1 pathway. It has been recently reported that in glioblastoma, exosomes increase the cancer cell's ability to survive radiation by increasing oncogenic cargo and decreasing tumour-suppressive cargo [48].

Despite the extensive range of studies that have been carried out in exosome-mediated functions in radiation therapy and oncology, there is a lack of knowledge regarding the role of exosomes derived from inside and outside of the radiation field in the initial and delayed induction of non-targeted effects of ionising radiation. We have therefore investigated the role of exosomes as possible molecular signalling mediators of radiation induced out-of-target effects/RIBE. Exosomal characteristics were investigated by qNano analysis, TEM, Western blot and Raman spectroscopy. In vitro functional effects were investigated by transfering exosomes derived from whole body irradiated (WBI) or partial body irradiated (PBI) mice after 24 h and 15 days post-irradiation to recipient mouse embryonic fibroblast (MEF) cells and evaluating their effect on cellular viability, DNA damage and calcium, reactive oxygen species and nitric oxide signalling compared to that of exosomes derived from unirradiated mice. Taken together, our results show that whole and partial-body irradiation increases the number of exosomes, instigating changes in exosome-treated MEF cells, depending on the source organ and time after exposure.

2. Results

2.1. Characterisation of Exosomes

Exosomes were extracted from plasma at 24 h and from organ samples (brain, liver, and heart) at 24 h and 15 days post different IR exposure conditions by the ultracentrifugation method. Exosome concentration and their size distribution, presence of exosome markers and the biochemical profile of exosomes were evaluated by qNano analysis, TEM, Western blot and Raman spectroscopy. Significant differences were particularly observed in exosome number between samples, as fully described below.

2.1.1. Characterisation of Exosomes by qNano

Exosome size and concentration of 2 Gy WBI and PBI mice organs (brain, liver, and heart) and plasma samples were assessed by using a tunable resistance pulse sensing (TRPS) system via qNano after 24 h and 15 days after irradiation (Figure 1).

Although exosome concentration levels varied for all organs and plasma, they were increased in the brain, liver, heart, and plasma of both 2 Gy X-ray irradiated WBI and PBI mice compared to the 0 Gy control groups after 24 h post-IR. Exosome concentrations in PBI mice organs showed a more dramatic increase compared to the WBI organs, while plasma obtained from WBI irradiated mice showed a higher level of exosomes compared to the plasma obtained from PBI mice.

For 15 days post-IR samples, exosome concentrations increased in the brain, liver, and heart of both 2 Gy WBI and PBI mice compared to the 0 Gy control groups. Taken together, the results show that ionising radiation can increase the yield of exosomes derived from 2 Gy mice organs after 24 h (Figure 1a–d) and 15 days (Figure 1e–g) following irradiation.

qNano analysis also showed that the exosome suspensions had a relatively narrow size distribution (70–130 nm) for all samples. Differences in the mean diameter of exosomes were significant for 2 Gy WBI and PBI brain samples compared to the unirradiated controls, whereas no significant difference in exosome size was observed between liver, heart, and plasma samples and their controls [30], indicating that ionising radiation can also alter the size distribution of exosomes in an organ-specific manner.

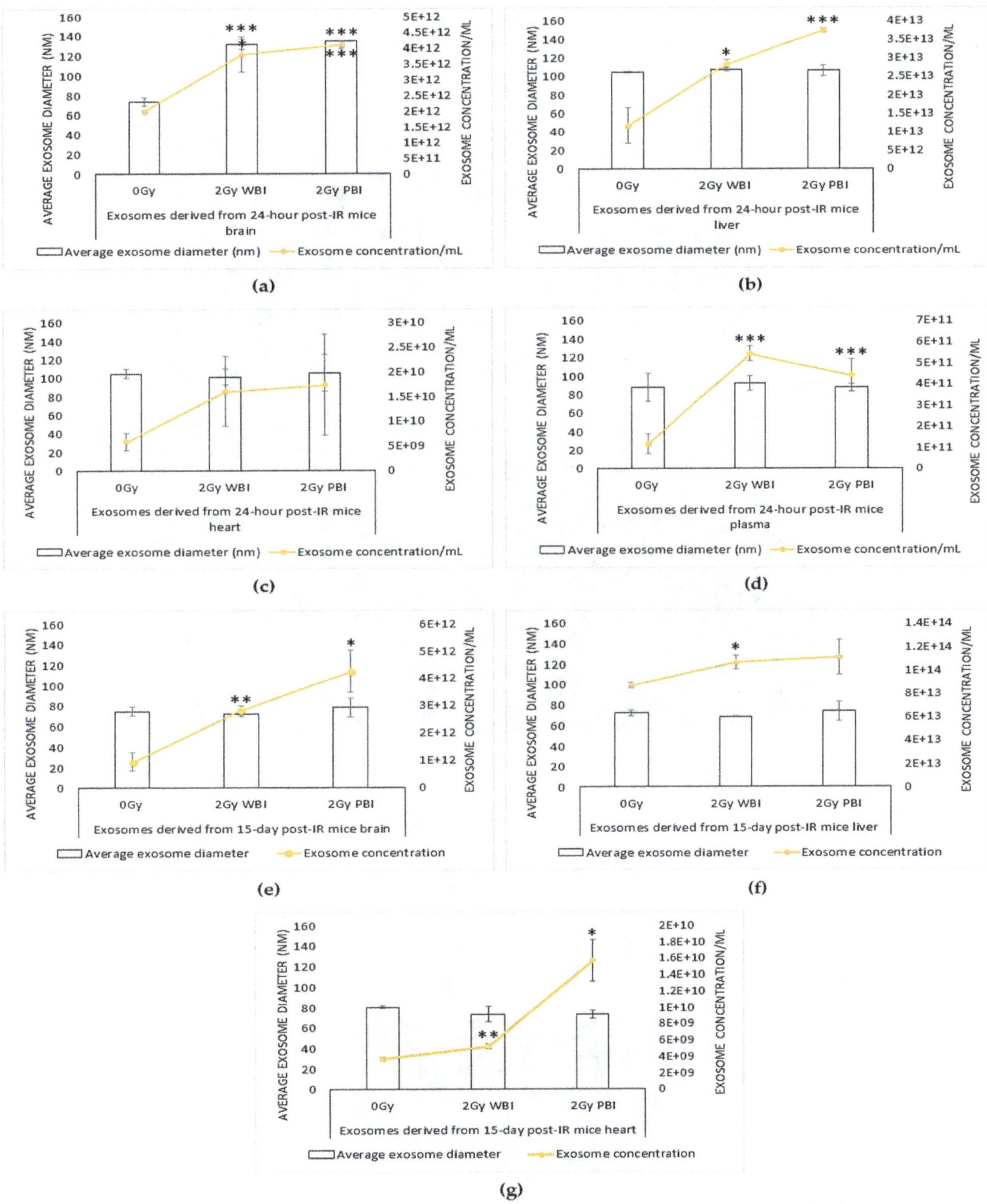

Figure 1. Concentration (exosome/mL) and size (nm) distribution of exosome suspensions obtained from (**a–d**) 24 h post ionising radiation (IR) and (**e–g**) 15 days post-IR 2 Gy whole body irradiated (WBI) and 2 Gy partial body irradiated (PBI) mouse compared to unirradiated mouse organs (brain, liver, heart) and plasma. Bars represent mean ± SD; significance was tested by Student's *t*-test (* $p < 0.05$, ** $p < 0.01$, *** $p < 0.001$).

2.1.2. Characterisation of Exosomes by Transmission Electron Microscopy (TEM)

Exosome suspensions were investigated by transmission electron microscopy for size. Negative staining of exosome suspensions showed vesicular structures in the anticipated size range and classical exosome morphology, as shown in Figure 2a.

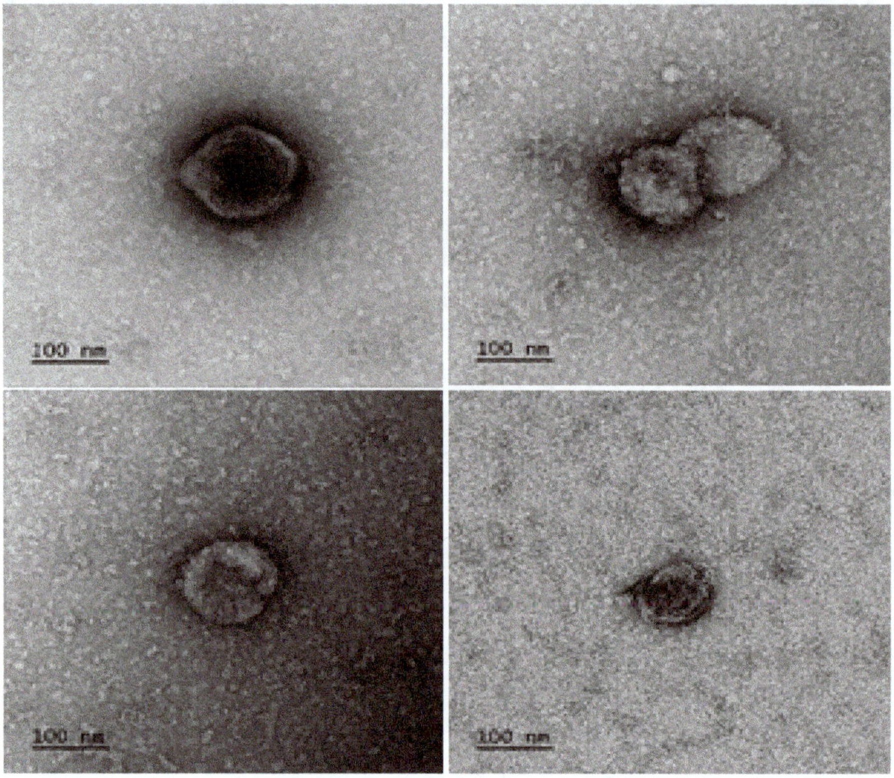

(a)

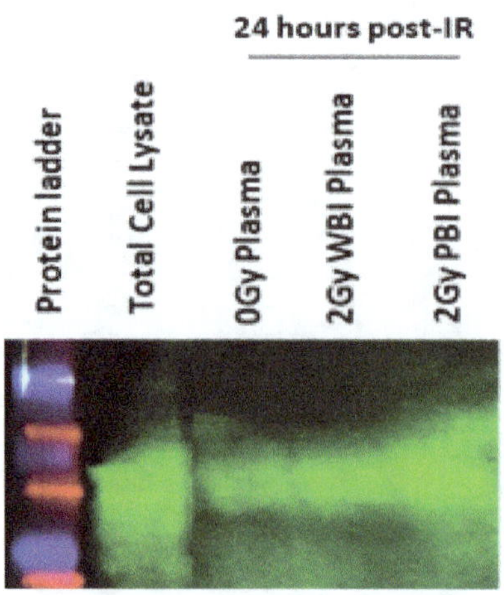

(b)

Figure 2. Confirmation of presence of exosomes. (**a**) TEM micrographs of exosomes. Representative images 1:10 or 1:100 diluted plasma exosome samples. (**b**) Western blot analysis of exosomes for CD63. Lane 1 protein ladder, lane 2: total cell lysate, lane 3: unirradiated (0 Gy) plasma sample, lane 4:2 Gy WBI plasma protein and lane 5:2 Gy PBI plasma protein sample.

2.1.3. Characterisation of Exosomes by Western Blot

The presence of exosomes in extracted samples was confirmed by Western blot analysis against exosome marker CD63. A representative Western blot analysis for plasma exosome samples is shown in Figure 2b. CD63 bands of samples were observed in the predicted molecular weight (26 kDa), which further confirms the presence of exosomes together with findings from qNano analysis and electron microscopy.

2.1.4. Characterisation of Exosomes by Raman Spectroscopy

Exosomes were also analysed by Raman spectroscopy, a label-free method based on light scattering which provides a biochemical profile of the sample [49,50]. Partial least squares discriminant analysis (PLSDA) showed good separation of the Raman spectral data from exosomes from brain, heart, and liver into control, PBI and WBI groups at 24 h and 15 days post-irradiation (Figure 3a,b). The spectral differences were based on changes in nucleic acid and protein features and these will be analysed in more detail in future work.

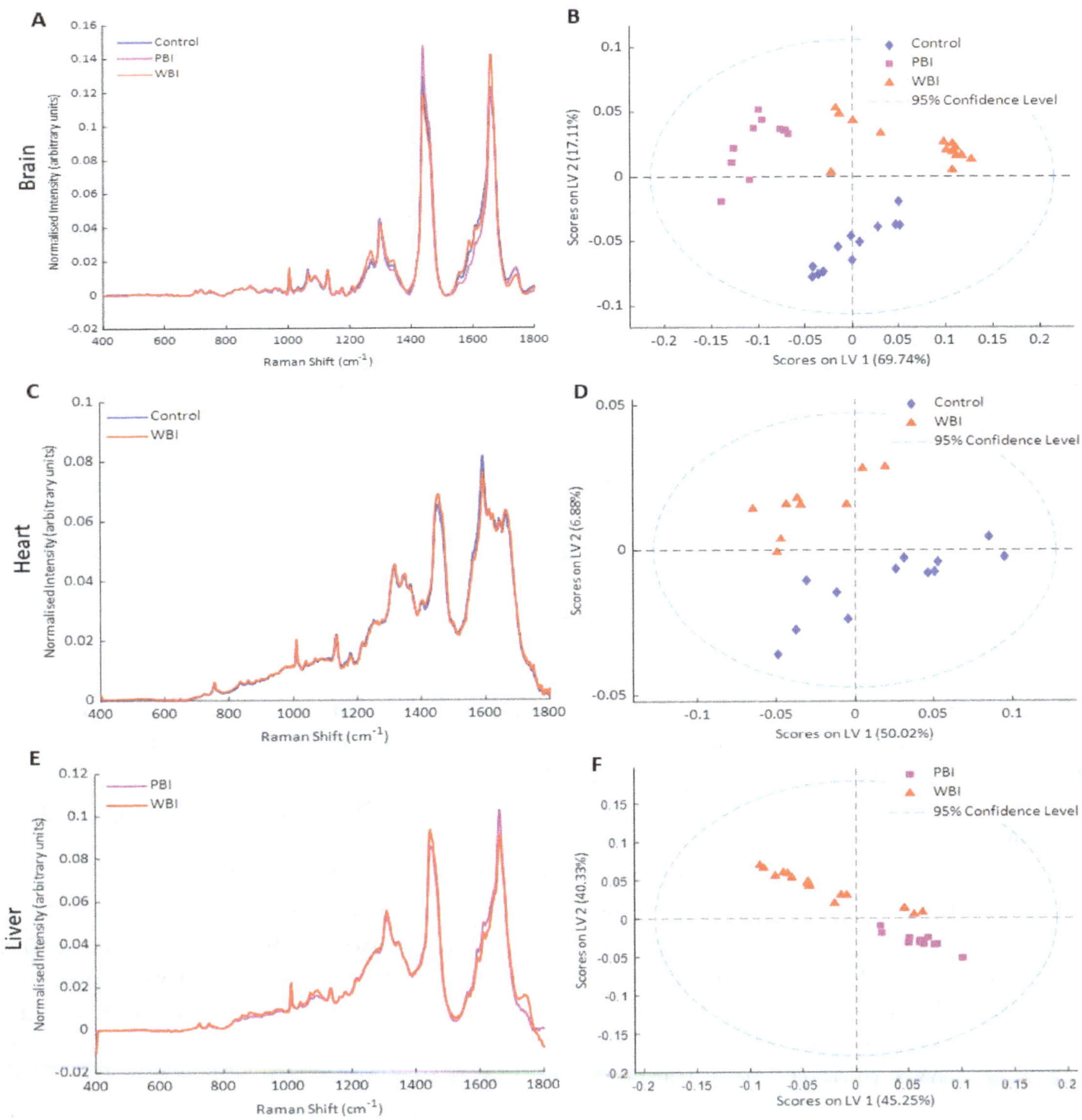

(a)

Figure 3. *Cont.*

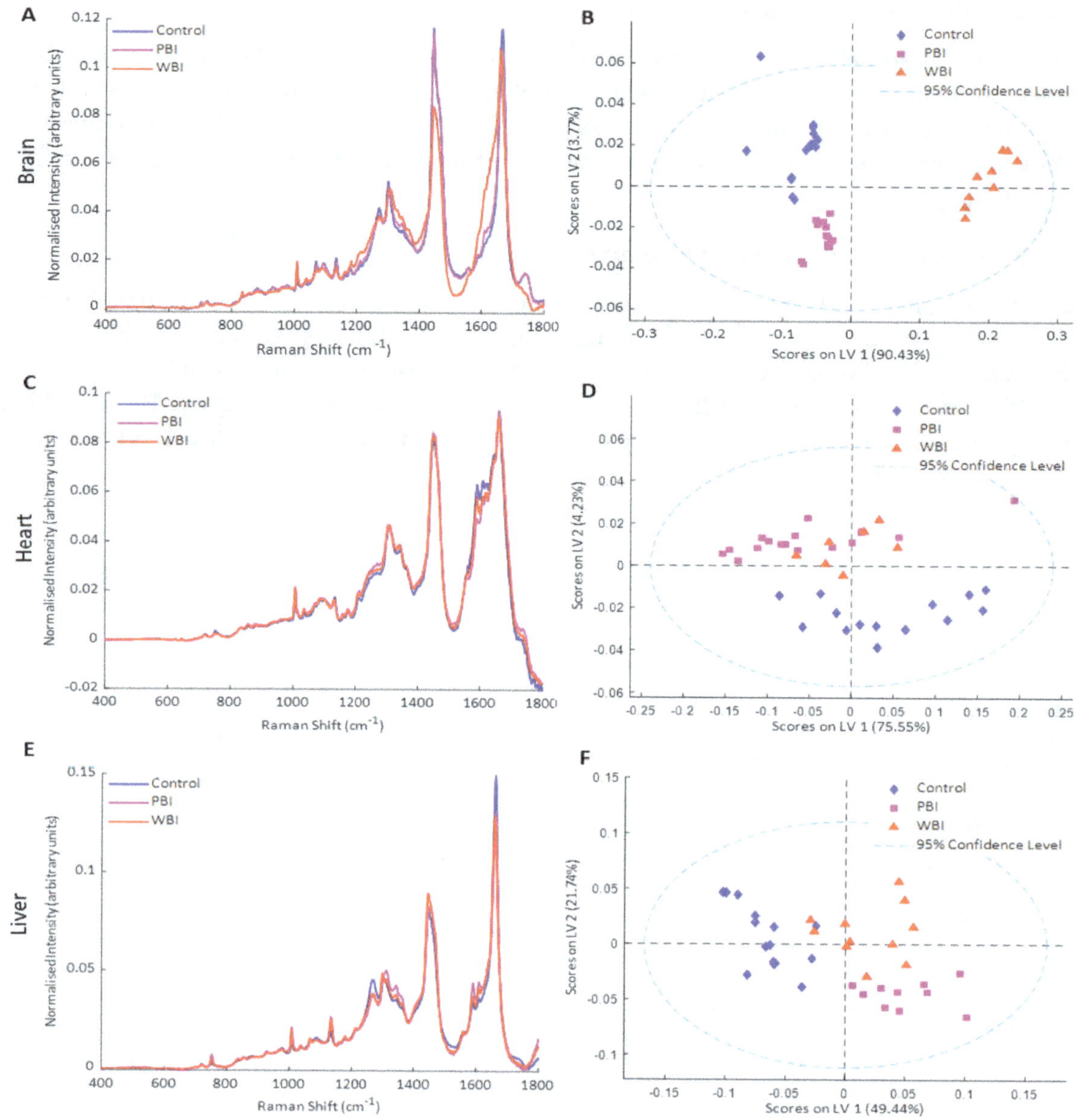

(**b**)

Figure 3. Raman spectroscopy of exosomes (**a**) 24 h post-IR: (**A**) mean Raman spectra from exosomes from brain of control, PBI and WBI mice at 24 h post irradiation, (**B**) PLSDA scatterplot of Raman spectral data from exosomes from control (blue), PBI (pink) and WBI (red) brain, (**C**) mean Raman spectra from exosomes from heart of control and WBI mice at 24 h post irradiation, (**D**) PLSDA scatterplot of Raman spectral data from exosomes from control (blue) and WBI (red) heart, (**E**) mean Raman spectra from exosomes from liver of PBI and WBI mice at 24 h post irradiation, (**F**) PLSDA scatterplot of Raman spectral data from exosomes from PBI (pink) and WBI (red) liver. (**b**) Fifteen days post-IR: (**A**) mean Raman spectra from exosomes from brain of control, PBI and WBI mice at 15 days post irradiation, (**B**) PLSDA scatterplot of Raman spectral data from exosomes from control (blue), PBI (pink) and WBI (red) brain, (**C**) mean Raman spectra from exosomes from heart of control and WBI mice at 15 days post irradiation, (**D**) PLSDA scatterplot of Raman spectral data from exosomes from control (blue) and WBI (red), (**E**) mean Raman spectra from exosomes from liver of control, PBI and WBI mice at 15 days post irradiation, (**F**) PLSDA scatterplot of Raman spectral data from exosomes from control (blue), PBI (pink) and WBI (red) liver.

2.2. *Effects of Exosomes from WBI and PBI Mice on Bystander MEF Cells*

2.2.1. Effects of Exosomes on Cell Viability

The effects of exosomes on cell viability were evaluated by using the MUSE Cell analyser. The results in Figure 4a show that 24 h post-IR, 2 Gy WBI brain, 2 Gy WBI liver, 2 Gy PBI liver and 2 Gy PBI heart exosomes significantly reduced the viability/survival of MEF cells. However, cells that received plasma exosomes did not show a significant change in the cell viability levels. High levels of cell death responses were observed in cells that received 15 days post-IR exosomes derived from organ samples, particularly the brain and liver, suggesting that exosomes can transfer long-lived signal-inducing radiation-induced genomic instability, as shown in Figure 4b. Moreover, a significant decrease in cell viability was also observed in cells treated with 15-day WBI and PBI exosomes compared to those that received 24-h WBI and PBI exosomes. Surprisingly, cells that received 0 Gy liver exosomes showed a high level of cell death compared to those that received WBI and PBI exosomes at the delayed time point (15-day exosomes), suggesting that delayed cell death can also be induced by exosomes derived from unirradiated liver cells. More investigations are needed to confirm this.

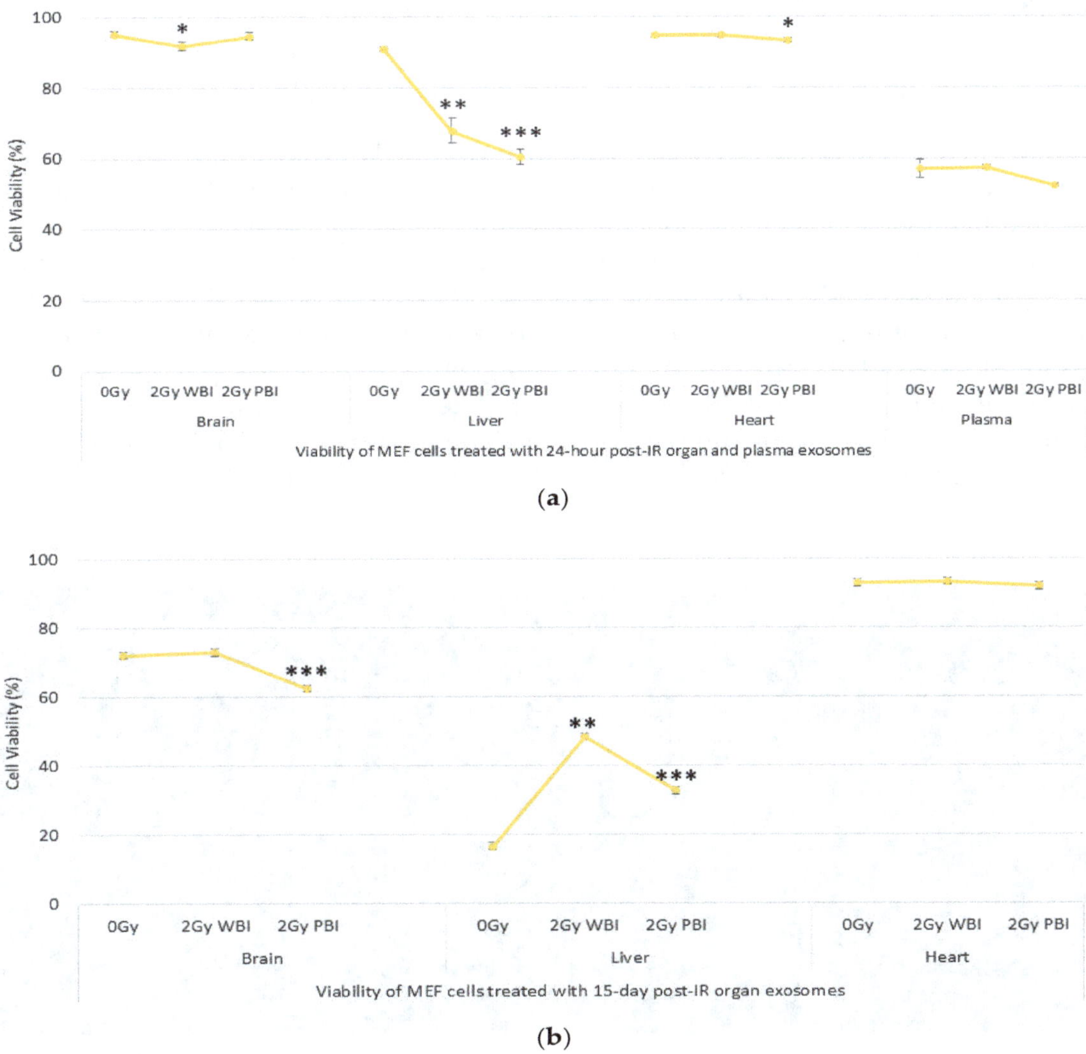

Figure 4. Viability of mouse embryonic fibroblast (MEF) cells treated with (**a**) 24 h post-IR and (**b**) 15 day post-IR exosomes obtained from organs and plasma of 2 Gy WBI or PBI mouse compared to unirradiated mouse organ and plasma exosomes. Data groups were obtained by triplícate measurements (* $p < 0.05$, ** $p < 0.01$, *** $p < 0.001$).

2.2.2. Effects of Exosomes on DNA Damage

DNA Damage in Comet Tail

Total DNA damage in the comet tail was measured in order to assess the role of exosomes obtained from 2 Gy WBI or PBI organs and plasma on the induction of DNA damage on the MEF recipient cells. As shown in Figure 5b–e, an increase in DNA damage was observed in 24 h post-IR brain, liver, heart, and plasma exosome-treated cells compared to unirradiated control groups. For brain and heart samples, significantly higher DNA damage was observed in MEF cells treated with 24 h 2 Gy WBI mice exosomes, while treatment with plasma exosomes from both 2 Gy WBI and PBI mice caused significantly higher DNA damage in MEF cells compared to unirradiated control treated MEF cells.

Similarly, as shown in Figure 5f–h, a significant increase in DNA damage was also observed in MEF cells treated with 15 days post-IR 2 Gy WBI and PBI brain, 2 Gy PBI liver, and 2 Gy WBI and PBI heart exosomes compared to their corresponding controls treated with unirradiated exosomes. In addition to the immediate bystander signal effects transferred by exosomes, exosome-transferred long-lived signals also have an ability to induce DNA damage responses in the treated cells.

γH2AX Immunostaining

In order to further evaluate DNA damage in terms of Double-strand breaks (DSBs)γH2AX immunostaining was carried out in MEF cells treated with 2 Gy PBI or WBI mice organ and plasma exosomes and unirradiated organ exosomes for both 24 h and 15 day time points. DSBs were significantly higher in MEF cells treated with all 24 h post-IR organ exosomes and plasma exosomes, as shown in Figure 6b–e, while the highest levels of DSBs were observed in MEF cells treated with 24 h post-IR brain exosomes, as shown in Figure 6a,b. The foci count was not significant for 15 days post-IR exosome-treated MEF cells (Figure 6f–h), suggesting that 15 day post-IR exosome-induced DNA DSBs could be faithfully or unfaithfully repaired. Alternatively, severely DSB damaged cells may have been removed from culture, as significantly delayed cell death was observed in the treated cells, so the γH2AX method was unable to detect the breaks in the 15 days post-IR exosome-treated samples.

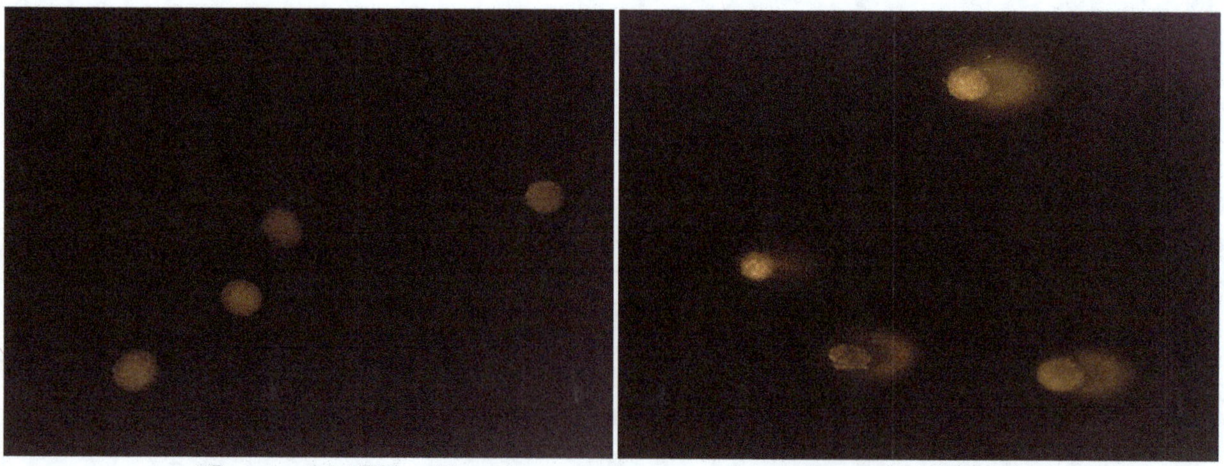

(a)

Figure 5. *Cont.*

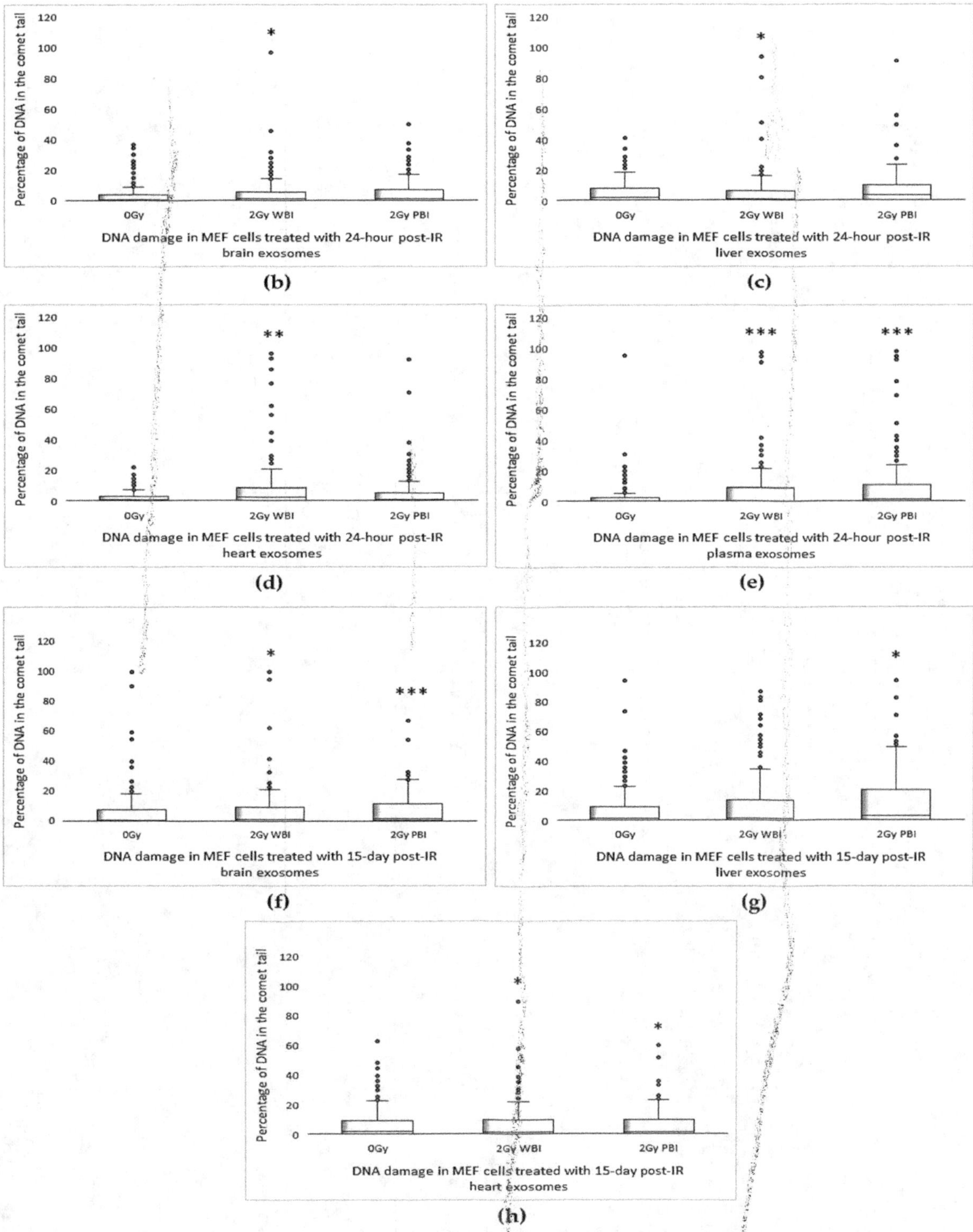

Figure 5. Comet assay showing the induction of DNA damage in MEF cells. (**a**) Representative fluorescent microscope images of untreated MEF cells and comet tails in treated MEF cells. (**b–e**) Induction of DNA damage in MEF cells treated with 24 h post-IR exosomes, (**f–h**) 15 day post-IR exosomes obtained from organs, and plasma of 2 Gy WBI or PBI mouse compared DNA damage in MEF cells treated with exosomes obtained from unirradiated mouse organ exosomes. Percentage of DNA in the comet tail was scored in 200 cells treated for each group. Statistical analysis was performed using the Mann–Whitney U Test (* $p < 0.05$, ** $p < 0.01$, *** $p < 0.001$).

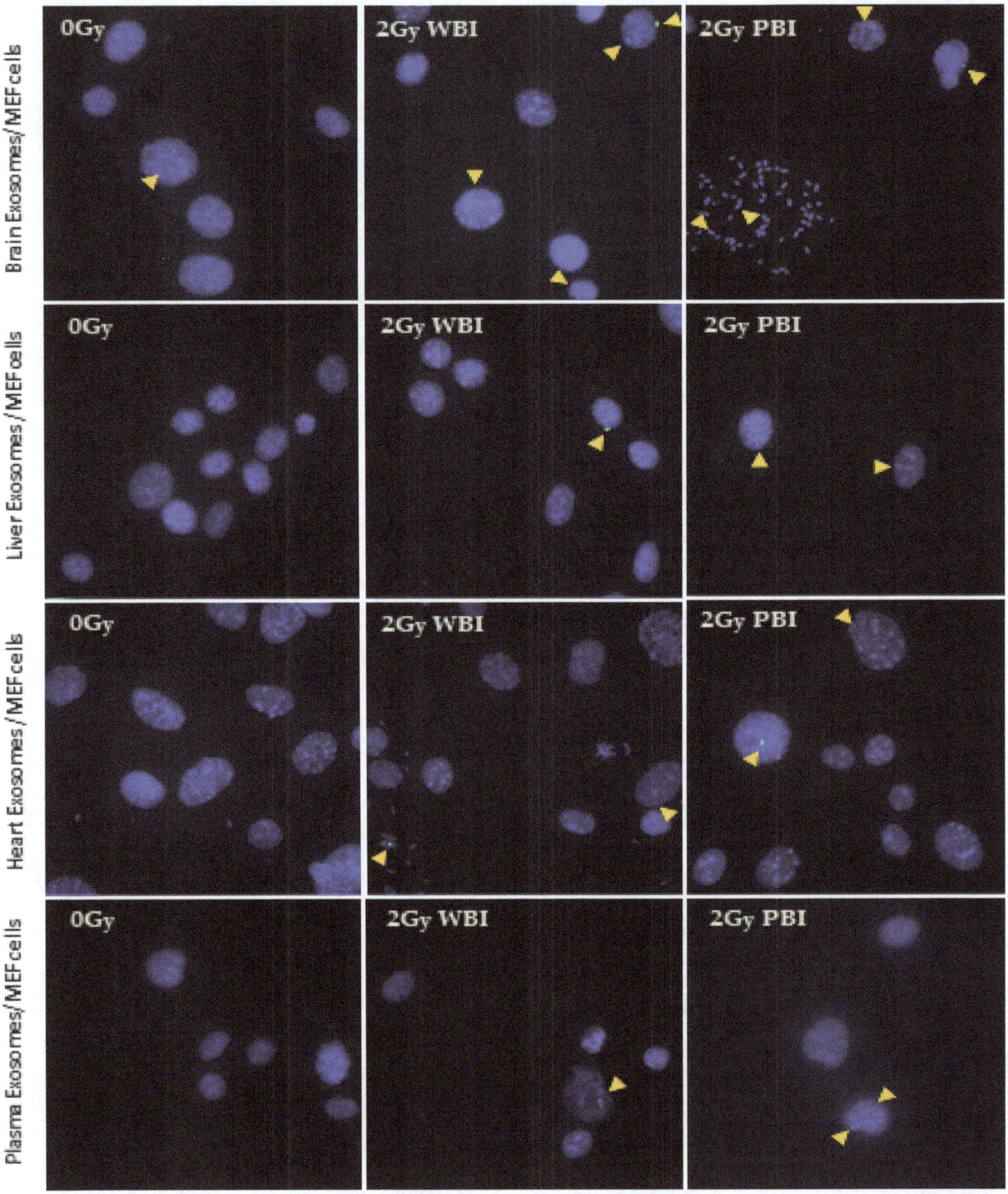

(a)

Figure 6. *Cont.*

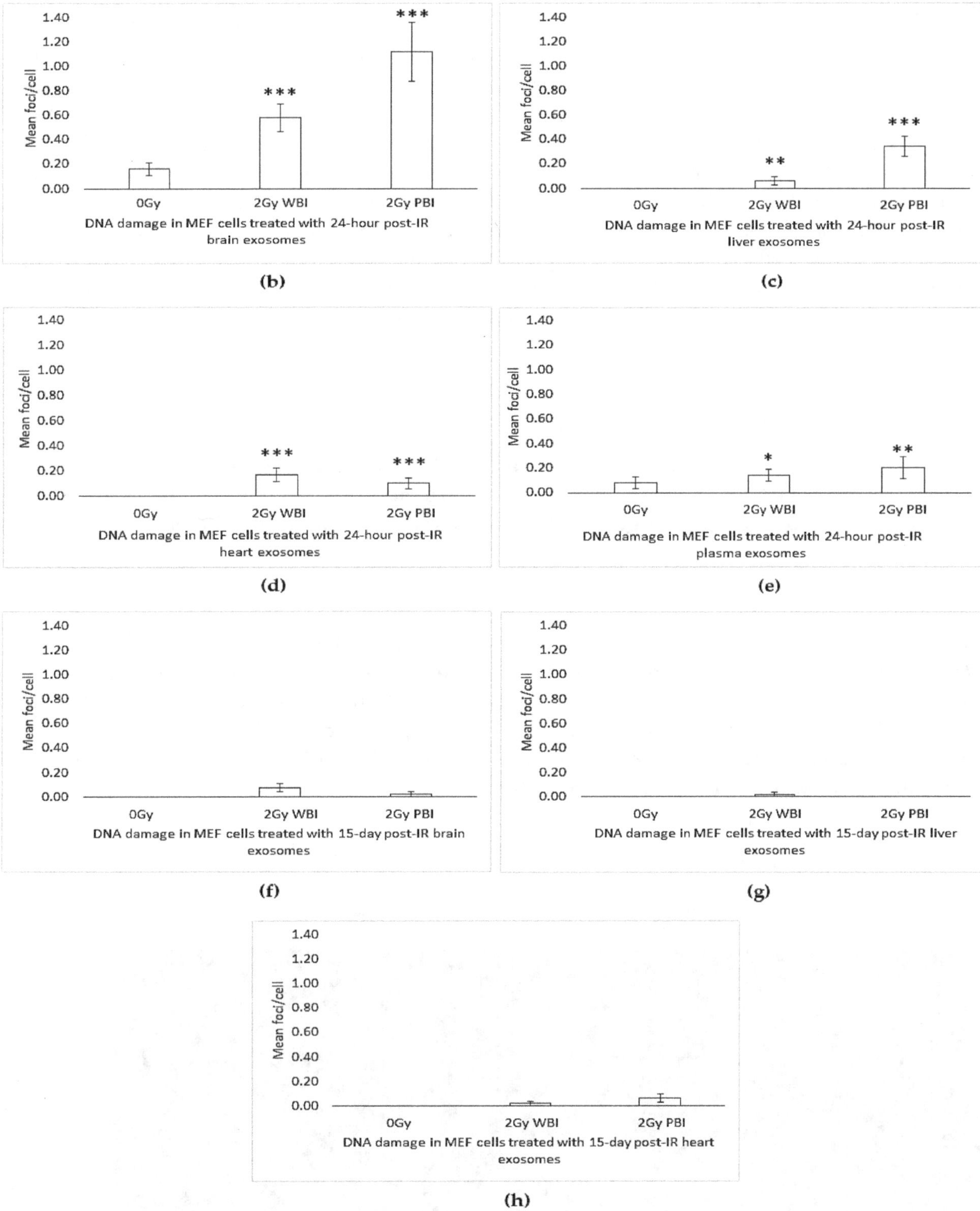

Figure 6. γH2AX foci formation in MEF cells after 24 h treatment with 24 h and 15 days post-IR organ and plasma of 2 Gy WBI or PBI mouse compared to unirradiated mouse organ exosomes treated cells. (**a**) Representative 63X fluorescent microscope images of MEF cells. Arrows indicate the location of γH2AX foci (Alexa488), cells were counterstained with 4′,6-Diamidine-2-phenylindole dihydrochloride (DAPI). (**b–e**) Bars represent the mean γH2AX foci formed per cell ± SEM after treatment with 24 h post-IR exosomes and (**f–h**) 15 day post-IR exosomes obtained from organs and plasma of 2 Gy WBI or PBI mouse compared to corresponding unirradiated mouse organ and plasma exosomes (* $p < 0.05$, ** $p < 0.01$, *** $p < 0.001$).

Chromosomal Aberrations

Chromosomal damage was assessed in metaphase chromosomes of MEF cells treated with 24 h post-IR exosomes and 15 day 2 Gy WBI and PBI mice post-IR exosomes. As shown in Figure 7b–e, chromosomal aberrations were increased significantly in MEF cells treated with 24 h post-IR WBI brain, heart, and plasma and PBI liver exosomes compared to the MEF cells treated with control organs and plasma exosomes. However, chromosomal aberrations do not show significant differences between treatment groups for MEF cells treated with 15 days post-IR exosomes, as shown in Figure 7f–h. The findings suggest that exosomes derived from organs and plasma are able to induce chromosome aberrations in the MEF cells only at early time point, while the exosomes' role in inducing chromosomal damage was reduced to the control level at the late time point post-exposure. This could be due to the ability of the cells to repair the damage at this time point of analysis.

2.2.3. Role of Exosomes as Signalling Mediators

To further investigate the role of exosomes as mediators of radiation induced damage, calcium, reactive oxygen species (ROS), and nitric oxide (NO) were monitored in real time in MEF cells exposed to media containing exosomes from organs and plasma as well as from unirradiated, WBI, and PBI mice.

Calcium Signalling

Increases in intracellular calcium were measured using the calcium-sensitive dyes Fluo 3 and Fura Red. Rapid calcium fluxes were induced in MEF recipient cells following addition of exosomes from brain, heart and liver from WBI and PBI mice but not from sham irradiated mice at 24 h (Figure 8a–c) and 15 days (Figure 8d–f) post-irradiation. Similarly, rapid calcium fluxes were induced in MEF recipient cells following addition of exosomes from plasma from WBI and PBI mice but not from sham irradiated mice at 24 h post irradiation (Figure 8g).

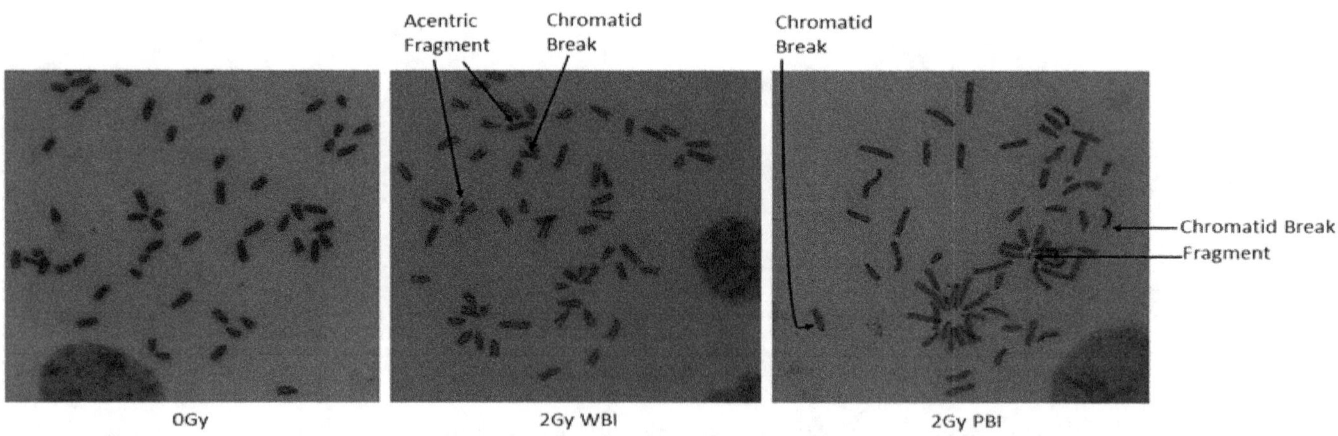

(a)

Figure 7. *Cont.*

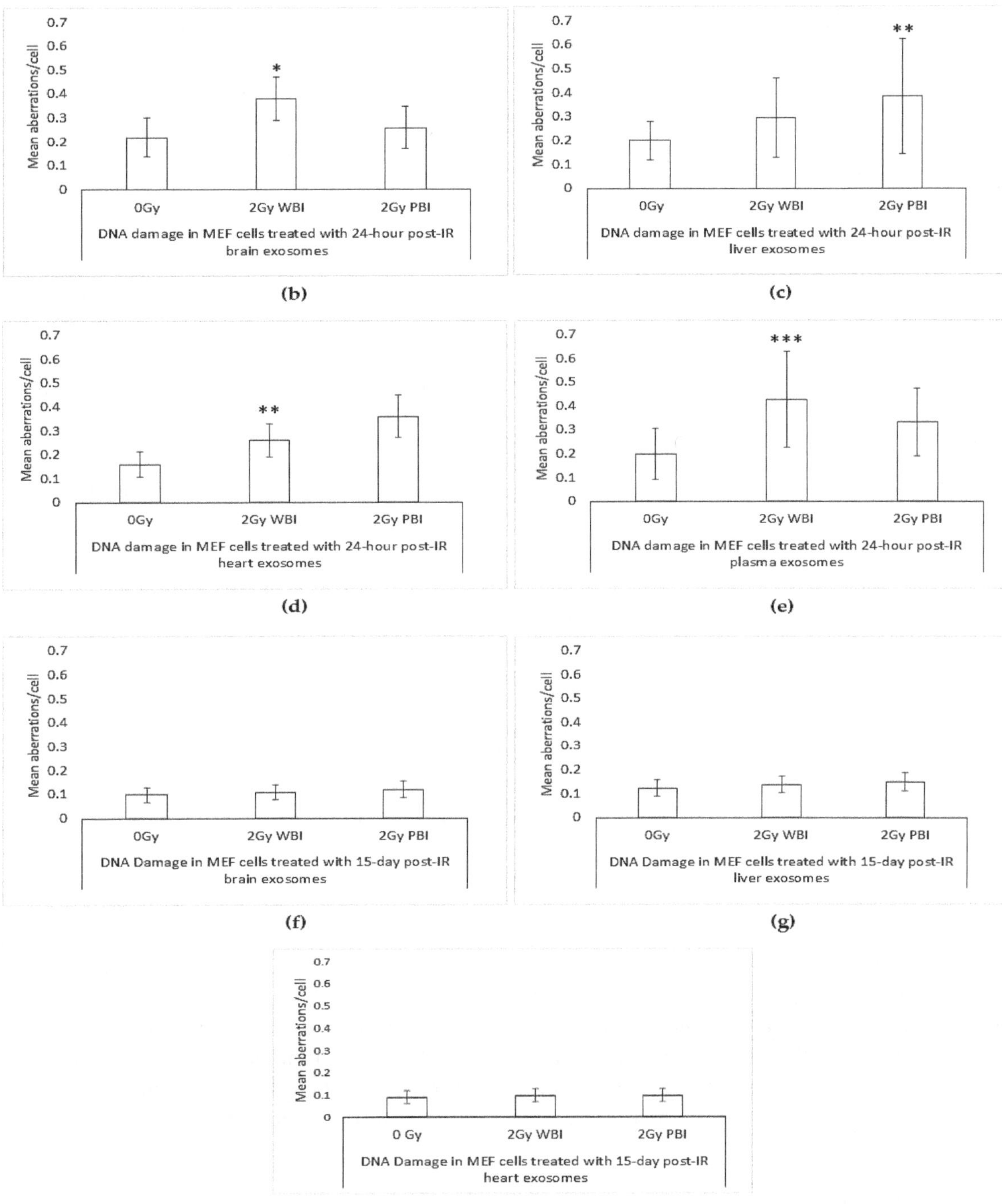

Figure 7. Chromosome analysis of MEF cells treated with 24 h and 15 days post-IR organ and plasma of 2 Gy WBI or PBI mouse compared to unirradiated mouse organ exosomes treated cells. (**a**) Normal and aberrant metaphase of MEF cells treated with exosomes derived from 24-h post-IR brain exosomes isolated from control (0 Gy), 2 Gy WBI and PBI mouse. Mean chromosomal aberrations/cell in MEF cells treated with (**b–e**) 24-h and (**f–h**) 15-day post-IR exosomes obtained from organs and plasma of 2 Gy WBI or PBI mouse compared to corresponding unirradiated mouse organ exosomes. Chromosomal aberrations (total) were scored in 100 metaphase spreads in cells treated with corresponding exosomes for 24 h. Bars represent mean chromosomal aberrations per cell ± SEM, significance was tested by Fisher's exact test. (* $p < 0.05$, ** $p < 0.01$, *** $p < 0.001$).

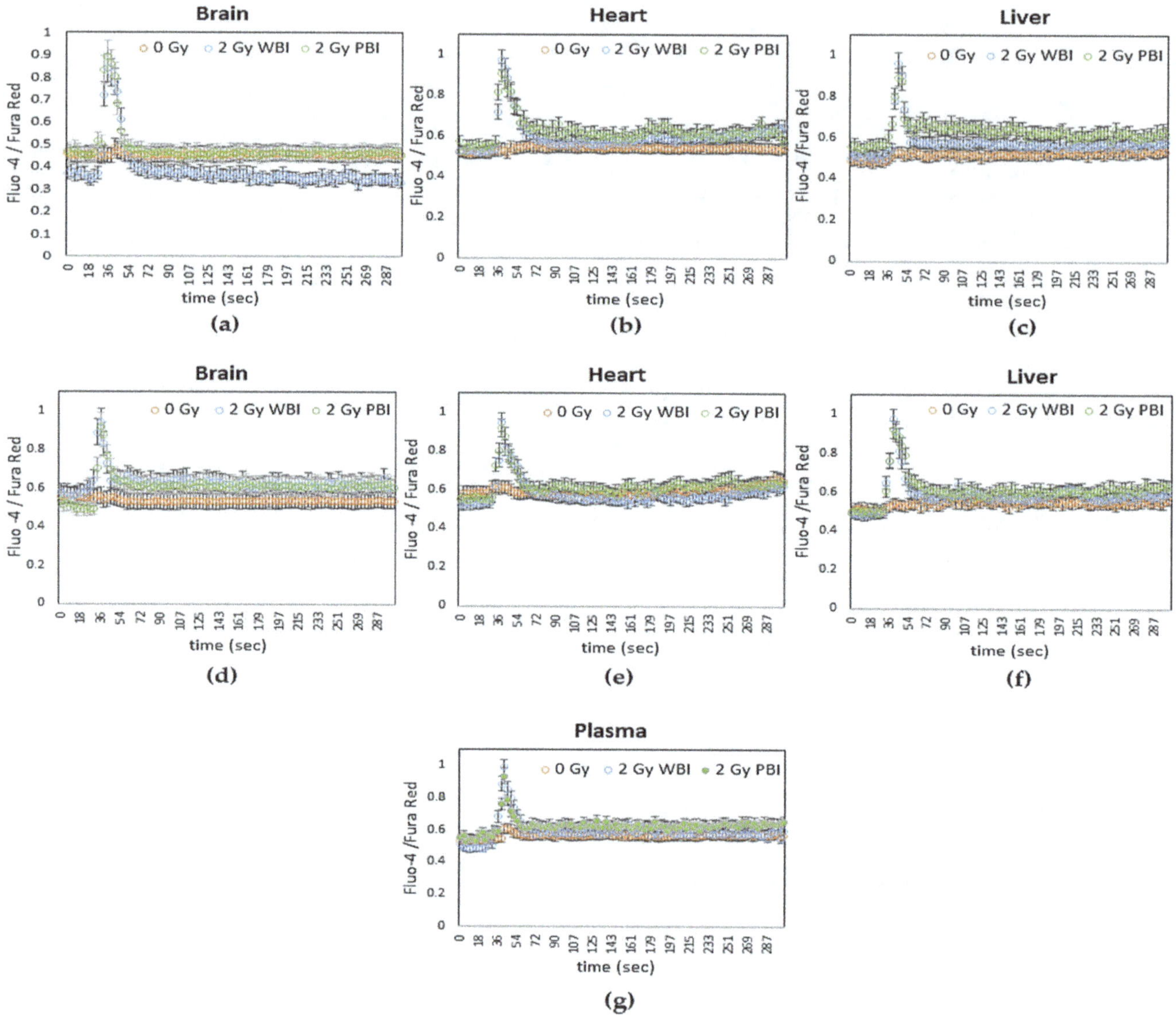

Figure 8. Intracellular calcium levels in MEF cells as indicated by the ratio of fluorescence emissions from the calcium sensitive dyes Fluo-4 and Fura Red after addition of media containing exosomes from (**a–c**) 24 h post-IR organs (**d–f**) 15 days post-IR organs and (**g**) 24 h post-IR plasma of 2 Gy WBI or PBI mice compared to corresponding unirradiated mouse organ and plasma exosomes.

ROS and NO Signaling

Production of ROS in MEF cells was measured using a fluorescent dye, CM-H2 DCFDA. The data were expressed as mean fluorescence intensity normalised to each respective control. Significant ROS production was observed within 5 min of addition of exosomes from brain, heart and liver from WBI and PBI mice at 24 h and 15 days post irradiation and within 5 min of addition of exosomes from plasma from WBI and PBI mice at 24 h post irradiation as shown in Figure 9a.

Similarly, NO levels were measured using a fluorescent dye, 4-amino-5-methylamino-2,7-ifluorofluorescein diacetate (DAF-FM). Again, the data were expressed as mean fluorescence intensity normalised to each respective control. Significant NO production was observed within 5 min of addition of exosomes from brain, heart and liver from WBI and PBI mice at 24 h and 15 days post-irradiation and within 5 min of addition of exosomes from plasma from WBI and PBI mice at 24 h post-irradiation (Figure 9b).

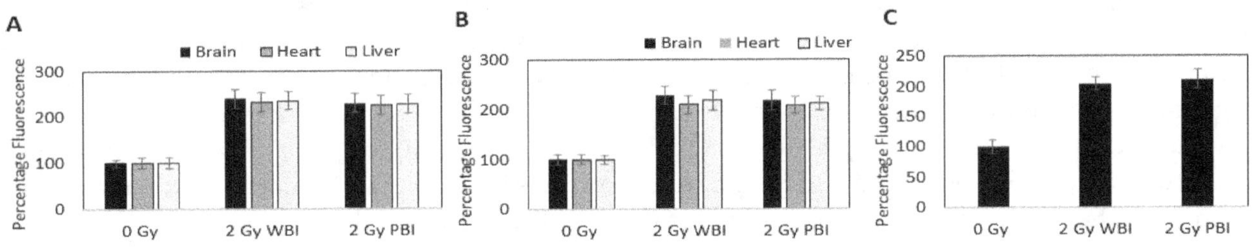

(a)

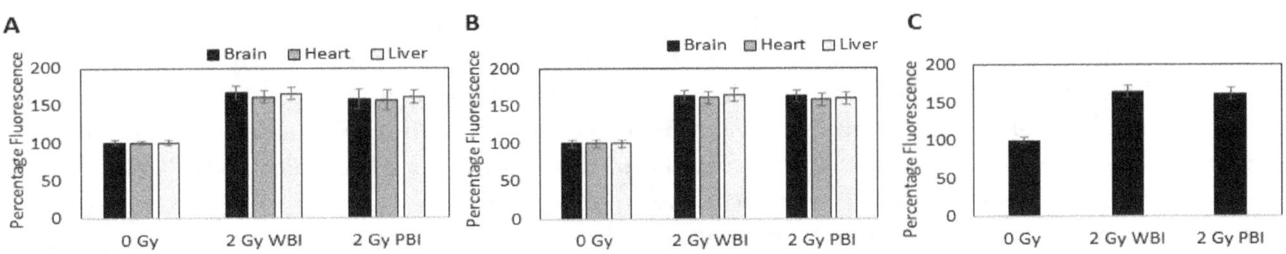

(b)

Figure 9. (**a**) Intracellular reactive oxygen species (ROS) levels in MEF cells as measured using CM-H2 DCFDA fluorescence after addition of media containing exosomes from brain, heart and liver from mice (**A**) 24 h and (**B**) 15 days after whole or partial body irradiation and (**C**) from plasma from mice 24 h after whole or partial body irradiation. Data are presented as mean ± SD after each sample was normalised to its respective control. (**b**) Intracellular NO levels in MEF cells as measured using DAF fluorescence after addition of media containing exosomes from brain, heart, and liver from mice (**A**) 24 h and (**B**) 15 days after whole or partial body irradiation and (**C**) from plasma from mice 24 h after whole or partial body irradiation. Data are presented as mean ± SD after each sample was normalised to its respective control.

3. Discussion

Exosome research is a new and growing field in radiobiology. To date, growing evidence supports the observation that ionising radiation can induce increased exosome release as well as changes in exosome content in in vitro models [32,34,51,52]. However, there is a lack of knowledge regarding the profile of exosomes released from directly irradiated versus abscopal organs and the systemic non-targeted effects of IR caused by those exosomes. Therefore, in the present study, we first investigated the organ responses to 2 Gy X-ray, a radiotherapeutic dose of ionising radiation, in terms of exosome profile including exosome concentration and size. Subsequently, we explored the role of these radiation-derived exosomes in the induction of cell death and DNA damaging effects in recipient MEF cells. Finally, the role of exosomes as mediators of radiation-induced damage was investigated by measuring calcium, ROS, and NO in exosome-exposed MEF cells.

Our data suggested that exosome yield can vary according to the organ type and whether exosomes were derived from directly irradiated organs (obtained from whole body irradiated animals) or abscopal organs (obtained from partial body irradiated animals). Differences in exosome concentrations between organs were also observed upon early (24 h post-IR) and delayed time (15 days post-IR) points.

Only significant change for size distribution of exosomes was observed in 24 h post-IR mice brain where WBI and PBI mice organs had an increased size distribution of exosomes compared to the unirradiated samples.

Although all three organs, the brain, liver, and heart, showed organ-specific exosome secretion levels, they exhibited a highly similar pattern of secretion when whole body and partial body irradiated sample organs were compared to their unirradiated counterparts at 24 h of post irradiation (Figure 1a–c). Both WBI and PBI organs showed a significantly increased level of exosome concentrations, particularly in the brain and liver, while exosome concentration levels were highest in PBI organs. One of the well-explained mechanisms of exosome release is linked to radiation-induced DNA damage and induction of the p53-related additional pathway of exosome biogenesis and secretion, which in turn leads to a significant increase in exosome release [52]. The same pattern of exosome secretion was also observed for all three organs at the delayed time point (Figure 1e–g).

Analysis of plasma samples from WBI and PBI mice, on the other hand, also showed significantly higher yield of exosomes for irradiated mice, albeit showing higher levels for WBI mice compared to PBI mice (Figure 1d). In addition, there was no alteration in size distribution between plasma samples.

Several studies support the involvement of exosomes in mediating RIBE in vivo and in vitro [29,30,37,53–55]. Bystander signals can be communicated through exosomes that can result in functional changes in target cells either by receptor-mediated interactions or by transfer of various bioactive molecules such as proteins, mRNA, miRNA or bioactive lipids carried by exosomes [56,57]. Therefore, in subsequent steps of our study, we interrogated whether WBI and PBI mice organ and plasma exosomes have an impact on the survival and DNA stability of recipient MEF cells.

It is evident from the literature that radiotherapeutic doses of X-ray can induce cell survival in cancer cell line models [34,48]. However, there has been a lack of information regarding impact of exosomes derived from both directly irradiated and abscopal organs on normal cells such as MEF cells, which was our model system in this study. Data showed that those exosomes from both early and delayed time points have an ability to reduce recipient MEF cell viability (Figure 4a,b). Albanese and co-authors showed that levels of TNFSF6 exfoliated on extracellular vesicles were increased following IR, suggesting a mechanism for abscopal and bystander effects after irradiation [58] which can also be one possible explanation for our findings.

Finally, we attempted to explore DNA damaging effects of exosomes in our model system as another aspect of bystander effects. Our findings show that both WBI and PBI organs and plasma induce DNA damage and chromosomal aberrations in the recipient MEF cells, where the extent of DNA damage was found to be organ and time specific. The amount of DNA in the comet tail showed significant results for both WBI and PBI organs when compared to MEF cells that received unirradiated organ and plasma exosomes as both early and delayed effects of IR, as depicted in Figure 5. Conversely, only early effects were significant when DSB foci (Figure 6b–e) or chromosomal aberrations (Figure 7b–e) were investigated. An absence of significant DNA DSBs or chromosomal aberrations in MEF cells that received delayed time point exosomes can be explained by the repair of DNA DSBs or removal of cells from the population (Figures 6f–h and 7f–h) as a delayed time point response to IR. It has been shown that the phosphorylation status of critical DNA damage repair proteins can be changed by exosomes released from breast cancer cells [59]. Moreover, exosomes from irradiated HNSCC cells were shown to enhance DNA repair in unirradiated recipient cells [34].

In addition, our study showed the involvement of exosomes in mediating radiation damage by increasing calcium levels (within 30 s) and inducing ROS and NO (within 5 min) following addition of exosomes from brain, heart and liver from WBI and PBI mice at 24 h and 15 days post-irradiation and from plasma from WBI and PBI mice at 24 h post irradiation. Calcium, ROS, and NO signalling has been shown previously in bystander cells exposed to media from irradiated cells [60–64], and to exosomes from irradiated cells [29].

4. Materials and Methods

4.1. Animal Breeding, Irradiation and Sample Collection

Animal studies were performed according to the European Community Council Directive 2010/63/EU, approved by the local Ethical Committee for Animal Experiments of the ENEA on 01/10/2017 with the project identification code No.004/2017- SEPARATE, and authorized by the Italian

Ministry of Health (n° 539/2018-PR). Eighty days of age C57 BL/6 female mice were either whole body irradiated (WBI) or partial body irradiated (PBI) with 2 Gy X-rays or sham irradiated (0 Gy).

For partial body exposure, the upper two-thirds of the adult mouse body was shielded, whilst exposing the lower one-third. At two different time points (24 h and 15 days) post-irradiation, animals were sacrificed by perfusion, washing out the blood and running physiological saline through the vascular system. The brains, livers and hearts were collected, snap frozen in liquid nitrogen, and stored at −80 °C for later exosome collection. Furthermore, at 24 h post-irradiation, animals were sacrificed, blood was collected and blood plasma were separated and snap frozen for later exosome extraction. All the samples were shipped to Oxford Brookes University (OBU) for exosome isolations as described below.

4.2. Exosome Isolation

The procedure of exosomes isolation from mouse organs (brain, liver and heart) were adapted from a protocol that has previously been established by Polanco et al. [65]. Briefly, organs (brain, liver and heart) were slowly defrosted, dissected and gently homogenised before being incubated in 7 mL of 20 units/mL papain (LS003119, Worthington, Lakewood, NJ, USA) in RPMI-1640 (R7388, Sigma, St Louis, MO, USA) for 20 min at 37 °C. Similarly, the reaction was stopped with 14 mL of ice-cold RPMI. The homogenised samples were then gently disrupted by pipetting with a 10 mL pipette, which was followed by a series of differential 4 °C centrifugations at 300× *g* for 10 min, 2000× *g* for 10 min, and 10,000× *g* for 30 min. However, the plasma samples were slowly defrosted, collected in falcon tubes and then similarly subjected to serial centrifugations at 300× *g* for 10 min, 2000× *g* for 10 min, and 10,000× *g* for 30 min. The supernatants from the 10,000× *g* centrifugations were passed through 0.45 μm and then 0.22 μm syringe filters, and then centrifuged at 120,000× *g* for 90 min at 4 °C to pellet exosomes. Exosome pellets were resuspended either in PBS or in exosome resuspension buffer (4478545, Invitrogen, Carlsbad, CA, USA) for downstream experiments.

4.3. Tunable Resistance Pulse Sensing (TRPS) via qNano

Exosome size and concentration were measured using TRPS via qNano machine (Izon Science™, Lyon, France), as described by Al-Mayah et al. [30], in which sample particles are driven through the nanopore (NP100 mm) by applying a combination of pressure and voltage.

Each particle causes a resistive pulse or blockade signal, which is detected and measured by the application software. Blockade magnitude is directly proportional to the volume of each particle [66], while blockade frequency is used to determine particle concentration [67]. Finally, magnitude and frequency values are converted to respective particle properties as size and concentration by normalizing against a known particle standard such as carboxylated polystyrene calibration nanoparticles.

4.4. Transmission Electron Microscopy (TEM): Morphological Analysis

In addition to qNano analysis and Western blotting to confirm existence of exosomes derived from organs and plasma, samples were morphologically analysed by electron microscopy [68]. In brief, 1:10 or 1:100 diluted exosome suspensions in PBS were incubated on formvar-coated and charged nickel grids (200 mesh) for 2 min. They were then fixed in 2.5% glutaraldehyde for 10 min, and then washed three times in 0.1 M phosphate buffer by dipping onto the surface of a water droplet and then stained with 8 μL of 2% aqueous uranyl acetate for 2 min. The stain was drawn off with cartridge paper to leave a thin negative stain. The grids were then examined and photographed under Jeol JEM-1400 Flash transmission electron microscope, with a Gatan OneView 16 Megapixel camera.

4.5. Western Blot

Western blotting has frequently been used for identifying and determining proteins, as one of the most commonly used methods in laboratories. It has also been utilised as a semi-quantitative technique in order to compare the expression of proteins in the cells and tissues [69].

Exosome proteins were extracted by using Total Exosome RNA and Protein Isolation kit (4478545, Invitrogen, Carlsbad, CA, USA). The protein content was measured using the modified Bradford assay using a Coomassie solution and Pre-Diluted Protein Assay Standards (23208, Thermo Scientific, Waltham, MA, USA). Exosome proteins were mixed with LDS sample buffer (NP0007, NuPAGE, Invitrogen, Carlsbad, CA, USA), and 30 μg from each sample were separated in a 4–20% polyacrylamide Mini-PROTEAN®TGX Stain-Free™ gels (456-8095, Bio-Rad, Hercules, CA, USA) and transferred onto Amersham™ Hybond™ PVDF membrane (10600090, GE Healthcare, Little Chalfont, UK). The membranes were blocked in 5% BSA and were incubated with the primary antibodies CD63 (ab 217345, Abcam, Cambridge, UK), in 5% BSA for 2 h at 1:1000 dilution. Membranes were washed three times with PBS-T which was followed by incubation with Goat Anti-Rabbit IgG H&L Alexa Fluor® 488 (ab150077, Abcam, Cambridge, UK) at 1:10,000 dilution for 1 h.

4.6. Raman Spectroscopy

A LabRam HR confocal Raman instrument (HORIBA, Northampton, UK) was used for spectral acquisition. Manual calibration of the grating was carried out using the 520.7 cm^{-1} Raman line of crystalline silicon. Dark current measurement and recording of the substrate and optics signal was also performed, for data correction. As a source, a 532 nm laser of ~12 mW power was focused by a 100 X (MPlanN, Olympus, NA = 0.9) objective onto the sample; and the resultant Raman signals were detected using a spectrograph with a 1200 g/mm grating coupled with a CCD. Raman spectra were acquired in the 400 to 1800 cm^{-1} region. Multiple calibration spectra of 1,4-bis(2-methylstyryl) benzene were recorded along with each sample acquisition. All spectra were subsequently wavenumber calibrated using in-house developed procedures in Matlab v.9.3 (Mathworks Inc., Natick, MA, USA). The instrument response correction was performed using the spectrum of NIST Standard Reference Material (SRM) no.2242.

Raman spectroscopic data were pre-processed (normalization, baseline subtraction, etc.) using in-house developed protocols within the Matlab (The Mathworks Inc.) environment and corrected spectra were subjected to Partial Least Squares Discriminant Analysis (PLSDA).

4.7. Cell Culture and Exosome Treatments

Mouse embryonic fibroblast (C57 BL/6) (ATCC®SCRC-1008™) cells (MEF-BL/6-1) were grown in ATCC-formulated Dulbecco's modified Eagle's medium (30-2002, ATCC, Manassas, VA, USA) in the presence of 15% Fetal Bovine Serum (FBS) and 1% penicillin (P4333, Sigma, St Louis, MO, USA) and 5% CO_2. A total of 1.5×10^6 MEF cells were treated with exosomes obtained from tissue corresponding to the 1:5 of the organ mass in15 mL for 24 h prior to experiments.

4.8. Cell Count and Viability

Cell count and viability were measured by using Muse™ Count and Viability Cell Dispersal Reagent (MCH100107, Merck, Millipore, Kenilworth, NJ, USA) and Muse™ Cell Analyzer (0500-3115, Merck, Millipore, Kenilworth, NJ, USA) as described by Laka et al. [70]. Briefly, 380 μL Muse™ Count and Viability reagent was added to 20 μL cell suspensions in PBS, and incubated for 5 min. Then, cell viability was measured in Muse™ Cell Analyzer according to the dilution factor, 20. Viability percentages were evaluated. Each sample was analysed in triplicates.

4.9. Alkaline Single Cell Gel Electrophoresis (Comet Assay)

Single-cell gel electrophoresis, or the comet assay, is a sensitive method to quantify total DNA damage (double-strand breaks, single-strand breaks and base damage) in individual cells [71,72]. The comet assay was carried out as described by Al-Mayah et al. [28]. Briefly, microscope slides were coated with 1% normal melting point agarose (NMPA) (A9539, Sigma, St Louis, MO, USA), and were allowed to dry overnight. The coated slides were then placed on a metal tray on ice. Twenty thousand cells were resuspended with 200 μL of 0.6% low melting point agarose (LMPA)

(BP165-25, Fisher Scientific, Pittsburgh, PA, USA) and placed immediately onto chilled pre-coated slides. The slides were then transferred to a Coplin jar, which was filled with cold alkaline lysis buffer (2.5 M NaCl, 100 mM EDTA pH 8.0, 10 mM Tris-HCl pH 7.6, and 1% Triton X-100, pH 10), and the jar was kept at 4 °C overnight.

The slides were then moved to a horizontal electrophoresis tank filled with electrophoresis buffer (0.3 M NaOH and 1 mM EDTA, (pH 13) at 4 °C for 40 min. The electrophoresis was run for 30 min, at 19 V, 300 A. Slides were neutralized with neutralizing buffer (0.4 M Tris-HCl, pH 7.5), washed with distilled water, and immediately stained with a 1:10,000 dilution of Diamond Nucleic Acid Dye (H1181, Promega, Madison, MA, USA). The slides were analysed using fluorescent microscopy and Comet Assay IV Image Analysis Software (Perceptive Instruments, Bury St Edmunds, UK). Tail intensities were evaluated for comparisons.

4.10. γH2AX Immunostaining

To investigate DNA damage in the exosome-treated cells, γH2AX Immunostaining assay was adapted from Zhang et al. [73]. Briefly, cells, fixed with 25% acetic acid in methanol, were dropped on slides and air-dried. Then cells were permeabilised with 0.2% Triton X-100 in PBS for 10 min. The cells were then blocked with 3% BSA in PBS for 1 h at room temperature (RT). Cells were incubated with γH2AX monoclonal antibody (ab26350, Abcam, Cambridge, UK) at 1:500 dilution overnight at 4 °C. Then, cells were washed three times with PBS-T and incubated with the secondary antibody conjugated with Alexa Fluor 488 (ab150113, Abcam, Cambridge, UK) for 1 h at RT at 1:1000 dilution. Cells were washed again three times with PBS-T and then mounted with anti-fade reagent containing DAPI.

4.11. Chromosome Analysis

Chromosomal preparation for Giemsa solid staining technique was carried out as described by [30]. Briefly, exosome-treated cells were incubated with 20 ng/mL demecolcine (D1925, Sigma, St Louis, MO, USA) for 1.5 h in a humidified 5% CO_2 incubator at 37 °C. Cells were collected and centrifuged 300× *g* for 10 min at RT. Cell pellet was re-suspended with 0.075 M KCL as hypotonic solution for 20 min at 37 °C. The hypotonic cell suspensions were centrifuged at 200× *g* for 10 min at RT after addition of few drops of 25% acetic acid in methanol 3:1 fixative. Next, the cell pellet was fixed twice with 25% acetic acid in methanol. Fixed cells were dropped onto clean slides, and stained using the Giemsa solid staining technique. Slides were mounted and at least 100 metaphases were analysed per group.

4.12. Live Cell Imaging

Intracellular calcium levels were determined using Fluo 3 and Fura Red (Invitrogen/Molecular Probes, BioSciences, Dublin, Ireland) and ROS and NO were followed in real time using the fluorescent probes 5-(and 6-)chloromethyl 2,7 dichlorodihydrofluorescein diacetate, acetyl ester (CM-H2 DCFDA) (Invitrogen/Molecular Probes) and 4-amino-5-methylamino-2,7-ifluorofluorescein diacetate (DAF-FM) (Invitrogen/Molecular Probes) respectively, as previously described (Lyng et al., 2006).

Briefly, MEF recipient cells were grown on 35 mm glass bottom culture dishes (Mat Tek Corporation, Ashland, MA, USA; # P35 G-0-20-C). Twenty-four hours after plating, cells were washed twice with a buffer containing 130 mM NaCl, 5 mM KCl, 1 mM Na_2 HPO_4, 1 mM $CaCl_2$ and 1 mM $MgCl_2$ (pH 7.4) and incubated with 3 μM Fluo 3 and 3 μM Fura Red acetoxymethyl esters for 1 h and with 5 μM CM-H2 DCFDA or DAF-FM for 30 min in the buffer at 37 °C. Subsequently, the cultures were washed three times with buffer. All dyes were excited at 488 nm and fluorescence emissions at 525 (Fluo 3, CM-H2 DCFDA and DAF-FM) and 660 nm (Fura Red) were recorded using a Zeiss LSM 510 confocal microscope. Exosomes were added after 60 s when a stable baseline had been established. All measurements were performed at room temperature.

4.13. Statistical Analysis

For exosome size, diameter and viability data, significance was tested by Student's t-test using raw data. Each experiment was carried out in triplicate. Analysis showed no significant inter-experimental variation; therefore, data from these experiments were pooled. For analysis of comet assays, statistical analysis was performed using the Mann–Whitney test, utilising the median of the raw data. Meanwhile, chromosomal analysis and γH2AX immunostaining data were subjected to Fisher's exact test. Data were considered statistically significant if p-value was lower than 0.05 (* $p < 0.05$, ** $p < 0.01$, *** $p < 0.001$).

5. Conclusions

In this study, we provide the first insights into the in vivo systemic effects of early and delayed effects of X-ray irradiation in terms of exosome profile and bystander effects. Taken together, the findings show that exosome yield is organ-specific and can be significantly increased in both directly irradiated and abscopal organs. On the other hand, changes in the levels of survival and DNA damage in MEF cells, receiving PBI or WBI exosomes, were not necessarily correlated with the increase in exosome yield in different radiation conditions or time points. Those manifested bystander effects in MEF cells also differed as early and delayed responses to ionizing radiation. The role of exosomes in mediating radiation damage was shown by rapid calcium fluxes and induction of ROS and NO in MEF recipient cells following addition of exosomes from WBI and PBI mice, but not from unirradiated mice. Altogether, these results draw attention to the content of those exosomes, which can be the key to understand the observed effects. Further studies such as miRNA expression profiles and proteomics will help to discover the molecules that are responsible for the observation of bystander effects in recipient cells.

Author Contributions: Conceptualization, M.K., F.L., S.T. and M.M.; methodology, S.T.C., A.M., S.P., M.M. and M.K.; software, A.M. and S.T.C.; validation, A.M., S.T.C. and M.K.; formal analysis, A.M., S.T.C., D.T., A.D., F.L. and M.K.; investigation, S.T.C., A.M., P.G., S.P., M.M., D.T., A.D., F.L. and M.K.; resources, A.S., M.M., F.L. and M.K.; data curation, S.T.C., A.M., F.L. and M.K.; writing—original draft preparation, S.T.C., A.M., D.T., A.D., A.D.M., F.L. and M.K.; writing—review and editing, S.T.C., A.M., D.T., A.D., A.D.M., F.L. and M.K.; visualization, S.T.C., A.M., F.L. and M.K.; supervision, M.K.; project administration, M.K., F.L., S.T., A.S. and M.M. All authors have read and agreed to the published version of the manuscript.

Acknowledgments: We confirm there was no any support given other than was covered by the author contribution and the funding sections.

Abbreviations

NTE	Non-targeted effects
RIBE	Radiation induced bystander effects
IR	Ionizing radiation
MEF	Mouse Embryonic Fibroblast
DSB	Doublestrand break
WBI	Whole body irradiation/irradiated
PBI	Partial body irradiation/irradiated

References

1. Kadhim, M.; Salomaa, S.; Wright, E.; Hildebrandt, G.; Belyakov, O.V.; Prise, K.M.; Little, M.P. Non-targeted effects of ionising radiation—Implications for low dose risk. *Mutat. Res.* **2013**, *752*, 84–98. [CrossRef] [PubMed]
2. Mothersill, C.; Rusin, A.; Seymour, C. Relevance of Non-Targeted Effects for Radiotherapy and Diagnostic Radiology; A Historical and Conceptual Analysis of Key Players. *Cancers* **2019**, *11*, 1236. [CrossRef] [PubMed]

3. Morgan, W.F. Non-targeted and delayed effects of exposure to ionizing radiation: I. Radiation-induced genomic instability and bystander effects in vitro. *Radiat. Res.* **2003**, *159*, 567–580. [CrossRef]
4. Azzam, E.I.; Little, J.B. The radiation-induced bystander effect: Evidence and significance. *Hum. Exp. Toxicol.* **2004**, *23*, 61–65. [CrossRef]
5. Kadhim, M.A.; Moore, S.R.; Goodwin, E.H. Interrelationships amongst radiation-induced genomic instability, bystander effects, and the adaptive response. *Mutat. Res.* **2004**, *568*, 21–32. [CrossRef]
6. Boyd, M.; Ross, S.C.; Dorrens, J.; Fullerton, N.E.; Tan, K.W.; Zalutsky, M.R.; Mairs, R.J. Radiation-induced biologic bystander effect elicited in vitro by targeted radiopharmaceuticals labeled with alpha-, beta-, and auger electron-emitting radionuclides. *J. Nucl. Med.* **2006**, *47*, 1007–1015.
7. Morgan, W.F.; Sowa, M.B. Non-targeted bystander effects induced by ionizing radiation. *Mutat. Res.* **2007**, *616*, 159–164. [CrossRef]
8. Ilnytskyy, Y.; Kovalchuk, O. Non-targeted radiation effects-an epigenetic connection. *Mutat. Res.* **2011**, *714*, 113–125. [CrossRef]
9. Pan, B.T.; Johnstone, R.M. Fate of the transferrin receptor during maturation of sheep reticulocytes in vitro: Selective externalization of the receptor. *Cell* **1983**, *33*, 967–978. [CrossRef]
10. Simpson, R.J.; Lim, J.W.; Moritz, R.L.; Mathivanan, S. Exosomes: Proteomic insights and diagnostic potential. *Expert Rev. Proteom.* **2009**, *6*, 267–283. [CrossRef]
11. Vlassov, A.V.; Magdaleno, S.; Setterquist, R.; Conrad, R. Exosomes: Current knowledge of their composition, biological functions, and diagnostic and therapeutic potentials. *Biochim. Biophys. Acta* **2012**, *1820*, 940–948. [CrossRef]
12. Klein-Scory, S.; Tehrani, M.M.; Eilert-Micus, C.; Adamczyk, K.A.; Wojtalewicz, N.; Schnölzer, M.; Hahn, S.A.; Schmiegel, W.; Schwarte-Waldhoff, I. New insights in the composition of extracellular vesicles from pancreatic cancer cells: Implications for biomarkers and functions. *Proteome Sci.* **2014**, *12*. [CrossRef]
13. Thery, C.; Boussac, M.; Veron, P.; Ricciardi-Castagnoli, P.; Raposo, G.; Garin, J.; Amigorena, S. Proteomic analysis of dendritic cell-derived exosomes: A secreted subcellular compartment distinct from apoptotic vesicles. *J. Immunol.* **2001**, *166*, 7309–7318. [CrossRef]
14. Gibbings, D.J.; Ciaudo, C.; Erhardt, M.; Voinnet, O. Multivesicular bodies associate with components of miRNA effector complexes and modulate miRNA activity. *Nat. Cell Biol.* **2009**, *11*, 1143–1149. [CrossRef]
15. Taylor, D.D.; Gercel-Taylor, C. MicroRNA signatures of tumor-derived exosomes as diagnostic biomarkers of ovarian cancer. *Gynecol. Oncol.* **2008**, *110*, 13–21. [CrossRef] [PubMed]
16. Balaj, L.; Lessard, R.; Dai, L.; Cho, Y.J.; Pomeroy, S.L.; Breakefield, X.O.; Skog, J. Tumour microvesicles contain retrotransposon elements and amplified oncogene sequences. *Nat. Commun.* **2011**, *2*, 180. [CrossRef] [PubMed]
17. Subra, C.; Laulagnier, K.; Perret, B.; Record, M. Exosome lipidomics unravels lipid sorting at the level of multivesicular bodies. *Biochimie* **2007**, *89*, 205–212. [CrossRef]
18. Stoorvogel, W.; Kleijmeer, M.J.; Geuze, H.J.; Raposo, G. The biogenesis and functions of exosomes. *Traffic* **2002**, *3*, 321–330. [CrossRef]
19. Gotz, J.; Chen, F.; van Dorpe, J.; Nitsch, R.M. Formation of neurofibrillary tangles in P301l tau transgenic mice induced by Aβ42 fibrils. *Science* **2001**, *293*, 1491–1495. [CrossRef]
20. Bolmont, T.; Clavaguera, F.; Meyer-Luehmann, M.; Herzig, M.C.; Radde, R.; Staufenbiel, M.; Lewis, J.; Hutton, M.; Tolnay, M.; Jucker, M. Induction of tau pathology by intracerebral infusion of amyloid-beta -containing brain extract and by amyloid-beta deposition in APP x Tau transgenic mice. *Am. J. Pathol.* **2007**, *171*, 2012–2020. [CrossRef]
21. Simons, M.; Raposo, G. Exosomes—vesicular carriers for intercellular communication. *Curr. Opin. Cell Biol.* **2009**, *21*, 575–581. [CrossRef] [PubMed]
22. Kucharzewska, P.; Christianson, H.C.; Welch, J.E.; Svensson, K.J.; Fredlund, E.; Ringner, M.; Morgelin, M.; Bourseau-Guilmain, E.; Bengzon, J.; Belting, M. Exosomes reflect the hypoxic status of glioma cells and mediate hypoxia-dependent activation of vascular cells during tumor development. *Proc. Natl. Acad. Sci. USA* **2013**, *110*, 7312–7317. [CrossRef]
23. Webber, J.; Steadman, R.; Mason, M.D.; Tabi, Z.; Clayton, A. Cancer exosomes trigger fibroblast to myofibroblast differentiation. *Cancer Res.* **2010**, *70*, 9621–9630. [CrossRef] [PubMed]

24. Rana, S.; Malinowska, K.; Zoller, M. Exosomal tumor microRNA modulates premetastatic organ cells. *Neoplasia* **2013**, *15*, 281–295. [CrossRef]
25. Montecalvo, A.; Larregina, A.T.; Shufesky, W.J.; Stolz, D.B.; Sullivan, M.L.; Karlsson, J.M.; Baty, C.J.; Gibson, G.A.; Erdos, G.; Wang, Z.; et al. Mechanism of transfer of functional microRNAs between mouse dendritic cells via exosomes. *Blood* **2012**, *119*, 756–766. [CrossRef]
26. Peinado, H.; Aleckovic, M.; Lavotshkin, S.; Matei, I.; Costa-Silva, B.; Moreno-Bueno, G.; Hergueta-Redondo, M.; Williams, C.; Garcia-Santos, G.; Ghajar, C.; et al. Melanoma exosomes educate bone marrow progenitor cells toward a pro-metastatic phenotype through MET. *Nat. Med.* **2012**, *18*, 883–891. [CrossRef]
27. Kosaka, N.; Takeshita, F.; Yoshioka, Y.; Hagiwara, K.; Katsuda, T.; Ono, M.; Ochiya, T. Exosomal tumor-suppressive microRNAs as novel cancer therapy: "exocure" is another choice for cancer treatment. *Adv. Drug Deliv. Rev.* **2013**, *65*, 376–382. [CrossRef]
28. Al-Mayah, A.H.; Irons, S.L.; Pink, R.C.; Carter, D.R.; Kadhim, M.A. Possible role of exosomes containing RNA in mediating nontargeted effect of ionizing radiation. *Radiat. Res.* **2012**, *177*, 539–545. [CrossRef]
29. Kumar Jella, K.; Rani, S.; O'Driscoll, L.; McClean, B.; Byrne, H.J.; Lyng, F.M. Exosomes Are Involved in Mediating Radiation Induced Bystander Signaling in Human Keratinocyte Cells. *Radiat. Res.* **2014**, *181*, 138–145. [CrossRef]
30. Al-Mayah, A.; Bright, S.; Chapman, K.; Irons, S.; Luo, P.; Carter, D.; Goodwin, E.; Kadhim, M. The non-targeted effects of radiation are perpetuated by exosomes. *Mutat. Res.* **2015**, *772*, 38–45. [CrossRef]
31. Diamond, J.M.; Vanpouille-Box, C.; Spada, S.; Rudqvist, N.P.; Chapman, J.R.; Ueberheide, B.M.; Pilones, K.A.; Sarfraz, Y.; Formenti, S.C.; Demaria, S. Exosomes Shuttle TREX1-Sensitive IFN-Stimulatory dsDNA from Irradiated Cancer Cells to DCs. *Cancer Immunol. Res.* **2018**, *6*, 910–920. [CrossRef] [PubMed]
32. Arscott, W.T.; Tandle, A.T.; Zhao, S.; Shabason, J.E.; Gordon, I.K.; Schlaff, C.D.; Zhang, G.; Tofilon, P.J.; Camphausen, K.A. Ionizing radiation and glioblastoma exosomes: Implications in tumor biology and cell migration. *Transl. Oncol.* **2013**, *6*, 638–648. [CrossRef]
33. Mutschelknaus, L.; Azimzadeh, O.; Heider, T.; Winkler, K.; Vetter, M.; Kell, R.; Tapio, S.; Merl-Pham, J.; Huber, S.M.; Edalat, L.; et al. Radiation alters the cargo of exosomes released from squamous head and neck cancer cells to promote migration of recipient cells. *Sci. Rep.* **2017**, *7*, 12423. [CrossRef]
34. Mutschelknaus, L.; Peters, C.; Winkler, K.; Yentrapalli, R.; Heider, T.; Atkinson, M.J.; Moertl, S. Exosomes Derived from Squamous Head and Neck Cancer Promote Cell Survival after Ionizing Radiation. *PLoS ONE* **2016**, *11*, e0152213. [CrossRef]
35. Boelens, M.C.; Wu, T.J.; Nabet, B.Y.; Xu, B.; Qiu, Y.; Yoon, T.; Azzam, D.J.; Twyman-Saint Victor, C.; Wiemann, B.Z.; Ishwaran, H.; et al. Exosome transfer from stromal to breast cancer cells regulates therapy resistance pathways. *Cell* **2014**, *159*, 499–513. [CrossRef]
36. Tang, Y.; Cui, Y.; Li, Z.; Jiao, Z.; Zhang, Y.; He, Y.; Chen, G.; Zhou, Q.; Wang, W.; Zhou, X.; et al. Radiation-induced miR-208a increases the proliferation and radioresistance by targeting p21 in human lung cancer cells. *J. Exp. Clin. Cancer Res.* **2016**, *35*, 7. [CrossRef]
37. Xu, S.; Wang, J.; Ding, N.; Hu, W.; Zhang, X.; Wang, B.; Hua, J.; Wei, W.; Zhu, Q. Exosome-mediated microRNA transfer plays a role in radiation-induced bystander effect. *RNA Biol.* **2015**, *12*, 1355–1363. [CrossRef]
38. Le, M.; Fernandez-Palomo, C.; McNeill, F.E.; Seymour, C.B.; Rainbow, A.J.; Mothersill, C.E. Exosomes are released by bystander cells exposed to radiation-induced biophoton signals: Reconciling the mechanisms mediating the bystander effect. *PLoS ONE* **2017**, *12*, e0173685. [CrossRef]
39. Ariyoshi, K.; Miura, T.; Kasai, K.; Fujishima, Y.; Nakata, A.; Yoshida, M. Radiation-Induced Bystander Effect is Mediated by Mitochondrial DNA in Exosome-Like Vesicles. *Sci. Rep.* **2019**, *9*, 9103. [CrossRef]
40. Frenz, M.B.; Mee, A.S. Diagnostic radiation exposure and cancer risk. *Gut* **2005**, *54*, 889–890. [CrossRef]
41. Shi, F.; Wang, X.; Teng, F.; Kong, L.; Yu, J. Abscopal effect of metastatic pancreatic cancer after local radiotherapy and granulocyte-macrophage colony-stimulating factor therapy. *Cancer Biol. Ther.* **2017**, *18*, 137–141. [CrossRef]
42. Wood, J.; Yasmin-Karim, S.; Mueller, R.; Viswanathan, A.N.; Ngwa, W. Single Radiotherapy Fraction with Local Anti-CD40 Therapy Generates Effective Abscopal Responses in Mouse Models of Cervical Cancer. *Cancers* **2020**, *12*, 1026. [CrossRef]

43. Pouget, J.P.; Georgakilas, A.G.; Ravanat, J.L. Targeted and Off-Target (Bystander and Abscopal) Effects of Radiation Therapy: Redox Mechanisms and Risk/Benefit Analysis. *Antioxid. Redox Signal.* **2018**, *29*, 1447–1487. [CrossRef] [PubMed]
44. Berrington de Gonzalez, A.; Darby, S. Risk of cancer from diagnostic X-rays: Estimates for the UK and 14 other countries. *Lancet* **2004**, *363*, 345–351. [CrossRef]
45. Barton, M.B.; Jacob, S.; Shafiq, J.; Wong, K.; Thompson, S.R.; Hanna, T.P.; Delaney, G.P. Estimating the demand for radiotherapy from the evidence: A review of changes from 2003 to 2012. *Radiother. Oncol.* **2014**, *112*, 140–144. [CrossRef]
46. Atun, R.; Jaffray, D.A.; Barton, M.B.; Bray, F.; Baumann, M.; Vikram, B.; Hanna, T.P.; Knaul, F.M.; Lievens, Y.; Lui, T.Y.; et al. Expanding global access to radiotherapy. *Lancet Oncol.* **2015**, *16*, 1153–1186. [CrossRef]
47. Jelonek, K.; Wojakowska, A.; Marczak, L.; Muer, A.; Tinhofer-Keilholz, I.; Lysek-Gladysinska, M.; Widlak, P.; Pietrowska, M. Ionizing radiation affects protein composition of exosomes secreted in vitro from head and neck squamous cell carcinoma. *Acta Biochim. Pol.* **2015**, *62*, 265–272. [CrossRef]
48. Mrowczynski, O.D.; Madhankumar, A.B.; Sundstrom, J.M.; Zhao, Y.; Kawasawa, Y.I.; Slagle-Webb, B.; Mau, C.; Payne, R.A.; Rizk, E.B.; Zacharia, B.E.; et al. Exosomes impact survival to radiation exposure in cell line models of nervous system cancer. *Oncotarget* **2018**, *9*, 36083–36101. [CrossRef] [PubMed]
49. Tatischeff, I.; Larquet, E.; Falcon-Perez, J.M.; Turpin, P.Y.; Kruglik, S.G. Fast characterisation of cell-derived extracellular vesicles by nanoparticles tracking analysis, cryo-electron microscopy, and Raman tweezers microspectroscopy. *J. Extracell. Vesicles* **2012**, *1*. [CrossRef]
50. Lee, W.; Nanou, A.; Rikkert, L.; Coumans, F.A.W.; Otto, C.; Terstappen, L.; Offerhaus, H.L. Label-Free Prostate Cancer Detection by Characterization of Extracellular Vesicles Using Raman Spectroscopy. *Anal. Chem.* **2018**, *90*, 11290–11296. [CrossRef]
51. Jabbari, N.; Nawaz, M.; Rezaie, J. Ionizing Radiation Increases the Activity of Exosomal Secretory Pathway in MCF-7 Human Breast Cancer Cells: A Possible Way to Communicate Resistance against Radiotherapy. *Int. J. Mol. Sci.* **2019**, *20*, 3649. [CrossRef]
52. Yu, X.; Harris, S.L.; Levine, A.J. The regulation of exosome secretion: A novel function of the p53 protein. *Cancer Res.* **2006**, *66*, 4795–4801. [CrossRef]
53. Jelonek, K.; Widlak, P.; Pietrowska, M. The Influence of Ionizing Radiation on Exosome Composition, Secretion and Intercellular Communication. *Protein Pept. Lett.* **2016**, *23*, 656–663. [CrossRef]
54. Szatmari, T.; Kis, D.; Bogdandi, E.N.; Benedek, A.; Bright, S.; Bowler, D.; Persa, E.; Kis, E.; Balogh, A.; Naszalyi, L.N.; et al. Extracellular Vesicles Mediate Radiation-Induced Systemic Bystander Signals in the Bone Marrow and Spleen. *Front. Immunol.* **2017**, *8*, 347. [CrossRef]
55. Szatmari, T.; Persa, E.; Kis, E.; Benedek, A.; Hargitai, R.; Safrany, G.; Lumniczky, K. Extracellular vesicles mediate low dose ionizing radiation-induced immune and inflammatory responses in the blood. *Int. J. Radiat. Biol.* **2019**, *95*, 12–22. [CrossRef]
56. Camussi, G.; Deregibus, M.C.; Bruno, S.; Cantaluppi, V.; Biancone, L. Exosomes/microvesicles as a mechanism of cell-to-cell communication. *Kidney Int.* **2010**, *78*, 838–848. [CrossRef]
57. Valadi, H.; Ekstrom, K.; Bossios, A.; Sjostrand, M.; Lee, J.J.; Lotvall, J.O. Exosome-mediated transfer of mRNAs and microRNAs is a novel mechanism of genetic exchange between cells. *Nat. Cell Biol.* **2007**, *9*, 654–659. [CrossRef] [PubMed]
58. Albanese, J.; Dainiak, N. Regulation of TNFRSF6 (Fas) expression in ataxia telangiectasia cells by ionizing radiation. *Radiat. Res.* **2000**, *154*, 616–624. [CrossRef]
59. Dutta, S.; Warshall, C.; Bandyopadhyay, C.; Dutta, D.; Chandran, B. Interactions between exosomes from breast cancer cells and primary mammary epithelial cells leads to generation of reactive oxygen species which induce DNA damage response, stabilization of p53 and autophagy in epithelial cells. *PLoS ONE* **2014**, *9*, e97580. [CrossRef]
60. Lyng, F.M.; Seymour, C.B.; Mothersill, C. Production of a signal by irradiated cells which leads to a response in unirradiated cells characteristic of initiation of apoptosis. *Br. J. Cancer* **2000**, *83*, 1223–1230. [CrossRef]
61. Lyng, F.M.; Seymour, C.B.; Mothersill, C. Initiation of apoptosis in cells exposed to medium from the progeny of irradiated cells: A possible mechanism for bystander-induced genomic instability? *Radiat. Res.* **2002**, *157*, 365–370. [CrossRef]

62. Lyng, F.M.; Maguire, P.; McClean, B.; Seymour, C.; Mothersill, C. The involvement of calcium and MAP kinase signaling pathways in the production of radiation-induced bystander effects. *Radiat. Res.* **2006**, *165*, 400–409. [CrossRef]
63. Lyng, F.M.; Howe, O.L.; McClean, B. Reactive oxygen species-induced release of signalling factors in irradiated cells triggers membrane signalling and calcium influx in bystander cells. *Int. J. Radiat. Biol.* **2011**, *87*, 683–695. [CrossRef]
64. Jella, K.K.; Moriarty, R.; McClean, B.; Byrne, H.J.; Lyng, F.M. Reactive oxygen species and nitric oxide signaling in bystander cells. *PLoS ONE* **2018**, *13*, e0195371. [CrossRef]
65. Polanco, J.C.; Scicluna, B.J.; Hill, A.F.; Gotz, J. Extracellular Vesicles Isolated from the Brains of rTg4510 Mice Seed Tau Protein Aggregation in a Threshold-dependent Manner. *J. Biol. Chem.* **2016**, *291*, 12445–12466. [CrossRef]
66. Vogel, R.; Willmott, G.; Kozak, D.; Roberts, G.S.; Anderson, W.; Groenewegen, L.; Glossop, B.; Barnett, A.; Turner, A.; Trau, M. Quantitative sizing of nano/microparticles with a tunable elastomeric pore sensor. *Anal. Chem.* **2011**, *83*, 3499–3506. [CrossRef]
67. Roberts, G.S.; Yu, S.; Zeng, Q.; Chan, L.C.; Anderson, W.; Colby, A.H.; Grinstaff, M.W.; Reid, S.; Vogel, R. Tunable pores for measuring concentrations of synthetic and biological nanoparticle dispersions. *Biosens. Bioelectron.* **2012**, *31*, 17–25. [CrossRef]
68. Grigor'eva, A.E.; Dyrkheeva, N.S.; Bryzgunova, O.E.; Tamkovich, S.N.; Chelobanov, B.P.; Ryabchikova, E.I. Contamination of exosome preparations, isolated from biological fluids. *Biomed. Khim.* **2017**, *63*, 91–96. [CrossRef]
69. Kurien, B.T.; Scofield, R.H. Western blotting. *Methods* **2006**, *38*, 283–293. [CrossRef]
70. Laka, K.; Makgoo, L.; Mbita, Z. Survivin Splice Variants in Arsenic Trioxide (As(2)O(3))-Induced Deactivation of PI3K and MAPK Cell Signalling Pathways in MCF-7 Cells. *Genes* **2019**, *10*, 41. [CrossRef]
71. Chandna, S. Single-cell gel electrophoresis assay monitors precise kinetics of DNA fragmentation induced during programmed cell death. *Cytom. A* **2004**, *61*, 127–133. [CrossRef] [PubMed]
72. Collins, A.R. The comet assay for DNA damage and repair: Principles, applications, and limitations. *Mol. Biotechnol.* **2004**, *26*, 249–261. [CrossRef]
73. Zhang, X.; Kluz, T.; Gesumaria, L.; Matsui, M.S.; Costa, M.; Sun, H. Solar Simulated Ultraviolet Radiation Induces Global Histone Hypoacetylation in Human Keratinocytes. *PLoS ONE* **2016**, *11*, e0150175. [CrossRef]

8

Analysis of Serum miRNA in Glioblastoma Patients: CD44-Based Enrichment of Extracellular Vesicles Enhances Specificity for the Prognostic Signature

Theophilos Tzaridis [1,2,3,†], Katrin S Reiners [1,*,†], Johannes Weller [2], Daniel Bachurski [4], Niklas Schäfer [2], Christina Schaub [2], Michael Hallek [4], Björn Scheffler [5], Martin Glas [6], Ulrich Herrlinger [2], Stefan Wild [7], Christoph Coch [1,7,†] and Gunther Hartmann [1,†]

1 Institute of Clinical Chemistry and Clinical Pharmacology, University of Bonn, 53127 Bonn, Germany; theophilos.tzaridis@ukbonn.de (T.T.); ccoch@uni-bonn.de (C.C.); Gunther.Hartmann@uni-bonn.de (G.H.)
2 Division of Clinical Neurooncology, Department of Neurology, Center of Integrated Oncology Aachen-Bonn-Cologne-Düsseldorf, Partner Site Bonn, University Hospital Bonn, 53127 Bonn, Germany; Johannes.weller@ukbonn.de (J.W.); Niklas.Schaefer@ukbonn.de (N.S.); Christina.Schaub@ukbonn.de (C.S.); Ulrich.Herrlinger@ukbonn.de (U.H.)
3 Tumor Initiation & Maintenance Program, NCI-Designated Cancer Center, Sanford Burnham Prebys Medical Discovery Institute, La Jolla, CA 92037, USA
4 Department I of Internal Medicine, Center for Integrated Oncology Aachen-Bonn-Cologne-Düsseldorf, Partner Site Cologne, CECAD Center of Excellence on "Cellular Stress Responses in Aging-Associated Diseases", Center for Molecular Medicine Cologne, University of Cologne, 50937 Cologne, Germany; Daniel.Bachurski@uk-koeln.de (D.B.); Michael.hallek@uk-koeln.de (M.H.)
5 DKFZ-Division Translational Neurooncology at the West German Cancer Center (WTZ), German Cancer Consortium (DKTK), DKFZ Heidelberg & Partner Site Univ Hospital Essen, 45147 Essen, Germany; b.scheffler@dkfz-heidelberg.de
6 Division of Clinical Neurooncology, Department of Neurology and West German Cancer Center (WTZ), German Cancer Consortium, University Hospital Essen, 45147 Essen, Germany; martin.glas@uk-essen.de
7 Miltenyi Biotec & Biomedicine GmbH, 51429 Bergisch Gladbach, Germany; StefanW@Miltenyi.com
* Correspondence: kreiners@uni-bonn.de
† These authors contributed equally to this work.

Abstract: Glioblastoma is a devastating disease, for which biomarkers allowing a prediction of prognosis are urgently needed. microRNAs have been described as potentially valuable biomarkers in cancer. Here, we studied a panel of microRNAs in extracellular vesicles (EVs) from the serum of glioblastoma patients and evaluated their correlation with the prognosis of these patients. The levels of 15 microRNAs in EVs that were separated by size-exclusion chromatography were studied by quantitative real-time PCR, followed by CD44 immunoprecipitation (SEC + CD44), and compared with those from the total serum of glioblastoma patients (n = 55) and healthy volunteers (n = 10). Compared to total serum, we found evidence for the enrichment of miR-21-3p and miR-106a-5p and, conversely, lower levels of miR-15b-3p, in SEC + CD44 EVs. miR-15b-3p and miR-21-3p were upregulated in glioblastoma patients compared to healthy subjects. A significant correlation with survival of the patients was found for levels of miR-15b-3p in total serum and miR-15b-3p, miR-21-3p, miR-106a-5p, and miR-328-3p in SEC + CD44 EVs. Combining miR-15b-3p in serum or miR-106a-5p in SEC + CD44 EVs with any one of the other three microRNAs in SEC + CD44 EVs allowed for a prognostic stratification of glioblastoma patients. We have thus identified four microRNAs in glioblastoma patients whose levels, in combination, can predict the prognosis for these patients.

Keywords: biomarkers; glioblastoma; extracellular vesicles; microRNA; immunoprecipitation; CD44

---

1. Introduction

Glioblastomas lacking isocitrate-dehydrogenase (*IDH*) 1 or 2 mutations (IDH-wildtype glioblastomas) are the most aggressive primary brain tumors, mainly occurring in adults, and exhibit a dismal prognosis [1]. First-line treatment includes surgical resection if feasible, followed by radiochemotherapy with temozolomide (TMZ) [2]. Despite a large number of studies on biomarkers for glioblastoma, only one has regularly been applied for prognostic stratification in routine clinical use: Promoter methylation of the *O-6-methylguanine-DNA methyltransferase* (*MGMT*) gene in tumor tissue. Promoter methylation of the *MGMT* gene leads to impaired DNA-repair mechanisms and not only correlates with superior survival, but also leads to a better response to TMZ treatment [3]. Based on this observation, novel data show an enhanced response of combination therapy with TMZ and lomustine (CCNU) for patients with tumors harboring *MGMT*-promoter methylation, and this intensified treatment does not compromise the patients' quality-of-life [4,5]. Due to the limited number of treatment options for recurrent disease [6], biomarkers capable of stratifying patients are crucial.

MicroRNAs (miRNAs) are non-coding single-stranded RNA molecules with a length of 21–25 nucleotides [7]. After initial processing in the nucleus by RNA polymerases II and III [8], the precursor miRNA (pre-miRNA) is transported into the cytoplasm and then processed by the cytoplasmic RNase III protein Dicer, thereby creating mature miRNA [7–9]. The purpose of the mature microRNAs is complex, and not yet fully understood, but usually involves the regulation (mostly repression) of gene expression by binding to either the 3′- or the 5′-untranslated region (UTR) of mRNAs [8,10,11]. microRNAs are loaded into an Argonaute protein (Ago) as part of the RNA-induced silencing complex (RISC) and are thereby protected from cytoplasmic nucleases [7].

The role of microRNAs in glioblastoma has already been extensively examined. They have been shown to be involved in invasiveness and proliferation (e.g., hsa-miR-21 or hsa-miR-15b) [12], to induce angiogenesis [13], and to modulate the innate immune system by polarizing it towards M2 macrophages and thereby inducing an immunosuppressive environment (hsa-miR-21 and hsa-miR-451) [13]. Interestingly, many of these functions require the trafficking of miRNA between cells, which occurs through either gap junctions or the exchange of extracellular vesicles (EVs) [13,14]. While some miRNAs are mainly found in miRNA-binding protein complexes, such as Ago-2, or high-density lipoproteins (HDL), other types of miRNAs are uniquely packaged into EVs [15], which makes them a potentially interesting biomarker in tumor diagnostics [16].

EVs are small particles formed by either direct budding of the plasma membrane (typically "large EVs") or by the fusion of a multivesicular endosome with the plasma membrane (typically "small EVs"), thus releasing the EVs into the extracellular space [17]. In a recent study, we identified relevant protein markers that are highly elevated in EVs from the serum of patients with glioblastoma, compared to those from healthy volunteers (HV), and these proteins are capable of detecting tumor progression [18]. Many studies have identified microRNAs in the biofluids (e.g., cerebrospinal fluid (CSF), serum and plasma) of patients with glioblastoma, which can be EV-independent [19] or contained in EVs [20]. While some reports have demonstrated the upregulation of microRNAs in patients with high-grade glioma compared to low-grade glioma [21], no report has thus far shown the potential of microRNAs in biofluids to stratify patients into different prognostic groups at critical time-points during treatment. Moreover, it is unknown whether the detection of any relevant microRNAs in the serum of glioma patients can be enhanced through specifically capturing tumor-derived EVs.

In this study, we evaluated suitable methods for the separation of EVs that could enrich the potentially glioblastoma-relevant microRNAs within the samples. In addition, we tested whether these microRNAs could serve as biomarkers to determine patient prognosis when purified from serum EVs (EV microRNAs) or as microRNAs directly from total serum (total serum microRNAs).

2. Results

2.1. Enrichment of Disease-Relevant microRNAs in Serum EV Samples Using Size-Exclusion Chromatography (SEC) Followed by CD44-Based Immunoprecipitation

Our study included 55 patients in total, 26 (47.3%) of which were treated within the multicenter Phase III CeTeG/NOA-09 trial [4] and 29 (52.7%) in the Division of Clinical Neurooncology of the University Hospital of Bonn. The median age at diagnosis was 56 years (range: 19–77 years) and the median overall survival was 2.35 years (range: 0.39–5.17 years). The vast majority of patients (53/55, i.e., 96.4%) had *IDH*-wildtype tumors and *MGMT* promoter methylation was observed in 33 (60%) cases (Table S1). Sampling was performed at the Q3 time point (i.e., the third quarter of the first year after diagnosis, 6–9 months after the initiation of primary radiotherapy and chemotherapy) for 54/55 patients and in the fifth month for one patient.

Consistent with previous results [21], serum from glioblastoma patients showed a much higher concentration of small EVs (size range: 87–166 nm) compared to HV when they were isolated by using size-exclusion chromatography (SEC, Figure 1A). In order to specifically isolate relevant glioblastoma-associated EVs, we subsequently performed immunoprecipitation using CD44 (SEC + CD44), which we had previously identified in a screen of serum EV proteins as being important in tumor progression [18]. Using this technique, we were able to capture a distinct EV subpopulation, as shown in Figure 1. Immunodetection was used to determine the presence of EV marker proteins flotillin-1 and CD9. Calnexin, which is a protein associated with the endoplasmic reticulum, serum protein albumin and apolipoprotein-A1 were absent or strongly reduced after EV isolation via SEC + CD44 (Figure 1C). Transmission electron microscopy (TEM) analysis showed that the CD44-enriched fraction mainly contained EVs with a size and shape typical of small EVs (Figure 1B). The TEM images show that lipoprotein contamination present after SEC (Figure 1B, left lane) is diminished after additional immunoprecipitation with CD44 beads (Figure 1B, middle lane). To allow the quantification of SEC + CD44 EVs, they were effectively diluted from capture-beads (visible as small black dots in middle lane pictures), as shown in Figure 1B, in the right lane.

Based on extensive screening of the literature, we selected 15 microRNAs that have been described to be present and relevant in the serum and plasma of glioblastoma patients (Table S2). Out of these 15 pre-screened microRNAs, we identified eight that consistently had Ct values lower than 36 in our glioblastoma patients. To determine if these miRNAs are more abundant in CD44-enriched EVs (SEC + CD44 EVs) compared to miRNAs isolated from total serum, we studied the levels of these microRNAs in the serum of glioblastoma patients ($n = 55$) at the Q3 time point. In total, we detected an enrichment of five out of the eight microRNAs tested in SEC + CD44-purified EVs compared to total serum (miR-21-3p, miR-106a-5p, miR-155-5p, and let-7a-5p with significantly higher levels and miR-486-5p with a non-significant trend toward a higher level, Figure 2A–H). Two out of eight miRNAs showed significantly higher levels in total serum compared to the purified EVs (miR-15b-3p and miR-23a-3p, Figure 2), indicating that these miRNAs are not enriched in SEC + CD44 EVs.

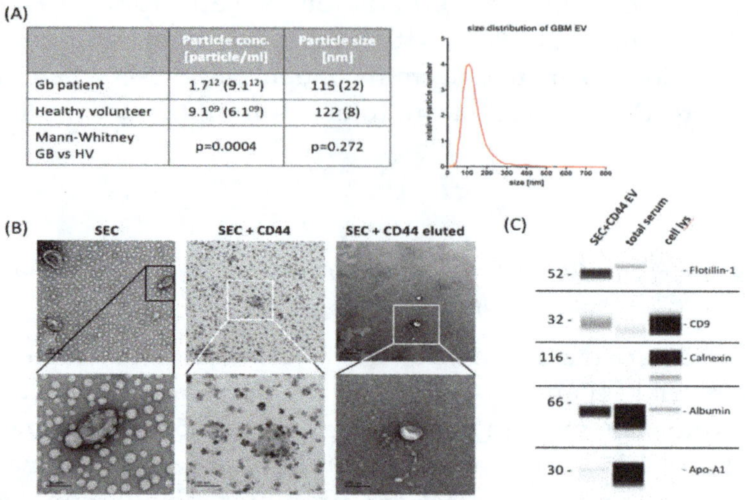

Figure 1. Yield and characteristics of serum-derived extracellular vesicles. (**A**) NTA of extracellular vesicles (EVs) isolated from 500 µL serum by size-exclusion chromatography (SEC) to determine the yield and particle size. The table states the mean particle concentration and mean modal size of 55 glioblastoma patients and 5 healthy volunteers (HV). Standard deviation is given in parentheses; histograms: exemplary size distribution of glioblastoma-EVs measured by NTA. (**B**) Transmission electron microscopy images showing the typical EV morphology. Scale bar represents, in all images, 200 nm in the upper row (30k × magnification) and 100 nm in the lower row (110k × magnification). Images in the left lane show EVs after SEC, middle lane images show CD44-captured SEC-EVs, and right lane images represent EVs that were diluted from CD44-beads. (**C**) Wes ProteinSimple immunodetection confirming the presence of the EV-markers flotillin-1 and CD9 and the absence of non-EV protein calnexin in SEC + CD44 EV preparations. Cell and total serum lysates were used as positive controls for the non-EV marker calnexin, apolipoprotein A1, and serum albumin, respectively. NTA = nanoparticle tracking analysis; cell lys = cell lysate of PBMCs (1:10) or glioblastoma cell line Gli36 (1:100); Apo-A1 = apolipoprotein A1.

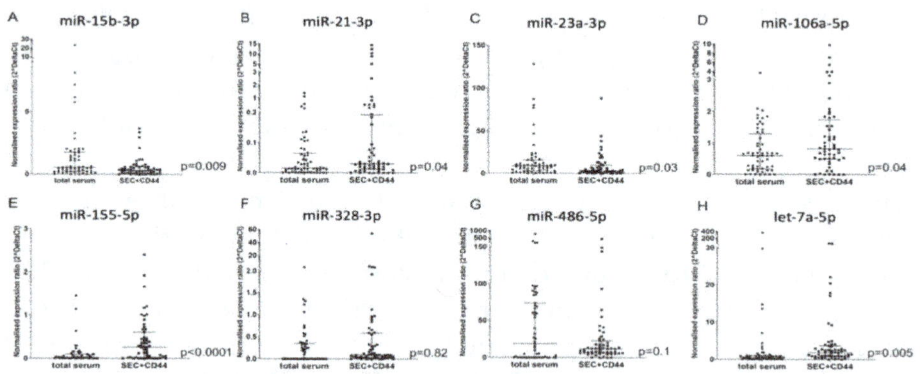

Figure 2. Comparison of microRNA detection in glioblastoma serum, depending on the source. (**A**) miR-15b-3p, (**B**) miR21-3p, (**C**) miR-23a-3p, (**D**) miR-106-5p, (**E**) miR-155-5p, (**F**) miR-328-3p, (**G**) miR-486-5p, (**H**) let-7a-5p. Depicted is the normalized expression ratio (2^deltaCt) of eight microRNAs (**A**–**H**) in total serum and in SEC + CD44-purified EVs from the serum of glioblastoma patients (n = 55). Expression levels were compared using a Wilcoxon rank sum test. Note that miR-15b-3p shows higher levels in total serum, as opposed to miR-21-3p and miR-106a-5p, which are upregulated in SEC + CD44 EVs. For miR-328-3p, no consistent up- or downregulation was observed between the two samples, which was partially because only 26/55 (47%) total serum samples were above the limit of detection. Bars depict the median value and the interquartile range.

When comparing SEC + CD44-purified EV microRNA from the serum of glioblastoma patients to HV, we saw significantly higher levels of miR-15b-3p, miR-21-3p, miR-155-5p, and let-7a-5p in glioblastoma patients ($p = 0.01$, $p = 0.001$, $p = 0.01$, and $p = 0.008$, respectively) and a non-significant trend for miR-23a-3p and miR-106a-5p ($p = 0.18$ and $p = 0.07$, respectively), while no significant difference was detected for miR-23a-3p, miR-328-3p, and miR-486-5p ($p = 0.45$ and $p = 0.48$ respectively, Figure 3A–H).

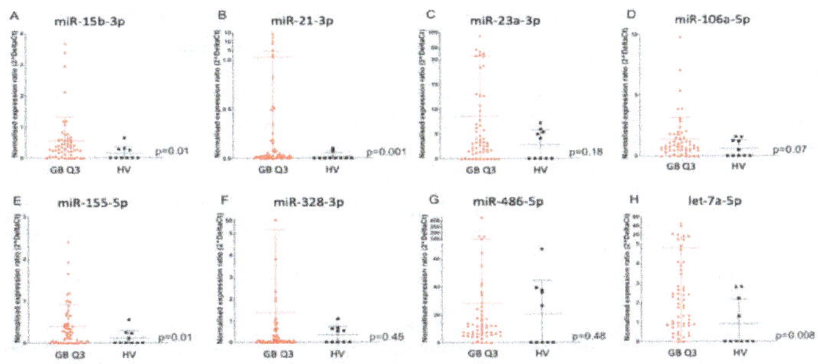

Figure 3. MicroRNA expression in SEC + CD44 EV. Normalized expression ratio (2^deltaCt) of eight microRNAs (**A**–**H**) in SEC + CD44 EV from glioblastoma patients ($n = 55$) versus HV ($n = 10$, average of two time points). (**A**) miR-15b-3p, (**B**) miR21-3p, (**C**) miR-23a-3p, (**D**) miR-106-5p, (**E**) miR-155-5p, (**F**) miR-328-3p, (**G**) miR-486-5p, (**H**) let-7a-5p. Expression levels were compared using the Mann–Whitney U test. Bars depict the median value and the interquartile range.

2.2. Four SEC+CD44-EV miRNAs and Total Serum miR-15b-3p Correlate with Survival in Glioblastoma Patients

To evaluate the prognostic potential of different microRNAs, we correlated the normalized expression ratio with the overall survival of glioblastoma patients ($n = 55$). miR-15b-3p, miR-21-3p, miR-106a-5p, and miR-328-3p in SEC + CD44 EVs showed a significant correlation with survival, with miR-15b-3p, miR-21-3p, and miR-328-3p exhibiting a negative correlation (high levels were associated with an inferior prognosis) and miR-106a-5p a positive correlation (high levels were associated with a better prognosis; Table 1A). While miR-15b-3p showed a weak correlation (Spearman $r = -0.27$), miR-21-3p, miR-106a-5p, and miR-328-3p showed intermediate correlation values (absolute Spearman $r > 0.34$). Interestingly, when considering total serum microRNAs, only miR-15b-3p correlated with inferior survival (Spearman $r = 0.4$) and miR-328-3p was not measurable in 29/55 (53%) patients (Table 1B). For all of the other microRNAs analyzed, no correlation with survival was found in either SEC + CD44 EVs or total serum (data not shown).

Interestingly, we found distinct differences of miR levels in SEC + CD44 EVs, depending on the MGMT promoter methylation status, with miR-15b-3p, miR-21-3p, and miR328-3p being enriched in MGMT-non-methylated and miR-106a-5p in MGMT-methylated patient samples (Figure S1). For total serum, we saw higher levels of miR-15b-3p in MGMT-non-methylated compared to MGMT-methylated patient samples (Figure S2). When only assessing the prognostic potential of the above-mentioned markers for MGMT-methylated patients, we observed a non-significant correlation with overall survival for miR-21-3p and miR-328-3p in SEC + CD44 EVs (data not shown).

The four microRNAs that were identified as putative prognostic markers were selected for dichotomous assessment based on the median of the normalized expression ratios and using Kaplan–Meier curves for data visualization. To define the subgroups, we calculated the median value of the normalized expression for each microRNA and split the patients into two groups, with one group including patients with expression values lower than the median and the other group with values equal to or higher than the median. In an analysis of single SEC + CD44 EV microRNAs, we saw a curve separation for all four miRNAs, which was less pronounced for miR-15b-3p and miR-328-3p (Figure 4A–D). Consistent with our correlation data, the subgroup of patients with levels of miR-106a-5p higher than the median had a clear survival benefit compared to those with lower levels, as opposed to miR-21-3p, where high levels of the miRNA defined a prognostically inferior group (Figure 4B,C).

Table 1. Correlation of the survival of 55 patients with microRNA levels from SEC + CD44 EV (**A**) or total serum (**B**) using the Spearman's rank correlation coefficient, r. Depicted are the Spearman r values showing either a positive or negative correlation, the p value, and the degree of significance (* $p < 0.05$, ** $p < 0.01$, *** $p < 0.001$, n.s. = not significant). For miR-328-3p in total serum, no correlation with survival was calculated because 53% of the values were non-measurable.

(A)				
SEC + CD44	**miR15b-3p**	**miR-21-3p**	**miR-328-3p**	**miR-106-5p**
Spearman R	−0.2746	−0.4372	−0.3407	0.3501
p (two-tailed)	0.043	0.00008	0.011	0.009
p value	*	***	*	**
(B)				
Total serum	**miR15b-3p**	**miR-21-3p**	**miR-328-3p**	**miR-106-5p**
Spearman R	-0.3947	0.0225	-	0.06753
p (two-tailed)	0.003	0. 78	-	0.62
p value	**	n.s.	-	n.s.

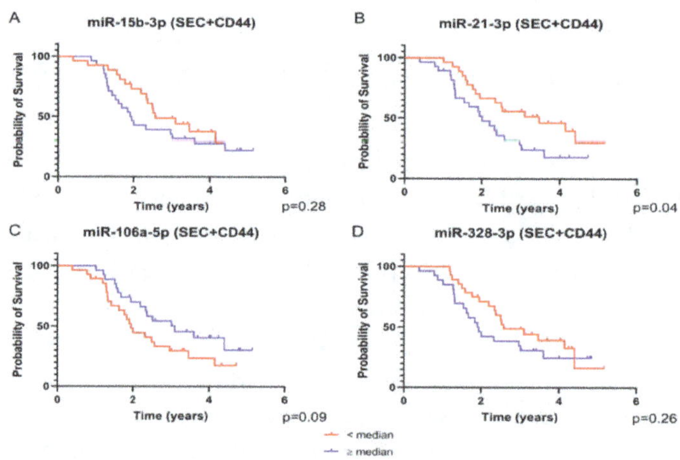

Figure 4. Survival analysis based on the expression of SEC + CD44 EV microRNA. Depicted are survival curves using a dichotomous assessment based on microRNA levels of SEC + CD44 EVs. ((**A**) miR-15b-3p, (**B**) miR-21-3p, (**C**) miR-106a-5p, (**D**) miR-328-3p) from glioblastoma patient serum ($n = 55$). Red color indicates patients with values equal to or higher than the median and blue color represents values lower than the median normalized expression ratio. Note that all four microRNAs were able to stratify the patients into different prognostic subgroups, albeit not reaching statistical significance. The p value for miR-21-3p is above the critical Benjamini–Hochberg value (0.03, Tables S3 and S4). Also note that for miR-106a-5p, higher values correlate with a better prognosis.

In analyses of single total serum microRNAs, we observed a clear curve separation for miR-15b-3p, as opposed to miR-21-3p and miR-106a-5p (Figure 5A–C). For miR-328-3p, no Kaplan–Meier curve was generated, because more than 50% of the values were below the detection limit.

2.3. Combination of SEC + CD44-EV- and Total Serum-microRNAs Allows Patient Stratification into Prognostically Relevant Subgroups

We next sought to determine whether a combination of different microRNAs results allows a more precise survival prediction compared to single microRNA analysis. Indeed, when we combined the analysis of miR-15b-3p in total serum with each of the other three EV-contained microRNAs, the prediction of prognosis was improved (based on the p-value determined) compared to the single microRNA analyses (Figure 6A–C).

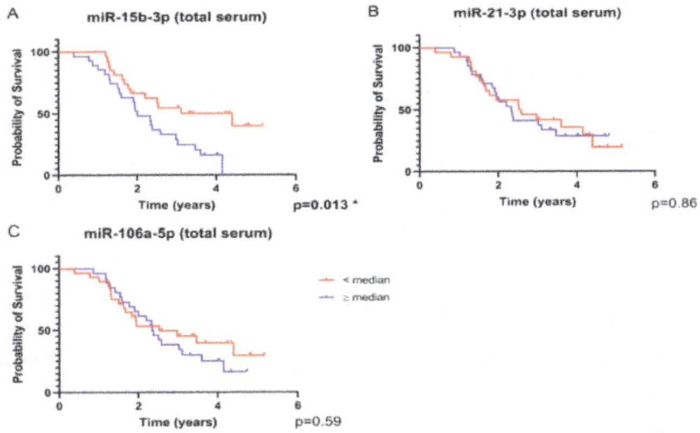

Figure 5. Survival analysis based on total serum microRNA expression. Depicted are survival curves using a dichotomous assessment based on microRNA levels ((**A**) miR-15b-3p, (**B**) miR-21-3p, (**C**) miR-106a-5p) from total serum of glioblastoma patients (n = 55). Red color indicates patients with values equal to or higher than the median and blue color represents values lower than the median normalized expression ratio. For miR-328-3p, no graph was generated because more than 50% of the values were non-measurable. Note that the separation of the curves based on miR-15b-3p is significant after Benjamini–Hochberg correction for multiple testing (Tables S3 and S4).

When combining markers tested only in SEC + CD44 EVs, three of these combinations, always including miR-106a-5p as one of the two microRNAs, yielded prognostically significant subgroups (Figure 7A–C). Because comparing multiple variables can give a bias and lead to a high risk of false significant results (type I error), we applied Benjamini–Hochberg correction [22]. Importantly, the combinations of microRNAs in Figures 6 and 7, as well as miR-15b-3p in total serum (Figure 5A), remained significant after this correction, as opposed to a single analysis of miR-21-3p (Figure 4B).

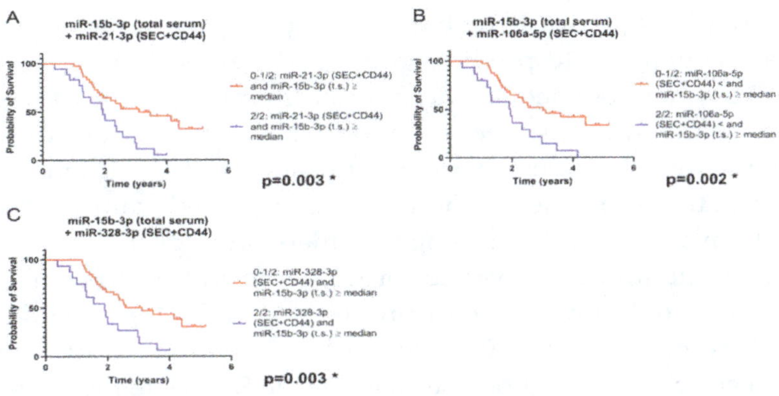

Figure 6. Survival analysis based on a combination of microRNA levels in total serum (t.s.) and/or SEC + CD44 EVs (combinations: miR-15b-3p in t.s. plus (**A**): miR-21-3p in SEC + CD44 EVs; (**B**): miR-106a-5p in SEC + CD44 EVs; (**C**): miR-328-3p in SEC + CD44 EVs). Depicted are survival curves using a dichotomous assessment from glioblastoma patients (n = 55) in both SEC + CD44 separated serum EVs and total serum. Blue color indicates patients whose microRNA profile fulfils both conditions, while red color indicates patients whose profile fulfils at most one out of two conditions. Note that the combination of miR-15b-3p in serum with each of these four microRNAs in SEC + CD44 EVs allowed for a prognostic stratification of this patient population, even after Benjamini–Hochberg correction (Table S1).

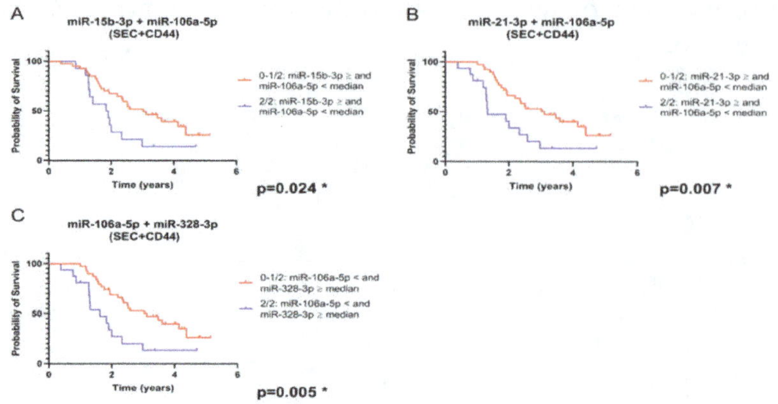

Figure 7. Survival analysis based on the expression of a combination of microRNAs isolated from SEC + CD44 EV (combinations: miR-106a-5p plus (**A**): miR-15b-3p; (**B**): miR-21-3p in SEC + CD44 EVs; (**C**): miR-328-3p). Depicted are survival curves using a dichotomous assessment based on a combination of microRNA levels from glioblastoma patients (n = 55) in SEC + CD44 separated serum EVs. Blue color indicates patients whose microRNA profile fulfils both conditions, while red color indicates patients whose profile fulfils at most one out of two conditions. Note that the combination of miR-106a-5p with each of the other three microRNAs allowed for a prognostic stratification of this patient population, even after Benjamini–Hochberg correction for multiple testing (Tables S3 and S4).

3. Discussion

In this study, we report on a new method for EV separation using the already established method of size-exclusion chromatography (by using qEV columns from IZON®), followed by immunoprecipitation with CD44-conjugated beads, thereby allowing specific enrichment of glioblastoma-associated EVs from patient serum. Using this novel method for EV separation, as well as total serum analyses, we identified a panel of four miRNAs suitable as biomarkers (alone or in combination) that allow for the prognostic stratification of glioblastoma patients at a relevant time point of the disease course.

Glioblastoma is a devastating disease which is currently lacking established biomarkers to aid in the prediction of prognosis and treatment decisions. Although progression-free survival (PFS) has been discussed as a surrogate parameter for overall survival in clinical trials of glioblastoma [23], it is mainly based on radiological findings [24,25], which are known to yield equivocal results [26]. Therefore, identifying biomarkers which would predict prognosis at an early time point during treatment is crucial. The median PFS is heterogeneous, and frequently ranges from 6 to 8 months [27,28]. In this study, we therefore chose the third quartile (6–9 months after the start of the adjuvant treatment) as a time for measuring microRNAs in serum and serum-derived EVs. Robust and feasible prognostic biomarkers at this critical time-point could therefore be highly relevant for the course of disease and help both patients and physicians in developing treatment strategies.

We have previously identified a key role for CD44, found on the surface of serum-EVs from glioblastoma patients, in the detection of tumor progression [18]. Due to a high upregulation of CD44 in glioblastoma patients compared to HV, we hypothesized that the immunoprecipitation of EVs with CD44 would lead to an enrichment of glioblastoma-specific EVs. The rationale for first separating the EVs using SEC was to ensure a purer sample of CD44-carrying EVs and minimize the risk of capturing EV-independent CD44, which is known to be present on various cell types, including T-cells and macrophages [29]. Furthermore, we intended to reduce the amount of soluble CD44 in the sample, which is shed by matrix-metalloproteases from the surface of tumor and non-tumor cells [30]. CD44 can also be found on EVs from different non-malignant cells, such as B- or T-cells [31,32], thus, we cannot exclude the possibility that some non-glioblastoma EVs are found in our SEC + CD44 EV fractions. However, the analysis of SEC-enriched serum-derived EVs from HV showed significantly lower CD44 levels compared to EVs from the serum of glioblastoma patients, indicating that CD44-positive EVs are enriched and highly relevant in glioblastoma patients [18]. This is in line with the intriguing hypothesis that glioblastoma is actually a systemic disease, as glioblastoma cells are known to affect non-tumor cells [33], suggesting that the EV-composition of glioblastoma patients differs from that of HV and thus could serve as a biomarker, irrespective of their origin.

The microRNAs selected for this project have all previously been studied in glioblastoma, either in total serum or EVs [19,20,34–36]. While some of these papers report that the majority of extracellular microRNAs are found bound to Ago [14], other studies claim that the majority of extracellular microRNAs are found inside small or large EVs [15,37,38]. Either way, it is clear that EV-associated microRNAs play a central role in intercellular communication and are capable of regulating key oncogenic processes, such as metastasis formation [16,39]. Ebrahimkhani et al. performed deep sequencing of exosomal microRNAs and discovered a panel of microRNAs that were not only upregulated in EVs from glioma patients, but were also capable of discriminating between *IDH*-mutated glioma and *IDH*-wildtype glioblastoma [34]. To account for differences between microRNAs found in total serum and glioblastoma-associated EVs, we examined the microRNA levels in both states and found a significant upregulation in four out of eight microRNAs, thus supporting our rationale for an enrichment of the relevant microRNAs through this isolation technique. Intriguingly, miR-328-3p, which was one of the microRNAs implicated in Ebrahimkhani et al. [34], did not show an increased concentration in EVs, yet its levels correlated with survival in our glioblastoma cohort. This finding underlines the importance of biological relevance rather than absolute quantitative changes in biomarker concentrations. miR-23, albeit having no prognostic significance in our study, might exude biological relevance based on its relatively strong increase in glioblastoma EVs. On the other hand, miR-15b-3p was found at higher levels in total serum and possessed the highest prognostic relevance using single microRNA analysis, while allowing for an even better stratification when combined with other microRNAs in SEC + CD44 EVs. Therefore, our data suggest that both targeted EV and total serum microRNAs should be studied for assessing the prognosis of glioblastoma patients. Notably, this is, to the best of our knowledge, the first study highlighting the prognostic potential of these biomarkers for glioblastoma patients.

The assessment of microRNA levels based on the MGMT methylation status showed that microRNAs indicating a non-favorable outcome (miR-15b-3p, miR-21-3p, and miR328-3p) were elevated in MGMT-non-methylated patient samples, while miR106a-5p showed lower levels compared to MGMT-methylated samples, thereby possibly highlighting a correlation of microRNAs with biologically aggressive tumors. While we only saw non-significant trends towards inferior survival for MGMT-methylated patients showing higher levels of miR-21-3p and miR-328-3p (data not shown), caution is warranted due to the low patient numbers.

All of the microRNAs from our prognostic panel have also been shown to carry important biological functions in glioblastoma, although their up- or downregulation in glioblastoma has not been conclusively resolved. miR-21-3p in EVs secreted by glioblastoma cells is known to promote oncogenesis, angiogenesis, and microglia activation [13] and has unanimously been described to be upregulated in the plasma of glioma patients [40], which is consistent with our data showing higher levels of miR-21-3p in glioblastoma-EVs compared to HV, as well as prognostic significance in SEC + CD44-purified EVs. The levels of miR-15b have also been shown to correlate with a high proliferation of glioma cells in vitro [12], but the levels in the serum of glioma patients were lower compared to HV [41]. Nevertheless, higher serum levels of miR-15b correlated with a higher WHO grade, thereby indicating that more aggressive tumors exhibit higher ratios of this microRNA compared to less proliferative tumors, which corresponds to the survival data in our study. Contradictory data have been published regarding the roles of miR-106a and miR-328 in glioma. High levels of miR-106a in glioma cells have been associated with reduced proliferation and increased apoptosis [12], but in other publications, have been associated with invasiveness [42]. Zhi and colleagues identified Fas-activated serine/threonine kinase (FASTK) as a direct target of miR-106a-5p and showed that a reduced expression of miR-106a-5p or increased expression of FASTK is significantly associated with poor survival in human astrocytoma patients [43], which is compatible with our data. While miR-328 has been shown to mediate the invasiveness of glioma cells via the downregulation of *Secreted Frizzled-related protein 1* (*SFRP1*) [44], Ebrahimkhani et al. reported that serum EVs from glioblastoma patients exhibited lower levels of miR-328 compared to HV [34]. Notably, the EVs used in this study were non-specifically purified from whole serum. Therefore, it would be tempting to speculate that a more specific separation of EVs using CD44 as a target antigen, and thereby enriching the population for glioblastoma EVs, would lead to a correlation of high miR-328 levels with a lower prognosis, as was the case in our study and as reported by Delic et al. [44].

In this study, we used serum as the biofluid source for miRNAs. While plasma used to be the preferred biomaterial for conducting EV studies, because coagulation activates platelets, resulting in an increased release of platelet-derived EVs [45], EVs and microRNAs are now increasingly being evaluated from the serum of glioblastoma patients [19,34]. Intriguingly, large-scale studies of different RNA subclasses, including microRNAs, in biofluids revealed only minor differences in the concentrations between serum and plasma [46,47]. These differences increased substantially if different RNA-extraction and EV-separation methods were applied, implying that both biofluids could yield similar results when established and validated protocols are used for purification. In our study, we used the miRNeasy kit by Qiagen for RNA extraction, as this method was identified as a suitable extraction method for high-yield and high-quality exosomal RNA [47]. There is still debate over what is the most suitable reference microRNA for calculating deltaCt. While some studies have used raw Ct values or exogenous non-human microRNAs as a reference, newer studies have discouraged the use of these conventions and instead recommend the use of at least one, and ideally two, microRNAs as a reference [48]. Based on these findings, we had previously performed screening for suitable reference microRNAs and identified miR-103 and miR-484 as two reliable housekeeper miRNAs (data not shown).

This is, to the best of our knowledge, the first study defining a microRNA signature (in EVs and cell-depleted serum) that shows a direct correlation with the survival of glioblastoma patients, thereby allowing prognostic stratification. Moreover, we introduce a novel approach for separating and enriching glioblastoma-specific EVs by combining size-exclusion chromatography and immunoprecipitation with CD44 as a unique target antigen. This optimized enrichment of glioblastoma-specific EVs leads to a more precise and sensitive prediction of prognosis compared to an unspecific serum-derived microRNA analysis. These encouraging data remain to be confirmed in larger prospective clinical trials, which could then pave the way for the introduction of an miRNA biomarker signature for clinicians, using methods that are feasible and time-efficient for routine diagnostic laboratories.

4. Materials and Methods

4.1. Ethical Approval

Studies on two cohorts of glioblastoma patients were separately approved by the Ethical Committee of the University of Bonn (protocol number for patients treated in Bonn: 182/08 and in the CeTeG/NOA-09 trial: 093/10) and on HV (Protocol number: 007/17).

4.2. Sample Collection

Serum was collected in 9 mL serum (S-Monovette, Sarstedt, Nuembrecht, Germany) tubes from HV and glioblastoma patients. For glioblastoma patients, blood was drawn at the time of an MRI visit in the third quartile period of their adjuvant treatment (i.e., 6–9 months after the initial diagnosis). For HV, blood was drawn at two different time-points with a time interval of three months. After a resting period of 30 min at room temperature (RT), the samples were centrifuged for 15 min at 2000× g at RT, followed by a further centrifugation step for 20 min at 3000× g at 6 °C. After filtration with a 0.45 μm filter, the serum was stored in aliquots at −80 °C.

4.3. EV Separation

EVs from serum samples were separated by size-exclusion chromatography (SEC) using the sepharose-based qEV columns (iZON Science, Christchurch, New Zealand), according to the manufacturer's recommendations. In short, 0.5 mL of serum was applied to the column and the EVs were eluted with Hank's balanced salt solution (HBSS). Next, 500 µL fractions were collected, fractions 8 to 10 were pooled, and a protease inhibitor (cOmplete, EDTA-free Protease Inhibitor Cocktail, Roche, Mannheim, Germany) was added to a final one-fold dilution. Subsequently, the combined fractions 8–10 (in total, 1.5 mL) were used for immunoprecipitation with 50 µL CD44-conjugated MicroBeads (CD44: Clone DB105) that were specifically designed for EV isolation and produced for this study. Separation of the EV-bound MicroBeads was performed with µColumns using the µMACS separator, according to the manufacturer's protocol (all Miltenyi Biotec, Bergisch Gladbach, Germany). EVs were eluted in 110 µL HBSS (ThermoFisher Scientific, Waltham, MA, USA) with protease inhibitor (Roche, Mannheim, Germany).

4.4. Nanoparticle Tracking Analysis (NTA)

As previously described, ZetaView Nanoparticle Tracking (Particle Metrix, Meerbusch, Germany) was used for NTA, according to the manufacturer's guidelines [49].

4.5. Transmission Electron Microscopy (TEM)

TEM was conducted based on the protocol previously described by Bachurski et al. [49]. Briefly, after loading 5 µL of an EV sample onto formvar-coated copper grids (Science Services, Munich, Germany), the EVs were fixed with 2% paraformaldehyde for 5 min, washed with PBS, fixed again for 5 min with 1% glutaraldehyde, washed with ddH_2O, and incubated with contrast dye (1.5% uranyl acetate) for 4 min. Images were captured with a Gatan OneView 4K camera (Gatan, Pleasanton, CA, USA) on a Jem-2100Plus microscope (JEOL) operating at 200 kV.

4.6. Wes^{TM} Simple Immunodetection

The presence of the EV markers CD9 and flotillin-1 and the absence of the endoplasmic-reticulum protein calnexin in the purified EV samples were confirmed using Wes^{TM} Simple Western technology with the Wes instrument (ProteinSimple, San Jose, CA, USA). In this study, 3 µL EVs were combined with 1 µL 0.1 × sample buffer and 1 µL 5 × fluorescent master mix for each lane. Analyses for flotillin-1 (clone 18/flotillin-1; BD Biosciences, Dilution: 1:100), calnexin (clone: C5C9; Cell Signaling Technology, Dilution: 1:80), apolipoprotein A1 (polylonal, R&D Systems, Concentration: 5 µg/mL), and human serum albumin polylonal (R&D Systems, Concentration: 5 µg/mL) were conducted under reducing (DTT-based buffer) conditions, while CD9 (clone D801A, Cell Signaling Technology, Dilution: 1:80) analysis was run under non-reducing conditions and using the 12–230 kDa Wes Separation Module. Anti-rabbit, anti-mouse, and anti-goat antibody detection modules (all from ProteinSimple) were used, according to the manufacturer's instructions. The default run conditions were changed to stocking-gel uptake: 22 s; sample uptake: 15 s; primary-antibody: 90 min; and secondary-antibody: 40 min incubation. Data analysis was performed with Compass software (ProteinSimple).

4.7. RNA Extraction and Quantitative Real-Time Polymerase Chain Reaction (qRT-PCR)

RNA extraction was performed using the Qiagen Micro-RNeasy kit, according to the manufacturer's protocol (Qiagen, Hilden, Germany). Following RNA extraction, equal volumes of the RNA samples were used for cDNA synthesis using a TaqMan Advanced miRNA cDNA Synthesis Kit, according to the manufacturer's protocol (ThermoFisher Scientific, Waltham, MA, USA). qRT-PCR was performed by using a QuantStudio 7 with TaqMan Advanced Control miRNA Assay (Table S2, both ThermoFisher Scientific, Waltham, MA, USA). miR-103a-3p and miR-484 were used as reference miRNAs (Table S2). The QuantStudio 7 PCR protocol included (1) enzyme activation at 95 °C for 20 s and (2) 40 cycles of denaturation (1 s at 95 °C) and annealing/extension (20 s at 60 °C).

4.8. Data Deposition

We have submitted all relevant data from our experiments to the EV-TRACK knowledgebase (EV-TRACK ID: 200051) [50].

4.9. Statistical Analysis

Statistical analysis was performed using GraphPad Prism software (Version 8.2.1, La Jolla, CA, USA). A Mann–Whitney *U* test was used to detect differences in the EV concentration, as well as differences in microRNAs between glioblastoma patients and HV. To compare levels of microRNAs isolated from total serum and SEC + CD44 EVs, a Wilcoxon signed-rank test was applied. Survival curves were generated using Kaplan–Meier plots and groups were evaluated using a log-rank test. The correlation of microRNA levels with survival was investigated using a nonparametric Spearman's rank correlation coefficient.

Since multiple microRNAs were tested, we corrected for multiple testing using the Benjamini–Hochberg procedure [22]. Out of the measured microRNAs used for dichotomous survival analysis, four out of four were measurable in SEC + CD44, but only three out of four in cell-depleted serum (miR-15b-3p, miR-21-3p, and miR-106a-5p, but not miR-328-3p). Therefore, we included seven groups in the single analysis and 21 in the two-fold combination analysis (Tables S3 and S4), leading to a total of $m = 28$ groups. After ranking the p-values (value i, range 1–28) and defining a false discovery rate (FDR = q) of 10%, we calculated the Benjamini–Hochberg critical value $q^*(i/m)$. If the log-rank p value was below the Benjamini–Hochberg critical value, the p values were considered statistically significant (*).

5. Conclusions

A panel of four microRNAs (miR-15b-3p, miR-21-3p, miR-106a-5p, and miR-328-3p) isolated from extracellular vesicles that were purified using size-exclusion chromatography and CD44-based immunoprecipitation in combination with total serum analysis allowed us to predict the prognosis of glioblastoma patients. Further analyses in larger prospective clinical trials are still warranted before these novel biomarkers become established in routine clinical practice.

Author Contributions: Conceptualization, T.T., G.H., S.W., C.C. and K.S.R.; methodology, T.T., S.W., C.C. and K.S.R.; visualization, T.T. and K.S.R.; validation, K.S.R.; formal analysis, T.T., J.W. and K.S.R.; investigation, T.T., G.H., U.H., C.C. and K.S.R.; resources, D.B., N.S., C.S., M.H., B.S., M.G., G.H. and S.W.; data curation, T.T., C.C. and K.S.R.; writing—original draft preparation, T.T. and K.S.R.; writing—review and editing, T.T., J.W., D.B., N.S., C.S., M.H., B.S., M.G., G.H., U.H., S.W., C.C. and K.S.R.; supervision, G.H., U.H., C.C. and K.S.R.; project administration, G.H.; funding acquisition, T.T. and G.H. All authors have read and agreed to the published version of the manuscript.

Acknowledgments: We thank Dagmar Christoph for excellent technical assistance and Meghan Campbell for her valuable input on the manuscript. We acknowledge Christiane Landwehr for her excellent trial and project management.

Abbreviations

MDPI	Multidisciplinary Digital Publishing Institute
LD	linear dichroism
EV	extracellular vesicles
HV	healthy volunteers
SEC	size-exclusion chromatography

References

1. Taylor, O.G.; Brzozowski, J.S.; A Skelding, K. Glioblastoma multiforme: An overview of emerging therapeutic targets. *Front. Oncol.* **2019**, *9*, 963. [CrossRef]
2. Stupp, R.; Mason, W.P.; Van Den Bent, M.J.; Weller, M.; Fisher, B.; Taphoorn, M.J.; Belanger, K.; Brandes, A.A.; Marosi, C.; Bogdahn, U.; et al. Radiotherapy plus concomitant and adjuvant temozolomide for glioblastoma. *N. Engl. J. Med.* **2005**, *352*, 987–996. [CrossRef]
3. Hegi, M.E.; Diserens, A.C.; Gorlia, T.; Hamou, M.F.; De Tribolet, N.; Weller, M.; Kros, J.M.; Hainfellner, J.A.; Mason, W.; Mariani, L.; et al. MGMT gene silencing and benefit from temozolomide in glioblastoma. *N. Engl. J. Med.* **2005**, *352*, 997–1003. [CrossRef]
4. Herrlinger, U.; Tzaridis, T.; Mack, F.; Steinbach, J.P.; Schlegel, U.; Sabel, M.; Hau, P.; Kortmann, R.-D.; Krex, D.; Grauer, O.; et al. Lomustine-temozolomide combination therapy versus standard temozolomide therapy in patients with newly diagnosed glioblastoma with methylated MGMT promoter (CeTeG/NOA–09): A randomised, open-label, phase 3 trial. *Lancet* **2019**, *393*, 678–688. [CrossRef]
5. Weller, J.; Tzaridis, T.; Mack, F.; Steinbach, J.P.; Schlegel, U.; Hau, P.; Krex, D.; Grauer, O.; Goldbrunner, R.; Bähr, O.; et al. Health-related quality of life and neurocognitive functioning with lomustine–temozolomide versus temozolomide in patients with newly diagnosed, MGMT-methylated glioblastoma (CeTeG/NOA-09): A randomised, multicentre, open-label, phase 3 trial. *Lancet Oncol.* **2019**, *20*, 1444–1453. [CrossRef]
6. Wick, W.; Osswald, M.; Wick, A.; Winkler, F. Treatment of glioblastoma in adults. *Ther. Adv. Neurol. Disord.* **2018**, *11*, 1756286418790452. [CrossRef]
7. Wahid, F.; Shehzad, A.; Khan, T.; Kim, Y.Y. MicroRNAs: Synthesis, mechanism, function, and recent clinical trials. *Biochim. Biophys. Acta* **2010**, *1803*, 1231–1243. [CrossRef]
8. O'Brien, J.; Hayder, H.; Zayed, Y.; Peng, C. Overview of MicroRNA Biogenesis, Mechanisms of Actions, and Circulation. *Front. Endocrinol. (Lausanne)* **2018**, *9*, 402. [CrossRef]
9. Knight, S.W.; Bass, B.L. A Role for the RNase III Enzyme DCR-1 in RNA Interference and Germ Line Development in Caenorhabditis elegans. *Science* **2001**, *293*, 2269–2271. [CrossRef]
10. Zhang, J.; Zhou, W.; Liu, Y.; Liu, T.; Li, C.; Wang, L. Oncogenic role of microRNA-532-5p in human colorectal cancer via targeting of the 5'UTR of RUNX3. *Oncol. Lett.* **2018**, *15*, 7215–7220. [CrossRef]
11. Huntzinger, E.; Izaurralde, E. Gene silencing by microRNAs: Contributions of translational repression and mRNA decay. *Nat. Rev. Genet.* **2011**, *12*, 99–110. [CrossRef] [PubMed]
12. Møller, H.G.; Rasmussen, A.P.; Andersen, H.H.; Johnsen, K.B.; Henriksen, M.; Duroux, M. A Systematic Review of MicroRNA in Glioblastoma Multiforme: Micro-modulators in the Mesenchymal Mode of Migration and Invasion. *Mol. Neurobiol.* **2013**, *47*, 131–144. [CrossRef]
13. Buruiană, A.; Florian, S.I.; Florian, I.A.; Timiș, T.L.; Mihu, C.M.; Miclăuș, M.; Oșan, S.; Hrapșa, I.; Cataniciu, R.C.; Farcaș, M.; et al. The Roles of miRNA in Glioblastoma Tumor Cell Communication: Diplomatic and Aggressive Negotiations. *Int. J. Mol. Sci.* **2020**, *21*, 1950. [CrossRef] [PubMed]
14. Arroyo, J.D.; Chevillet, J.R.; Kroh, E.M.; Ruf, I.K.; Pritchard, C.C.; Gibson, D.F.; Mitchell, P.S.; Bennett, C.F.; Pogosova-Agadjanyan, E.L.; Stirewalt, D.L.; et al. Argonaute2 complexes carry a population of circulating microRNAs independent of vesicles in human plasma. *Proc. Natl. Acad. Sci. USA* **2011**, *108*, 5003–5008. [CrossRef]
15. Bayraktar, R.; Van Roosbroeck, K.; Calin, G.A. Cell-to-cell communication: microRNAs as hormones. *Mol. Oncol.* **2017**, *11*, 1673–1686. [CrossRef]
16. Valadi, H.; Ekstrom, K.; Bossios, A.; Sjöstrand, M.; Lee, J.J.; Lötvall, J. Exosome-mediated transfer of mRNAs and microRNAs is a novel mechanism of genetic exchange between cells. *Nat. Cell Biol.* **2007**, *9*, 654–659. [CrossRef]
17. Théry, C.; Ostrowski, M.; Segura, E. Membrane vesicles as conveyors of immune responses. *Nat. Rev. Immunol.* **2009**, *9*, 581–593. [CrossRef]
18. Tzaridis, T.; Reiners, K.; Herrlinger, U.; Gunther, H.; Scheffler, B.; Glas, M.; Coch, C. CBMT-17. Novel Approach of Utilising Serum/Plasma ev and Cell-Free RNA for Treatment Monitoring in Glioblastoma Patients. *Neuro Oncol.* **2018**, *20*, vi35–vi36. [CrossRef]
19. Kros, J.M.; Mustafa, D.M.; Dekker, L.J.; Smitt, P.A.S.; Luider, T.M.; Zheng, P.-P. Circulating glioma biomarkers. *Neuro Oncol.* **2015**, *17*, 343–360. [CrossRef]

20. Santangelo, A.; Imbrucè, P.; Gardenghi, B.; Belli, L.; Agushi, R.; Tamanini, A.; Munari, S.; Bossi, A.M.; Scambi, I.; Benati, N.; et al. A microRNA signature from serum exosomes of patients with glioma as complementary diagnostic biomarker. *J. Neuro Oncol.* **2018**, *136*, 51–62. [CrossRef]
21. Wang, H.; Jiang, D.; Li, W.; Xiang, X.; Zhao, J.; Yu, B.; Wang, C.; He, Z.; Zhu, L.; Yang, Y. Evaluation of serum extracellular vesicles as noninvasive diagnostic markers of glioma. *Theranostics* **2019**, *9*, 5347–5358. [CrossRef]
22. Benjamini, Y.; Hochberg, Y. Controlling the False Discovery Rate: A Practical and Powerful Approach to Multiple Testing. *J. R. Stat. Soc.* **1995**, *57*, 289–300. [CrossRef]
23. Han, K.; Ren, M.; Wick, W.; Abrey, L.; Das, A.; Jin, J.; Reardon, D.A. Progression-free survival as a surrogate endpoint for overall survival in glioblastoma: A literature-based meta-analysis from 91 trials. *Neuro Oncol.* **2014**, *16*, 696–706. [CrossRef]
24. Chukwueke, U.N.; Wen, P.Y. Use of the Response Assessment in Neuro-Oncology (RANO) criteria in clinical trials and clinical practice. *CNS Oncol.* **2019**, *8*, CNS28. [CrossRef]
25. Ellingson, B.M.; Wen, P.Y.; Cloughesy, T.F. Modified Criteria for Radiographic Response Assessment in Glioblastoma Clinical Trials. *Neurotherapeutics* **2017**, *14*, 307–320. [CrossRef]
26. Galldiks, N.; Kocher, M.; Langen, K.-J. Pseudoprogression after glioma therapy: An update. *Expert Rev. Neurother.* **2017**, *17*, 1109–1115. [CrossRef]
27. Kelly, C.; Majewska, P.; Ioannidis, S.; Raza, M.H.; Williams, M. Estimating progression-free survival in patients with glioblastoma using routinely collected data. *J. Neuro Oncol.* **2017**, *135*, 621–627. [CrossRef]
28. Majewska, P.; Ioannidis, S.; Raza, M.H.; Tanna, N.; Bulbeck, H.; Williams, M. Postprogression survival in patients with glioblastoma treated with concurrent chemoradiotherapy: A routine care cohort study. *CNS Oncol.* **2017**, *6*, 307–313. [CrossRef]
29. Senbanjo, L.T.; Chellaiah, M.A. CD44: A Multifunctional Cell Surface Adhesion Receptor Is a Regulator of Progression and Metastasis of Cancer Cells. *Front. Cell Dev. Biol.* **2017**, *5*, 18. [CrossRef]
30. Stamenkovic, I.; Yu, Q. Shedding Light on Proteolytic Cleavage of CD44: The Responsible Sheddase and Functional Significance of Shedding. *J. Investig. Dermatol.* **2009**, *129*, 1321–1324. [CrossRef]
31. Buschow, S.I.; Van Balkom, B.W.M.; Aalberts, M.; Heck, A.J.R.; Wauben, M.; Stoorvogel, W. MHC class II-associated proteins in B-cell exosomes and potential functional implications for exosome biogenesis. *Immunol. Cell Biol.* **2010**, *88*, 851–856. [CrossRef] [PubMed]
32. Perez-Hernandez, D.; Gutiérrez-Vázquez, C.; Jorge, I.; López-Martín, S.; Ursa, A.; Sánchez-Madrid, F.; Vázquez, J.; Yáñez-Mó, M. The Intracellular Interactome of Tetraspanin-enriched Microdomains Reveals Their Function as Sorting Machineries toward Exosomes. *J. Biol. Chem.* **2013**, *288*, 11649–11661. [CrossRef] [PubMed]
33. Rak, J.; Milsom, C.; Yu, J. Vascular determinants of cancer stem cell dormancy-do age and coagulation system play a role? *APMIS* **2008**, *116*, 660–676. [CrossRef]
34. Ebrahimkhani, S.; Vafaee, F.; Hallal, S.; Wei, H.; Lee, M.Y.T.; Young, P.E.; Satgunaseelan, L.; Beadnall, H.; Barnett, M.H.; Shivalingam, B.; et al. Deep sequencing of circulating exosomal microRNA allows non-invasive glioblastoma diagnosis. *NPJ Precis. Oncol.* **2018**, *2*, 28. [CrossRef]
35. Niyazi, M.; Pitea, A.; Mittelbronn, M.; Steinbach, J.; Sticht, C.; Zehentmayr, F.; Piehlmaier, D.; Zitzelsberger, H.; Ganswindt, U.; Rödel, C.; et al. A 4-miRNA signature predicts the therapeutic outcome of glioblastoma. *Oncotarget* **2016**, *7*, 45764–45775. [CrossRef]
36. Wei, Z.; Batagov, A.O.; Schinelli, S.; Wang, J.; Wang, Y.; El Fatimy, R.; Rabinovsky, R.; Balaj, L.; Chen, C.C.; Hochberg, F.; et al. Coding and noncoding landscape of extracellular RNA released by human glioma stem cells. *Nat. Commun.* **2017**, *8*, 1145. [CrossRef]
37. Gallo, A.; Tandon, M.; Alevizos, I.; Illei, G.G. The Majority of MicroRNAs Detectable in Serum and Saliva Is Concentrated in Exosomes. *PLoS ONE* **2012**, *7*, e30679. [CrossRef]
38. Hunter, M.P.; Ismail, N.; Zhang, X.; Aguda, B.D.; Lee, E.J.; Yu, L.; Xiao, T.; Schafer, J.; Lee, M.-L.T.; Schmittgen, T.D.; et al. Detection of microRNA Expression in Human Peripheral Blood Microvesicles. *PLoS ONE* **2008**, *3*, e3694. [CrossRef]
39. Kosaka, N.; Iguchi, H.; Hagiwara, K.; Yoshioka, Y.; Takeshita, F.; Ochiya, T. Neutral Sphingomyelinase 2 (nSMase2)-dependent Exosomal Transfer of Angiogenic MicroRNAs Regulate Cancer Cell Metastasis. *J. Biol. Chem.* **2013**, *288*, 10849–10859. [CrossRef]

40. Ilhan-Mutlu, A.; Wagner, L.; Woehrer, A.; Furtner, J.; Widhalm, G.; Marosi, C.; Preusser, M. Plasma MicroRNA-21 Concentration May Be a Useful Biomarker in Glioblastoma Patients. *Cancer Investig.* **2012**, *30*, 615–621. [CrossRef]
41. Yang, C.; Wang, C.; Chen, X.; Chen, S.; Zhang, Y.; Zhi, F.; Wang, J.; Li, L.; Zhou, X.; Li, N.; et al. Identification of seven serum microRNAs from a genome-wide serum microRNA expression profile as potential noninvasive biomarkers for malignant astrocytomas. *Int. J. Cancer* **2013**, *132*, 116–127. [CrossRef] [PubMed]
42. Lavon, I.; Zrihan, D.; Granit, A.; Einstein, O.; Fainstein, N.; Cohen, M.A.; Cohen, M.A.; Zelikovitch, B.; Shoshan, Y.; Spektor, S.; et al. Gliomas display a microRNA expression profile reminiscent of neural precursor cells. *Neuro Oncol.* **2010**, *12*, 422–433. [CrossRef] [PubMed]
43. Zhi, F.; Zhou, G.; Shao, N.; Xia, X.; Shi, Y.; Wang, Q.; Zhang, Y.; Wang, R.; Xue, L.; Wang, S.; et al. miR-106a-5p Inhibits the Proliferation and Migration of Astrocytoma Cells and Promotes Apoptosis by Targeting FASTK. *PLoS ONE* **2013**, *8*, e72390. [CrossRef]
44. Delic, S.; Lottmann, N.; Stelzl, A.; Liesenberg, F.; Wolter, M.; Götze, S.; Zapatka, M.; Shiio, Y.; Sabel, M.C.; Felsberg, J.; et al. MiR-328 promotes glioma cell invasion via SFRP1-dependent Wnt-signaling activation. *Neuro Oncol.* **2014**, *16*, 179–190. [CrossRef] [PubMed]
45. Ambrose, A.; Alsahli, M.A.; Kurmani, S.A.; Goodall, A.H. Comparison of the release of microRNAs and extracellular vesicles from platelets in response to different agonists. *Platelets* **2018**, *29*, 446–454. [CrossRef]
46. Murillo, O.D.; Thistlethwaite, W.; Rozowsky, J.; Subramanian, S.L.; Lucero, R.; Shah, N.; Jackson, A.R.; Srinivasan, S.; Chung, A.; Laurent, C.D.; et al. exRNA Atlas Analysis Reveals Distinct Extracellular RNA Cargo Types and Their Carriers Present across Human Biofluids. *Cell* **2019**, *177*, 463–477.e15. [CrossRef]
47. Srinivasan, S.; Yeri, A.; Cheah, P.S.; Chung, A.; Danielson, K.M.; De Hoff, P.; Filant, J.; Laurent, C.D.; Laurent, L.D.; Magee, R.; et al. Small RNA Sequencing across Diverse Biofluids Identifies Optimal Methods for exRNA Isolation. *Cell* **2019**, *177*, 446–462.e16. [CrossRef]
48. Occhipinti, G.; Giulietti, M.; Principato, G.; Piva, F. The choice of endogenous controls in exosomal microRNA assessments from biofluids. *Tumor Biol.* **2016**, *37*, 11657–11665. [CrossRef]
49. Bachurski, D.; Schuldner, M.; Nguyen, P.-H.; Malz, A.; Reiners, K.S.; Grenzi, P.C.; Babatz, F.; Schauss, A.C.; Hansen, H.P.; Hallek, M.; et al. Extracellular vesicle measurements with nanoparticle tracking analysis—An accuracy and repeatability comparison between NanoSight NS300 and ZetaView. *J. Extracell. Vesicles* **2019**, *8*, 1596016. [CrossRef]
50. Van Deun, J.; EV-TRACK Consortium; Mestdagh, P.; Agostinis, P.; Akay, Ö.; Anand, S.; Anckaert, J.; Martinez, Z.A.; Baetens, T.; Beghein, E.; et al. EV-TRACK: Transparent reporting and centralizing knowledge in extracellular vesicle research. *Nat. Methods* **2017**, *14*, 228–232. [CrossRef]

9

Exosomes in Prostate Cancer Diagnosis, Prognosis and Therapy

Tomasz Lorenc [1,*], Katarzyna Klimczyk [2,3], Izabela Michalczewska [2,3], Monika Słomka [2,3], Grażyna Kubiak-Tomaszewska [2,3] and Wioletta Olejarz [2,3]

1 1st Department of Clinical Radiology, Medical University of Warsaw, 02-004 Warsaw, Poland
2 Department of Biochemistry and Pharmacogenomics, Faculty of Pharmacy, Medical University of Warsaw, 02-097 Warsaw, Poland; kasiaklimczyk6@gmail.com (K.K.); izabela1356@gmail.com (I.M.); slomka.monika.94@gmail.com (M.S.); grazyna.kubiak-tomaszewska@wum.edu.pl (G.K.-T.); wolejarz@wum.edu.pl (W.O.)
3 Centre for Preclinical Research, Medical University of Warsaw, 02-097 Warsaw, Poland
* Correspondence: tomasz.lorenc@wum.edu.pl

Abstract: Prostate cancer (PCa) is the second most common cause of cancer-related mortality among men in the developed world. Conventional anti-PCa therapies are not effective for patients with advanced and/or metastatic disease. In most cases, cancer therapies fail due to an incomplete depletion of tumor cells, resulting in tumor relapse. Exosomes are involved in tumor progression, promoting the angiogenesis and migration of tumor cells during metastasis. These structures contribute to the dissemination of pathogenic agents through interaction with recipient cells. Exosomes may deliver molecules that are able to induce the transdifferentiation process, known as "epithelial to mesenchymal transition". The composition of exosomes and the associated possibilities of interacting with cells make exosomes multifaceted regulators of cancer development. Extracellular vesicles have biophysical properties, such as stability, biocompatibility, permeability, low toxicity and low immunogenicity, which are key for the successful development of an innovative drug delivery system. They have an enhanced circulation stability and bio-barrier permeation ability, and they can therefore be used as effective chemotherapeutic carriers to improve the regulation of target tissues and organs. Exosomes have the capacity to deliver different types of cargo and to target specific cells. Chemotherapeutics, natural products and RNA have been encapsulated for the treatment of prostate cancers.

Keywords: extracellular vesicle; precision oncology; cancer biomarker; prostate cancer

---

1. Introduction

Prostate cancer (PCa) is the most common solid malignancy, with a high mortality in men [1]. In many cases, successful treatment of prostate cancer is difficult due to the late detection and rate of metastasis [2]. Importantly, the tumors of many patients with prostate cancer become refractory to androgen therapy and progress to metastatic castration-resistant disease [3]. An effective treatment course of prostate cancer patients requires predictive biomarkers in metastatic castration-resistant prostate cancer that support individual therapy [4]. Liquid biopsies, circulating tumor cells, exosomes and circulating nucleic acids have been developed as minimally invasive assays to monitor PCa patients [5]. Exosomes are extracellular vesicles (EVs), which may serve as novel tools for various therapeutic approaches, including drug delivery [6], anti-tumor therapy, pathogen vaccination, immune-modulatory and regenerative therapies. They are secreted by cells and detected in various biological fluids, and they can serve as biomarkers for cancer diagnosis, prognosis and therapy [7]. The presence of EVs in urine was discovered in 2004 [8]. It is believed that urinary EVs originate from epithelial cells of the urogenital system, which includes the organs involved in reproduction and urine excretion. Urine has additional advantages relating to cancers of the urogenital system, since the composition of urine directly reflects changes in the function of associated organs. Blood-specific prostate antigen (PSA) remains the most widely used biomarker in the detection of early prostate cancer, but new biomarkers, like exosomal miRNAs, have been proposed to increase specificity and distinguish aggressive from non-aggressive PCas [9]. It has been shown that urinary markers can aid in the decision-making process regarding whether to carry out a prostate biopsy and in the design of a therapeutic strategy [10]. Urinary exosomes and their cargo, especially miR-21 and miR-375, have become an emerging source of biomarkers in the detection and prognosis of PCa [11]. Moreover, the expression of serum exosomal miRNAs induced by radiotherapy may have potential value as prognostic and predictive biomarkers PCa [12]. Exosomes are also promising carriers of drugs and other therapeutic molecules targeting prostate cancer, but there are still several challenges relating to their use as drug carriers [13].

2. Structure and Function of Exosomes

Exosomes are small (from 30 to 120 nm in diameter) extracellular vesicles (EVs) (Figure 1). Their lipid bilayer membrane, with a width of 5 nm, protects them from the negative action of RNases and proteases. Exosomes have a longer retention in circulation in comparison to polymersomes or liposomes [14]. This characteristic originates from transmembrane protein (CD47-SIRPα), which prevents exosomes from being phagocytosed [15]. Moreover, the double membrane structures contain various cargo, such as miRNAs, proteins (e.g., tetraspanin CD63, CD81, CD82, CD53 and CD37, as well as cystolic proteins), lipids (e.g., sphingomyelin, cholesterol and generally saturated fats) and viral particles [16], and their presence depends on the origin cells and organism's health conditions. The proteins in exosomes include endosomal, plasma and nuclear proteins [17]. Therefore, in comparison to cellular proteins in prostate cancer, those in exosomes have a higher level of glycosylation [18]. Moreover, proteins enriched in exosomes include those relevant for individual exosomal biogenesis pathways and for exosome secretion. The Minimal Information for Studies of Extracellular Vesicles 2018 highlights three categories of markers that must be found in isolated extracellular exosomes. They include at least one transmembrane/lipidbound protein or cytosolic protein and one negative protein marker [19]. The cargo present in extracellular vesicles can be transferred and alter signaling pathways in recipient cells [13]. For example, exosomes from metastatic prostate cancer cells showed high contents of miR-21 and miR-141, which are responsible for the regulation of osteoclastogenesis and osteoblastogenesis. Cancer-derived exosomes can promote epithelial–mesenchymal transition (EMT) via miRNAs. They play an important role in the conversion from benign to malignant cancers [13] and in the regulation of the response to docetaxel, such as miR-34 in prostate cancer cells and cell-derived exosomes targeting Bcl-2. This shows the great influence of extracellular vesicles on the drug resistance of prostate cancer cells.

Exosomes are released by the exocytosis of multivesicular bodies (MVBs), developed from early and then late endosomes [20]. Those naturally occurring membrane particles mediate intercellular communication by delivering molecular information between cancer and stromal cells, especially cancer-associated fibroblast (CAFs) [14]. Cancer cell-derived EVs cause very diverse effects and may depend on their target cells, which appear to take them up in several ways, such as receptor/lipid raft endocytosis, phagocytosis, micropinocytosis or fusion with the plasma membrane [17]. Some exosomes can reduce the anti-cancer immune response, interact with specific membrane receptors [9] and promote a suitable microenvironment, while others may cause drug resistance and the failure of antibody therapy involving RNA species and protein delivery [21]. Recent studies point out that exosomes released from the tumor microenvironment can regulate (also by tethering TGFβ) a proliferation, a reduction of apoptosis, a promotion of angiogenesis and, finally, an evasion of immune surveillance (Figure 2). Moreover, exosomes can provide candidate biomarkers for prostate cancer, contribute to tumor progression and, after a loss of environment homeostasis, promote tumor metastasis [18].

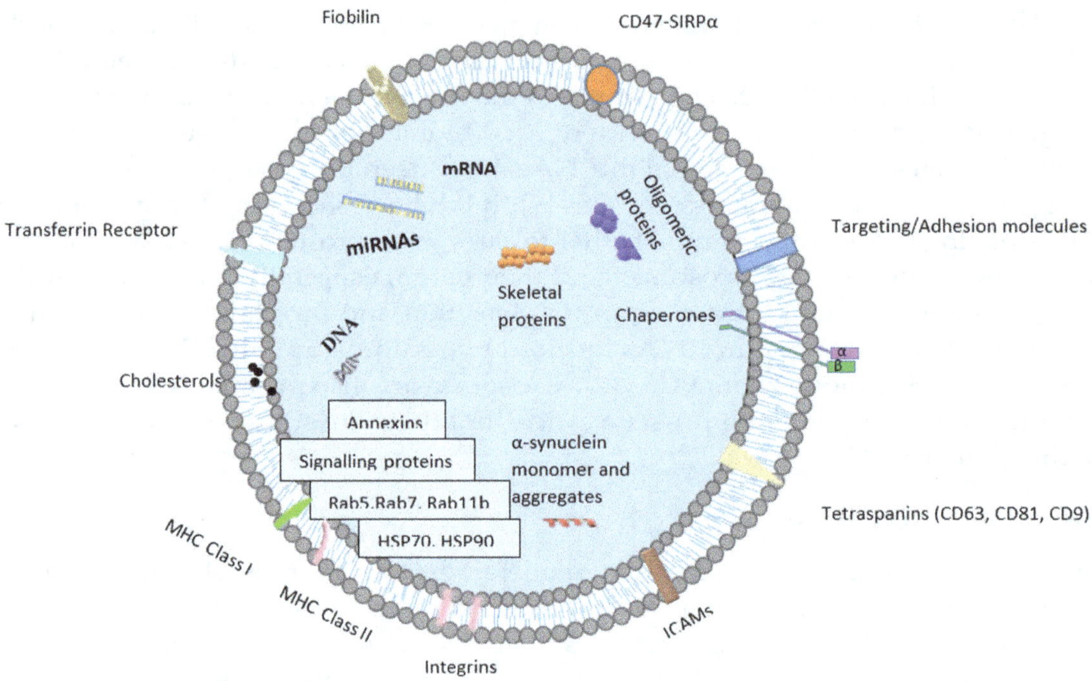

Figure 1. Schematic diagram of an exosome.

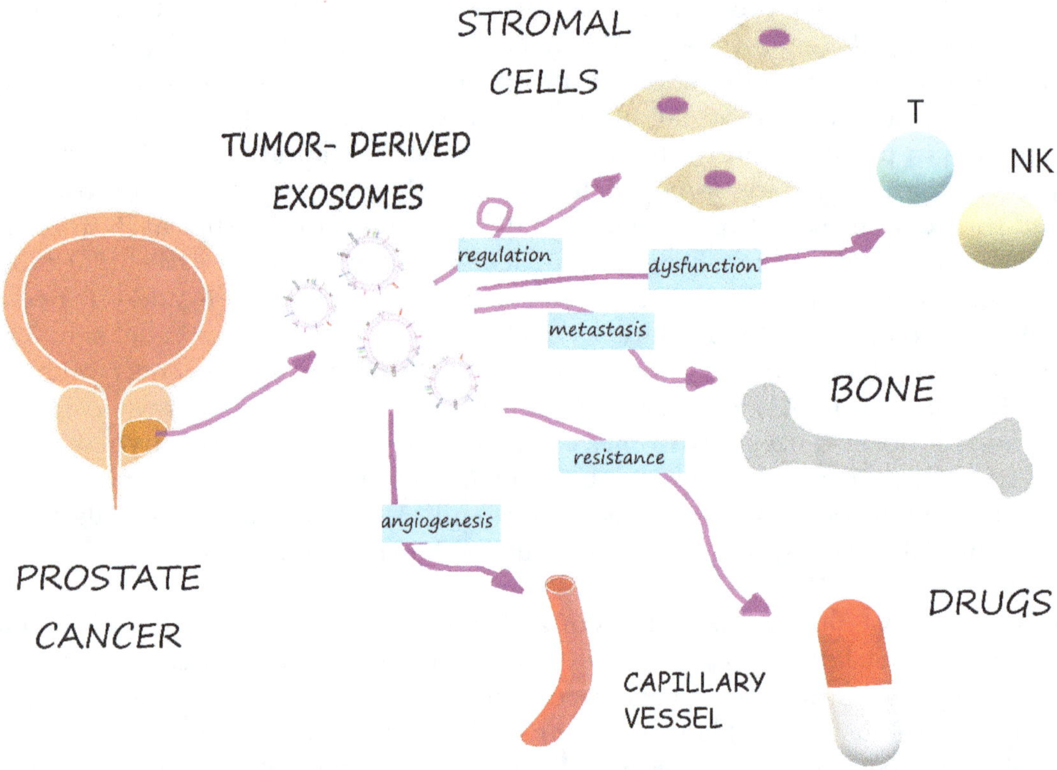

Figure 2. Role of prostate cell-derived exosomes in cancer progression. Exosomes regulate stromal cells, impair immune cells and alter the microenvironment, which could lead to tumor growth and metastasis. Exosomes are also responsible for drug resistance.

3. Tumor-Derived Exosomes in Cancer Progression

3.1. Tumorigenesis

Tumorigenesis or carcinogenesis is the process of an uncontrolled multiplication of cells or deregulated apoptotic cell death, which leads to the formation of cancer. Exosomes derived from prostate cancer cells, which have become the common object of scientists' interest, are implicated in creating a premetastatic niche [13]. It is a microenvironment that is especially favorable to cancer cells composed of different cell types, like fibroblasts, lymphocytes, epithelial cells, matrix molecules, factors, such as growth factors, and cytokines [22]. Exosomes influence the immune system in a premetastatic niche through various mechanisms. Prostate cells are involved in producing tumor-derived secreted factors (TDSFs), including VEGF, TNF-α and interleukins in response to the local inflammatory niche. First, TDSFs stimulate the recruitment of myeloid cells and immune cells to the pre-metastatic niche. Furthermore, the expression of inflammatory factors is upregulated by stromal cells under the influence of TDSFs. Moreover, tumor-derived exosome content enhances the pre-metastatic niche [23]. Tumor exosomes possess the capacity to support the migration of immune cells, like neutrophils, macrophages and regulatory T to secondary sites, thereby reducing immune response against tumors and inhibiting antigen-presenting cells, such as dendritic cells. These nanoparticles could impair the function of T-cells and NK-cells via the blocking, proliferation, activation and provision of apoptosis [23,24].

3.2. *Tumor Progression*

Recent research suggests that exosomes isolated from the prostate cancer microenvironment are an important factor in the progression of this type of cancer. The increase in the mass of prostate tumors may be the result of tumor stem cell proliferation, which possess a self-renewing ability [13,25]. Prostate cancer exosomes, as the carriers of many lipids, proteins and RNAs, can affect the proliferation, angiogenesis and survival of cancer cells, as well as their ability to avoid immune surveillance [26]. Exosomes, which form in the tumor microenvironment, transport, among other things, miRNAs (miRs, short noncoding RNAs that are responsible for the regulation of gene expression). miR-20a, miR-21 and miR-125b cause the inhibition of apoptosis and survival, promoting the effects of cancer cells. miRNA-221 and miR-222 are responsible for cancer growth. miR-92a and miR-17-92 are involved in the promotion of angiogenesis, as a result of an increased proliferation and migration of endothelial cells. It has been shown that miR-210 is responsible for the induction of metastasis by EMT promotion. miR-21, miR-100 and miR-139 induce fibroblast migration by increasing the expression of MMP-2, MMP-9, MMP-13 and RANKL. miR-21, miR375 and miR-141 overcome low androgen conditions during distant metastasis. miR-1290 and miR-375 are connected with a poor patient prognosis. miR-409 downregulates tumor suppressors, like RSU1 and STAG2, and promotes cancer cell tumorigenesis. The ANXA6/LRP1/TSP1 complex causes tumor growth induction. miR-126 and miR-146a are involved in tumor suppression [27–30]. Moreover, exosomes, with a structure rich in lipids, especially cholesterol and sphingomyelin, can modulate the lipid composition of the target cells, disturbing their homeostasis. A study on prostate cancer has shown that the accumulation of cholesterol esters in prostate cancer cells is associated with tumor progression and metastasis. This was confirmed by studies, using synthetic exosomes carried out by the Lombardo Group, which showed that exosomal lipids can increase the tumor aggressiveness, metastatic progression and drug resistance of pancreatic cancer cells [31].

It is well known that exosomes secreted by prostate cancer cells under hypoxic conditions increase the invasiveness, mobility and EMT in native PC cells and promote the transformation of fibroblasts into myofibroblasts. Moreover, under these conditions, lipid accumulation (triglycerides, bis-monoacylglycerolphosphate, ceramides, cholesterol, etc.) increase in PCs, which leads to increased exosome secretion by hypoxic tumor cells [32].

Exosomes also participate in the regulation of the progression of prostatic tumors, as the carriers of numerous proteins. These vesicular structures, as carriers of TGF-β, can induce the above-mentioned transformation of fibroblasts into myofibroblasts (CAFs) by the activation of TGF-β/Smad3 signaling or independent SMAD signaling pathways and promote neoangiogenesis. Webber found that exosomal TGFβ1 induces a highly aggressive myofibroblast phenotype, with a high proangiogenic activity [13,33]. ITGA3, ITGB1, ITGB4 and ITGB3 are exosomal proteins involved in the progression of prostate cancers. These integrins are responsible for the promotion of the migration and invasion of epithelial cells by the activation of Src phosphorylation in recipient cells. The matrix metalloproteinases MMP-9 and MMP-14 promote cancer cell protection from apoptosis and intensify their mobility through the stimulation of ERK1/2 phosphorylation. Subsequently, they prepare the metastatic site [26,27]. The next group, ligands for the NKG2D and Fas receptors, causes tumor immune evasion by receptor downregulation and silences the cytotoxic activity of NK cells and CD8+ cells. Caveolin-1 increases the survival of cancer cells and their independence from androgens, and they also promote distant metastasis by the positive regulation of fatty acid synthase activity. Hypoxia-inducible factor 1α (HIF-1α) promotes the initiation and progression of metastasis by promoting the loss of E-cadherin. The tetraspanin–integrin complex increases the adhesion of exosomes to the right cells. The epidermal growth factor receptor (EGFR) induces tumor angiogenesis by the activation of the autocrine VEGF/VEGFR-2 pathway in endothelial cells. Tyrosine-protein kinase Met induces metastasis and promotes a phenotype resistant to castration therapy. Other exosomal proteins, c-Src tyrosine kinase, IGF-1R and FAK, induce angiogenesis by the stimulation of the VEGF transcription within the tumor microenvironment [34–39].

It has also been shown that tumor-derived exosomes enhance interleukin-6 production in myeloid-derived suppressor cells through the activation of Toll-like receptor 2 via the membrane-linked heat shock protein, which promotes the autocrine phosphorylation of Stat3 and enhances the effect of the immunosuppression of the immune system and the promotion of prostate cancer [40].

Exosomes secreted by prostate cancer cells are also involved in tumor metastasis in bone, which is common in patients with this type of cancer. Recent studies have shown that prostate cancer cell-derived exosomes mediate cell–cell communication in osteoblastic metastasis. On the other hand, osteoblast-derived exosomes may regulate prostate cancer cell proliferation at the original site of tumor development [41].

Exosomes secreted by tumor cells affect the stromal cells by stimulating the formation of pro-proliferative and proangiogenic phenotypes in these cells. It has been shown that exosomes secreted by cancer-associated fibroblasts increase the ability of prostate cancer cells to proliferate and survive in low-oxygen and low-nutrient environments by inhibiting mitochondrial oxidative phosphorylation and increasing anaerobic glycolysis [33,42].

3.3. Angiogenesis

Exosomes obtained from prostate cancer stem cells support tumorigenesis by promoting angiogenesis, which, as a process of developing a new vasculature, is crucial for tumor growth and migrations and is the main cause of metastasis and malignancy. A recent report has shown that some conditions, like hypoxia or acidosis, enhance secreting exosomes in bodily fluids, and these exosomes cause angiogenesis more frequently [30]. Studies have also reported that the expression of E-cadherin and carbonic anhydrase 9 in exosomes could contribute to the angiogenesis process [43]. In the case of prostate cancer, the process of vascularization is supported by the transfer of sphingomyelin and CD147 via exosomes into endothelial cells [44]. Newly formed blood vessels are created to facilitate tumor growth by transporting TDSFs and circulating tumor cells (CTCs) into secondary tissues, thereby beginning vascular leakage [23]. Prostate cancer cell-derived exosomes, via the activation of TGF-β\SMAD3 signaling, lead to the transformation of fibroblasts into myofibroblasts, referred to as CAFs. The exosomes derived from these special kinds of cells are able to cause an explosive growth of prostate cancer cells by transferring the miRNAs (miR-21 and miR-409) into neighboring epithelia. miR-21 suppresses the expression of APAF1 (apoptotic peptidase activating factor 1) and PDCD4 (programmed cell death 4) to inhibit apoptosis and make cancer cells resistant to chemotherapeutics [13]. Increased angiogenesis and vascular permeability promote metastasis.

3.4. Metastasis

Tumor metastasis is a complicated process, including vascular leakiness and an alteration of the microenvironment, in which exosomes are also involved. Initially, exosomes begin an epithelial–mesenchymal transition (EMT) via miRNAs by losing their junction and adhesion ability. Thus, epithelial tumor cells obtain mesenchymal cell properties and are responsive to malignancy [45].

Exosomes support the formation of a pre-metastatic niche, and cells could then be found at the vascularized organs, but metastasis cannot be developed randomly, but rather only in preferential sites under the direction of exosomes. In the process of this expansion, the special kind of CTCs (metastases-initiating cells (MICs)) are involved [46]. MICs can act on other cells by secreting exosomes that reprogram adjacent stromal cells to create a more favorable tumor microenvironment in order to support cancer growth and progression. As mentioned above, cancer-derived exosomes determine organotropism. Prostate cancer possesses an affinity with bone. Exosomes obtained from prostate cancer cells by osteoclast fusion and differentiation support transmission to this destination [47]. Furthermore, exosomes derived therefrom, through the activation of RANKL, FOXM1 and c-Myc, support EMT [46]. As described previously, cancer cell-derived exosomes transfer various substances, including integrins, which are responsible for organotropism. Integrin σ3 and β1 lead to the migration and dissemination of epithelial cells. In addition, integrin avb6, expressed on the surface, is responsible for the metastatic phenotype [48]. The tumor microenvironment contributes to the regulation of prostate cancer progression through proliferation, angiogenesis and metastasis, and it also regulates immunity.

3.5. Tumor Immune Escape

The immune system is able to recognize transformed cells and eliminate them. Thus, tumors use diverse mechanisms to escape from immune-mediated surveillance. Exosomes derived from prostate cancer cells impair the cytotoxic function of lymphocytes and induce the apoptosis of CD8+ T cells [23]. In the first step, exosomes activate the T cell receptor, and the expression of Fas on the T cell is then upregulated. Subsequently, FasL induces apoptosis directly via receptor CD 95/APO1 or indirectly using dendritic cells [49]. Fas-mediated apoptosis, as an immune-evasive mechanism, may lead to tumorigenesis, but it may also be responsible for drug resistance [49]. Some other mechanisms and molecules may also contribute to T cell apoptosis. Among the substances carried by exosomes is the programmed death ligand 1 (PDL-1). Exosomal PDL-1 plays the same role as tumor PDL-1. PDL-1 binds to its receptor, PD-1, expressed on the surface of activated T and B cells or macrophages, and enables T cell apoptosis [50]. On the other hand, cancer-derived exosomes can not only cause the dysfunction of T and NK cells by blocking the activation and proliferation or induction of apoptosis, but also inhibit antigen-presenting cells and antitumor immune response [23]. While exosomes are characterized by heterogeneity and a risk of developing metastasis, those nanoparticles could be used as diagnostic and prognostic biomarkers for various cancers and as promising drug carriers.

4. The Potential of Exosomes in Prostate Cancer Diagnosis and Monitoring

EVs may be key biomarkers in the early diagnosis of prostate cancer and in a personalized approach to treatment and to future patients' prognosis [26]. EVs in the blood and urine of prostate cancer patients contain unique prostate-cancer-specific contents, which are biomarkers of prostate cancer and cancer metastasis [51,52]. Exosomes present many different proteins on their surface. These proteins can act as epitopes, which are recognized using different mono- or polyclonal antibodies. A dedicated assay has been established to determine if the exosomes in blood plasma can serve as markers for prostate cancer [51]. Additionally, several EV-derived proteins have been investigated as potential cancer biomarkers in clinical settings [53]. Several urinary exosome proteins showed a high sensitivity and specificity for prostate cancer as individual biomarkers, and combining them in a multi-panel test has the potential for a full differentiation of prostate cancer from non-disease controls [52]. Liu et al. showed that exosomes from prostate cancer are highly enriched with PSA, representing characteristics of the original PCa cells [54]. Current evidence indicates that the strongest candidates for intercellular communications in PCa are exosomal RNAs [55]. The EV-derived RNAs have also been assessed in terms of their potential as cancer biomarkers in clinical samples. Yang et al. confirmed, through meta-analysis, that plasma exosomal miRNAs have a high diagnostic value for prostate cancer patients [56]. Hessvik et al. identified 36 exosomal miRNAs as biomarker candidates for PCa in clinical studies [57]. In prostate cancer, plasma vesicles, isolated using the precipitation-based ExoQuick method, identified miR-1290 and miR-375 as potential prognostic biomarkers in castration-resistant prostate cancer (CRPC), since their level correlates with a poorer overall survival ($p < 0.004$) [58,59]. Joncas et al. demonstrated that the exosomal androgen receptor splice variant (AR-V7) is correlated with lower sex steroid levels and with a poor prognosis in CRPC patients [60]. For prostate cancer, the expression of the AR-V7 RNA in CTCs was identified as a predictive marker for response to enzalutamide (anti-androgen) and abiraterone (anti-androgen and CYP17 inhibitor) [61]. Whether the AR-V7 transcript can also be measured in plasma EVs and serve as a biomarker was investigated by Del Re et al. [62]. Using the exoRNeasy purification method, EV-RNA was extracted, and the AR-V7 transcript was detected, preferentially in patients resistant to enzalutamide or abiraterone treatment. This study, as well as other studies, shows the biomarker potential of plasma-derived EV-RNA and the advantages with respect to the cost, ease of workflow and likelihood that all (heterogeneous) tumors are represented by a specific EV population. Communication between a tumor and its environment in PCa via extracellular vesicle (EV) RNAs is a potential mechanism of bone metastasis [63]. It has been shown that exosomal miR-375 significantly promoted osteoblast activity [64]. Krishn et al. proved that prostate cancer sheds the $\alpha v\beta 3$ integrin in vivo through exosomes. The exosomal $\alpha v\beta 3$ integrin has been shown to promote aggressive phenotypes in many types of cancers and may be clinically useful as a non-invasive biomarker to follow prostate cancer progression [65]. Another paper has reported survivin as a plasma-derived EV biomarker through the isolation of total EVs by the ultracentrifugation-based method for prostate cancer [66]. It has been shown that exosomes released by irradiated prostate cancer cells are enriched in B7-H3 protein (CD276), which has been identified as a diagnostic marker [67]. Furthermore, the serum exosomes of prostate cancer patients undergoing radiotherapy had increased levels of HSP72, which plays a key role in the stimulation of pro-inflammatory immune responses [68]. Additionally, urinary exosomes are a promising non-invasive biomarker, with a potential use in the diagnosis, prognosis and monitoring of prostate cancer [69].

5. The Potential of Exosomes in Prostate Cancer Therapy

5.1. Exosomes as Drug Carriers for Prostate Cancer Therapy

Extracellular vesicles have biophysical properties, such as stability, biocompatibility, permeability, low toxicity and low immunogenicity, which are key to successful drug delivery systems [70]. They have an enhanced circulation stability and bio-barrier permeation ability, and they can therefore be used as effective chemotherapeutics carriers to improve the regulation of target tissues and organs [71]. Chemotherapeutics, natural products and RNA were combined for the treatment of prostate, breast, pancreatic, and lung cancers, as well as glioblastoma [14]. Exosomes have the capacity to deliver different types of cargo and to target specific cells (Figure 3). They have been tested for the delivery of different therapeutic agents in in vitro and in vivo experiments [72]. EVs can be used as carriers to deliver therapeutic agents to tumor cells, leading to an effective tumor cell killing, while minimizing the side effects of the drugs [73]. It has been shown that mesenchymal stromal cells are able to package and deliver active drugs through their membrane microvesicles (MVs) [74]. Additionally, Tian et al. demonstrated that exosomes modified by targeting ligands can be used therapeutically for the delivery of doxorubicin to tumors [75]. Additionally, Qi et al. confirmed that drug-loaded exosomes enhanced cancer cell targeting under an external magnetic field and suppressed tumor growth [76]. Saari et al. confirmed that cancer cell-derived EVs can be used as effective carriers of Paclitaxel to autologous prostate cancer cells by increasing its cytotoxicity [21]. The simultaneous application of either radiation technology or nuclear medicine with exosomes are promising tools for the realization of the enhancement of targeting strategies using radiation technology [77].

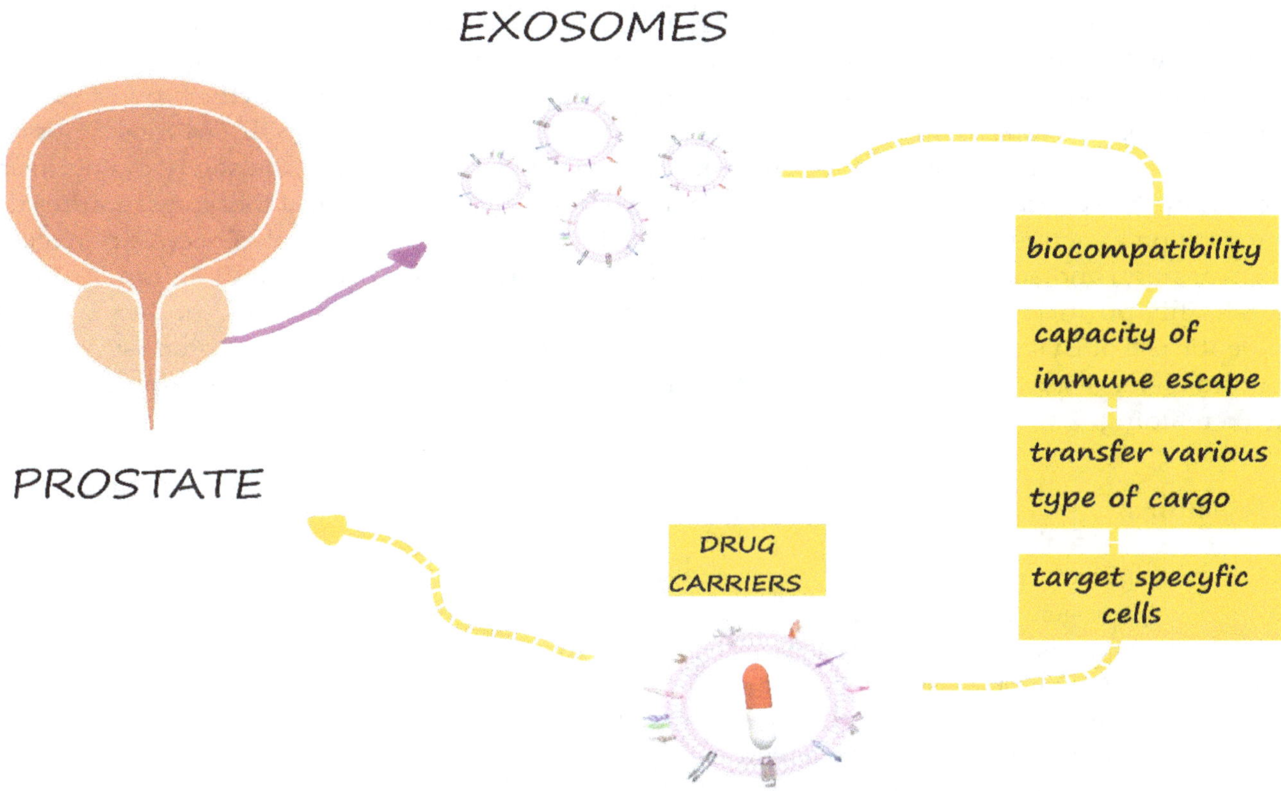

Figure 3. Exosomes are promising drug carriers in prostate cancer therapy.

5.2. Precision Therapy

Exosomes have been shown to be crucial for the development of drug resistance in patients with prostate tumor. In 2017, Del Re et al. assessed AR-V7 as a predictor of resistance to hormonal therapy by highly sensitive digital droplet polymerase chain reaction in plasma-derived exosomal RNA. They found that both the median progression-free survival (20 vs. 3 months; $p < 0.001$) and overall survival (8 months vs. not reached; $p < 0.001$) were significantly longer in AR-V7-negative vs. AR-V7-positive patients [62]. Exosome-derived microRNAs also contribute to PCa chemoresistance [78] and can act as surrogate biomarkers of tumor response to taxanes [79]. It has been observed that the transfer of exosomes (in particular, MDR-1/P-gp) from docetaxel-resistant cell lines to the DU145, 22Rv1, and LNCap PCa cell lines induces an acquired resistance to this drug [80]. In the same view, Kawakami et al. reported that β4 (ITGB4) and VCL in exosomes could be useful markers of PCa progression, which is correlated with taxane resistance [81]. Interestingly, high serum exosomal P-glycoprotein levels are associated with resistance to docetaxel, but not to cabazitaxel, thus representing a potential biomarker for guiding the decision-making process of PCa patients [82].

There is currently also a list of challenges that remain in the era of individualized or person-specific targeting of cancer. EVs are being pursued as intercellular vectors for RNA-based therapy (both miRNAs and siRNAs), with a documented efficacy in animal models of disease. For example, the in vivo suppression of prostate cancer has been achieved through the exosome delivery of the tumor suppressor miRNA, miR-143, in mice [83].

6. Concluding Remarks and Future Directions

Increasingly, studies confirm the potential of exosomes as therapeutic vehicles for cancer treatment. Exosomes can transfer cargos with both an immunoregulatory potential and genetic information. They can decrease tumor cell invasion, migration and proliferation by enhancing immune response, cell death and sensitivity to chemotherapy. Exosomes have a high potential for both as diagnostic and therapeutic agents in immune therapy, vaccination trials and regenerative medicine. The native structure and unique cellular functions of exosomes give them a great potential as natural drug/gene delivery vehicles, but developing efficient and reliable isolation methods is necessary to fully utilize their potential [84]. The use of exosomes in clinical applications must be further investigated in order to establish standards for exosome characterization and manipulation [85]. The most important thing is to choose optimal methods for engineering exosomes and ensuring the safety of engineered exosomes in clinical trials [86]. Furthermore, the translation of EVs into clinical therapies requires their categorization as active drug components or drug delivery vehicles [6]. It is also necessary to clarify the composition and action mechanism of the various substances in exosomes and determine how to obtain highly purified exosomes and the right dosage for their clinical use [15]. Another important therapeutic use of EVs could be cancer vaccination [87]. In brief, exosomes are very promising tools for the future theranostics of prostate cancer, but their use first requires solutions for the many challenging issues remaining.

Abbreviations

APAF1	apoptotic peptidase-activating factor 1
AR-V7	androgen receptor variant 7
CAFs	cancer-associated fibroblast
CRPC	castrate-resistant prostate cancer
CTCs	circulating tumor cells
EGFR	epidermal growth factor receptor
EMT	epithelial–mesenchymal transition
EV	extracellular vesicles
FASL	ligand Fas
HIF-1α	hypoxia-inducible factor 1α
HSP	heat shock proteins
ITGB4	integrin β4
MICs	metastases-initiating cells
miR	micro RNA
MMP	matrix metalloproteinase
MVBs	multivesicular bodies
PCa	prostate cancer
PDCD4	programmed cell death 4
PDL-1	programmed death ligand 1
PSA	prostate specific antigen
RANKL	receptor activator for nuclear factor κB ligand
RSU1	Ras suppressor protein 1
siRNA	small interfering RNA
STAG2	cohesin subunit SA-2
TDSFs	tumor-derived secreted factors
TGFβ	transforming growth factor β
VCL	vinculin

References

1. Siegel, R.L.; Miller, K.D.; Jemal, A. Cancer statistics, 2019. *CA Cancer J. Clin.* **2019**, *69*, 7–34. [CrossRef]
2. Kim, S.J.; Kim, S.I. Current treatment strategies for castration-resistant prostate cancer. *Korean J. Urol.* **2011**, *52*, 157–165. [CrossRef] [PubMed]
3. Ingrosso, G.; Detti, B.; Scartoni, D.; Lancia, A.; Giacomelli, I.; Baki, M.; Carta, G.; Livi, L.; Santoni, R. Current therapeutic options in metastatic castration-resistant prostate cancer. *Semin. Oncol.* **2018**, *45*, 303–315. [PubMed]

4. Kretschmer, A.; Tilki, D. Biomarkers in prostate cancer—Current clinical utility and future perspectives. *Crit. Rev. Oncol. Hematol.* **2017**, *120*, 180–193. [CrossRef] [PubMed]
5. Lu, Y.T.; Delijani, K.; Mecum, A.; Goldkorn, A. Current status of liquid biopsies for the detection and management of prostate cancer. *Cancer Manag. Res.* **2019**, *11*, 5271–5291. [CrossRef]
6. Lener, T.; Gimona, M.; Aigner, L.; Borger, V.; Buzas, E.; Camussi, G.; Chaput, N.; Chatterjee, D.; Court, F.A.; Del Portillo, H.A.; et al. Applying extracellular vesicles based therapeutics in clinical trials—An ISEV position paper. *J. Extracell. Vesicles* **2015**, *4*, 30087. [CrossRef]
7. Jiang, L.; Gu, Y.; Du, Y.; Liu, J. Exosomes: Diagnostic Biomarkers and Therapeutic Delivery Vehicles for Cancer. *Mol. Pharm.* **2019**, *16*, 3333–3349. [CrossRef] [PubMed]
8. Pisitkun, T.; Shen, R.F.; Knepper, M.A. Identification and proteomic profiling of exosomes in human urine. *Proc. Natl. Acad. Sci. USA* **2004**, *101*, 13368–13373. [CrossRef] [PubMed]
9. Filella, X.; Foj, L. Prostate Cancer Detection and Prognosis: From Prostate Specific Antigen (PSA) to Exosomal Biomarkers. *Int. J. Mol. Sci.* **2016**, *17*, 1784. [CrossRef]
10. Fujita, K.; Nonomura, N. Urinary biomarkers of prostate cancer. *Int. J. Urol.* **2018**, *25*, 770–779. [CrossRef]
11. Foj, L.; Ferrer, F.; Serra, M.; Arevalo, A.; Gavagnach, M.; Gimenez, N.; Filella, X. Exosomal and Non-Exosomal Urinary miRNAs in Prostate Cancer Detection and Prognosis. *Prostate* **2017**, *77*, 573–583. [CrossRef] [PubMed]
12. Malla, B.; Aebersold, D.M.; Pra, A.D. Protocol for serum exosomal miRNAs analysis in prostate cancer patients treated with radiotherapy. *J. Transl. Med.* **2018**, *16*, 223. [CrossRef] [PubMed]
13. Pan, J.; Ding, M.; Xu, K.; Yang, C.; Mao, L.J. Exosomes in diagnosis and therapy of prostate cancer. *Oncotarget* **2017**, *8*, 97693–97700. [CrossRef] [PubMed]
14. Pullan, J.E.; Confeld, M.I.; Osborn, J.K.; Kim, J.; Sarkar, K.; Mallik, S. Exosomes as Drug Carriers for Cancer Therapy. *Mol. Pharm.* **2019**, *16*, 1789–1798. [CrossRef] [PubMed]
15. Di, C.; Zhang, Q.; Wang, Y.; Wang, F.; Chen, Y.; Gan, L.; Zhou, R.; Sun, C.; Li, H.; Zhang, X.; et al. Exosomes as drug carriers for clinical application. *Artif. Cells Nanomed. Biotechnol.* **2018**, *46*, S564–S570. [CrossRef]
16. Qin, J.; Xu, A.Q. Functions and application of exosomes. *Acta Pol. Pharm.* **2014**, *71*, 537–543.
17. Guo, W.; Gao, Y.; Li, N.; Shao, F.; Wang, C.; Wang, P.; Yang, Z.; Li, R.; He, J. Exosomes: New players in cancer (Review). *Oncol. Rep.* **2017**, *38*, 665–675. [CrossRef]
18. Liu, C.M.; Hsieh, C.L.; Shen, C.N.; Lin, C.C.; Shigemura, K.; Sung, S.Y. Exosomes from the tumor microenvironment as reciprocal regulators that enhance prostate cancer progression. *Int. J. Urol.* **2016**, *23*, 734–744. [CrossRef]
19. Thery, C.; Witwer, K.W.; Aikawa, E.; Alcaraz, M.J.; Anderson, J.D.; Andriantsitohaina, R.; Antoniou, A.; Arab, T.; Archer, F.; Atkin-Smith, G.K.; et al. Minimal information for studies of extracellular vesicles 2018 (MISEV2018): A position statement of the International Society for Extracellular Vesicles and update of the MISEV2014 guidelines. *J. Extracell. Vesicles* **2018**, *7*, 1535750. [CrossRef]
20. Panigrahi, G.K.; Deep, G. Exosomes-based biomarker discovery for diagnosis and prognosis of prostate cancer. *Front. Biosci.* **2017**, *22*, 1682–1696.
21. Saari, H.; Lazaro-Ibanez, E.; Viitala, T.; Vuorimaa-Laukkanen, E.; Siljander, P.; Yliperttula, M. Microvesicle- and exosome-mediated drug delivery enhances the cytotoxicity of Paclitaxel in autologous prostate cancer cells. *J. Control. Release* **2015**, *220*, 727–737. [CrossRef] [PubMed]
22. Egeblad, M.; Nakasone, E.S.; Werb, Z. Tumors as organs: Complex tissues that interface with the entire organism. *Dev. Cell* **2010**, *18*, 884–901. [CrossRef] [PubMed]
23. Guo, Y.; Ji, X.; Liu, J.; Fan, D.; Zhou, Q.; Chen, C.; Wang, W.; Wang, G.; Wang, H.; Yuan, W.; et al. Effects of exosomes on pre-metastatic niche formation in tumors. *Mol. Cancer* **2019**, *18*, 39. [CrossRef] [PubMed]
24. Liu, Y.; Xiang, X.; Zhuang, X.; Zhang, S.; Liu, C.; Cheng, Z.; Michalek, S.; Grizzle, W.; Zhang, H.G. Contribution of MyD88 to the tumor exosome-mediated induction of myeloid derived suppressor cells. *Am. J. Pathol.* **2010**, *176*, 2490–2499. [CrossRef] [PubMed]
25. Alzahrani, F.A.; El-Magd, M.A.; Abdelfattah-Hassan, A.; Saleh, A.A.; Saadeldin, I.M.; El-Shetry, E.S.; Badawy, A.A.; Alkarim, S. Potential Effect of Exosomes Derived from Cancer Stem Cells and MSCs on Progression of DEN-Induced HCC in Rats. *Stem Cells Int.* **2018**, *2018*, 8058979. [CrossRef] [PubMed]
26. Vlaeminck-Guillem, V. Extracellular Vesicles in Prostate Cancer Carcinogenesis, Diagnosis, and Management. *Front. Oncol.* **2018**, *8*, 222. [CrossRef]

27. Giusti, I.; Dolo, V. Extracellular vesicles in prostate cancer: New future clinical strategies? *Biomed. Res. Int.* **2014**, *2014*, 561571. [CrossRef]
28. Umezu, T.; Ohyashiki, K.; Kuroda, M.; Ohyashiki, J.H. Leukemia cell to endothelial cell communication via exosomal miRNAs. *Oncogene* **2013**, *32*, 2747–2755. [CrossRef]
29. Wang, Z.; Tan, Y.; Yu, W.; Zheng, S.; Zhang, S.; Sun, L.; Ding, K. Small role with big impact: miRNAs as communicators in the cross-talk between cancer-associated fibroblasts and cancer cells. *Int. J. Biol. Sci.* **2017**, *13*, 339–348. [CrossRef]
30. Mashouri, L.; Yousefi, H.; Aref, A.R.; Ahadi, A.M.; Molaei, F.; Alahari, S.K. Exosomes: Composition, biogenesis, and mechanisms in cancer metastasis and drug resistance. *Mol. Cancer* **2019**, *18*, 75. [CrossRef] [PubMed]
31. Brzozowski, J.S.; Jankowski, H.; Bond, D.R.; McCague, S.B.; Munro, B.R.; Predebon, M.J.; Scarlett, C.J.; Skelding, K.A. Lipidomic profiling of extracellular vesicles derived from prostate and prostate cancer cell lines. *Lipids Health Dis.* **2018**, *17*, 211. [CrossRef] [PubMed]
32. Deep, G.; Schlaepfer, I.R. Aberrant Lipid Metabolism Promotes Prostate Cancer: Role in Cell Survival under Hypoxia and Extracellular Vesicles Biogenesis. *Int. J. Mol. Sci.* **2016**, *17*, 1061. [CrossRef] [PubMed]
33. Webber, J.P.; Spary, L.K.; Sanders, A.J.; Chowdhury, R.; Jiang, W.G.; Steadman, R.; Wymant, J.; Jones, A.T.; Kynaston, H.; Mason, M.D.; et al. Differentiation of tumour-promoting stromal myofibroblasts by cancer exosomes. *Oncogene* **2015**, *34*, 290–302. [CrossRef] [PubMed]
34. Bellezza, I.; Aisa, M.C.; Palazzo, R.; Costanzi, E.; Mearini, E.; Minelli, A. Extracellular matrix degrading enzymes at the prostasome surface. *Prostate Cancer Prostatic Dis.* **2005**, *8*, 344–348. [CrossRef] [PubMed]
35. Di Vizio, D.; Sotgia, F.; Williams, T.M.; Hassan, G.S.; Capozza, F.; Frank, P.G.; Pestell, R.G.; Loda, M.; Freeman, M.R.; Lisanti, M.P. Caveolin-1 is required for the upregulation of fatty acid synthase (FASN), a tumor promoter, during prostate cancer progression. *Cancer Biol. Ther.* **2007**, *6*, 1263–1268. [CrossRef]
36. Wu, M.; Wang, G.; Hu, W.; Yao, Y.; Yu, X.F. Emerging roles and therapeutic value of exosomes in cancer metastasis. *Mol. Cancer* **2019**, *18*, 53. [CrossRef]
37. Conigliaro, A.; Cicchini, C. Exosome-Mediated Signaling in Epithelial to Mesenchymal Transition and Tumor Progression. *J. Clin. Med.* **2018**, *8*, 26. [CrossRef]
38. Tian, W.; Liu, S.; Li, B. Potential Role of Exosomes in Cancer Metastasis. *Biomed. Res. Int.* **2019**, *2019*, 4649705. [CrossRef]
39. Lopez, T.; Hanahan, D. Elevated levels of IGF-1 receptor convey invasive and metastatic capability in a mouse model of pancreatic islet tumorigenesis. *Cancer Cell* **2002**, *1*, 339–353. [CrossRef]
40. Chalmin, F.; Ladoire, S.; Mignot, G.; Vincent, J.; Bruchard, M.; Remy-Martin, J.P.; Boireau, W.; Rouleau, A.; Simon, B.; Lanneau, D.; et al. Membrane-associated Hsp72 from tumor-derived exosomes mediates STAT3-dependent immunosuppressive function of mouse and human myeloid-derived suppressor cells. *J. Clin. Invest.* **2010**, *120*, 457–471. [CrossRef]
41. Li, F.X.; Liu, J.J.; Xu, F.; Lin, X.; Zhong, J.Y.; Wu, F.; Yuan, L.Q. Role of tumor-derived exosomes in bone metastasis. *Oncol. Lett.* **2019**, *18*, 3935–3945. [CrossRef] [PubMed]
42. Vander Heiden, M.G.; Cantley, L.C.; Thompson, C.B. Understanding the Warburg effect: The metabolic requirements of cell proliferation. *Science* **2009**, *324*, 1029–1033. [CrossRef] [PubMed]
43. Tang, M.K.S.; Yue, P.Y.K.; Ip, P.P.; Huang, R.L.; Lai, H.C.; Cheung, A.N.Y.; Tse, K.Y.; Ngan, H.Y.S.; Wong, A.S.T. Soluble E-cadherin promotes tumor angiogenesis and localizes to exosome surface. *Nat. Commun.* **2018**, *9*, 2270. [CrossRef] [PubMed]
44. Naito, Y.; Yoshioka, Y.; Yamamoto, Y.; Ochiya, T. How cancer cells dictate their microenvironment: Present roles of extracellular vesicles. *Cell Mol. Life Sci.* **2017**, *74*, 697–713. [CrossRef]
45. Roma-Rodrigues, C.; Mendes, R.; Baptista, P.; Fernandes, A. Targeting Tumor Microenvironment for Cancer Therapy. *Int. J. Mol. Sci.* **2019**, *20*, 840. [CrossRef]
46. Shiao, S.L.; Chu, G.C.; Chung, L.W. Regulation of prostate cancer progression by the tumor microenvironment. *Cancer Lett.* **2016**, *380*, 340–348. [CrossRef]
47. Karlsson, T.; Lundholm, M.; Widmark, A.; Persson, E. Tumor Cell-Derived Exosomes from the Prostate Cancer Cell Line TRAMP-C1 Impair Osteoclast Formation and Differentiation. *PLoS ONE* **2016**, *11*, e0166284. [CrossRef]

48. Bijnsdorp, I.V.; Geldof, A.A.; Lavaei, M.; Piersma, S.R.; van Moorselaar, R.J.; Jimenez, C.R. Exosomal ITGA3 interferes with non-cancerous prostate cell functions and is increased in urine exosomes of metastatic prostate cancer patients. *J. Extracell. Vesicles* **2013**, *2*. [CrossRef]
49. Ichim, T.E.; Zhong, Z.; Kaushal, S.; Zheng, X.; Ren, X.; Hao, X.; Joyce, J.A.; Hanley, H.H.; Riordan, N.H.; Koropatnick, J.; et al. Exosomes as a tumor immune escape mechanism: Possible therapeutic implications. *J. Transl. Med.* **2008**, *6*, 37. [CrossRef]
50. Chen, G.; Huang, A.C.; Zhang, W.; Zhang, G.; Wu, M.; Xu, W.; Yu, Z.; Yang, J.; Wang, B.; Sun, H.; et al. Exosomal PD-L1 contributes to immunosuppression and is associated with anti-PD-1 response. *Nature* **2018**, *560*, 382–386. [CrossRef]
51. Tavoosidana, G.; Ronquist, G.; Darmanis, S.; Yan, J.; Carlsson, L.; Wu, D.; Conze, T.; Ek, P.; Semjonow, A.; Eltze, E.; et al. Multiple recognition assay reveals prostasomes as promising plasma biomarkers for prostate cancer. *Proc. Natl. Acad. Sci. USA* **2011**, *108*, 8809–8814. [CrossRef] [PubMed]
52. Overbye, A.; Skotland, T.; Koehler, C.J.; Thiede, B.; Seierstad, T.; Berge, V.; Sandvig, K.; Llorente, A. Identification of prostate cancer biomarkers in urinary exosomes. *Oncotarget* **2015**, *6*, 30357–30376. [CrossRef] [PubMed]
53. Soekmadji, C.; Corcoran, N.M.; Oleinikova, I.; Jovanovic, L. Extracellular vesicles for personalized therapy decision support in advanced metastatic cancers and its potential impact for prostate cancer. *Prostate* **2017**, *77*, 1416–1423. [CrossRef] [PubMed]
54. Liu, T.; Mendes, D.E.; Berkman, C.E. Functional prostate-specific membrane antigen is enriched in exosomes from prostate cancer cells. *Int. J. Oncol.* **2014**, *44*, 918–922. [CrossRef] [PubMed]
55. Soekmadji, C.; Russell, P.J.; Nelson, C.C. Exosomes in prostate cancer: Putting together the pieces of a puzzle. *Cancers* **2013**, *5*, 1522–1544. [CrossRef]
56. Yang, B.; Xiong, W.Y.; Hou, H.J.; Xu, Q.; Cai, X.L.; Zeng, T.X.; Ha, X.Q. Exosomal miRNAs as Biomarkers of Cancer: A Meta-Analysis. *Clin. Lab.* **2019**. [CrossRef]
57. Hessvik, N.P.; Sandvig, K.; Llorente, A. Exosomal miRNAs as Biomarkers for Prostate Cancer. *Front. Genet.* **2013**, *4*, 36. [CrossRef]
58. Huang, X.; Yuan, T.; Tschannen, M.; Sun, Z.; Jacob, H.; Du, M.; Liang, M.; Dittmar, R.L.; Liu, Y.; Liang, M.; et al. Characterization of human plasma-derived exosomal RNAs by deep sequencing. *BMC Genomics* **2013**, *14*, 319. [CrossRef]
59. Huang, X.; Yuan, T.; Liang, M.; Du, M.; Xia, S.; Dittmar, R.; Wang, D.; See, W.; Costello, B.A.; Quevedo, F.; et al. Exosomal miR-1290 and miR-375 as prognostic markers in castration-resistant prostate cancer. *Eur. Urol.* **2015**, *67*, 33–41. [CrossRef]
60. Joncas, F.H.; Lucien, F.; Rouleau, M.; Morin, F.; Leong, H.S.; Pouliot, F.; Fradet, Y.; Gilbert, C.; Toren, P. Plasma extracellular vesicles as phenotypic biomarkers in prostate cancer patients. *Prostate* **2019**, *79*, 1767–1776. [CrossRef]
61. Antonarakis, E.S.; Lu, C.; Wang, H.; Luber, B.; Nakazawa, M.; Roeser, J.C.; Chen, Y.; Mohammad, T.A.; Chen, Y.; Fedor, H.L.; et al. AR-V7 and resistance to enzalutamide and abiraterone in prostate cancer. *N. Engl. J. Med.* **2014**, *371*, 1028–1038. [CrossRef] [PubMed]
62. Del Re, M.; Biasco, E.; Crucitta, S.; Derosa, L.; Rofi, E.; Orlandini, C.; Miccoli, M.; Galli, L.; Falcone, A.; Jenster, G.W.; et al. The Detection of Androgen Receptor Splice Variant 7 in Plasma-derived Exosomal RNA Strongly Predicts Resistance to Hormonal Therapy in Metastatic Prostate Cancer Patients. *Eur. Urol.* **2017**, *71*, 680–687. [CrossRef] [PubMed]
63. Probert, C.; Dottorini, T.; Speakman, A.; Hunt, S.; Nafee, T.; Fazeli, A.; Wood, S.; Brown, J.E.; James, V. Communication of prostate cancer cells with bone cells via extracellular vesicle RNA; a potential mechanism of metastasis. *Oncogene* **2019**, *38*, 1751–1763. [CrossRef] [PubMed]
64. Li, S.L.; An, N.; Liu, B.; Wang, S.Y.; Wang, J.J.; Ye, Y. Exosomes from LNCaP cells promote osteoblast activity through miR-375 transfer. *Oncol. Lett.* **2019**, *17*, 4463–4473. [CrossRef]
65. Krishn, S.R.; Singh, A.; Bowler, N.; Duffy, A.N.; Friedman, A.; Fedele, C.; Kurtoglu, S.; Tripathi, S.K.; Wang, K.; Hawkins, A.; et al. Prostate cancer sheds the alphavbeta3 integrin in vivo through exosomes. *Matrix Biol.* **2019**, *77*, 41–57. [CrossRef]
66. Khan, S.; Jutzy, J.M.; Valenzuela, M.M.; Turay, D.; Aspe, J.R.; Ashok, A.; Mirshahidi, S.; Mercola, D.; Lilly, M.B.; Wall, N.R. Plasma-derived exosomal survivin, a plausible biomarker for early detection of prostate cancer. *PLoS ONE* **2012**, *7*, e46737. [CrossRef]

67. Lehmann, B.D.; Paine, M.S.; Brooks, A.M.; McCubrey, J.A.; Renegar, R.H.; Wang, R.; Terrian, D.M. Senescence-associated exosome release from human prostate cancer cells. *Cancer Res.* **2008**, *68*, 7864–7871. [CrossRef]
68. Hurwitz, M.D.; Kaur, P.; Nagaraja, G.M.; Bausero, M.A.; Manola, J.; Asea, A. Radiation therapy induces circulating serum Hsp72 in patients with prostate cancer. *Radiother Oncol.* **2010**, *95*, 350–358. [CrossRef]
69. Erozenci, L.A.; Bottger, F.; Bijnsdorp, I.V.; Jimenez, C.R. Urinary exosomal proteins as (pan-)cancer biomarkers: Insights from the proteome. *FEBS Lett.* **2019**, *593*, 1580–1597. [CrossRef]
70. Kim, S.M.; Kim, H.S. Engineering of extracellular vesicles as drug delivery vehicles. *Stem Cell Investig.* **2017**, *4*, 74. [CrossRef]
71. Lin, Y.; Lu, Y.; Li, X. Biological characteristics of exosomes and genetically engineered exosomes for the targeted delivery of therapeutic agents. *J. Drug Target.* **2019**, *28*, 129–141. [CrossRef] [PubMed]
72. Johnsen, K.B.; Gudbergsson, J.M.; Skov, M.N.; Pilgaard, L.; Moos, T.; Duroux, M. A comprehensive overview of exosomes as drug delivery vehicles—Endogenous nanocarriers for targeted cancer therapy. *Biochim. Biophys. Acta* **2014**, *1846*, 75–87. [CrossRef] [PubMed]
73. Tang, K.; Zhang, Y.; Zhang, H.; Xu, P.; Liu, J.; Ma, J.; Lv, M.; Li, D.; Katirai, F.; Shen, G.X.; et al. Delivery of chemotherapeutic drugs in tumour cell-derived microparticles. *Nat. Commun.* **2012**, *3*, 1282. [CrossRef]
74. Pascucci, L.; Cocce, V.; Bonomi, A.; Ami, D.; Ceccarelli, P.; Ciusani, E.; Vigano, L.; Locatelli, A.; Sisto, F.; Doglia, S.M.; et al. Paclitaxel is incorporated by mesenchymal stromal cells and released in exosomes that inhibit in vitro tumor growth: A new approach for drug delivery. *J. Control. Release* **2014**, *192*, 262–270. [CrossRef] [PubMed]
75. Tian, Y.; Li, S.; Song, J.; Ji, T.; Zhu, M.; Anderson, G.J.; Wei, J.; Nie, G. A doxorubicin delivery platform using engineered natural membrane vesicle exosomes for targeted tumor therapy. *Biomaterials* **2014**, *35*, 2383–2390. [CrossRef] [PubMed]
76. Qi, H.; Liu, C.; Long, L.; Ren, Y.; Zhang, S.; Chang, X.; Qian, X.; Jia, H.; Zhao, J.; Sun, J.; et al. Blood Exosomes Endowed with Magnetic and Targeting Properties for Cancer Therapy. *ACS Nano* **2016**, *10*, 3323–3333. [CrossRef]
77. Bakht, M.K.; Oh, S.W.; Hwang, D.W.; Lee, Y.S.; Youn, H.; Porter, L.A.; Cheon, G.J.; Kwak, C.; Lee, D.S.; Kang, K.W. The Potential Roles of Radionanomedicine and Radioexosomics in Prostate Cancer Research and Treatment. *Curr. Pharm. Des.* **2017**, *23*, 2976–2990. [CrossRef]
78. Li, J.; Yang, X.; Guan, H.; Mizokami, A.; Keller, E.T.; Xu, X.; Liu, X.; Tan, J.; Hu, L.; Lu, Y.; et al. Exosome-derived microRNAs contribute to prostate cancer chemoresistance. *Int. J. Oncol.* **2016**, *49*, 838–846. [CrossRef]
79. Kharaziha, P.; Chioureas, D.; Rutishauser, D.; Baltatzis, G.; Lennartsson, L.; Fonseca, P.; Azimi, A.; Hultenby, K.; Zubarev, R.; Ullen, A.; et al. Molecular profiling of prostate cancer derived exosomes may reveal a predictive signature for response to docetaxel. *Oncotarget* **2015**, *6*, 21740–21754. [CrossRef]
80. Corcoran, C.; Rani, S.; O'Brien, K.; O'Neill, A.; Prencipe, M.; Sheikh, R.; Webb, G.; McDermott, R.; Watson, W.; Crown, J.; et al. Docetaxel-resistance in prostate cancer: Evaluating associated phenotypic changes and potential for resistance transfer via exosomes. *PLoS ONE* **2012**, *7*, e50999. [CrossRef]
81. Kawakami, K.; Fujita, Y.; Kato, T.; Mizutani, K.; Kameyama, K.; Tsumoto, H.; Miura, Y.; Deguchi, T.; Ito, M. Integrin beta4 and vinculin contained in exosomes are potential markers for progression of prostate cancer associated with taxane-resistance. *Int. J. Oncol.* **2015**, *47*, 384–390. [CrossRef] [PubMed]
82. Kato, T.; Mizutani, K.; Kameyama, K.; Kawakami, K.; Fujita, Y.; Nakane, K.; Kanimoto, Y.; Ehara, H.; Ito, H.; Seishima, M.; et al. Serum exosomal P-glycoprotein is a potential marker to diagnose docetaxel resistance and select a taxoid for patients with prostate cancer. *Urol. Oncol.* **2015**, *33*, 385.e15–385.e20. [CrossRef] [PubMed]
83. Kosaka, N.; Iguchi, H.; Yoshioka, Y.; Hagiwara, K.; Takeshita, F.; Ochiya, T. Competitive interactions of cancer cells and normal cells via secretory microRNAs. *J. Biol. Chem.* **2012**, *287*, 1397–1405. [CrossRef] [PubMed]
84. Li, X.; Corbett, A.L.; Taatizadeh, E.; Tasnim, N.; Little, J.P.; Garnis, C.; Daugaard, M.; Guns, E.; Hoorfar, M.; Li, I.T.S. Challenges and opportunities in exosome research-Perspectives from biology, engineering, and cancer therapy. *APL Bioeng.* **2019**, *3*, 011503. [CrossRef]
85. Gilligan, K.E.; Dwyer, R.M. Engineering Exosomes for Cancer Therapy. *Int. J. Mol. Sci.* **2017**, *18*, 1122. [CrossRef]

86. Rahbarghazi, R.; Jabbari, N.; Sani, N.A.; Asghari, R.; Salimi, L.; Kalashani, S.A.; Feghhi, M.; Etemadi, T.; Akbariazar, E.; Mahmoudi, M.; et al. Tumor-derived extracellular vesicles: Reliable tools for Cancer diagnosis and clinical applications. *Cell Commun. Signal.* **2019**, *17*, 73. [CrossRef] [PubMed]
87. Giulietti, M.; Santoni, M.; Cimadamore, A.; Carrozza, F.; Piva, F.; Cheng, L.; Lopez-Beltran, A.; Scarpelli, M.; Battelli, N.; Montironi, R. Exploring Small Extracellular Vesicles for Precision Medicine in Prostate Cancer. *Front. Oncol.* **2018**, *8*, 221. [CrossRef] [PubMed]

10

Platelets Extracellular Vesicles as Regulators of Cancer Progression

Magdalena Żmigrodzka *, Olga Witkowska-Piłaszewicz and Anna Winnicka

Department of Pathology and Veterinary Diagnostics, Institute of Veterinary Medicine, Warsaw University of Life Sciences (WULS-SGGW), Nowoursynowska 159c, 02-787 Warsaw, Poland; olga_witkowska_pilaszewicz@sggw.edu.pl (O.W.-P.); anna_winnicka@sggw.edu.pl (A.W.)

* Correspondence: magdalena_zmigrodzka@sggw.edu.pl

Abstract: Extracellular vesicles (EVs) are a diverse group of membrane-bound structures secreted in physiological and pathological conditions by prokaryotic and eukaryotic cells. Their role in cell-to-cell communications has been discussed for more than two decades. More attention is paid to assess the impact of EVs in cancer. Numerous papers showed EVs as tumorigenesis regulators, by transferring their cargo molecules (miRNA, DNA, protein, cytokines, receptors, etc.) among cancer cells and cells in the tumor microenvironment. During platelet activation or apoptosis, platelet extracellular vesicles (PEVs) are formed. PEVs present a highly heterogeneous EVs population and are the most abundant EVs group in the circulatory system. The reason for the PEVs heterogeneity are their maternal activators, which is reflected on PEVs size and cargo. As PLTs role in cancer development is well-known, and PEVs are the most numerous EVs in blood, their feasible impact on cancer growth is strongly discussed. PEVs crosstalk could promote proliferation, change tumor microenvironment, favor metastasis formation. In many cases these functions were linked to the transfer into recipient cells specific cargo molecules from PEVs. The article reviews the PEVs biogenesis, cargo molecules, and their impact on the cancer progression.

Keywords: extracellular vesicles; exosomes; ectosomes; neoplasia

1. Introduction

The number of research work and scientific papers that discuss the involvement of cell-derived extracellular vesicles (EVs) in multiple physiological and pathological processes has increased rapidly during the last two decades. EVs might have an influence on target cells by delivering ligands and signaling complexes, and transferring mRNA and transcription factors that cause the epigenetic reprograming of recipient cells. EVs are submicron spherical membrane bound structures, that are generated by different prokaryotic (termed as membrane vesicles) and eukaryotic cells [1–3]. EVs nomenclature take into account their cellular origin and size. Their size ranges between 10 nm to 5 μm and comprises three heterogeneous populations of vesicles—exosomes (EXSMs), ectosomes (ECTSMs) also named microparticles (MPs), and apoptotic bodies (ABs) [4,5]. EVs actively secreted form parental cells with a diameter of 10 to 100 nm are named EXSMs, and those with a diameter ranging between 100 nm to 1 μm are ECTSMs. Lipid bilayer membrane protects their cargo from enzymes like proteases and ribonucleases [6]. The largest of EVs are ABs (with diameter 1–5 μm) represented by clumps of material generated during the late stage of cell apoptosis [5–7].

During activation, maturation, proliferation, stress, aging, or apoptosis, cells shed EVs into the extracellular space [8]. Their presence in a number of body fluids including—urine, synovial fluid, bronchoalveolar lavage fluid, saliva, and bile was confirmed [7,9–11]. In the bloodstream, EVs are released by—erythrocytes, leukocytes, platelets (PEVs), megakaryocytes, and endothelial cells [10,12].

In addition, EVs are also secreted by cancer cells known as tumor-derived extracellular vesicles (TEVs) [4,12]. In both healthy subjects and those with a variety of pathologies, peripheral blood is a rich source of EVs, where the most abundant population are PEVs. Their percentage ranges between 70 to 90% of all EVs in the plasma of healthy individuals [13–15].

In 1967, Peter Wolf described "platelet dust"—a subcellular material derived from thrombocytes in the plasma and serum of healthy individuals [16,17]. This was a milestone in medicine research, allowing further examinations evaluating PEVs involvement in physiological and pathological processes. PEVs share many functional features with PLTs. These tiny fragments smaller than platelets (PLTs) were secreted during PLT activation and were known to be crucial in coagulation and clot formation [16,18]. Despite the fact that PLTs play a crucial role in hemostasis, PEVs coagulation capacity is several dozen higher than PLTs [19]. Platelets microparticles (PMPs) are enriched in tissue factor (TF), coagulation factors, and dozens of them expose about 3-fold higher phosphatidylserine (PS) concentration on the outer membrane than PLTs [20]. The coagulation process initiated by TF connection with coagulation factor VII, activates coagulation cascade. Activated PLTs, PMPs PS + offer a catalytic surface for the coagulation and binding of consecutive clotting factors. Moreover, in healthy individuals, the presence of integrin αIIbβ3 (CD41/CD61) on PMPs supports fibrin clot formation [21]. In various bleeding disorders, abnormalities in PMPs functions and their reduced number in blood were reported [22]. On the other hand, their increased amount was presented in thrombotic state and other pathologies [23]. PLTs of patients described by Castaman are unable to shed PMPs, conversely to patients with Scott syndrome in which the PMPs number is adequate, but the incorrect translocation of PS impairs prothrombinase activity, and causes hemorrhagic diathesis [22]. Patients with immune thrombocytopenia have higher PEVs level than healthy individuals, which might be an evolutionary way to prevent blood loss and maintain tissue integrity [24]. Additionally, contemporary papers showed that PEVs might be a potential biomarker or prognostic factor in other pathologies—inflammatory, cardiovascular, and autoimmune diseases, solid tumors and hematological malignancies [14,25].

In this review, the role of PEVs in the cancerogenesis, tumor growth, and metastasis formation in distant organs is reported. Furthermore, the possible evaluation of PEVs as markers for cancer detection, and effectiveness of anticancer treatment is discussed.

2. EVs Biogenesis and Elimination

Based on the current knowledge, the mechanism of EVs formation and secretion to the extracellular space vary, depending on the EXSMs or ECTSMs descent. The EXSM definition was originally used for microparticles secreted from variety of cultured cells, thereafter, Johnstone and colleagues in 1987 explained the mechanism of transferrin receptor loss during reticulocytes maturation via secretion of nanosize vesicles; for this term EXSMs is used [26]. The latest research confirmed that the pathways of EVs biogenesis might differ between the parental cells types and EVs secretion, which does not seem to be accidental [1,27].

2.1. ECTSMs Formation

The blebbing of the plasma was documented in apoptosis during ABs formation, but it was confirmed as well in ECTSMs biogenesis. Changes in lipid components affect the rearrangement within plasma membrane. This process is initiated by an increased level of intracellular calcium ions. It causes activation of floppase and scramblase enzymes and inhibition of flippase (Figure 1) [1,8]. The membrane phospholipids—PS and phosphatidyl-ethanolamine, are vertically translocated from the inner leaflet to the outer cell membrane surface. The rearrangement breaks the bonds between cytoskeleton and cell membrane phospholipids. Partial degradation of actin filaments leads to restructuring of the cytoskeleton filaments, which favor formation of ECTSMs [1,8,10].

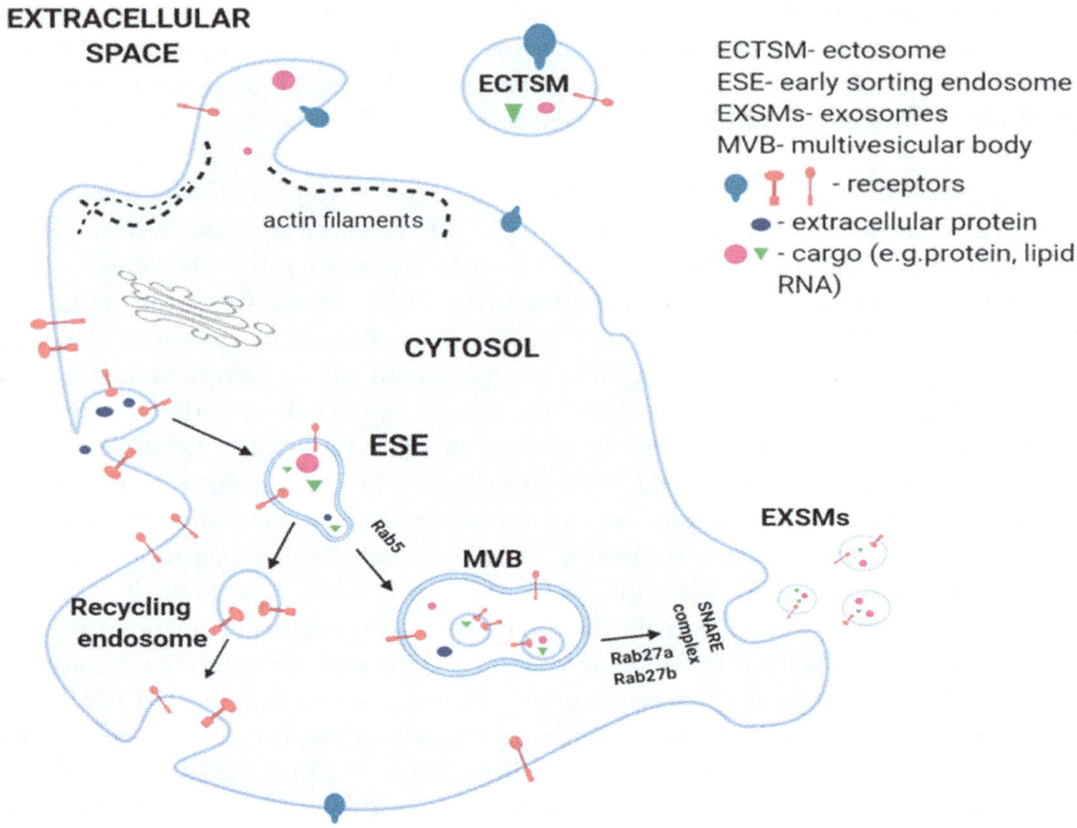

Figure 1. Extracellular vesicle biogenesis and secretion. The exosomes (EXSMs) generation begins with the membrane bulging into the lumen of the ESE. Part of them form a part of the plasma membrane (recycling endosome), others are converted into multi vesicular body (MVB). Members of the Rab family, Rab27a and Rab27b, are involved in MVB transport and fusion with cell membrane. Transmembrane protein complex SNARE enables the MVB to dock with the cell membrane that leads to release of EXSMs to extracellular space. Ectosome (ECTSM) are formed directly by cell membrane blebbing. This process is initiated with an increase in intracellular calcium that causes the activation of enzymes—floppase and scramblase and the inhibition of flippase. This causes the rearrangement of phospholipids in the cell membrane, as well as results in breaking bonds between cytoskeleton and partial degradation of actin filaments. During formation of EXSMs and ECTSMs, mRNA and miRNA that are located in cytoplasm are randomly entered.

The fast phospholipid membrane remodeling and PS exposure are relevant for PLTs physiological procoagulant response in hemostasis. PMPs formation in the circulation could result from PLTs activation via multiple agonists, high shear stress or apoptosis [20,28]. In the high shear rate, the loss of membrane integrity is initiated through the dislocated connection between the membrane glycoprotein Ib receptor (CD42b) and PLTs cytoskeleton, which began PMPs formation [20]. Natural PLTs activators, such as thrombin or collagen, induce PMPs formation via transmembrane integrin receptor gpIIb/IIIa (CD41/CD61) or tetraspanin 29 [29]. Altogether, these observations become the starting point for subsequent works assessing, how different types of PLTs activators induce PMPs formation, and how they affect the heterogeneity of PEVs population. Noticeably, a research conducted in 2017 confirmed that PS negative tubular PMPs population with structural similarities to filopodia could be formed during PLTs activation. Lack of PS expression on their surface implied that during their formation, there is no PS translocation [30].

2.2. EXSMs Formation

EXSMs generation begins with the inward bulging of the plasma membrane by endocytosis into the cytoplasm lumen. It leads to forming early sorting endosomes (ESEs) (Figure 1) [1].

Part of ESEs is returned into plasma membrane, other under the Rab5 control are changed into late endosomes or multivesicular bodies (MVBs) [1,10]. During this process, proteins and antigens are packaged into intraluminal vesicles (ILVs) and the budding of the ESEs membrane transform into MVBs [31]. Four protein subunits of the endosomal sorting complex required for transport (ESCRT) machinery are involved in this process. ESCTR-III is essential for the scission of ILVs into MVBs lumen. Cargo clustering and membrane budding can occur by ESCRT-dependent or -independent machinery [1]. ESCRT-0 recognizes ubiquitinated proteins (cargo) by the hepatocyte growth factor-regulated tyrosine kinase substrate (Hrs), in association with clathrin. This complex helps ESCRT I and II to connect with ESCRT 0 and ubiquitinated cargo, on the part of the endosomal membrane, where it will finally pullulate. ESCRT III connects with the complex and ultimately bud ILVs into the endosome [32]. The MVBs fuse with the plasma membrane to secrete the ILVs as exosomes or absorb with lysosomes for their degradation [1]. Members of Rab family, Rab27a and Rab27b, are essential mediators in transport of MVBs and its fusion with cell membrane. Transmembrane protein complex SANRE enables dock EXSMs with cell membrane that leads to the release of EXSMs to extracellular space (exocytosis). Increased concentration of calcium ions is one of the EXSMs secretion regulators [33]. Targeting selected Rabs via specific inhibitors modulates their structure or secretory function and becomes a new promising strategy of limiting EXSMs formation, both by PLTs and cancer cells. Wang et al. showed in a pre-clinical study, that elevated number of PMPs in patients with sepsis after intravenous administration of small GTPase inhibitor NSC23766, reduced PMPs secretion for about 87% [34,35].

Aatonen et al. showed that PMPs and platelet derived EXSMs (PdEXSMs) biogenesis is also observed by non-activated PLTs [36]. Examination potency of various agonists on EVs formation confirmed that Ca^{2+} ionophore is the strongest agonist, these include—thrombin, collagen, LPS, TRAP-6, and the weakest one is ADP [36]. Moreover, authors considered that Ca^{2+} ionophore causes vesiculation in unselective way or fragmentation and ABs formation, and should be advisedly used as agonist. The strongest PdEXSMs activators are thrombin and collagen or collagen-related peptide XL. Interestingly, the proteins cargo in PdEXSMs derived from stimulated PLTs was richer than from resting PLTs [36]. Nowadays, the utility of EXSMs as a new diagnostic cancer marker is extensively studied. Recent work performed by Lea et al. showed an increased number of EXSMs with PS expression in peritoneal fluid and plasma of patients with ovarian carcinoma [37]. It confirmed that, when PS is routinely used as a PMPs marker, it is also present on cancer derived EXSMs and causes a possibility to exploit these results in early diagnostic tests of women with ovarian malignancies [37].

2.3. EVs Elimination and Impact of Storage Conditions on PEVs Number

The PEVs rapid clearance from circulation varied depending on their molecular content, and the induction signal in different species [38,39]. As they have pro-coagulant and pro-inflammatory nanosize structures, their rapid turnover is essential for prevention of thrombotic diseases. PMPs turnover in rabbits is less than ten minutes, compared to people where PMPs were shown in circulation for more than 3 h [20,38,39]. Flumenhaft found that mice PMPs are eliminated from bloodstream within 30 min [40]. PEVs could be phagocytized after their opsonization with thrombospondin or complement components C3b [40]. The PS on the PMPs outer leaflet of the plasma membrane is recognized by macrophages and it originates a signal to remove them. Moreover, the role of lactadherin (LA) in clearance of EVs from circulation is discussed [41]. LA secreted by macrophages and adipocytes is also detected on the circulating PMPs. An increased PEVs level was observed in lactadherin–deficient mice, which could suggest the role of LA as a one of "eat-me" signals for phagocytosis [41]. Dasgupta et al. showed that developmental endothelial locus–1 in endothelial cells mediates PS-positive PMPs elimination via endocytosis [17,42]. Shorter half-life of ECTSMs, compared to EXSMs in blood, might arise from the higher concentration of membrane lipids in ECTSMs and activity of phospholipase A2 in serum [43]. Furthermore, EXSMs elimination via IgM immunoglobulins binding to lipid lysophosphatidylcholine was reported and liver macrophages were shown to be crucial elements of EXSMs clearance [43,44].

In EVs analysis, preanalytical steps standardization is crucial for the minimization of false results of PEVs number and their quality tests. Different anticoagulants could activate PLTs during blood collection and storage. Wisgrill et al. confirmed that the EVs number and their functionality is stable in sodium citrate for 8 h in room temperature (RT), after blood samples collection [45]. In EDTA, routinely used in clinical practice, PMPs and erythrocytes' derived EVs count is stable for 48 h in RT [45]. Thus, it could be an alternative when the collected samples are stored before analysis [45].

3. Content of Platelet Extracellular Vesicles

Physiological or pathological processes in parental cells define their EVs cargo and biological properties. As described above, the PEVs formation, membrane composition and specific markers expression on the outer membrane leaflet depends on the PLTs activators (Ca^{2+} ionophore, adenosine diphosphate, thrombin, collagen, epinephrine) [20]. Most of the EVs circulating in plasma are classified as PEVs based on their surface receptors. Nevertheless, heterogeneity of PEVs surface receptors starts discussion about EVs derived from megakaryocytes (Mk-EVs), as a part of PEVs subpopulation [46]. The EVs phenotyping conception to distinguish PEVs from Mk-EVs involves the usefulness of cluster of differentiation (CD) CD41/CD61 as a constitutive marker for both PLTs and Mks, while CD62P and CD107a act as a PLTs activation markers [47]. Flaumenhaft et al. showed that mouse and human Mk-EVs are PS/CD41/CD61 positive and CD62/CD107a negative [46]. In support of this finding, after irradiation of bone marrow, the CD61 positive EVs population largely disappears from mice circulation, whereas CD62P remains unchanged [48]. A study by Brisson et al. showed that small PMPs population—PS negative and CD41 positive, is a result of cell membrane shedding without PS redistribution. Moreover, PMPs could contain organelles like mitochondria and dense granules [30]. EVs are identified based on their size and expression of characteristic surface markers. PS expression is an a ECTSMs marker, when the presence of tetraspanin CD63 is used for EXSMs identification. During PLTs activation, both ECTSMs and EXSMs are CD63 positive but the CD63 expression is higher on EXSMs. It could be useful for determining the purity of the EXSMs population [30,49]. A characteristic of PEVs is the diversity of their surface markers and cargo. PEVs display a wide array of bioactive molecules like adhesion molecules, chemo- and cytokines, apoptosis regulators, miRNAs. They also harbor a broad spectrum of coagulation factors, enzymes, complement proteins, and bioactive lipids (Table 1). PEVs express glycoprotein (gp) IIb/IIIa, Ib, IIa, as well as P-selectin and a lysosome-associated glycoprotein-1 (LAMP-1). C-type lectin domain family 1-member B (CLEC-2) and gp VI expression was documented on Mk-EVs [17]. PdEXSMs are substantial with proteins from α granules, whereas ECTSMs are substantial with lipid mediators and mitochondrial proteins [17,50].

Molecules presented on PEVs were involved in triggering receptors on the target cells or regulating them via bioactive molecules, signaling molecules or a plethora of genetic material including miRNAs [51]. PEVs can interact with donor cells in multiple ways—(i) stimulation via signaling complex, using specific PEVs surface receptors and lipids; (ii) transfer membrane receptors and adhesion molecules; (iii) horizontal transfer of heterogeneous proteins, miRNAs, bioactive lipids, and other factors including infectious particles (prions) or even organelles (mitochondria) [3].

Table 1. Comparison of the PEVs cargo and their function. Biologically active molecules, receptors, enzymes, chemokines were categorized based on their functions of the PEVs, but there are no discrepancies detailed for some molecules.

	Function or Category Name							
	Clotting	**Enzymes**	**Adhesion Molecules**	**Bioactive Lipids**	**Programmed Cell Death**	**Growth Factors**	**Chemokines /Cytokines**	**Immune Response**
PEVs Cargo (Ref)	TF [52–54]	12-LO [55]	CD41/61 [56–59] CD31 [49,59]	PS [60]	caspase-3 [58]	TGF β1 [50]	CXCR4-(PF-4) [57,61]	CD 154 [32,62]
	FVa, FVIII [60,63]	heparynase [64]	CD62P [57,59,65,66]	AA [67,68]	CD95 [57]	PDGF bFGF [64]	IL-1β [69]	C5b-9 [70]
	PAR-1 [57]	PDI [71]	fibrinogen, vWF, vitronectin [65]	LPA [70]	caspase-9 [72]	VEGF [64]	CCL5, CCL23 [50,73]	CD55, CD59 [52]
	TFPI [74]	NADPH oxidase [75]	CD42a, CD42b [49,59]	TXA2 [76]			CX3CR1 [73]	Factor H [52]

PEVs are able to transfer receptors expressed on their surface (i.e., CD41, CD61, CD184, CD62P, PAR-1) to recipient cells (monocytes, myeloblasts, hematopoietic stem cells) and induce their adhesion or proliferation [3,27]. PEVs functional gpIIb/IIIa (CD41/CD61) transferred to neutrophils, activated NF-κB, in response to GM-CSF and enhanced inflammation [77]. Tang and colleagues showed that PEVs transfer arachidonate 12-lipoxygenase to mast cells, which increased synthesis of one of the negative regulator of inflammation lipoxin A4 (LXA4) [27,55]. Thus, PEVs play both positive and negative role in inflammation response, depending on the target cell.

PEVs are rich in sphingosine 1-phosphate, metalloproteinases, heparyanase, PDI, and arachidonic acid (AA) [3]. Transfer of AA by PMPs to monocytes and endothelial cells induced by prostanoids and cyclooxygenase 2 synthesis enhances these cells interactions [3]. Treatment of human umbilical vein endothelial cells (HUVECs) with PEVs showed intensification of angiogenesis and cell proliferation versus activated charcoal treated PEVs (removed nonpolar lipids), where a reduction of these effects was observed. This experiment showed that PEVs lipid components were involved in HUVECs stimulation [27,78]. The horizontal transfer of non-coding RNAs via EVs regulates gene expression by post-transcriptional repression. miRNA from parental cells encapsulated in EVs was protected from ribonuclease activity in circulation [51]. In human's, several PMPs miRNAs were detected, e.g., miR-19, miR-21, miR-22, miR-126, miR-133, miR-146, miR-185, miR-223, and miR-320b [3]. Moreover, it was confirmed that PEVs miRNA was transferred to macrophages, endothelial, and cancer cells. In macrophages, MiR-126–3p transferred from PEVs led to decreased ATF3 and ATP1B1 expression and protein synthesis [27]. Recently presented data support the notion of PEVs tumor microenvironment infiltration and interaction with cancer cells via the mechanisms described above.

4. PEVs in Cancer Progression

PEVs are highly interesting group of EVs because of their percentage participation in bloodstream, as well as their increased number in patients with cancer, such as glioblastoma, gastric, lung and skin cancer, and other diseases. This makes them potentially useful as a diagnostic marker [79]. It is known that PLTs facilitate cancer metastasis. Moreover, the number of papers that discuss PEVs contribution in cancerogenesis increased recently [51,80]. EVs as cell-to-cell messenger molecules can start phenotypic and functional changes in donor cells, by reaching the recipient cells and delivering EVs content. PEVs are also discussed as potentially early markers of disease progression.

4.1. PEVs in Tumor Angiogenesis

The cancer cells without blood circulation can grow up to 2 mm^3 in diameter, forming a tumor and then stop and undergo apoptosis or necrosis [81]. Growth of the vascular network is pivotal for the cancer cells survival, proliferation, as well as metastatic spread of cancer [81]. Angiogenesis is essential for formation of a new vascular network that supplies nutrients, oxygen, and immune cells, and also removes waste products of cellular metabolism. Therefore, angiogenesis is a critical factor in the progression of cancer. The tumor microenvironment (TME) consists of diverse cellular populations, including tumor cells, endothelial cells, fibroblasts, infiltrating immune cells (monocytes, macrophages, neutrophils, mast cells, T cells), extracellular matrix, and newly formed blood vessels [79]. The PEVs interaction with TME components could reveal their functions in cancer progression. Newly stirring blood vessels permit tumor growth, which is critical in cancer progression. Interestingly, Happonen et al. demonstrated a mechanism of PEVs transfer to human aortic endothelial cells (HAECs) and HUVECs [82]. PS-positive PEVs are taken up by phagocytosis via tyrosine kinase receptor Axl, and its ligand protein Gas6 on endothelial cells [82]. Janowska-Wieczorek et al. used lung cancer cell lines to elucidate PEVs importance in cancer angiogenesis [83]. After PEVs stimulation of IL-8 (about 35-fold), vascular endothelial growth factor (VEGF) (3-fold) and scatter factor (4-fold) mRNA expression increased in the A549 cell line [83].

PEVs delivery of bioactive molecules like cytokines or microRNA to recipient cells could regulate tumor growth [84]. miRNAs are small non-coding RNAs that regulate gene expression post-transcriptionally. Anene et al. demonstrated regulatory angiogenesis miRNAs transfer from PEVs to HUVECs cells during co-culturing on extracellular matrix gel [85]. A robust capillary-like structure formation and simultaneously decreased synthesis of anti-angiogenic thrombospondin-1 (THBS-1) was observed. miRNA Let-7 a from PEVs was delivered to HUVECs and targeted THBS-1 mRNA to induce angiogenic responses of HUVECs [85]. Blood vessel formation is controlled by a balance between localized production of pro- and anti-angiogenic molecules and changes in THBS-1 concentration is the key determinant of this "angiogenic switch" [85]. PEVs ability to bind TF and the platelet-activating factor potentiates their pro- angiogenic competence even more [86].

Pan et al. demonstrated that after incubation, PEVs with HUVECs cells miR-223 level in endothelial cells increased, which promoted glycation end-product-induced vascular endothelial cell apoptosis via targeting insulin-like growth factor 1 receptor [87]. Another work showed that HUVECs cells preferentially uptake miR-223 from PEVs generated by thrombin-activated PLTs [4,88]. This leads to the formation of functional Argonaute 2 (Ago2) miR-223 complexes. These complexes are able to regulate gene expression and protein level for ephrin A1 and F-box/WD repeat-containing protein 7 in HUVECs cells and conduces apoptosis [4].

Increased angiogenesis in TME could be a result of metalloproteinase-1 (MMP-1) transfer, as well as increased MMP-9, VEGF, and IL-8 mRNA expression in lung cancer cells lines, after co-incubation with PEVs [83]. Moreover, PEVs molecules from α granules like VEGF, platelet-derived growth factor and fibroblast growth factor are a component of their cargo with pro-angiogenic properties.

4.2. PEVs in Migration, Invasion, and Metastasis

A key for distant metastases formation is cancer cells passage through the newly formed vascular walls in primary tumor, surviving in the circulation, and finally proliferation at the distant tissue. In solid tumors, vasculature is highly permeable, allowing the possibility to PEVs infiltration to TEM and contact with cells. A great number of studies indicate the PEVs involvement in cancer progression and some discuss their anti-cancer properties. Michael et al. showed that PEVs have the ability to infiltrate murine and human tumors [84]. This ability creates conditions for the horizontal transfer of miRNA-24, which targets mitochondrial mt-Nd2, and Snora75. This entails mitochondrial dysfunction and results in an increased cancer cell apoptosis [84].

Bakewell and colleagues showed that platelets gpIIb/IIIa antagonists minimize formation of distant metastasis from B16 melanoma cells in bones, due to the inhibition of the interaction between cancer cells and PLTs [89]. Lung cancer cell line A549 increases adhesiveness to the fibrinogen and HUVECs, after receiving CD41 from PEVs. PEVs chemoattract lung cell lines from 2.5 to 7-fold more than the control [83]. Moreover, evaluation of PEVs interaction with lung cancer cell lines confirmed the activation of mitogen-activated protein kinase (MAPK) MAPK p42/44 and AKT, signaling pathways participating in proliferative responses [83]. Murine lung cells covered by PEVs injected intravenously into mince resulted in significant increase metastasis formation in lungs and bone marrow [83]. Transfer onto the surface of donor cells CD184-, a chemokine receptor type 4 from PEVs and respond to stromal cell-derived factor 1, which is rich niche in bone marrow in the murine model, confirmed their high metastatic potential [83]. Moreover, activation of cyclin D2 by PEVs in lung cancer cell lines could change the phenotype of cancer cells into a more invasive phenotype. Similar observations were made in human squamous carcinoma or breast cancer cell lines in murine in vivo model [90].

Interestingly, Gasperi and colleagues confirmed the modulatory influence of polyunsaturated fatty acids (PUFAs) diet, especially the $\omega3$ and $\omega6$ on cellular processes in carcinogenesis [62]. The PUFAs $\omega3$ cancer preventive activity is well known, in contrast to high concentration of $\omega6$ in diet, which correlates with higher risk of breast and prostate cancer [62,91]. Their role in cancerogenesis is related to changes in fatty acids compositions of membrane rafts in cells membranes. PEVs contains miR-126 and miRNA-223, which are important players in tumorigenesis. VEGF-dependent proliferation of endothelial cells is stimulated by miRNA-126, while miRNA-223 inhibit formation of new blood vessels by targeting endothelial $\beta1$ integrin [92]. Gasperi et al. examined the influence of increased level of PUFAs $\omega6$

on both PEVs formation and their cargo [62]. The newly formed PEVs had an increased amounts of miRNA-123 and miRNA233. Breast cell line BT549 blocked its cell cycle and decreased cell migration after internalizing PEVs [62].

A Tang et al. study revealed an important PEVs role in the epithelial-to-mesenchymal transition of ovarian epithelial cancer cell line (SKOV3). miR-939 transfer leads to enhanced invasion and cancer progression [93,94]. Tropomyosin 3 (TPM3) contributes cancerogenesis in thyroid papillary carcinoma and esophageal squamous cell carcinoma by fusing neurotrophic receptor tyrosine kinase 1 and PDGF receptors [95]. Yao et al. demonstrated increased TPM3 mRNA in PLTs and revealed their transfer by PEVs into breast cancer cells and promotion of an invasion [94]. Moreover, in patients with distant metastases, compared to subjects without metastases TPM3 mRNA in PLTs was significantly increased [94].

Another interesting issue is the ability of cancer cells to educate PLTs. Zarà et al. demonstrated that breast cancer cell lines—highly aggressive MDM-MB-231 and MCF7 could educate PLTs to produce PEVs in an amount similar to that after thrombin activation [96]. Next, those PEVs were co-cultured with cancer cells to investigate if the newly formed PEVs impact cells. Only in the MDM-MB-231 cell line, authors observed cells activation and phosphorylation of p38MAPK and myosin light chain. Moreover, increased migration and invasion was noted. This experiment showed that PEVs can novel paracrine-positive feedback mechanism initiated by MDA-MB-231 to escalate their invasive phenotype [96].

PEVs formed by PLTs during apoptosis-like process show surface gpIIb/IIIa, and PS and stimulate their own phagocytic removal by monocytes, moreover, they are able to change macrophages into M2 macrophages [97,98]. In contrast to effect on endothelial cells, after PEVs miR-223 transfer into gastric cell line SGC7901, increased proliferation and invasion in vitro, as well as decreased apoptosis, was observed. This showed that horizontal miRNA transfer via PEVs could have diverse effect contingent on donor cells [4,99]. Another noteworthy experimental work showed that peripheral blood mononuclear cells (PBMCs) isolated from patients with B-precursor acute lymphoblastic leukemia had increased apoptotic markers CD95, active caspse-3, and an increased number of apoptotic cells, after two days of co-culturing with PEVs [100].

Cancer cells transmigration from circulation into the tissues is mediated likewise by tissue-specific enzymes, the majority of which belongs to the MMP family. Dashevsky et al. confirmed transfer of MMP-2 and its' increased secretion from Cl-1 cells after co-culturing with PEVs. Interesting observation was made when Cl-1 cells were incubated with PEVs lysate. Values of MMP-2 concentration and secretion were similar to that after cells co-culturing with PEVs. It suggests that the transfer of MMP-2 is not dependent only on PEVs internalization. The other possible candidates for increased MMP-2 value might be free miRNA from PEVs lysate or lysophosphatidic acid (LPA) as an MMP-2 activator presented on PLTs and in prostate cancer cells [101].

Natural killer (NK) cells efficiently recognize and kill circulating tumor cells of almost any origin, but their effectiveness in TME is discussed. PEVs miR-183 transfer into NK cells suppressed activator adapter DAP12 and suppressed their cytolytic functions in tumor-associated NK cells [102]. PEVs could also horizontally transfer functional miR-126–3p into primary human macrophages. The PEVs dose-dependent down regulation of miR-126–3p targets CCL4, CSF1, and TNF was observed. Decreased secretion of cytokines/chemokines was correlated with reprogramming into phagocytic macrophages [88,103]. The role of TF in angiogenesis and metastasis formation is well documented, therefore, the role of TF-positive PEVs in tumor growth seems clear. Another interesting aspect of PEVs as a potentially important immune checkpoint in cancer biology is a presence of PS on PEVs surface. PEVs as an abundant source of PS might be a possible ligand for PS receptor (PSR) on the immune cells. Activation of PS–PSR pathway leads to the inhibition of innate and adaptive immune response in TME, as well as in circulation [104]. The new oncotherapy strategies examined the PSR inhibitors as a new anticancer target, but only a highly selective inhibition strategy could be applied in

the cancer treatment. Table 2 summarize PEVs pivotal role in crosstalk between PLTs and other cells, particularly with cancer cells (Table 2).

Table 2. The role of PEVs in cell-to-cell communication. PEVs secreted from activated PLTs transfer to target cells and their cargo promotes phenotypic changes and novel functions in donor cells.

Target Cell	PEVs Derived Factors/Molecules	Functional Changes (References)
A549, CRL 2066, CRL 2062, HTB 183, HTB 177 lung CCL; LCC * CCL	CD41, CD61 CD184	(+) adhesion to fibrinogen and HUVECs [83] (+) metastatic potential [83] (+) mRNA expression of angiogenic factors (MMP-9, VEGF, IL-8) [83] (+) proliferation and chemoinvasion [83]
HUVECs	miRNA Let-7a miRNA-223	(−) synthesis THBS-1 anti-angiogenic molecule [85] (+) apoptosis by IGF-1 [4,87]
MC-38 colon CCL, LCC * CCL	miRNA-24	(+) apoptosis [84]
BT549 breast CCL	miRNA-123 miRNA-233	(−) migration [62] (−) cell cycle [62]
SKOV3 ovarian CCL	miRNA-939	(+) invasion via TPM3 [94] (+) progression [94]
MDM-MB-231 breast CCL		(+) invasion [96] (+) migration [94]
SGC7901 gastric CCL	miRNA-223	(+) proliferation and invasion [4,105] (−) apoptosis [4,105]
PBMCs from patients with ALL	CD95 Caspase-3	(+) apoptosis [100]
Cl-1 prostate CCL	MMP-2 miRNA?	(+) migration [101]
macrophages	PS, gpIIb/IIIa miR-126-3p	polarization into macrophage M2 [97] (−)CCL4, CSF1, TNF [88]
NK cells	miR-183	(−) cytolysis [102]

Abbreviations: CCL-cancer cell line; * murine cell line; (+) increase; (−) decrease.

5. The Potential of PEVs as Diagnostics Cancer Biomarkers

PEVs number in blood was raised about twice in myeloproliferative neoplasms, compared to healthy controls, up to four times in oral cancer and colorectal subjects and more than ten times in breast cancer patients [86,106]. The highest concentration of PEVs, more than 30-fold, was noticed in patients with IV stage of gastric cancer. In each group, the highest PEVs concertation were demonstrated in advanced cancer stages and in patients with distal metastases [86,106–108].

Investigation in patients with non-small cell lung cancer (NSCLC) categorized based on disease progression, showed the significantly higher number of circulating EVs from activated or apoptotic PLTs and from endothelial apoptotic cells, compared to healthy subjects. Changes in EVs levels in different stages of NSCLC showed that serial measurements of circulating PEVs are valuable prognostic biomarkers, mainly in the advanced stages of NSCLC [109].

PEVs as source of anionic phospholipids and TF on their surface are one of the important factors of procoagulant activity. Data demonstrated by Ren et al. showed the significantly increased number of EVs and PEVs in patients with oral squamous cell carcinoma (OSCC) in peripheral blood. PEVs level was also positively correlated with clinical stage and with fibrinogen concentration and patients hypercoagulable state [107]. Mege and colleagues showed correlations between increased PEVs number and the stage of the disease in patients with pancreatic cancer and colorectal cancer. They suggested that PEVs concentration in blood could be a useful marker for evaluation of the disease progression in these types of neoplasia [110].

Yenigürbüz et al. described another aspect of increased PEVs number in patients with neoplasia. Thromboembolism is one of the complications during induction of therapy in pediatric acute lymphoblastic leukemia (ALL) patients [111]. Children with ALL have increased levels of ABs, PEVs, endothelial-derived, and tissue factor-positive microvesicles during induction therapy. Further studies are needed to confirm the PEVs contribution in thromboembolism during the induction therapy period in children with ALL [111]. Similar observations were made in adult patients with myeloproliferative neoplasia, where the number of TF positive PMPs and endothelial derived EVs was significantly increased, which might also play a role in thrombotic complications in that group of subjects [112]. Tjon-Kon-Fat et al. demonstrated that tumor educated PLTs are a source of prostate cancer biomarkers [113,114]. In this context it seems to be interesting to evaluate the presence and role of EVs generated from tumor-educated PLTs.

6. The Potential of PEVs in Cancer Therapy

The paradigm of using nanoparticulate pharmaceutics as delivery vectors was established over the past decade [56]. To use EVs as drug transporters, their pharmacokinetics should be analyzed. Mice models of EXSMs distribution showed that the route of administration, EXSMs origin, and concentration critically influenced their biodistribution [115]. In the mice model, after intraperitoneal and subcutaneous administration of EXSMs, they preferentially localized in the pancreas and gastrointestinal tract. Whereas, intravenous administrated EXSMs were detected in the spleen and the liver [116]. In addition, EXSMs loaded with therapeutic anti-miRNA could be transferred locally into tumor or systemically. Other therapeutic strategies in cancer therapy were elimination of EXSMs from blood or prevention of EXSMs fusion with target cell [117,118]. Various strategies of using EXSMs in anticancer therapy are characterized in the literature, but more research is still needed.

In an elegant study, Kailashiya et al. documented that doxorubicin-loaded PEVs (doxo-PEVs) were taken by HL60, K562 cells (leukemia cell lines), and blast cells, in whole blood harvested from patients with newly diagnosed leukemia. Doxo-PEVs were uptaken by cells via P-selectin ligands and integrins. Moreover, doxo-PEVs transfer into leukemia cells was higher, compared to free doxorubicin, which could be used to increase the effectiveness of the therapy and minimize the side effects of drugs [56]. Gasperi et al. showed that PEVs with miR-126 and with miR-223 increased sensitivity of BT549 cells to the cisplatin chemotherapy [62].

PEVs drug-loaded could be a natural vectors-targeted medications. Engineering them from autologous platelets in large quantity and storing for several days, seems to be a new biocompatible and non-immunogenic new-generation medicine. However, to make PEVs applicable and efficacious in clinical treatments, some of their underlying functions still need to be better researched and understood.

7. Summary

PEVs biogenesis depends on different signals that control their formation from PLTs. The role of PEVs in various physiological conditions, like hemostasis, or pathological like inflammation or atherosclerosis was confirmed. This review focused on the PEVs participation in cancerogenesis. A better understanding of the biology of PEVs and the mechanisms that allow them to function as mediators in cell-to-cell communication in cancer growth, could become a contribution to the development of new therapeutic strategies, which could also be applicable in cancer. Moreover, determining the number of PEVs and their cargo becomes a useful diagnostic marker or prognostic factor for the different clinical stages in a variety of neoplasia. Knowledge about the formation of distinct PEVs types dependent on PLTs activators could lead to the development of specific techniques for PEVs-mediated drug delivery to cancer cells, or to TME, to modulate their immune response or angiogenesis.

Author Contributions: Conceptualization, M.Ż. and O.W.-P., Writing-Original Draft Preparation, M.Ż., O.W.-P.; Writing-Review & Editing, M.Ż., A.W. and O.W.-P. Supervision, M.Ż. All authors have read and agreed to the published version of the manuscript.

Abbreviations

ABs	apoptotic bodies
ALL	acute lymphoblastic leukemia
CD	cluster of differentiation
CCL	cancer cell line
CLEC-2	C-type lectin domain family 1-member B
ECTSMs	ectosomes
ESCRT	endosomal sorting complex required for transport
ESE	early sorting endosomes
EVs	extracellular vesicles
EXSMs	exosomes
HAECs	human aortic endothelial cells
HUVECs	human umbilical vein endothelial cells
ILVs	intraluminal vesicles
LA	lactadherin
LAMP-1	lysosome-associated glycoprotein-1
LPS	lipopolysaccharide
LXA4	lipoxin A4
MAPK	mitogen-activated protein kinase
Mk-EVs	EVs derived from megakaryocytes
MMP	metalloproteinase
MVBs	multivesicular bodies
MPs	microparticles
NSCLC	non-small cell lung cancer
OSCC	oral squamous cell carcinoma
PBMCs	peripheral blood mononuclear cells
PdEXSMs	platelet derived exosomes
PEVs	platelets extracellular vesicles
PLTs	platelets
PMPs	platelets microparticles
PUFAs	polyunsaturated fatty acids
RT	room temperature
TEVs	tumor derived extracellular vesicles
TF	tissue factor
TGF-β1	transforming growth factor beta 1
THBS-1	thrombospondin-1
TLR-4	toll-like receptor 4
TME	tumor microenvironment
TPM3	tropomyosin 3
VEGF	vascular endothelial growth factor
vWf	von Willebrand factor

References

1. Van Niel, G.; D'Angelo, G.; Raposo, G. Shedding light on the cell biology of extracellular vesicles. *Nat. Rev. Mol. Cell Biol.* **2018**, *4*, 213–228. [CrossRef]
2. Dauros Singorenko, P.; Chang, V.; Whitcombe, A.; Simonov, D.; Hong, J.; Phillips, A.; Swift, S.; Blenkiron, C. Isolation of membrane vesicles from prokaryotes: A technical and biological comparison reveals heterogeneity. *J. Extracell. Vesicles* **2017**, *1*, 1324731. [CrossRef] [PubMed]

3. Dovizio, M.; Bruno, A.; Contursi, A.; Grande, R.; Patrignani, P. Platelets and extracellular vesicles in cancer: Diagnostic and therapeutic implications. *Cancer Metastasis Rev.* **2018**, *37*, 455–467. [CrossRef] [PubMed]
4. Wojtukiewicz, M.Z.; Sierko, E.; Hempel, D.; Tucker, S.C.; Honn, K. Platelets and cancer angiogenesis nexus. *Cancer Metastasis Rev.* **2017**, *2*, 249–262. [CrossRef] [PubMed]
5. Navarro-Tableros, V.; Gomez, Y.; Camussi, G.; Brizzi, M.F. Extracellular vesicles: New players in lymphomas. *Int. J. Mol. Sci.* **2018**, *21*, 41. [CrossRef] [PubMed]
6. Lorenc, T.; Klimczyk, K.; Michalczewska, I.; Słomka, M.; Kubiak-Tomaszewska, G.; Olejarz, W. Exosomes in prostate cancer diagnosis, prognosis and therapy. *Int. J. Mol. Sci.* **2020**, *21*, 2118. [CrossRef]
7. Meldolesi, J. Extracellular vesicles, news about their role in immune cells: Physiology, pathology and diseases. *Clin. Exp. Immunol.* **2019**, *13*, 318–327. [CrossRef]
8. Stahl, P.D.; Raposo, G. Extracellular vesicles: Exosomes and microvesicles, integrators of homeostasis. *Physiology* **2019**, *3*, 169–177. [CrossRef]
9. Frydrychowicz, M.; Kolecka-Bednarczyk, A.; Madejczyk, M.; Yasar, S.; Dworacki, G. Exosomes—Structure, biogenesis and biological role in non-small-cell lung cancer. *Scand. J. Immunol.* **2015**, *81*, 2–10. [CrossRef]
10. Żmigrodzka, M.; Guzera, M.; Miśkiewicz, A.; Jagielski, D.; Winnicka, A. The biology of extracellular vesicles with focus on platelet microparticles and their role in cancer development and progression. *Tumour. Biol.* **2016**, *11*, 14391–14401. [CrossRef]
11. Van der Pol, E.; Böing, A.N.; Gool, E.L.; Nieuwland, R. Recent developments in the nomenclature, presence, isolation, detection and clinical impact of extracellular vesicles. *J. Thromb. Haemost.* **2016**, *14*, 48–56. [CrossRef] [PubMed]
12. Menck, K.; Bleckmann, A.; Wachter, A.; Hennies, B.; Ries, L.; Schulz, M.; Balkenhol, M.; Pukrop, T.; Schatlo, B.; Rost, U.; et al. Characterisation of tumour-derived microvesicles in cancer patients' blood and correlation with clinical outcome. *J. Extracell. Vesicles* **2017**, *1*, 1340745. [CrossRef] [PubMed]
13. Laroche, M.; Dunois, C.; Vissac, A.M.; Amiral, J. Update on functional and genetic laboratory assays for the detection of platelet microvesicles. *Platelets* **2017**, *3*, 235–241. [CrossRef] [PubMed]
14. Italiano, J.E., Jr.; Mairuhu, A.T.; Flaumenhaft, R. Clinical relevance of microparticles from platelets and megakaryocytes. *Curr. Opin. Hematol.* **2010**, *6*, 578–584. [CrossRef] [PubMed]
15. Berckmans, R.J.; Nieuwland, R.; Böing, A.N.; Romijn, F.P.; Hack, C.E.; Sturk, A. Cell-derived microparticles circulate in healthy humans and support low grade thrombin generation. *Thromb. Haemost.* **2001**, *4*, 639–646.
16. Wolf, P. The nature and significance of platelet products in human plasma. *Br. J. Haematol.* **1967**, *3*, 269–288. [CrossRef]
17. Melki, I.; Tessandier, N.; Zufferey, A.; Boilard, E. Platelet microvesicles in health and disease. *Platelets* **2017**, *3*, 214–221. [CrossRef]
18. Van der Pol, E.; Harrison, P. From platelet dust to gold dust: Physiological importance and detection of platelet microvesicles. *Platelets* **2017**, *3*, 211–213. [CrossRef]
19. Sinauridze, E.I.; Kireev, D.A.; Popenko, N.Y.; Pichugin, A.V.; Panteleev, M.A.; Krymskaya, O.V.; Ataullakhanov, F.I. Platelet microparticle membranes have 50- to 100-fold higher specific procoagulant activity than activated platelets. *Thromb. Haemost.* **2007**, *97*, 425–434.
20. Aatonen, M.; Grönholm, M.; Siljander, P.R. Platelet-derived microvesicles: Multitalented participants in intercellular communication. *Semin. Thromb. Hemost.* **2012**, *1*, 102–113. [CrossRef]
21. Zubairova, L.D.; Nabiullina, R.M.; Nagaswami, C.; Zuev, Y.F.; Mustafin, I.G.; Litvinov, R.I.; Weisel, J.W. Circulating microparticles alter formation, structure, and properties of fibrin clots. *Sci. Rep.* **2015**, *5*, 17611. [CrossRef] [PubMed]
22. Castaman, G.; Li, Y.-F.; Battistin, E.; Rodeghiero, F. Characterization of a novel bleeding disorder with isolated prolonged bleeding time and deficiency of platelet microvesicle generation. *Br. J. Haematol.* **1997**, *96*, 458–463. [CrossRef]
23. Chen, Y.; Xiao, Y.; Lin, Z.; Xiao, X.; He, C.; Bihl, J.C.; Zhao, B.; Ma, X.; Chen, Y. The role of circulating platelets microparticles and platelet parameters in acute ischemic stroke patients. *J. Stroke Cereb. Dis.* **2015**, *10*, 2313–2320. [CrossRef] [PubMed]
24. Álvarez-Román, M.T.; Fernández-Bello, I.; Jiménez-Yuste, V.; Martín-Salces, M.; Arias-Salgado, E.G.; Rivas Pollmar, M.I.; Justo Sanz, R.; Butta, N.V. Procoagulant profile in patients with immune thrombocytopenia. *Br. J. Haematol.* **2016**, *5*, 925–934. [CrossRef] [PubMed]

25. Kim, H.K.; Song, K.S.; Park, Y.S.; Kang, Y.H.; Lee, Y.J.; Lee, K.R.; Kim, H.K.; Ryu, K.W.; Bae, J.M.; Kim, S. Elevated levels of circulating platelet microparticles, VEGF, IL-6 and RANTES in patients with gastric cancer: Possible role of a metastasis predictor. *Eur. J. Cancer* **2003**, *2*, 184–191. [CrossRef]
26. Johnstone, R.M.; Adam, M.; Hammond, J.R.; Orr, L.; Turbide, C. Vesicle formation during reticulocyte maturation. Association of plasma membrane activities with released vesicles (exosomes). *J. Biol. Chem.* **1987**, *19*, 9412–9420.
27. Edelstein, L.C. The role of platelet microvesicles in intercellular communication. *Platelets* **2017**, *3*, 222–227. [CrossRef]
28. Schoenwaelder, S.M.; Yuan, Y.; Josefsson, E.C.; White, M.J.; Yao, Y.; Mason, K.D.; O'Reilly, L.A.; Henley, K.J.; Ono, A.; Hsiao, S.; et al. Two distinct pathways regulate platelet phosphatidylserine exposure and procoagulant function. *Blood* **2009**, *3*, 663–666. [CrossRef]
29. Dale, G.L.; Remenyi, G.; Friese, P. Tetraspanin CD9 is required for microparticle release from coated-platelets. *Platelets* **2009**, *20*, 361–366. [CrossRef]
30. Brisson, A.R.; Tan, S.; Linares, R.; Gounou, C.; Arraud, N. Extracellular vesicles from activated platelets: A semiquantitative cryo-electron microscopy and immuno-gold labeling study. *Platelets* **2017**, *3*, 263–271. [CrossRef]
31. Mashouri, L.; Yousefi, H.; Aref, A.R.; Ahadi, A.M.; Molaei, F.; Alahari, S.K. Exosomes: Composition, biogenesis, and mechanisms in cancer metastasis and drug resistance. *Mol. Cancer* **2019**, *75*. [CrossRef] [PubMed]
32. Meldolesi, J. Exosomes and ectosomes in intercellular communication. *Curr. Biol.* **2018**, *8*, R435–R444. [CrossRef] [PubMed]
33. Hessvik, N.P.; Llorente, A. Current knowledge on exosome biogenesis and release. *Cell Mol. Life Sci.* **2018**, *2*, 193–208. [CrossRef] [PubMed]
34. Catalano, M.; O'Driscoll, L. Inhibiting extracellular vesicles formation and release: A review of EV inhibitors. *J. Extracell. Vesicles* **2019**, *9*, 1703244. [CrossRef]
35. Wang, Y.; Zhang, S.; Luo, L.; Norstrom, E.; Braun, O.O.; Morgelin, M.; Thorlacius, H. Platelet-derived microparticles regulates thrombin generation via phophatidylserine in abdominal sepsis. *J. Cell Physiol.* **2018**, *2*, 1051–1060. [CrossRef]
36. Aatonen, M.T.; Ohman, T.; Nyman, T.A.; Laitinen, S.; Grönholm, M.; Siljander, P.R. Isolation and characterization of platelet-derived extracellular vesicles. *J. Extracell. Vesicles* **2014**. [CrossRef]
37. Lea, J.; Sharma, R.; Yang, F.; Zhu, H.; Ward, E.S.; Schroit, A.J. Detection of phosphatidylserine-positive exosomes as a diagnostic marker for ovarian malignancies: A proof of concept study. *Oncotarget* **2017**, *9*, 14395–14407. [CrossRef]
38. Rand, M.L.; Wang, H.; Bang, K.W.; Packham, M.A.; Freedman, J. Rapid clearance of procoagulant platelet-derived microparticles from the circulation of rabbits. *J. Thromb. Haemost.* **2006**, *7*, 1621–1623. [CrossRef]
39. Rank, A.; Nieuwland, R.; Crispin, A.; Grützner, S.; Iberer, M.; Toth, B.; Pihusch, R. Clearance of platelet microparticles in vivo. *Platelets* **2011**, *2*, 111–116. [CrossRef]
40. Flaumenhaft, R. Formation and fate of platelet microparticles. *Blood Cells Mol. Dis.* **2006**, 2, 182–187. [CrossRef]
41. Abdel-Monem, H.; Dasgupta, S.K.; Le, A.; Prakasam, A.; Thiagarajan, P. Phagocytosis of platelet microvesicles and beta2- glycoprotein I. *Thromb. Haemost.* **2010**, *2*, 335–341.
42. Dasgupta, S.K.; Le, A.; Chavakis, T.; Rumbaut, R.E.; Thiagarajan, P. Developmental endothelial locus-1 (Del-1) mediates clearance of platelet microparticles by the endothelium. *Circulation* **2012**, *13*, 1664–1672. [CrossRef] [PubMed]
43. Record, M.; Silvente-Poirot, S.; Poirot, M.; Wakelam, M.J.O. Extracellular vesicles: Lipids as key components of their biogenesis and functions. *J. Lipid. Res.* **2018**, *8*, 1316–1324. [CrossRef] [PubMed]
44. Charoenviriyakul, C.; Takahashi, Y.; Morishita, M.; Matsumoto, A.; Nishikawa, M.; Takakura, Y. Cell type-specific and common characteristics of exosomes derived from mouse cell lines: Yield, physicochemical properties, and pharmacokinetics. *Eur. J. Pharm. Sci.* **2017**, *1*, 316–322. [CrossRef]
45. Wisgrill, L.; Lamm, C.; Hartmann, J.; Preiβing, F.; Dragostis, K.; Bee, A.; Hell, L.; Thaler, J.; Ay, C.; Pabinger, I.; et al. Peripheral blood microvesicles secretion is influenced by storage time, temperature, and anticoagulants. *Cytometry A* **2016**, *7*, 663–672. [CrossRef]

46. Flaumenhaft, R.; Dilks, J.R.; Richardson, J.; Alden, E.; Patel-Hett, S.R.; Battinelli, E.; Klement, G.L.; Sola-Visner, M.; Italiano, J.E., Jr. Megakaryocyte-derived microparticles: Direct visualization and distinction from platelet-derived microparticles. *Blood* **2009**, *5*, 1112–1121. [CrossRef]
47. Vajen, T.; Mause, S.F.; Koenen, R.R. Microvesicles from platelets: Novel drivers of vascular inflammation. *Thromb. Haemost.* **2015**, *2*, 228–236. [CrossRef]
48. Rank, A.; Nieuwland, R.; Delker, R.; Kohler, A.; Toth, B.; Pihusch, V. Cellular origin of platelet-derived microparticles In Vivo. *Thromb. Res.* **2010**, *126*, e255–e259. [CrossRef]
49. Heijnen, H.F.; Schiel, A.E.; Fijnheer, R.; Geuze, H.J.; Sixma, J.J. Activated platelets release two types of membrane vesicles: Microvesicles by surface shedding and exosomes derived from exocytosis of multivesicular bodies and alpha-granules. *Blood* **1999**, *11*, 3791–3799. [CrossRef]
50. Dean, W.L.; Lee, M.J.; Cummins, T.D.; Schultz, D.J.; Powell, D.W. Proteomic and functional characterisation of platelet microparticle size classes. *Thromb. Haemost.* **2009**, *102*, 711–718. [CrossRef]
51. Menter, D.G.; Kanikarla-Marie, P.; Lam, M.; Davis, J.S.; Kopetz, S. Platelet microparticles: Small payloads with profound effects on tumor growth. *Noncoding RNA Investig.* **2017**, *15*. [CrossRef] [PubMed]
52. Sadallah, S.; Eken, C.; Martin, P.J.; Schifferli, J.A. Microparticles (ectosomes) shed by stored human platelets downregulate macrophages and modify the development of dendritic cells. *J. Immunol.* **2011**, *11*, 6543–6552. [CrossRef] [PubMed]
53. Falati, S.; Liu, Q.; Gross, P.; Merrill-Skoloff, G.; Chou, J.; Vandendries, E.; Celi, A.; Croce, K.; Furie, B.C.; Furie, B. Accumulation of tissue factor into developing thrombi in vivo is dependent upon microparticle Pselectin glycoprotein ligand 1 and platelet P-selectin. *J. Exp. Med.* **2003**, *11*, 1585–1598. [CrossRef] [PubMed]
54. Diamant, M.; Nieuwland, R.; Pablo, R.F.; Sturk, A.; Smit, J.W.; Radder, J.K. Elevated numbers of tissue-factor exposing microparticles correlate with components of the metabolic syndrome in uncomplicated type 2 diabetes mellitus. *Circulation* **2002**, *19*, 2442–2447. [CrossRef]
55. Tang, K.; Liu, J.; Yang, Z.; Zhang, B.; Zhang, H.; Huang, C.; Ma, J.; Shen, G.X.; Ye, D.; Huang, B. Microparticles mediate enzyme transfer from platelets to mast cells: A new pathway for lipoxin a4 biosynthesis. *Biochem. Biophys. Res. Commun.* **2010**, *3*, 432–436. [CrossRef]
56. Kailashiya, J.; Gupta, V.; Dash, D. Engineered human platelet-derived microparticles as natural vectors for targeted drug delivery. *Oncotarget* **2019**, *56*, 5835–5846. [CrossRef]
57. Baj-Krzyworzeka, M.; Majka, M.; Pratico, D.; Ratajczak, J.; Vilaire, G.; Kijowski, J.; Reca, R.; Janowska-Wieczorek, A.; Ratajczak, M.Z. Platelet-derived microparticles stimulate proliferation, survival, adhesion, and chemotaxis of hematopoietic cells. *Exp. Hematol.* **2002**, *5*, 450–459. [CrossRef]
58. Gelderman, M.P.; Simak, J. Flow cytometric analysis of cell membrane microparticles. *Methods Mol. Biol.* **2008**, *484*, 79–93.
59. Abid, H.M.N.; Meesters, E.W.; Osmanovic, N.; Romijn, F.P.; Nieuwland, R.; Sturk, A. Antigenic characterization of endothelial cellderived microparticles and their detection ex vivo. *J. Thromb. Haemost.* **2003**, *11*, 2434–2443. [CrossRef]
60. Thiagarajan, P.; Tait, J.F. Collagen-induced exposure of anionic phospholipid in platelets and platelet-derived microparticles. *J. Biol. Chem.* **1991**, *36*, 24302–24307.
61. Rozmyslowicz, T.; Majka, M.; Kijowski, J.; Murphy, S.L.; Conover, D.O.; Poncz, M.; Ratajczak, J.; Gaulton, G.N.; Ratajczak, M.Z. Platelet- and megakaryocyte-derived microparticles transfer CXCR4 receptor to CXCR4-null cells and make them susceptible to infection by X4-HIV. *AIDS* **2003**, *1*, 33–42. [CrossRef] [PubMed]
62. Gasperi, W.; Vangapandu, C.; Savini, I.; Ventimiglia, G.; Adoro, G.; Catani, M.V. Polyunsaturated fatty acids modulate the delivery of platelet microvesicle-derived microRNAs into human breast cancer cell lines. *J. Nutr. Biochem.* **2019**, *74*, 108242. [CrossRef] [PubMed]
63. Gilbert, G.E.; Sims, P.J.; Wiedmer, T.; Furie, B.; Furie, B.C.; Shattil, S.J. Platelet-derivedmicroparticles express high affinity receptors for factor VIII. *J. Biol. Chem.* **1991**, *26*, 17261–17268.
64. Brill, A.; Dashevsky, O.; Rivo, J.; Gozal, Y.; Varon, D. Platelet-derived microparticles induce angiogenesis and stimulate post-ischemic revascularization. *Cardiovasc. Res.* **2005**, *1*, 30–38. [CrossRef]
65. Fox, J.E.; Austin, C.D.; Boyles, J.K.; Steffen, P.K. Role of the membrane skeleton in preventing the shedding of procoagulant-rich microvesicles from the platelet plasma membrane. *J. Cell Biol.* **1990**, *2*, 483–493. [CrossRef] [PubMed]

66. Podor, T.J.; Singh, D.; Chindemi, P.; Foulon, D.M.; McKelvie, R.; Weitz, J.I.; Austin, R.; Boudreau, G.; Davies, R. Vimentin exposed on activated platelets and platelet microparticles localizes vitronectin and plasminogen activator inhibitor complexes on their surface. *J. Biol. Chem.* **2002**, *9*, 7529–7539. [CrossRef]
67. Barry, O.P.; Praticò, D.; Savani, R.C.; FitzGerald, G.A. Modulation of monocyte-endothelial cell interactions by platelet microparticles. *J. Clin. Investig.* **1998**, *1*, 136–144. [CrossRef]
68. Barry, O.P.; Kazanietz, M.G.; Praticò, D.; FitzGerald, G.A. Arachidonic acid in platelet microparticles up-regulates cyclooxygenase-2-dependent prostaglandin formation via a protein kinase C/mitogen-activated protein kinase-dependent pathway. *J. Biol. Chem.* **1999**, *11*, 7545–7556. [CrossRef]
69. Boilard, E.; Nigrovic, P.A.; Larabee, K.; Watts, G.F.M.; Coblyn, J.S.; Weinblatt, M.E.; Massarotti, E.M.; Remold-O'Donnell, E.; Farndale, R.W.; Ware, J.; et al. Platelets amplify inflammation in arthritis via collagen-dependent microparticle production. *Science* **2010**, *5965*, 580–583. [CrossRef]
70. Sims, P.J.; Faioni, E.M.; Wiedmer, T.; Shattil, S.J. Complement proteins C5b-9 cause release of membrane vesicles from the platelet surface that are enriched in the membrane receptor for coagulation factor Va and express prothrombinase activity. *J. Biol. Chem.* **1988**, *34*, 18205–18212.
71. Raturi, A.; Miersch, S.; Hudson, J.W.; Mutus, B. Platelet microparticleassociated protein disulfide isomerase promotes platelet aggregation and inactivates insulin. *Biochim. Biophys. Acta* **2008**, *12*, 2790–2796. [CrossRef] [PubMed]
72. Böing, A.N.; Hau, C.M.; Sturk, A.; Nieuwland, R. Platelet microparticles contain active caspase 3. *Platelets* **2008**, *19*, 96–103. [CrossRef] [PubMed]
73. Mause, S.F.; von Hundelshausen, P.; Zernecke, A.; Koenen, R.R.; Weber, C. Platelet microparticles: A transcellular delivery system for RANTES promoting monocyte recruitment on endothelium. *Arter. Thromb. Vasc. Biol.* **2005**, *7*, 1512–1518. [CrossRef] [PubMed]
74. Maroney, S.A.; Haberichter, S.L.; Friese, P.; Collins, M.L.; Ferrel, J.P.; Dale, G.L.; Mast, A.E. Active tissue factor pathway inhibitor is expressed on the surface of coated platelets. *Blood* **2007**, *5*, 1931–1937. [CrossRef] [PubMed]
75. Gambim, M.H.; do Carmo, A.; Marti, L.; Veríssimo-Filho, S.; Lopes, L.R.; Janiszewski, M. Platelet-derived exosomes induce endothelial cell apoptosis through peroxynitrite generation: Experimental evidence for a novel mechanism of septic vascular dysfunction. *Crit. Care* **2007**, *5*, R107. [CrossRef] [PubMed]
76. Pfister, S.L. Role of platelet microparticles in the production of thromboxane by rabbit pulmonary artery. *Hypertension* **2004**, *2*, 428–433. [CrossRef]
77. Salanova, B.; Choi, M.; Rolle, S.; Wellner, M.; Luft, F.C.; Kettritz, R. Beta2-integrins and acquired glycoprotein IIb/IIIa (GPIIb/IIIa) receptors cooperate in NF-kappaB activation of human neutrophils. *J. Biol. Chem.* **2007**, *38*, 27960–27969. [CrossRef]
78. Kim, H.K.; Song, K.S.; Chung, J.H.; Lee, K.R.; Lee, S.N. Platelet microparticles induce angiogenesis in vitro. *Br. J. Haematol.* **2004**, *3*, 376–384. [CrossRef]
79. Saber, S.H.; Ali, H.E.A.; Gaballa, R.; Gaballah, M.; Ali, H.I.; Zerfaoui, M.; Abd Elmageed, Z.Y. Exosomes are the driving force in preparing the soil for the metastatic seeds: Lessons from the prostate cancer. *Cells* **2020**, *3*, 564. [CrossRef]
80. Doyle, L.M.; Wang, M.Z. Overview of extracellular vesicles, their origin, composition, purpose, and methods for exosome isolation and analysis. *Cells* **2019**, *7*, 727. [CrossRef]
81. Nishida, N.; Yano, H.; Nishida, T.; Kamura, T.; Kojiro, M. Angiogenesis in cancer. *Vasc. Health Risk Manag.* **2006**, *3*, 213–219. [CrossRef] [PubMed]
82. Happonen, K.E.; Tran, S.; Mörgelin, M.; Prince, R.; Calzavarini, S.; Angelillo-Scherrer, A.; Dählback, B. The Gas6-Axl protein interaction mediates endothelial uptake of platelet microparticles. *J. Biol. Chem.* **2016**, *20*, 10586–10601. [CrossRef] [PubMed]
83. Janowska-Wieczorek, A.; Wysoczynski, M.; Kijowski, J.; Marquez-Curtis, L.; Machalinski, B.; Ratajczak, J.; Ratajczak, M.Z. Microvesicles derived from activated platelets induce metastasis and angiogenesis in lung cancer. *Int. J. Cancer* **2005**, *5*, 752–760. [CrossRef] [PubMed]
84. Michael, J.V.; Wurtzel, J.G.T.; Mao, G.F.; Rao, A.K.; Kolpakov, M.A.; Sabri, A.; Hoffman, N.E.; Rajan, S.; Tomar, D.; Madesh, M.; et al. Platelet microparticles infiltrating solid tumors transfer miRNAs that suppress tumor growth. *Blood* **2017**, *5*, 567–580. [CrossRef]
85. Anene, C.; Graham, A.M.; Boyne, J.; Roberts, W. Platelet microparticle delivered microRNA-Let-7a promotes the angiogenic switch. *Biochim. Biophys. Acta Mol. Basis Dis.* **2018**, *8*, 2633–2643. [CrossRef]

86. Lazar, S.; Goldfinger, L.E. Platelet microparticles and miRNA transfer in cancer progression: Many targets, modes of action, and effects across cancer stages. *Front. Cardiovasc. Med.* **2018**, *5*, 13. [CrossRef]
87. Pan, Y.; Liang, H.; Liu, H.; Li, D.; Chen, X.; Li, L.; Zhang, C.Y.; Zen, K. Platelet-secreted microRNA-223 promotes endothelial cell apoptosis induced by advanced glycation end products via targeting the insulin-like growth factor 1 receptor. *J. Immunol.* **2014**, *1*, 437–446. [CrossRef]
88. Laffont, B.; Corduan, A.; Rousseau, M.; Duchez, A.C.; Lee, C.H.C.; Boilard, E.; Provost, P. Platelet microparticles reprogram macrophage gene expression and function. *Thromb. Haemost.* **2016**, *2*, 311–323.
89. Bakewell, S.J.; Nestor, P.; Prasad, S.; Tomasson, M.H.; Dowland, N.; Mehrorta, M.; Scarborough, R.; Kanter, J.; Abe, K.; Phillips, D.; et al. Platelet and osteoclast beta3 integrins are critical for bone metastasis. *Proc. Natl. Acad. Sci. USA* **2003**, *24*, 14205–14210. [CrossRef]
90. Liu, S.C.; Bassi, D.E.; Zhang, S.Y.; Holoran, D.; Conti, C.J.; Klein-Szanto, A.J. Overexpression of cyclin D2 is associated with increased in vivo invasiveness of human squamous carcinoma cells. *Mol. Carcinog.* **2002**, *3*, 131–139. [CrossRef]
91. Shahidi, F.; Ambigaipalan, P. Omega-3 polyunsaturated fatty acids and their health benefits. *Ann. Rev. Food Sci. Technol.* **2018**, *9*, 345–381. [CrossRef] [PubMed]
92. Shi, L.; Fisslthaler, B.; Zippel, N.; Frömel, T.; Hu, J.; Elgheznawy, A.; Heide, H.; Popp, R.; Fleming, I. MicroRNA-223 antagonizes angiogenesis by targeting β1 integrin and preventing growth factor signaling in endothelial cells. *Circ. Res.* **2013**, *113*, 1320–1330. [CrossRef] [PubMed]
93. Tang, M.L.; Jiang, L.; Lin, Y.Y.; Wu, X.L.; Wang, K.; He, Q.Z.; Wang, X.P.; Li, W.P. Platelet microparticle mediated transfer of miR-939 to epithelial ovarian cancer cells promotes epithelial to mesenchymal transition. *Oncotarget* **2017**, *8*, 97464–97475. [CrossRef] [PubMed]
94. Yao, B.; Qu, S.; Hu, R.; Gao, W.; Jin, S.; Ju, J.; Zhao, Q. Delivery of platelet TPM3 mRNA into breast cancer cells via microvesicles enhances metastasis. *FEBS Open Biol.* **2019**, *12*, 2159–2169. [CrossRef] [PubMed]
95. Yu, S.B.; Gao, Q.; Lin, W.W.; Kang, M.Q. Proteomic analysis indicates the importance of TPM3 in esophageal squamous cell carcinoma invasion and metastasis. *Mol. Med. Rep.* **2017**, *15*, 1236–1242. [CrossRef] [PubMed]
96. Zarà, M.; Guidetti, G.F.; Boselli, D.; Villa, C.; Canobbio, I.; Seppi, C.; Visconte, C.; Canio, J.; Torti, M. Release of prometastatic platelet-derived microparticles induced by breast cancer cells: A novel positive feedback mechanism for metastasis. *TH Open* **2017**, *2*, e155–e163. [CrossRef]
97. Burnouf, T.; Goubran, H.A.; Chou, M.L.; Devos, D.; Radosevic, M. Platelet microparticles: Detection and assessment of their paradoxical functional roles in disease and regenerative medicine. *Blood Rev.* **2014**, *4*, 155–166. [CrossRef]
98. Vasina, E.M.; Cauwenberghs, S.; Feijge, M.A.; Heemskerk, J.W.; Weber, C.; Koenen, R.R. Microparticles from apoptotic platelets promote resident macrophage differentiation. *Cell Death Dis.* **2011**, *9*, e211. [CrossRef]
99. Li, J.; Guo, Y.; Liang, X.; Sun, M.; Wang, G.; De, W.; Wu, W. MicroRNA-223 functions as an oncogene in human gastric cancer by targeting FBXW7/hCdc4. *J. Cancer Res. Clin. Oncol.* **2012**, *5*, 763–774. [CrossRef]
100. Yaftian, M.; Yari, F.; Ghasemzadeh, M.; Fallah, A.V.; Haghighi, M. Induction of apoptosis in cancer cells of pre-B ALL patients after exposure to platelets, platelet-derived microparticles and soluble CD40 ligand. *Cell J.* **2018**, *1*, 120–126.
101. Dashevsky, O.; Varon, D.; Brill, A. Platelet-derived microparticles promote invasiveness of prostate cancer cells via upregulation of MMP-2 production. *Int. J. Cancer* **2009**, *8*, 1773–1777. [CrossRef] [PubMed]
102. Sadallah, S.; Schmied, L.; Eken, C.; Charoudeh, H.N.; Amicarella, F.; Schifferli, J.A. Platelet-derived ectosomes reduce NK cell function. *J. Immunol.* **2016**, *5*, 1663–1671. [CrossRef] [PubMed]
103. Laffont, B.; Corduan, A.; Plé, H.; Duchez, A.C.; Cloutier, N.; Boilard, E.; Provost, P. Activated platelets can deliver mRNA regulatory Ago2 microRNA complexes to endothelial cells via microparticles. *Blood* **2013**, *2*, 253–261. [CrossRef] [PubMed]
104. Park, M.; Kang, K.W. Phosphatidylserine receptor-targeting therapies for the treatment of cancer. *Arch. Pharm. Res.* **2019**, *7*, 617–628. [CrossRef] [PubMed]
105. Li, B.; Antonyak, M.A.; Zhang, J.; Cerione, R.A. RhoA triggers a specific signaling pathway that generates transforming microvesicles in cancer cells. *Oncogene* **2012**, *45*, 4740–4749. [CrossRef]
106. Zhang, W.; Qi, J.; Zhao, S.; Shen, W.; Dai, L.; Han, W.; Huang, M.; Wang, Z.; Ruan, C.; Wu, D.; et al. Clinical significance of circulating microparticles in Ph- myeloproliferative neoplasms. *Oncol. Lett.* **2017**, *2*, 2531–2536. [CrossRef]

107. Ren, J.G.; Man, Q.W.; Zhang, W.; Li, C.; Xiong, X.P.; Zhu, J.Y.; Wang, W.M.; Sun, Z.J.; Jia, J.; Zhang, W.F.; et al. Elevated level of circulating platelet-derived microparticles in oral cancer. *J. Dent. Res.* **2016**, *1*, 87–93. [CrossRef]
108. Dymicka-Piekarska, V.; Gryko, M.; Lipska, A.; Korniluk, A.; Siergiejko, E.; Kemona, H. Platelet-derived microparticles in patients with colorectal cancer. *J. Cancer Ther.* **2012**, *6*, 898–901. [CrossRef]
109. Wang, C.C.; Tseng, C.C.; Chang, H.C.; Huang, K.T.; Fang, W.F.; Chen, Y.M.; Yang, C.T.; Hsiao, C.C.; Lin, M.C.; Ho, C.K.; et al. Circulating microparticles are prognostic biomarkers in advanced non-small cell lung cancer patients. *Oncotarget* **2017**, *44*, 75952–75967. [CrossRef]
110. Mege, D.; Panicot-Dubois, L.; Ouaissi, M.; Robert, S.; Sielezneff, I.; Sastre, B.; Digant-George, F.; Dubois, C. The origin and concentration of circulating microparticles differ according to cancer type and evolution: A prospective single-center study. *Int. J. Cancer* **2016**, *4*, 939–948. [CrossRef]
111. Yenigürbüz, F.D.; Kızmazoğlu, D.; Ateş, H.; Erdem, M.; Tufekci, O.; Yilmaz, S.; Oren, H. Analysis of apoptotic, platelet-derived, endothelial-derived, and tissue factor-positive microparticles of children with acute lymphoblastic leukemia during induction therapy. *Blood Coagul. Fibrinolysis* **2019**, *4*, 149–155. [CrossRef] [PubMed]
112. Ball, S.; Nugent, K. Microparticles in Hematological Malignancies: Role in Coagulopathy and Tumor Pathogenesis. *Am. J. Med. Sci.* **2018**, *3*, 207–214. [CrossRef] [PubMed]
113. Tjon-Kon-Fat, L.A.; Lundholm, M.; Schroder, M.; Wurdinger, T.; Thellenberg-Karlsson, C.; Widmark, A.; Wikstrom, P.; Nilsson, R.J.A. Platelets harbor prostate cancer biomarkers and the ability to predict therapeutic response to abiraterone in castration resistant patients. *Prostate* **2018**, *1*, 48–53. [CrossRef] [PubMed]
114. Boerrigter, E.; Groen, L.N.; Van Erp, N.P.; Verhaegh, G.W.; Schalken, J.A. Clinical utility of emerging biomarkers in prostate cancer liquid biopsies. *Expert Rev. Mol. Diagn.* **2020**, *2*, 219–230. [CrossRef]
115. Tarasov, V.V.; Svistunov, A.A.; Chubarev, V.N. Extracellular vesicles in cancer nanomedicine. *Semin. Cancer Biol.* **2019**. [CrossRef]
116. Wiklander, O.P. Extracellular vesicle in vivo biodistribution is determined by cell source, route of administration and targeting. *J. Extracell. Vesicles* **2015**, *4*, 26316. [CrossRef]
117. Dilsiz, N. Role of exosomes and exosomal microRNAs in cancer. *Future Sci. OA* **2020**, *4*, FSO465. [CrossRef]
118. Zhao, X.; Wu, D.; Ma, X.; Wang, J.; Hou, W.; Zhang, W. Exosomes as drug carriers for cancer therapy and challenges regarding exosome uptake. *Biomed. Pharmacother.* **2020**, *128*, 110237. [CrossRef]

11

Measuring Extracellular Vesicles by Conventional Flow Cytometry: Dream or Reality?

Donatella Lucchetti [1], Alessandra Battaglia [2], Claudio Ricciardi-Tenore [1], Filomena Colella [1], Luigi Perelli [1], Ruggero De Maria [1], Giovanni Scambia [3], Alessandro Sgambato [4,*] and Andrea Fattorossi [3]

1 Department of Translational Medicine and Surgery, Università Cattolica del Sacro Cuore, 00168 Rome, Italy; dnlucchetti@gmail.com (D.L.); c.ricciarditenore@hotmail.it (C.R.-T.); colella.filomena@gmail.com (F.C.); luigi.perelli19934@libero.it (L.P.); demariaruggero@gmail.com (R.D.M.)
2 Department of Life Science and Public Health, Università Cattolica del Sacro Cuore, 00168 Rome, Italy; alessandra.battaglia@unicatt.it
3 Laboratory of Cytometry and Immunology, Department of Obstetrics and Gynecology, Università Cattolica del Sacro Cuore, Fondazione Policlinico Universitario A. Gemelli IRCCS, 00168 Rome, Italy; giovanni.scambia@policlinicogemelli.it (G.S.); andrea.fattorossi@guest.policlinicogemelli.it (A.F.)
4 Centro di Riferimento Oncologico della Basilicata (IRCCS-CROB), Rionero in Vulture (PZ), 85028 Potenza, Italy
* Correspondence: alessandro.sgambato@crob.it

Abstract: Intense research is being conducted using flow cytometers available in clinically oriented laboratories to assess extracellular vesicles (EVs) surface cargo in a variety of diseases. Using EVs of various sizes purified from the HT29 human colorectal adenocarcinoma cell line, we report on the difficulty to assess small and medium sized EVs by conventional flow cytometer that combines light side scatter off a 405 nm laser with the fluorescent signal from the EVs general labels Calcein-green and Calcein-violet, and surface markers. Small sized EVs (~70 nm) immunophenotyping failed, consistent with the scarcity of monoclonal antibody binding sites, and were therefore excluded from further investigation. Medium sized EVs (~250 nm) immunophenotyping was possible but their detection was plagued by an excess of coincident particles (swarm detection) and by a high abort rate; both factors affected the measured EVs concentration. By running samples containing equal amounts of Calcein-green and Calcein-violet stained medium sized EVs, we found that swarm detection produced false double positive events, a phenomenon that was significantly reduced, but not totally eliminated, by sample dilution. Moreover, running highly diluted samples required long periods of cytometer time. Present findings raise questions about the routine applicability of conventional flow cytometers for EV analysis.

Keywords: extracellular vesicles; exosomes; flow cytometry; immunophenotyping; swarm detection

---

1. Introduction

Extracellular vesicles (EVs) are membrane-surrounded structures released in the intercellular environment and blood stream by a large variety of cells. EVs shuttle lipids, proteins, RNA, DNA, and other metabolites between cells and tissues. They diverge into two main subgroups according to their biogenesis and release mechanism: microvesicles (150–1000 nm in diameter, MVs) shed from the plasma membrane, and exosomes, which are generally smaller in size (30–150 nm in diameter, EXOs) originating from the endosome as intraluminal vesicles enclosed within multivesicular bodies [1]. EVs are central in regulating multiple physiological processes—e.g., tissue repair, stem cell maintenance and coagulation—and pathophysiological processes—e.g., cancer, neurodegenerative diseases and viral infections [2]—because of their ability to transfer biological content. Since EVs are found in accessible body fluids and express a variety of bioactive molecules of the cells of origin, intense research is being conducted to understand EVs' potential as biomarkers for personalized medicine, and to develop relatively simple and fast methods to assess EVs in translational studies using high-throughput technologies.

Indeed, a number of techniques are potentially suited to assess individual EVs, including electron microscopy, resistive pulse sensing, nanoparticle tracking analysis, dynamic light scatter (DLS), and flow cytometry [3]. However, only the latter technique is able to combine high-throughput and adequate speed allowing EVs evaluation in translational studies and in a routine clinical setting. Several custom-constructed flow cytometers or last generation modified cell sorter, with optimized fluidics and flow cell design, have been developed to detect extremely small particles [4,5]. However, these instruments are not optimally suited for other more common applications in clinical settings, mostly cell immunophenotyping. Paradoxically, it is in the clinical setting that EVs are currently most extensively investigated by flow cytometry in a variety of pathological processes.

The current generation of commercial flow cytometers include highly complex and sensitive instruments, which are optimized to assess lymphocytes and other similar sized cells. Commercially available flow cytometers routinely measure light scatter in the forward scatter (FSC) and right angle, or side scatter (SSC), directions, and the two parameters combined provide a good foundation to begin cell population analysis. To identify a particle, the scattered light must exceed the triggering threshold, which must be set to exclude the optical and electronic noise. This is easily accomplished when analyzing micrometer-sized particles, such as cells and the largest EVs, e.g., apoptotic bodies. Conversely, smaller EVs generate scatter signals that may be extremely low and fall within the range of the optical and electronic noise; because the intensity of the scattered light attenuates exponentially with size (the sixth power of particle size) [6], these EVs remain hidden in the background.

In the cytometry of EVs with conventional flow cytometers, FSC is generally less used than SSC, as only particles with a diameter larger than the typical 488 nm wavelength excitation provided by the standard blue laser preferentially scatter (in fact diffract) light in the "forward" direction. SSC is better suited to identify particles with diameters smaller than the wavelength of the incident laser light, because SSC is a measure of mostly refracted and reflected light. However, SSC signal intensity also depends on the ratio between the particle size and the wavelength of the incident laser light. To improve resolution, conventional flow cytometers have been developed to measure the SSC off the violet laser (hereafter referred to as VSSC) instead of blue laser SSC (hereafter referred to as BSSC), as the 405 nm wavelength of the violet laser compared with the 488 nm wavelength of the blue laser is closer to EVs size [7].

In the present study, we explored the feasibility to identify small/medium sized EVs (size range 70 to 300 nm) by a conventional flow cytometer, equipped with blue and violet laser excitation sources, designed to be used in clinical setting for a wide range of applications (CytoFLEX S, Beckman Coulter, Milano, Italy) [8].

We demonstrate that immunophenotyping of small sized EVs (in the 100 nm size range) is in fact not feasible, not even when the staining involves very abundant surface molecules, showing that conventional flow cytometry is inadequate for these EVs assessment. Additionally, we show that the flow cytometry analysis of medium sized EVs (around 250 nm) is plagued by an excess of coincident particles (swarm detection)—a limitation that must be considered when formulating schemes for EVs studies using conventional flow cytometers.

2. Results

2.1. EVs Generation and Flow Cytometer Set Up

EVs were obtained from the HT29 (human colorectal adenocarcinoma) cell line. EVs isolation and characterization are detailed in [9] and Supplementary Information) and shown in Figure S1.

The window of analysis was determined by VSSC and fluorescence parameters using fluorescent Megamix-Plus FSC polystyrene microbeads (Figure S2). The superiority of VSSC over BSSC in terms of resolution is shown in Figure S2. During apoptotic cell death, apoptotic exosome-like vesicles similar in size and in certain surface marker expression—e.g., CD63—are released in culture supernatant [10], so we collected supernatant for EVs purification only when cell culture viability exceeded 90% by trypan blue exclusion.

2.2. *Identification of Small and Medium Sized EVs by VSSC and Calceins*

In the cytometry of EVs, fluorescence is less affected by background noise than light scatter. Thus, the poor resolution of dim light scatter signals generated by EVs can be improved by adding a fluorescent label as parameter. Among several available fluorescent labels potentially useful for EVs staining [11], we chose Calceins (Calcein-green and Calcein-violet) [12]. Calcein probes are best suited to avoid possible interferences related to staining of cell membrane fragments and non-intact EVs following ultracentrifugation procedures [13], because they require hydrolysis by intracellular esterases to become fluorescent, and, therefore, identify only metabolically active, intact vesicles, which can transform the non-fluorescent dye into the fluorescent form [12]. Moreover, these probes have little to no spectral overlap with each other and with the other fluorochromes we used, minimizing compensation and spreading error [14].

Unless otherwise stated, we adopted 1.5 μg/mL of EVs (by Bradford assay) in a solution with either Calcein-green or Calcein-violet at final concentration of 1 μM in filtered PBS. Most experiments were performed using Calcein-green; thus, Calcein will refer to Calcein-green, unless otherwise stated (Figure 1).

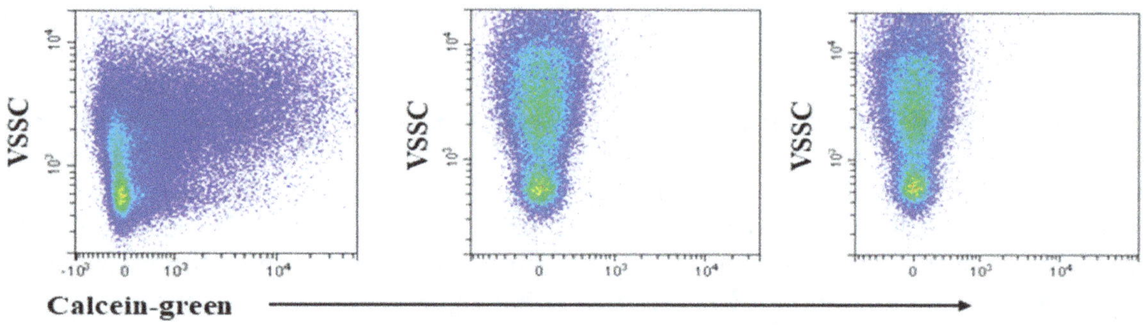

Figure 1. Calcein stains intact extracellular vesicles (EVs). Violet side scatter (VSSC)/Calcein-green fluorescence profile of EVs incubated with Calcein-green at 37 °C (left and right panels) and 4 °C (middle panel). The low temperature prevented the non-fluorescent Calcein from being converted into the green fluorescent form inside EVs, indicating that free dye does not contribute to the observed fluorescence pattern. Triton-X-100 treatment abrogates Calcein fluorescence at 37 °C (right panel). Using Calcein-violet produced identical results.

To confirm previous observation [12] that Calceins only stain intact EVs, an EVs sample was incubated with the dyes at 4 °C, as this temperature inhibits internal esterase activity and, therefore, prevents the non-fluorescent Calcein form conversion to the fluorescent form. Data in the left and middle panel of Figure 1 were generated by staining samples with Calcein at 37 °C and 4 °C, respectively. EVs remained non-fluorescent when incubated at low temperature. Subsequently, as any disruption of plasma membrane, obtained with detergents, leads to leakage of the dye from particle, we treated Calcein-stained EVs with Triton-X-100. In accordance with a previous report [12], positive events disappeared following Triton-X-100 treatment, thus excluding the contribution of free esterases to Calcein signal (Figure 1, right panel).

Calcein staining was dim to moderate, yet the distinction between positive and negative events, remained discernible by visual inspection (Figure 1, left panel). These findings seem at odds with those of De Rond et al. [15] that deemed Calcein as scarcely sensitive. We can only conjecture as to why our findings differ from previous results because the lack of a detailed description of the gating strategy and particle size in the De Rond study [15] make a comparison difficult.

After demonstrating that the inclusion of Calcein as a second parameter was viable to identify EVs, we applied this procedure to purified medium and small sized EVs

The left panel of Figure 2a shows that, in purified medium sized EVs preparations, most Calcein positive events generated a VSSC signal that roughly corresponded to that of the 100 nm microbeads (Figure S2a, left panel). Thus, owing to the medium sized EVs distribution of 210 ± 49 nm (Figure S1), it can be inferred that these EVs generated a scatter signal with intensity comparable to that of the 100 nm polystyrene beads. This is well in line with the notion that the inner refractive index of EVs is less than 1.4, which is considerably lower than that of polymer beads (generally ~1.6) [16].

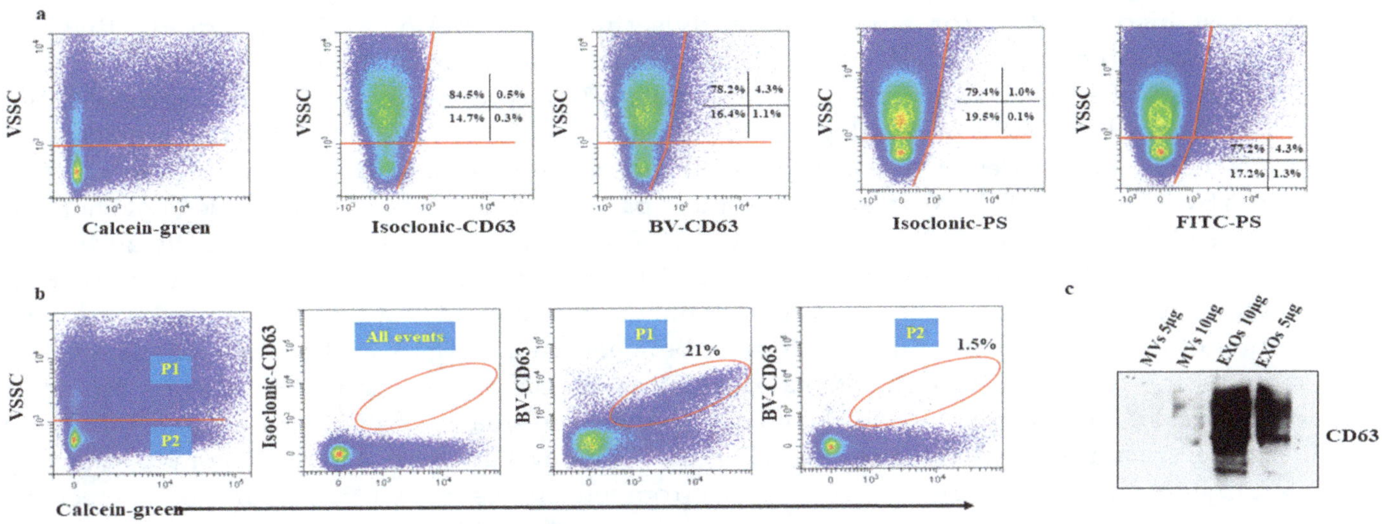

Figure 2. Immunostaining of purified medium and small sized EVs, (**a**) and (**b**), respectively. Both in (**a**) and (**b**), two populations can be visually discerned based on the intensity of the VSSC signal (horizontal red line). (**a**) Left panel, in this representative experiment the percentage of events with high VSSC signal intensity was 84%. Second panel from left, to demonstrate specificity of staining and set the quad markers, the binding of brilliant violet-conjugated CD63 (BV-CD63) moAb was blocked by pre-incubation with unlabelled CD63 moAb prior to staining with the BV-CD63 moAb (isoclonic control). Third panel from left, only a marginal amount of events reacted with the BV-CD63 moAb, consistent with the non-exosomes (EXOs) nature of these particles. Fourth panel from left, to demonstrate specificity of staining and set the quad markers, the binding of FITC (Fluorescein isothiocyanate)-conjugated anti- Phosphatidylserine (PS) moAb was blocked by pre-incubation with unlabelled anti-PS moAb prior to staining with the anti-PS moAb (isoclonic control). Right panel, a detectable proportion of medium sized EVs expressed PS demonstrating that immunophenotyping EVs of that size is feasible. (**b**) Left panel, in this representative experiment, the percentage of events with high VSSC (P1) and low (P2) signal intensity was 43% and 57%, respectively. Middle panel, isoclonic control. Right and far right panels, BV-CD63 staining in P1 and P2 populations, respectively. BV-CD63 moAb reacted only with events in the P1 region, consistent with the EXOs nature of these particles. Values in regions show percentages of positive events. We addressed the issue of possible BV-CD63 fluorescence quenching by Calcein because of the spectral characteristics of BV fluorochrome and run side by side small sized EVs stained with only BV-CD63 or double stained with BV-CD63 and Calcein. (**c**) Representative western bot assay showing the preferential expression of CD63 by small sized EVs. Red lines = analysis gates; red circles = EVs population analyzed.

A Calcein positive particle population (~15% in six independent experiments) with a clearly lower VSSC signal intensity was also visible (Figure 2a left panel). We hypothesized that these events reflected contaminating small sized EVs deriving from imperfect ultracentrifugation procedures, which remained undetected by the DLS analysis because of their low frequency. EVs are defined according to their modality of generation, rather than size. However, microvesicles (MVs) are generally larger than exosomes (EXOs), thus, we inferred that small sized EVs samples were mostly EXOs and medium sized EVs samples were mostly MVs. It is known that CD63 and CD9, two tetraspanins, are specially enriched in EXOs membranes and scarcely expressed by MVs [17,18], and Western blot analysis of small and medium sized EVs samples showed that only the former exhibited a strong reactivity for CD63 (Figure 2c). Thus, Calcein-stained medium sized EVs samples were incubated with monoclonal antibodies (moAbs) to CD63 and CD9, to determine the origin of particles with lower VSSC signal intensity in medium sized EVs samples. Virtually no particle stained positive for CD63 (Figure 2a, second panel from left). Although these findings suggests a non-EXOs nature of these particles, this issue will be discussed again below. The moAb to CD63 marginally stained particles in the high VSSC region—a finding consistent with the low CD63 expression by MVs [17,18].

MVs stained positive for anti-Phosphatidylserine (PS) antibody, demonstrating that the immunophenotyping of medium sized EVs is possible [19] (right panel, Figure 2).

Running purified small sized EVs, we observed that a large proportion of Calcein positive particles was located in the high VSSC region (Figure 2b, left panel), which is a region that corresponds to that of the ~250 nm sized particles (Figure 2a, left panel) and therefore, visibly exceeds the size of these EVs (68 ± 7 nm by DLS, Figure S1). We reasoned that these particles could be aggregates, possibly generated by high-speed centrifugation [20]. In the cytometry of cells, the aggregate issue is usually addressed by checking the height, area, and width of the FSC and/or SSC pulse. In the cytometry of EVs, the exceedingly small pulse makes this conventional doublet discrimination impossible, because the magnitude of difference in any pulse parameter between an aggregate and a single event is small, and requires the resolution of a linear scale, which is unfeasible, as scatter parameters require a logarithmic scale. To determine the origin of larger particles in small sized EVs samples, we incubated small sized EVs samples with the moAb to CD63 and CD9, and analyszd the staining in relationship with VSSC signal intensity. Figure 2b, third panel from left, shows that CD63 staining was detectable only in particles with high VSSC signal intensity (Figure 2b, right panel), particles with low VSSC signal intensity remained non-fluorescent. These data supports our hypothesis that particles with high VSSC signal intensity were aggregates. However, it remains unclear why such a large proportion of aggregates remained undetected by DLS analysis. Although it is held that DLS fails to detect small proportions of aggregates in otherwise homogeneous EVs preparations [21], the proportion of aggregates we observed in our experiments is too large to remain undetected. We hypothesize that the cytometer preferentially detected aggregates due to their larger size, while most non-aggregated particles are below the VSSC threshold, and are simply not "seen" by the instrument. An alternative, not mutually exclusive explanation, is that the larger particles are coincident events (swarm detection), an issue that is addressed below.

2.3. Particles in the Low VSSC Signal Region Do Not Stain for Tetraspanins Because of an Insufficient Number of Binding Sites

As outlined above, findings in Figure 2 indicated that the nature of the events in the low VSSC region in both the small and medium sized EVs samples could not be assessed by staining for the typical small sized EVs (mostly EXOs) markers. The analyses of medium sized EVs indicated that ~250 nm EVs were positioned in the high VSSC region (Figure 2a), and the intensity of the scattered light attenuates exponentially with size [21], so we infer that the EVs size in the low VSSC region should be considerably lower, possibly around 70 nm, in line with DLS data. Studies showed that the number of antigenic sites, which can react with a fluorochrome-labelled moAb on the surface of an EV in this size range, is around 10/15 epitopes, which is too scarce to generate a fluorescence signal detectable by a commercial flow cytometer [22–24]. Thus, we hypothesized that the absence of EVs immunolabeling in the low VSSC region reflected insufficient target epitopes. To verify this hypothesis, we mimicked fluorescence generated by the anti-tetraspanins fluorochrome-conjugated moAbs using FITC-conjugated mouse IgG (Immunoglobulin G) bound to 50 nm–sized anti-mouse IgG microbeads. The experiments were performed at various microbeads/FITC-conjugated mouse IgG ratios. No fluorescence signal was ever detected in the low VSSC region (Figure 3 left and middle panel). To insure that the amount of FITC-conjugated mouse IgG was enough to produce a detectable fluorescence signal in the presence of particles carrying a suitable number of binding sites, the lowest amount of FITC-conjugated mouse IgG, which was incubated with the 50 nm microbeads, was incubated with larger (3μm) microbeads coated with anti-mouse IgG. In this condition, FITC-conjugated mouse IgG generated a brilliant fluorescence signal (Figure 3 right panel). We also ascertained by fluorescence microscopy that FITC-conjugated mouse IgG made the 50 nm–sized anti-mouse IgG microbeads fluorescent (Figure 3, lower panels).

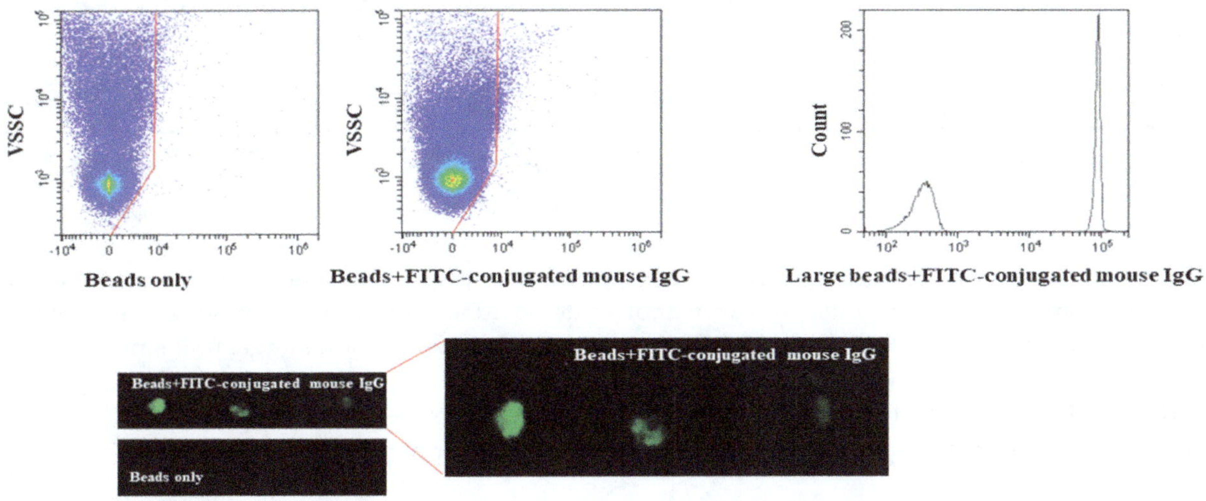

Figure 3. Fifty nm-sized anti-mouse IgG microbeads do not bind enough FITC-conjugated mouse IgG to generate a measurable fluorescence signal. Anti-mouse IgG microbeads background fluorescence (**top left panel**) and anti-mouse IgG microbeads coated with FITC-conjugated mouse IgG fluorescence (**top middle panel**). The same amount of FITC-conjugated mouse IgG generates a brilliant fluorescence signal when incubated with larger-sized (3 μm) microbeads coated with anti-mouse IgG (**top right panel**). Fluorescence microscopy analysis of anti-mouse IgG microbeads carrying FITC-conjugated mouse IgG (**lower panel**). Red lines in top panel = analysis gates.

Thus, assessing the surface cargo of small sized EVs is a difficult task using conventional flow cytometers, because the fluorescence signal may remain hidden in the background.

2.4. Swarming and Abort Rate Affect EVs Detection

Unlike cells, EVs are not prone to align and traverse the laser path in a single row and separated from each other because of their small size. Therefore, swarm detection is the most common cause for spurious or artifactual results in the cytometry of EVs. As outlined above, we found that immunophenotyping of small sized EVs (in the size range of most EXOs) was essentially impossible. In a clinical setting, immunophenotyping is the most common approach to identify the EVs populations of interest in bodily fluids. Therefore, small sized EVs were excluded from further investigation, and we focused on medium sized EVs to investigate how swarm detection affected their analysis.

Studies suggested that the occurrence of swarming detection can be verified by serial sample dilution experiments [24–26]; upon dilution, the number of particles simultaneously traversing the laser beam should progressively reduce to a single particle, and the number of counted events should then decrease linearly in response to further dilution. Concomitantly, the intensity of fluorescence signal generated by the many particles coincidentally traversing the laser beam should progressively decline upon dilution and finally remain stable, which is consistent with analysis of single particles.

Thus, we quantitated changes in measured medium sized EVs concentrations and fluorescence intensity at various sample dilutions. Concentration (events/l) was calculated using the volumetric measurement featured in the CytoFLEX S cytometer (CytoFLEX S, Beckman Coulter, Milano, Italy).

A relation between particle concentrations measured by the cytometer and sample dilutions was noted, which was more evident at higher dilutions, leading to hypothesize the occurrence of detection as single event (Figure 4, left panel). However, the fluorescence signal intensity never levelled completely (Figure 4 center panel). Collectively, these findings suggest that dilution reduces but does not fully prevent swarm detection.

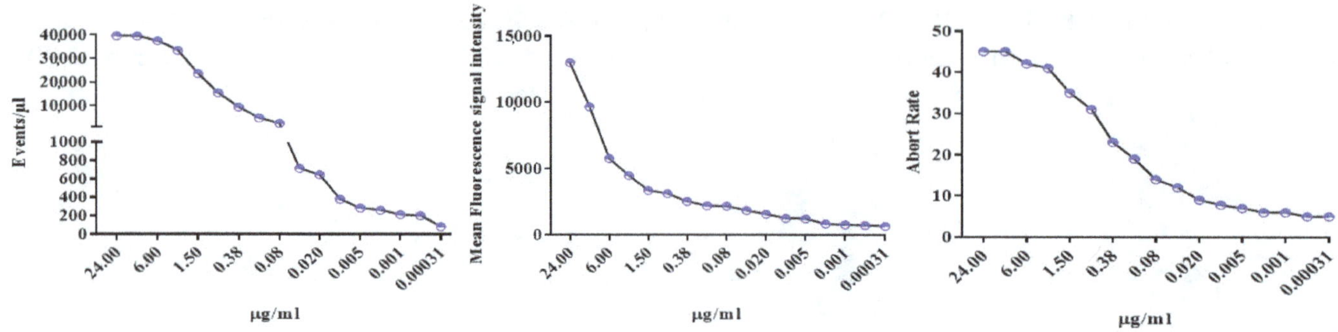

Figure 4. Four-fold dilutions experiments comparing changes in estimated Calcein-stained MVs concentration, fluorescence intensity and abort rate yielded not univocal results. **Left panel**, the measured MVs concentration decreased in proportion to the dilution suggesting single particle detection. **Middle** and **right panel**, fluorescence intensities and the abort rate, respectively, almost levelled only at the last four dilutions as it is the case for single particle detection. Data are from one experiment representative for the other four experiments conducted.

As a result of swarming, the flow cytometry of EVs is plagued by a high abort rate; the electronics is still processing the previous pulse and aborts the new event that, consequently, is lost to analysis. Thus, we hypothesized that if EVs were detected as single events as a consequence of dilution, the abort rate would be concomitantly reduced. Figure 4, right panel, shows that the abort rate declined upon dilution, a finding that concurs with data shown in the left and center panels, supporting the progressive decrease of swarm detection. Of note, minimizing swarm detection required a progressive increase of flow cytometry time; at the highest dilution a single run required ~40 min (including washing to prevent carry-over) to collect at least 5000 Calcein positive events.

Subsequently, we approached the issue of swarm detection using a model that included Calcein-green MVs mixed with Calcein-violet medium sized EVs (final concentration in the tube 0.325 µg/mL each) at a 1:1 ratio just before the flow cytometry run. Under these experimental conditions, any double-stained event undoubtedly denotes more than one EVs coincidentally traversing the laser beam. Data in Figure 5a demonstrate the presence of a sizeable number of double positive events in samples containing Calcein-green and Calcein-violet medium sized EVs, consistent with swarm detection. Consistent with the concomitant presence of several particles in the flow chamber, the VSCC signal of Calcein-green/Calcein-violet double positive EVs was higher than that of either Calcein-green or Calcein-violet single positive EVs.

To provide further experimental evidence of swarm detection and its impact, medium sized EVs samples, containing Calcein-green medium sized EVs mixed with Calcein-violet medium sized EVs, were run at higher flow rates to increase the sample fluid stream size and to boost coincidences. Thus, the samples were run at 10 µL/min—the lowest flow rate allowed by the instrument—30 µL/min and 60 µL/min. Consistent with a progressive increase of particles, simultaneously illuminated by the laser beam, the percentage of the double stained population increased in step with the flow rate (Figure 5b left panel) [23–25]. Additionally, the abort rate was enhanced, which further indicated the concomitant presence of several particles simultaneously traversing the laser beam, which were too close together to be processed as single events (Figure 5b right panel). Interestingly, swarm detection and the concomitant high abort rate reduced the number of particles detected. In the experiment depicted in Figure 5, the number of particles/μL was 25.2×10^3, 17.3×10^3 and 10.6×10^3 at 10 μL/min, 30 μL/min and 60 μL/min, respectively.

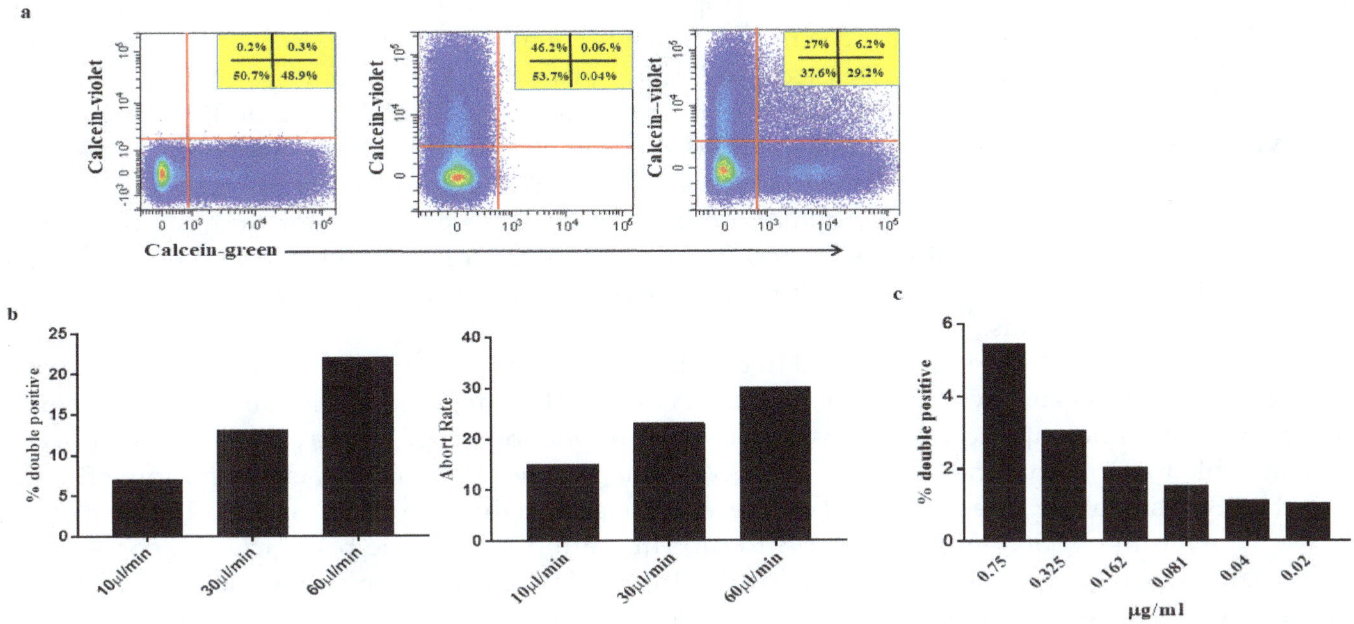

Figure 5. Two aliquots of the same medium sized EVs preparation were separately labelled, one with Calcein-green and the other with Calcein-violet, kept refrigerated and mixed at 1:1 ratio just before analysis. (**a**) Flow cytometry profile (green and violet fluorescence) of medium sized EVs marked individually with Calcein-green (left plot) and Calcein-violet (middle plot). The two Calceins stained medium sized EVs with comparable efficiency and the fluorescence signal of each dye did not interfere with the other. In the mixture (right plot), the two dyes identified single stained particles (either Calcein-green or Calcein-violet) and double-stained particles. Double-stained particles are visible in the upper right quadstat gate. These particles are indicative of coincident events, as there should not be double positive events when the two markers in comparison (Calcein-green and Calcein-violet) are not present on the same event. Samples were run at 10 μL/min. Values in panels show percentages of positive events. (**b**) Same sample as in (**a**), but run at the different flow rates featured in the cytometer. The frequency of double stained events (left panel) increased in step with the flow rate indicating that several particles were simultaneously traversing the laser beam as a consequence of the enlarged size of sample fluid stream. Consistent with the presence of several particles in the laser beam illuminated simultaneously, the abort ratio also increased (right panel) in step with the flow rate. Data are from one experiment representative for the other three experiments conducted. (**c**) The mixture of Calcein-green and Calcein-violet medium sized EVs were diluted to further explore the effect of high dilution on swarm recognition. Samples were run at 10 μL/min. The percentage of double positive events declined with dilution and close to zero at the lowest concentrations. Red lines = analysis gates.

After demonstrating swarm detection impact on EVs analysis, we used the Calcein-green/Calcein-violet model, to test again how dilution may reduce coincidences contributing to overcome the problem. To this end, the Calcein-green/Calcein-violet medium sized EVs mixture was progressively diluted yielding a final concentration of 0.04 μg/tube, and the samples were run at the lowest flow rate allowed by the instrument. The proportion of double positive events declined with dilution and approximated to zero (Figure 5c), comforting data obtained in the dilution experiments carried out using Calcein-green stained medium sized EVs (Figure 4). Confirming the preceding experiments (Figure 4), minimizing double positive events entailed a long flow cytometer time ($\geq$ 40 min).

3. Discussion

In the era of precision medicine, considerable interest using flow cytometry exists to assess both quantitative and phenotypic EVs characteristics in patients suffering from a variety of diseases in the hope to reveal clinically relevant EVs subpopulations. Moreover, the presence of EVs in readily accessible sources e.g., blood, urine, and cerebrospinal fluid, makes EV flow cytometric characterization particularly appealing to monitor diseases in longitudinal studies. Unfortunately, EVs size, which is well below that of whole cells, pushes flow cytometry to the edge of its lower reliability limits.

Within recent years, several dedicated flow cytometric approaches have been developed and refined for single EVs counting and for studying EVs heterogeneity in terms of surface markers [4]. Such methods generally demand extensive operator expertise for sample preparation and, most importantly, require dedicated instrumentation for sample acquisition and data analysis, which is unfeasible for most clinically oriented flow cytometry facilities using standard nonspecialized instruments.

In the present study, we investigated in depth whether small/medium-sized EVs (50 to 250 nm size range) can be detected by a conventional flow cytometer (CytoFLEX S, Beckman Coulter) designed primarily for common applications, mostly cell immunophenotyping. Using calibrated microbeads, we were able to show that the VSSC, i.e., the SSC signal off the 405 nm laser, is much more effective than the BSSC signal, i.e., the SSC signal off the 488 nm laser in detecting EVs, as already reported [7]. However, even the VSSC signal fells short to discriminate EVs over the background generated by particulate, which survived extensive buffer filtering, and by electronic noise of the system. In fact, EVs recognition was only possible by the concomitant assessment of VSSC and the fluorescence signal produced by the general fluorescent Calcein dyes.

Perhaps the most important application of flow cytometry in the EVs study is the purported ability to evaluate their surface cargo by staining with fluorochrome-conjugated monoclonal antibodies. We found that immunophenotyping of small sized EVs, i.e., below the 100 nm range, in our experience, was impossible. Our data show that the immunofluorescent signal generated by EVs of that size becomes measurable only when generated by multiple particles that traverse the laser beam at the same time and emit enough fluorescence to be detected, whereas single particles remain underneath the fluorescence sensitivity. Notably, these conclusions were drawn by staining for markers expressed at the highest possible level on the surface of the particle, emphasizing that they will extend to all the less expressed surface markers. While we cannot exclude that effective staining may depend on type/source of EVs and the markers being evaluated, the present findings cast doubts on the view that conventional flow cytometer is a convenient tool to explore such small sized EVs. This view is in line with the early literature indicating that a single EV of that size carries around 10 epitopes; one would not expect this particle to bind enough fluorescent moAb molecules to be detected by conventional flow cytometers [22,23]. Thus, great caution must be exerted when interpreting immunophenotyping data on small sized EVs. Additional uncertainties exist performing small EVs immunophenotyping; early work showed that the attachment of large labels to antibody molecules resulted in reduced antibody binding even to surface antigens of whole cells [27]. Notably, some of the commonly used fluorophores (PE—240 kDa, PE-Cy7—255 kDa) exceed the antibody size itself and attenuation by bulky fluorophores in multicolor flow cytometry has been reported [28].

The conclusion that assessing surface cargo of EVs in the EXOs size range (50–70 nm) by conventional flow cytometry is questionable seems to be in contrast with some earlier data, in which anti-CD63 moAb reportedly stained EXOs [29]. Comparisons are difficult because of several factors that might have affected the results, such as the modality to set the boundary between negative and positive events, the use of an isotype rather than of a more reliable isoclonic control, and the type of instrument. However, differences in the capabilities of the Apogee A 50 dedicated cytometer used in that study [24], and the CytoFLEX S conventional cytometer in our study, to resolve artificial particles of ~100 nm size, do not seem to be relevant; the Apogee A 50 depicts large angle light scatter (LALS) compared with fluorescence profiles of FITC-labelled latex beads included in our study, which closely resemble plots depicting VSSC/FITC-labelled beads. Moreover, a comparison of our results with those by Van der Pol et al. [24], suggests that the sensitivity of the scatter signals of the Apogee A 50 dedicated cytometer and the CytoFLEX S conventional cytometer is similar, as the Apogee A 50 dedicated cytometer was able to detect single 102 nm polystyrene beads [30], as we did with the CytoFLEX S. However, it should be emphasized that our conclusions, which are based on 50 nm beads as a model for small particles immunophenotyping, cannot be generalized, as different fluorescent dyes and/or different excitation sources and/or different detection devices might provide different results. This issue, as well as other related questions, will be addressed in future studies. In this scenario, in a recent paper, the CytoFLEX was reported to detect EVs as small as 60 nm, triggering with VSSC and immunophenotyping [8]. While it is possible that the surface antigen density on the EVs used in that study was much higher than that of the EVs we tested here, it should be noted that the swarming issue was not addressed, so that it cannot be excluded that multiple small sized EVs coincidentally traversing the laser spot may have contributed to the detected fluorescence signal.

Immunophenotyping was possible in larger EVs, in the 250 nm size range. However, their assessment was plagued by swarming—a finding in line with several earlier publications [24–26]. We also confirmed that swarming could be minimized by extensive sample dilution [24–26]. Unfortunately, in translational studies, particle concentration in a given sample is most often unknown, and quite hard to predict. Thus, one should measure each given sample at various dilutions, which is unfeasible in clinically oriented laboratories. Moreover, minimizing coincidences by running high diluted samples comes at the cost of very long cytometry times (up to more than half an hour in our experience) to collect enough events of interest—a constraint hardly compatible with applications in clinical research. Notably, a (unknown) portion of double stained EVs reported in immunophenotyping studies may in fact represent coincident events, because overlooked swarm detection injects 'polluted' events into the cytometry data [22].

In conclusion, our study emphasizes the inherent difficulties in investigating the surface cargo of small sized EVs (in the EXOs size range ~50–70 nm). Conventional flow cytometry does detect surface cargo of relatively large sized EVs (in the MVs size range ~200–300 nm). However, we must be aware of the intrinsic limitations due to swarm detection.

A limitation of the present study is that data were generated using CytoFLEX S as example of a commercially available last generation flow cytometer designed for a wide spectrum of applications, which does not necessitate labor intensive manual hardware adjustments and calibrations. We do not have direct knowledge of the performance of other commercial instruments offering the same features to assess EVs, and we are not aware of side-by-side comparative tests performed using the same EVs preparation. This latter point is quite an important one, as knowing what size and level of fluorescence a given platform is able to detect will allow for comparisons amongst different instruments, and even of the same producer, which is an important issue for standardization, and, therefore, reproducibility. We are now designing studies to allow the conversion of scattering intensity in arbitrary units to diameter distribution and of fluorescence intensity to MESF (Molecules of Equivalent Soluble Fluorochrome) to better define the detection limits of each instrument [31].

4. Methods

4.1. Reagents

Calcein-AM green (referred to as Calcein-green) and Calcein-AM violet (referred to as Calcein-violet) were purchased from Life Technologies. Both dyes were solubilized in DMSO to produce a 100x stock solution (1 mM). The nonionic detergent Triton X-100 was purchased from Thermofisher. Mouse anti-human brilliant violet-conjugated CD63 (BV-CD63, clone H5C6) and mouse anti-human R-phycoerythrin (PE)-conjugated CD9 (PE-CD9, clone HI9a) monoclonal antibodies (moAbs), as well as unconjugated identical (isoclonic control) CD63 and CD9 moAbs were purchased from BioLegend, Inc. Fluorochrome-conjugated moAbs were centrifuged before using to minimize the antibody aggregates commonly found in commercial preparations. Fifty nm sized microbeads carrying goat anti-mouse IgG (H+L) F (ab')2 fragments were purchased from Miltenyi Biotec (referred to as anti-mouse IgG microbeads). These beads are made of very dense material to be paramagnetic, a feature that implies a refraction index at least comparable to that of polystyrene beads. The VSSC threshold we used in all experiments included events producing a signal one and half decade lower than that generated by the smallest polystyrene beads (100 nm) included in the Megamix-Plus FSC mix (Supplementary Figure S2a and Supplementary Information), so we speculated that the Miltenyi 50 nm beads might become visible if made fluorescent. FITC Mouse IgG1, κ Isotype Control (clone MOPC-21) was purchased from BD Biosciences. Megamix-Plus FSC consisting in 100 nm, 300 nm, 500 nm, and 900 nm microbeads were purchased from BioCytex. Rainbow Calibration Particles consisting in a series of beads with eight different predefined levels of fluorescence intensity, including nonfluorescent beads, were purchased from BD Biosciences. VersaComp 3μm beads were purchased from Beckman Coulter.

4.2. EVs Staining by Calcein-Green and Calcein-Violet, and Anti-Tetraspanins moAbs

Calcein-green and Calcein-violet are essentially colorless and non-fluorescent until hydrolyzed. Once inside the EVs, the lipophilic blocking groups are cleaved by nonspecific esterases. As a result, Calceins are trapped inside the EV and emit a green or a violet fluorescence signal (emission max 516 nm and 450 nm, respectively) following excitation with a blue (488 nm) or violet (405 nm) laser. After some preliminary experiments, the incubation time was set at 20 min at 37 °C, final concentration 1 μM All staining steps were carried out in filtered PBS.

In immunophenotyping experiments, EVs, which had been previously loaded with Calcein, were incubated with moAbs to CD63, CD9, Anti-Phosphatidylserine for 30 min at room temperature (RT). All moAbs were used at concentration of 1.25 $\mu g/mL^{-1}$. All moAbs were used at 1:20 final dilution. To assess moAb non-specific binding, Calcein-loaded MVs and EXOs were incubated with a five-fold excess of identical unlabelled moAb (isoclonic control) before being reacted with the fluorochrome-conjugated moAb. Under these conditions, all specific binding sites of the fluorochrome-conjugated moAb are blocked by the large excess of unconjugated antibody and non-specific staining can be measured. Summary of reagents used for flow cytometry experiment are shown in Table 1.

To avoid false positive events, moAbs were centrifuged using 0.22 μm centrifugal filter tubes and a fixed angle single speed centrifuge (~750× *g*) at RT for 2 min. The flow through was collected and used for immunophenotyping experiments. In each staining session, the moAbs were run in filtered PBS alone to check signal from antibody aggregates that survived centrifugation.

4.3. Microbead Experiments

The 50 nm anti-mouse IgG microbeads served to mimic the low amount of binding sites available on an EXO-sized particle. These microbeads are designed to bind the largest possible number of moAb molecules per surface unit. The microbeads (5 to 10 μL directly from original bottle) were incubated with FITC-conjugated mouse IgG (final dilution 1:20 in filtered PBS) for 30′ at 4 °C in rotation. For comparison, the same amount of FITC-conjugated mouse IgG was incubated with 5 μL of larger sized beads (VersaComp 3 μm beads, Beckman Coulter, CytoFLEX S, Beckman Coulter, Milano, Italy), coated with anti-mouse IgG.

Table 1. Summary of reagents.

Characteristic[s] Measured	Analyte	Analyte Detector	Reporter	Isotype	Clone	Final Concentration	Manufacturer	Catalogue. Number	Lot Number
Intracellular Esterase activity	Vesicles esterases	Calcein-green	Green-fluorescent calcein	NA	NA	1 µM	Life Technologies	C3100MP	1837717
		Calcein-violet	Violet-fluorescent calcein	NA	NA	1 µM	Life Technologies	C34858	2018203
Cell surface protein	Human CD63	Anti-human CD63 antibody	Brilliant Violet 421™	Mouse $IgG1_k$	H5C6	1.25 $\mu g/mL^{-1}$	BioLegend	353030	B275650
	Human CD63	Anti-human CD63 antibody	NA	Mouse $IgG1_k$	H5C6	5 $\mu g/mL^{-1}$	BD Pharmingen	556019	VP036
Membrane Phospholipids	Phosphatidylserine	Anti-Phosphatidylserine Antibody	Alexa Fluor 488	Mouse IgG1	1H6	1.25 $\mu g/mL^{-1}$	Millipore	16-256	2926493
		Anti-Phosphatidylserine Antibody	NA	Mouse IgG1	1H6	5 $\mu g/mL^{-1}$	Millipore	05-719	2867579
Mouse IgG1 Fluorescein-conjug Antibody	Miltenyi Microbeads	50 nm-microbeads	Fluorescein	Mouse $IgG1_k$	N/A	Dilution 1/20	R&D system	IC002F	1171
Anti-mouse IgG MicroBeads, human	NA	NA	NA	Mouse IgG1	NA	5 µl from the bottle	Miltenyi Biotec	130-057-501	5171201227

NA= not applicable.

4.4. Flow Cytometry

EVs were analysed using a CytoFLEX S instrument equipped with blue (488 nm) and violet laser (405 nm) excitation sources. The instrument is able to collect SSC off the blue laser (BSSC) and the violet laser (VSSC). The set-up of the instrument is a critical point for EVs analysis, so the flow cytometer was first calibrated using the 8-Peak Rainbow Beads. The linear regression equation between the predefined fluorescence intensity values of the fluorescent microbeads and the instrument's response in histogram channel values was then computed at several photodiode gains. The best gain was then determined for each fluorescent channel and used throughout all experiments. The Megamix-Plus FSC beads emitting a FITC-like fluorescent light of different sizes (100, 300, 500, and 900 nm), used throughout all experiments, served to establish the best photodiode gains for BSSC and VSSC, which produced the smallest coefficient of variation. Megamix-Plus FSC beads were also used for daily standardization. All signals were collected in log area mode, except the VSSC, which was also collected in log height mode and served as threshold parameter. Time delays between lasers were optimized and controlled by the standard daily QC start-up procedure.

To ensure system and reagent cleanliness, the sample line was washed with sterile distilled water filtered through 100 nm filter before each flow cytometry run, and the same fluid was used as sheath fluid. The sheath fluid tank was thoroughly rinsed with the filtered water. In between samples, the sample line was flushed by boosting filtered PBS and Coulter Clenz® Cleaning Agent (Beckman Coulter, CytoFLEX S, Beckman Coulter, Milano, Italy) and clean distilled water to minimize carry-over.

Unless otherwise stated, EVs samples were run at the lowest rate allowed by the instrument (10 μL/min) to maintain the diameter of the sample stream as small as possible. Data were acquired and analysed by the CytExpert 2.2™ software (version 2.2, CytoFLEX S, Beckman Coulter, Milano, Italy). EVs number was measured using the cell-counting feature of the instrument that relies on a calibrated peristaltic pump for sample delivery.

To tighten the pulse window and thus reducing the background for small-particle analyses, the event rate setting feature implemented in the CytoFLEX S instrument was set to "high", according to manufacturer's instructions.

4.5. Western Blot Analysis

The EVs (microvesicles or exosomes) were lysed using lysis buffer (50 mmol/L Tris-HCl pH 7.2, 5 mmol/L MgCl2, 50 mmol/L NaCl, 0.25%, 0.1% SDS, and 1% Triton X-100) containing protease inhibitors (2 mmol/L phenyl methyl sulfonyl fluoride, 10 mg/mL aprotinin, and 2 mmol/L Na3VO4, 100 mmol/L NaF). Protein concentration was assessed using the Bradford method (Bradford protein assay kit II, Bio-Rad, Hercules, CA, USA), with BSA used as a standard. EVs extracted proteins (5–10 μg) were resolved by SDS PAGE (Sodium Dodecyl Sulfate PolyAcrylamide Gel Electrophoresis) 10% under reducing or non-reducing conditions, and were transferred to PVDF blotting membranes (GE Healthcare, Solingen, Germany) and analyzed using the enhanced chemiluminescence kit for Western blotting detection (Advansta, WesternBright TM ECL, Bering Drive San Jose, CA, USA). CD63 primary monoclonal antibody was used following suppliers' (dilution, 1:500; sc-5275; Santa Cruz Biotechnology, Inc., Dallas, TX, USA). This antibody recognizes the glycosylated forms of CD63 (30–60 kDa).

4.6. Fluorescence Microscopy

The 50 nm anti-mouse IgG microbeads (5 μL directly from original bottle) were incubated with FITC-conjugated mouse IgG (final dilution 1:20 in filtered PBS) for 30′ at 4 °C in rotation. All images were collected with a Nikon ECLIPSE TE2000-S (Chiyoda, Tokyo, Japan) inverted microscope equipped with: 20× objective (Numerical aperture 0.4), filters TRITC, FITC, and UV, and a Nikon Digital Camera DXM1200F (Chiyoda, Tokyo, Japan). The images were analyzed with NIS Elements BR 2.10 software.

Author Contributions: Authors' individual contributions: Conceptualization, D.L. and A.F.; Methodology, C.R.-T. and F.C.; Formal Analysis, L.P.; Resources, R.D.M. and G.S.; Writing–Original Draft Preparation, A.F. and D.L.; Writing–Review A.B. and A.S. All authors have read and agreed to the published version of the manuscript.

References

1. Lane, R.E.; Korbie, D.; Hill, M.M.; Trau, M. Extracellular vesicles as circulating cancer biomarkers: Opportunities and challenges. *Clin. Transl. Med.* **2019**, *7*, 14. [CrossRef] [PubMed]
2. Meldolesi, J. Extracellular vesicles, news about their role in immune cells: Physiology, pathology and diseases. *Clin. Exp. Immunol.* **2019**, *196*, 318–327. [CrossRef]
3. Lucchetti, D.; Fattorossi, A.; Sgambato, A. Extracellular Vesicles in Oncology: Progress and Pitfalls in the Methods of Isolation and Analysis. *Biotechnol. J.* **2019**, *14*, e1700716. [CrossRef] [PubMed]
4. Zhu, S.; Ma, L.; Wang, S.; Chen, C.; Zhang, W.; Yang, L.; Hang, W.; Nolan, J.P.; Wu, L.; Yan, X. Light-scattering detection below the level of single fluorescent molecules for high-resolution characterization of functional nanoparticles. *ACS Nano* **2014**, *8*, 10998–11006. [CrossRef]
5. Morales-Kastresana, A.; Musich, T.A.; Welsh, J.A.; Telford, W.; Demberg, T.; Wood, J.; Bigos, M.; Ross, C.D.; Kachynski, A.; Dean, A.; et al. High-fidelity detection and sorting of nanoscale vesicles in viral disease and cancer. *J. Extracell. Vesicles* **2019**, *8*, 1597603. [CrossRef] [PubMed]
6. Chýlek, P. Absorption and scattering of light by small particles. By C. F. Bohren and d. R. Huffman. *Appl. Opt.* **1986**, *25*, 3166. [PubMed]
7. McVey, M.J.; Spring, C.M.; Kuebler, W.M. Improved resolution in extracellular vesicle populations using 405 instead of 488 nm side scatter. *J. Extracell. Vesicles* **2018**, *7*. [CrossRef]
8. Brittain, G.C.; Chen, Y.Q.; Martinez, E.; Tang, V.A.; Renner, T.M.; Langlois, M.A.; Gulnik, S. A Novel Semiconductor-Based Flow Cytometer with Enhanced Light-Scatter Sensitivity for the Analysis of Biological Nanoparticles. *Sci. Rep.* **2019**, *9*, 1–13. [CrossRef]
9. Lucchetti, D.; Calapà, F.; Palmieri, V.; Fanali, C.; Carbone, F.; Papa, A.; De Maria, R.; De Spirito, M.; Sgambato, A. Differentiation Affects the Release of Exosomes from Colon Cancer Cells and Their Ability to Modulate the Behavior of Recipient Cells. *Am. J. Pathol.* **2017**, *187*, 1633–1647. [CrossRef]
10. Park, S.J.; Kim, J.M.; Kim, J.; Hur, J.; Park, S.; Kim, K.; Shin, H.-J.; Chwae, Y.J. Molecular mechanisms of biogenesis of apoptotic exosome-like vesicles and their roles as damage-associated molecular patterns. *Proc. Natl. Acad. Sci. USA* **2018**, *11*, E11721–E11730. [CrossRef]
11. Tsien, R.Y. Fluorescent probes of cell signaling. *Annu. Rev. Neurosci.* **1989**, *12*, 227–253. [CrossRef] [PubMed]
12. Gray, W.D.; Mitchell, A.J.; Searles, C.D. An accurate, precise method for general labeling of extracellular vesicles. *MethodsX* **2015**, *2*, 360–367. [CrossRef] [PubMed]
13. Nordin, J.Z.; Lee, Y.; Vader, P.; Mäger, I.; Johansson, H.J.; Heusermann, W.; Wiklander, O.P.; Hällbrink, M.; Seow, Y.; Bultema, J.; et al. Ultrafiltration with size-exclusion liquid chromatography for high yieldisolation of extracellular vesicles preserving intact biophysical and functional properties. *Nanomedicine* **2015**, *11*, 879–883. [CrossRef] [PubMed]
14. Roederer, M. Spectral compensation for flow cytometry: Visualization artifacts, limitations, and caveats. *Cytometry* **2001**, *45*, 194–205. [CrossRef]
15. de Rond, L.; Libregts, S.; Rikkert, L.G.; Hau, C.M.; van der Pol, E.; Nieuwland, R.; van Leeuwen, T.G.; Coumans, F. Refractive index to evaluate staining specificity of extracellular vesicles by flow cytometry. *J. Extracell. Vesicles* **2019**, *8*, 1643671. [CrossRef]
16. Coumans, F.; Brisson, A.R.; Buzas, E.I.; Dignat-George, F.; Drees, E.; El-Andaloussi, S.; Emanueli, C.; Gasecka, A.; Hendrix, A.; Hill, A.; et al. Methodological Guidelines to Study Extracellular Vesicles. *Circ. Res.* **2017**, *120*, 1632–1648. [CrossRef]
17. Escola, J.-M.; Kleijmeer, M.J.; Stoorvogel, W.; Griffith, J.; Yoshie, O.; Geuze, H.J. Selective enrichment of tetraspan proteins on the internal vesiclesof multivesicular endosomes and exosomes secreted by humanB-lymphocytes. *J. Biol. Chem.* **1998**, *273*, 20121–20127. [CrossRef]
18. Chutipongtanate, S.; Greis, K.D. Multiplex Biomarker Screening Assay for Urinary Extracellular Vesicles Study: A Targeted Label-Free Proteomic Approach. *Sci. Rep.* **2018**, *9*, 15039. [CrossRef]
19. Arraud, N.; Linares, R.; Tan, S.; Gounou, C.; Pasquet, J.M.; Mornet, S.; Brisson, A.R. Extracellular vesicles from blood plasma: Determination of their morphology, size, phenotype and concentration. *J. Thromb. Haemost.* **2014**, *12*, 614–627. [CrossRef]

20. Linares, R.; Tan, S.; Gounou, C.; Arraud, N.; Brisson, A.R. High-speed centrifugation induces aggregation of extracellular vesicles. *J. Extracell. Vesicles* **2015**, *4*, 29509. [CrossRef]
21. Maas, S.L.; de Vrij, J.; van der Vlist, E.J.; Geragousian, B.; van Bloois, L.; Mastrobattista, E.; Schiffelers, R.M.; Wauben, M.H.; Broekman, M.L.; Nolte-'t Hoen, E.N. Possibilities and limitations of current technologies for quantification of biological extracellular vesicles and synthetic mimics. *J. Control. Release* **2015**, *200*, 87–96. [CrossRef] [PubMed]
22. Nolan, J.P.; Jones, J.C. Detection of platelet vesicles by flow cytometry. *Platelets* **2017**, *28*, 256–262. [CrossRef] [PubMed]
23. Welsh, J.A.; Holloway, J.A.; Wilkinson, J.S.; Englyst, N.A. Extracellular Vesicle Flow Cytometry Analysis and Standardization. *Front. Cell Dev. Biol.* **2017**, *5*, 78. [CrossRef] [PubMed]
24. Van der Pol, E.; van Gemert, M.J.; Sturk, A.; Nieuwland, R.; van Leeuwen, T.G. Single vs swarm detection of microparticles and exosomes by flow cytometry. *J. Thromb. Haemost.* **2012**, *10*, 919–930. [CrossRef] [PubMed]
25. Kormelink, T.G.; Arkesteijn, G.J.; Nauwelaers, F.A.; van den Engh, G.; Nolte-'t Hoen, E.N.; Wauben, M.H. Prerequisites for the analysis and sorting of extracellular vesicle subpopulations by high-resolution flow cytometry. *Cytom. Part A* **2016**, *89*, 135–147. [CrossRef] [PubMed]
26. Libregts, S.F.W.M.; Arkesteijn, G.J.A.; Németh, A.; Nolte-'t Hoen, E.N.M.; Wauben, M.H.M. Flow cytometric analysis of extracellular vesicle subsets in plasma: Impact of swarm by particles of non-interest. *J. Thromb. Haemost.* **2018**, *16*, 1423–1436. [CrossRef] [PubMed]
27. Kent, S.P.; Ryan, K.H.; Siegel, A.L. Steric hindrance as a factor in the reaction of labeled antibody with cell surface antigenic determinants. *J. Histochem. Cytochem.* **1978**, *26*, 618–621. [CrossRef] [PubMed]
28. De Vita, M.; Catzola, V.; Buzzonetti, A.; Fossati, M.; Battaglia, A.; Zamai, L.; Fattorossi, A. Unexpected interference in cell surface staining by monoclonal antibodies to unrelated antigens. *Cytom. Part B Clin. Cytom.* **2015**, *88*, 352–354. [CrossRef]
29. Kibria, G.; Ramos, E.K.; Lee, K.E.; Bedoyan, S.; Huang, S.; Samaeekia, R.; Athman, J.J.; Harding, C.V.; Lötvall, J.; Harris, L.; et al. A rapid, automated surface protein profiling of single circulating exosomes in human blood. *Sci. Rep.* **2016**, *6*, 36502. [CrossRef]
30. Van der Pol, E.; Coumans, F.A.; Grootemaat, A.E.; Gardiner, C.; Sargent, I.L.; Harrison, P.; Sturk, A.; van Leeuwen, T.G.; Nieuwland, R. Particle size distribution of exosomes and microvesicles determined by transmission electron microscopy, flow cytometry, nanoparticle tracking analysis, and resistive pulse sensing. *J. Thromb. Haemost.* **2014**, *12*, 1182–1192. [CrossRef]
31. Welsh, J.A.; Horak, P.; Wilkinson, J.S.; Ford, V.J.; Jones, J.C.; Smith, D. FCMPASS Software Aids Extracellular Vesicle Light Scatter Standardization. *Cytom. Part A* **2000**, *97*, 569–581. [CrossRef] [PubMed]

12

The Biological Function and Therapeutic Potential of Exosomes in Cancer: Exosomes as Efficient Nanocommunicators for Cancer Therapy

Jeong Uk Choi [1], In-Kyu Park [2], Yong-Kyu Lee [3] and Seung Rim Hwang [4,5,*]

1 College of Pharmacy, Chonnam National University, 77 Yongbong-ro, Buk-gu, Gwangju 61186, Korea; cju0667@jnu.ac.kr
2 Department of Biomedical Sciences, Chonnam National University Medical School, 322 Seoyang-ro, Hwasun 58128, Korea; pik96@jnu.ac.kr
3 Department of Chemical and Biological Engineering, Korea National University of Transportation, 50 Daehak-ro, Chungju, Chungbuk 27469, Korea; leeyk@ut.ac.kr
4 College of Pharmacy, Chosun University, 309 Pilmun-daero, Dong-gu, Gwangju 61452, Korea
5 Department of Biomedical Sciences, Graduate School, Chosun University, 309 Pilmun-daero, Dong-gu, Gwangju 61452, Korea
* Correspondence: srhwang@chosun.ac.kr;

Abstract: Cancer therapeutics must be delivered to their targets for improving efficacy and reducing toxicity, though they encounter physiological barriers in the tumor microenvironment. They also face limitations associated with genetic instability and dynamic changes of surface proteins in cancer cells. Nanosized exosomes generated from the endosomal compartment, however, transfer their cargo to the recipient cells and mediate the intercellular communication, which affects malignancy progression, tumor immunity, and chemoresistance. In this review, we give an overview of exosomes' biological aspects and therapeutic potential as diagnostic biomarkers and drug delivery vehicles for oncotherapy. Furthermore, we discuss whether exosomes could contribute to personalized cancer immunotherapy drug design as efficient nanocommunicators.

Keywords: exosome; cancer; nanocommunicator; diagnostic biomarker; drug delivery vehicle; personalized cancer immunotherapy

1. Introduction

Oncology drugs constitute the largest therapeutics section approved by the Center for Drug Evaluation and Research, a division of the United States Food and Drug Administration [1]. Once cancer metastasizes from the primary tumor to new sites at the time of detection, the survival rate of cancer patients decreases substantially, posing a threat to overall health [2]. In recent times, an early diagnosis of cancer via timely screening using liquid biopsy tools such as circulating tumor cells (CTCs) or circulating tumor DNA (ctDNA) as well as extracellular vesicles (EVs) has received attention, and the survival rate of cancer patients has increased with the development of treatment strategies [3,4].

Anticancer drug approval trends have changed since cancer chemotherapeutic agents were first developed in the 1940s. Cancer chemotherapeutics show efficacy as well as side effects as they not only interfere with the growth or division of cancer cells but also normal cells. Cancer cells even evade anticancer drugs by mediating a cellular efflux of drugs or by reducing target gene expression. They not only receive external signals but also transmit signals to form new blood vessels in cancer tissues. EVs secreted by cancer cells transfer signals for cancer progression and metastasis [5].

As technical limitations associated with cancer chemotherapy have been recognized, and the understanding of cancer biology has improved, the paradigm of research and development has shifted towards targeted therapy using monoclonal antibodies and small molecules that target the signaling process of cancer cells. Recently, synthetic small molecules such as a FGFR (fibroblast growth factor receptor) inhibitor (erdafitinib; Janssen Pharmaceutica, Beerse, Belgium), CSF1R (colony-stimulating factor-1 receptor)/RTK (receptor tyrosine kinase)/FLT3 (FMS-like tyrosine kinase 3) inhibitor (pexidartinib; Daiichi Sankyo, Tokyo, Japan), and exportin 1 inhibitor (selinexor; Karyopharm Therapeutics, Newton, MA, USA), and biologics such as an antibody-drug conjugate (ADC) targeting nectin-4 (enfortumab vedotin; Astellas Pharma, Tokyo, Japan), and ADC targeting CD79B (polatuzumab vedotin; Genentech/Roche, South San Francisco, CA, USA) have been approved. Interestingly, entrectinib (Genentech/Roche) simultaneously targets c-ROS oncogene 1 (ROS1), ALK (anaplastic lymphoma kinase) RTKs, and tropomyosin receptor kinase proteins encoded by neurotrophic tyrosine receptor kinase (NTRK) genes; it was approved as a biomarker-based treatment for ROS1-positive and NTRK fusion-positive cancer. In the same vein, engineered EVs that carry small interfering RNA (siRNA) or short hairpin RNA specifically targeting oncogenic KRAS mutation showed their therapeutic potential in pancreatic ductal adenocarcinoma mouse models [6].

Despite the remarkable performance of targeted anticancer drugs, limitations have been associated with the targeting of markers on the cancer cell surface, because cancer cells are genetically unstable, and surface proteins in cancer cells change dynamically during disease progression [7]. The necessity of developing new therapeutic approaches has emerged due to the genetic heterogeneity between cancer cells and drug resistance mechanisms [8]. Since 2010, cancer immunotherapy drugs including CTLA4 (cytotoxic T-lymphocyte-associated protein 4) inhibitors, PD-1 (programmed cell death protein 1) inhibitors, PD-L1 (programmed cell death ligand 1) inhibitors, and CAR (chimeric antigen receptor)-T cell therapy have received a lot of attention. As cancer cells proliferate by evading the immune system, cancer immunotherapy drugs interfere with the evasion mechanism or stimulate immune cells to attack tumor cells [9].

However, cancer therapeutics encounter barriers against transport to target sites owing to the elevated levels of solid stress, vascular network formation, interstitial fluid pressure, and density of extracellular matrix (ECM) in the tumor microenvironment [10]. Nanocarriers can enhance the permeability and retention of their cargo drugs in solid tumor tissues. Certain cancer therapeutics need to be delivered to intracellular targets such as the cytosol or nucleus to elicit their proper action [11]. Interestingly, nanosized EVs called exosomes, transfer their cargo nucleic acids and proteins to the recipient cells via the cellular uptake of vesicles; this contributes to the intercellular communication between tumor cells and bone marrow stromal cells. Recurrent mutations or specific alterations of niches within hematopoietic cells of the bone marrow regulating the production of blood and immune cells play roles in malignancy progression and chemoresistance [12]. In that exosomes orchestrate immune cells in the tumor microenvironment through cell-to-cell signaling, they have been tested for cancer immunotherapy in clinical trials. Dendritic cell (DC)-derived exosomes (DEXs) showed modest efficacy in patients with metastatic melanoma and non-small cell lung cancer (NSCLC) [13,14]. Autologous tumor-derived exosomes (TEXs) in combination with the GM-CSF (granulocyte-macrophage colony-stimulating factor) could induce antitumor T lymphocyte response in colorectal cancer [15]. TEXs can also provide diagnostic biomarkers, because they circulate in biological fluids and the exosomal components enclosed in lipid membrane vesicles reflect the characteristics of the cells of origin in the tumor tissue.

In this review, we aimed to provide a comprehensive overview of the biological aspects and potential therapeutic applications of exosomes in cancer. Here, we also discuss whether exosomes could contribute to personalized cancer immunotherapy drug design as efficient nanocommunicators.

2. Biologic Aspects of Exosomes and Cancer

2.1. Exosome Biogenesis

Cells release EVs enclosed in lipid membranes into the extracellular environment. Exosomes, microvesicles (MVs)/microparticles, and apoptotic bodies form a subgroup of EVs. They have been defined by their biogenesis, size, or constituent molecules. The process of exosome biogenesis starts with endocytic membrane transport through which the cell surface proteins can be recycled [16]. The perimeter membrane of endocytic vesicles buds inward during endosome maturation from the early endosome to the late endosome [17]. Further invagination of the endosomal membrane into the endosomal compartment forms intraluminal vesicles (ILVs) in the multivesicular body (MVB). Subsequently, the MVB is either fused with the lysosome for degradation or release its contents in the form of exosomes by merging with the plasma membrane [18]. This process of exosome formation is different from that of MV/microparticle formation that takes place via outward budding directly from the plasma membrane (Figure 1) [18].

The most well-known mechanism for packaging receptors internalized from the cell surface and other exosomal cargo proteins in the late endosome membrane depend on the endosomal sorting complex required for transport (ESCRT) machinery [19]. Cytosolic protein complexes composed of ESCRT-0, ESCRT-I, ESCRT-II, and ESCRT–III, together with accessory proteins, participate in binding ubiquitinated cargo and sculpting MVB vesicles [20]. The addition of a regulatory ubiquitin protein to the substrate is a reversible post-translational modification catalyzed by a ubiquitin-activating enzyme, ubiquitin-conjugating enzyme, and ubiquitin ligase [21]. Even though it is debatable whether the contents of MVBs are released into the extracellular medium or enter the lysosome under certain circumstances, the tagging of misfolded or damaged proteins with ubiquitin plays a role in maintaining intracellular protein levels for cell cycle regulation and is also associated with the oncogenic processes [22]. For example, mutations in genes encoding components for ubiquitin ligase activity lead to the development of renal cell carcinoma and breast cancer [23,24].

Alternatively, ESCRT-independent pathways are supported by MVB formation even in the depletion of key subunits of ESCRTs [25]. Sphingolipids and cholesterol that are enriched in detergent-resistant membrane domains may be involved in ubiquitin-independent protein sorting [26]. The sphingolipid ceramide also triggers the formation of ILVs in the late endosome that are destined for secretion as exosomes [27]. Observation of human lymphoblastoid cells via immunoelectron microscopy demonstrated low cholesterol labeling in the lysosome but high cholesterol labeling in the MVB and exosomes [28]. Despite the possibility of ubiquitin-independent exosomal cargo sorting, certain ESCRT components are involved in exosome formation. Apoptosis-linked gene 2-interacting protein X (ALIX), an ESCRT accessory protein, contributes to the sorting of transferrin receptor into the late endosome membrane and interacts with syntenin-linking syndecan-mediated signaling [29,30]. Hepatocyte growth factor-regulated tyrosine kinase substrate (HRS), an ESCRT-0 protein, is related to exosomal secretion and antigen-presenting activity in DCs [31].

The pathways of packaging RNAs into exosomes are still unclear. Specific linear sequence motifs that are shared by exosomal RNAs may function as *cis*-acting elements that target RNAs to exosomes [32]. GW-bodies containing protein components of RNA-induced silencing complex congregate with the endosome and MVB, where microRNA (miRNAs) are enriched. The exosome-like vesicles secreted by MVBs are rich in GW182, which modulates miRNA loading or gene silencing [33].

2.2. Exosome–Cell Interaction and Biodistribution of Exosomes

After being secreted by the original cell into the extracellular space, exosomes circulate in body fluids or are distributed into the tissue ECM [34,35]. Owing to their nanosize, they even penetrate the nasal mucosa and bypass the blood-brain barrier [36]. Labeling and tracking exosomes using fluorescence or bioluminescence helps us understand exosome–cell interaction and the biodistribution of exosomes [37].

Exosomes with lipid bilayer structures can be taken up into the recipient cell via membrane fusion, clathrin-mediated endocytosis, caveolin-dependent endocytosis, macropinocytosis, or phagocytosis, leading to the delivery of exosomal contents to the cytosolic space of the recipient cell (Figure 2) [38–41]. Exosomal cargo is then released by the acidification of the endo/lysosome compartment in the recipient cell [42]. Receptor-ligand interactions between cell surface receptors and exosomal ligands may also occur based on specific cell types, and mediate antigen presentation, cell signaling, the release of soluble factor, disease progression, and immune surveillance [43].

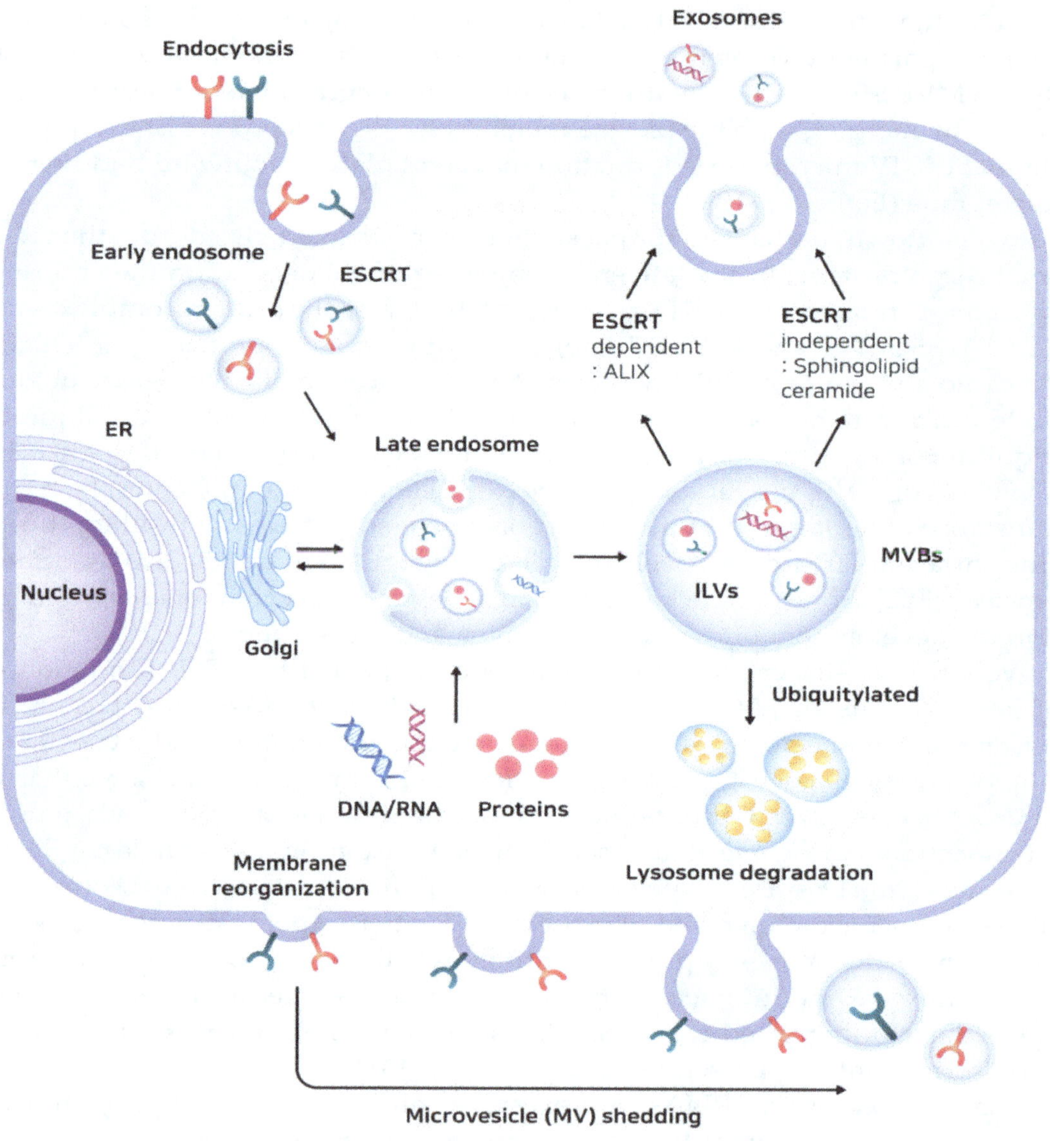

Figure 1. The process of exosome biogenesis. The perimeter membrane of endocytic vesicles buds inward during endosome maturation from the early endosome to the late endosome. Further invagination of the endosomal membrane forms intraluminal vesicles (ILVs) in the multivesicular body (MVB). Subsequently, the MVB is fused with the lysosome or release its contents in the form of exosomes (top right). This process of exosome biogenesis is different from that of microvesicle (MV) shedding (bottom). Receptors internalized from the cell surface and other exosomal cargo proteins are packed in the late endosome either by endosomal sorting complex required for transport (ESCRT)-dependent or ESCRT-independent pathway. ER: endoplasmic reticulum; ILV: intraluminal vesicle; MVB: multivesicular body; MV: microvesicle; ESCRT: endosomal sorting complex required for transport; ALIX: apoptosis-linked gene 2-interacting protein X.

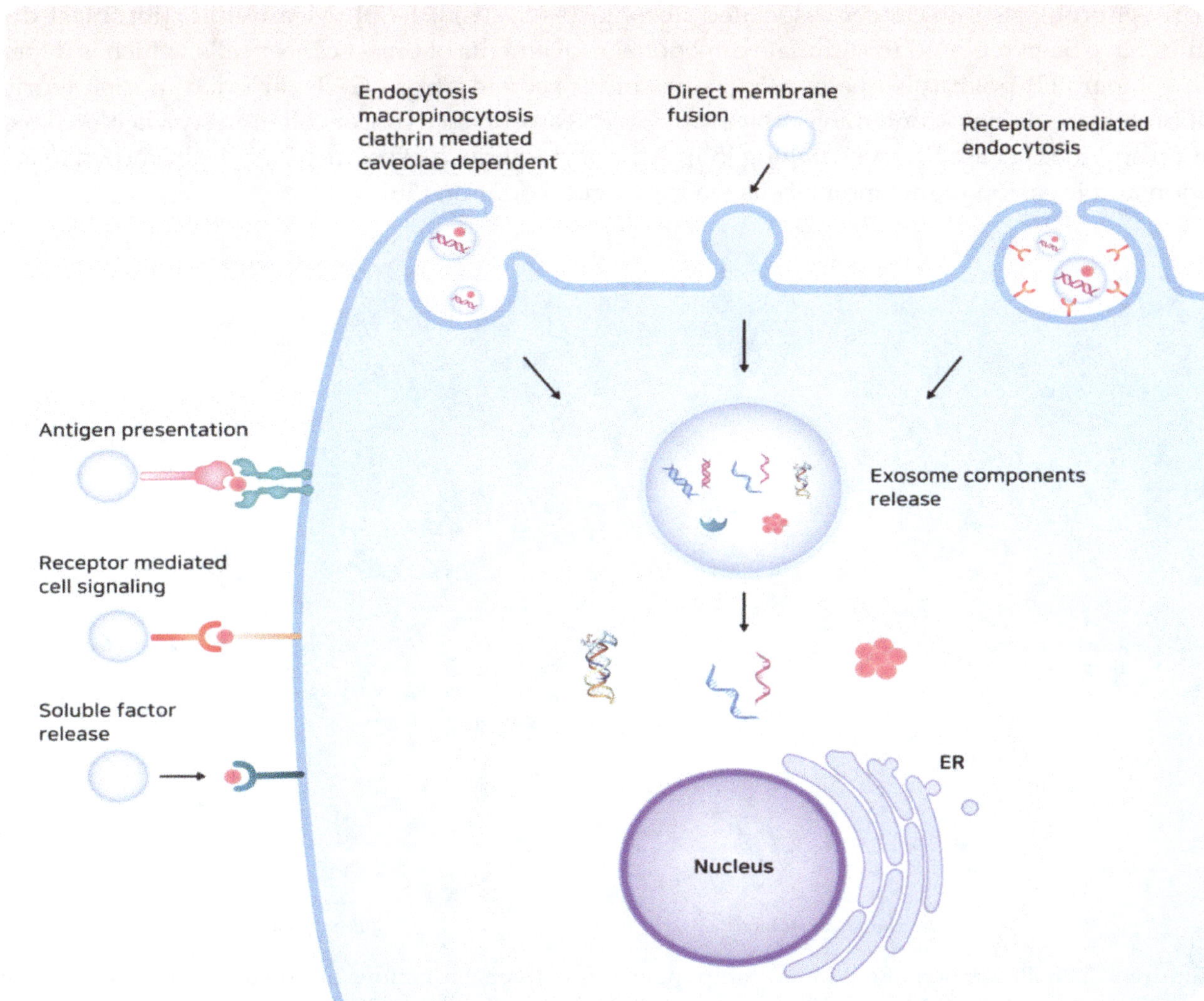

Figure 2. The illustration of exosome–cell interaction. Exosomes can be taken up into the recipient cell via direct membrane fusion or endocytosis, leading to the delivery of exosomal contents such as DNAs, messenger RNAs, long non-coding RNAs, enzymes, and signaling peptides or proteins to the cytosolic space of the recipient cell. Receptor-ligand interactions between cell surface receptors and exosomal ligands may also occur.

The diverse functions of exosomes are governed by the delivery of exosomal components, including lipids, nucleic acids, and proteins such as tetraspanins, adhesion molecules, antigen-presenting molecules, transmembrane receptors, MVB formation proteins, membrane trafficking proteins, cytoskeletal proteins, enzymes, signaling proteins, and heat shock proteins (Figure 3) [44]. Finally, clearance of exosomes from the body might take place via the liver, spleen, and kidneys with the mononuclear phagocytic system [45].

2.3. The Biological Functions of Exosomes in Cancer

Exosomes function as unique intercellular communicators and debris managers for cellular homeostasis [46]. Delivery of exosomal cargo mediates cell motility, immune responses, and reprogramming of the tumor microenvironment. Whether exosomes promote cancer progression and escape from immunosurveillance depends on the type of the cells of origin and malignancy at the time of exosome release [47,48]. TEXs have autocrine and paracrine roles in cancer progression [49]. At the site of the primary lesion, they carry fibronectin and proteinases, including membrane type 1-matrix metalloproteinase (MMP) and MMP2, and facilitate adhesion and invasiveness of cancer cells [50]. Delivery of miRNAs via TEXs transforms fibroblasts into cancer-associated fibroblasts (CAFs) [51–53]. Meanwhile, fibroblast-derived exosomes have been reported to stimulate directional movements of breast cancer cells, which is dependent on Wnt-planar cell polarity signaling [54]. Exosomes secreted from CAFs are rich in disintegrin and metalloproteinase domain-containing protein 10, which can enhance cancer cell motility via Notch receptor activation and the GTPase RhoA signalling [55]. Adipocyte-derived exosomes have been shown to increase migration and invasion of melanoma cells via fatty acid oxidation [56].

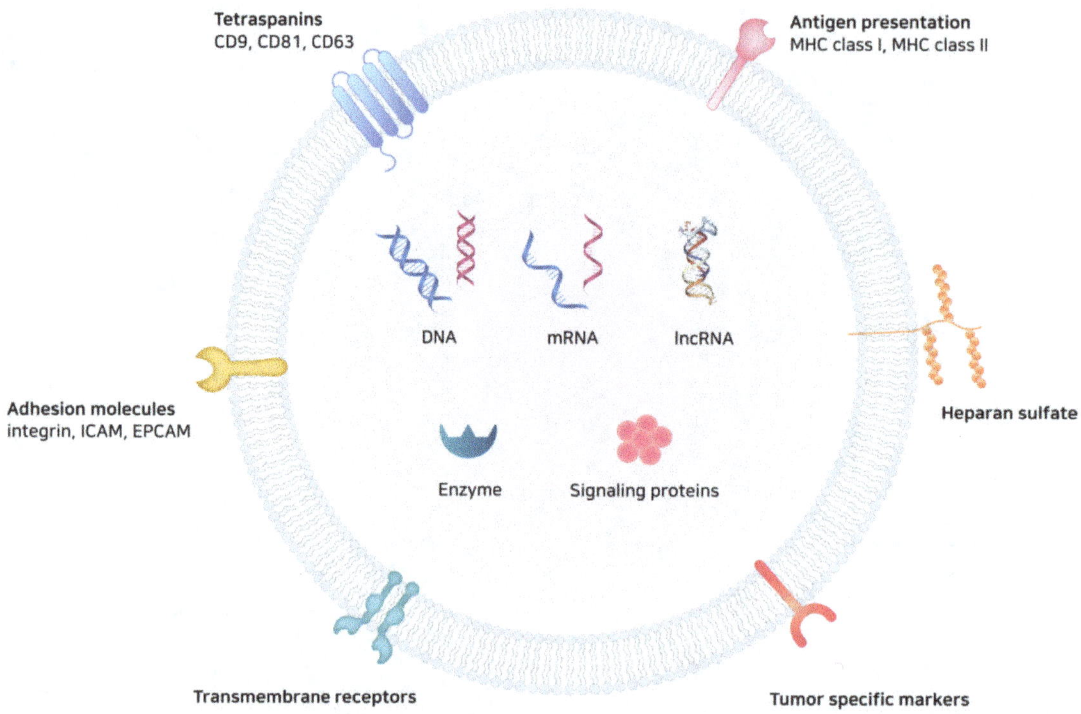

Figure 3. The illustration of exosomal components. The diverse functions of exosomes are governed by the delivery of exosomal cargo proteins and nucleic acids to the recipient cells. Exosomal components include lipids, nucleic acids, tetraspanins, adhesion molecules, antigen-presenting molecules, transmembrane receptors, MVB formation proteins, membrane trafficking proteins, enzymes, signaling proteins, etc. mRNA: messenger RNAs; lncRNA: long non-coding RNA; ICAM: intercellular adhesion molecule; EpCAM: epithelial cell adhesion molecule; MHC: major histocompatibility complex.

Exosomes also regulate angiogenesis and vascular permeability [57]. A mucin-type podoplanin glycoprotein, which is upregulated in certain types of cancer and incorporated into exosomes, reprograms exosomal proteins and promotes lymphangiogenesis [58]. Uptake of leukemia-derived exosomes containing miR-92a by endothelial cells enhanced endothelial tube formation [59]. Exosomes released by metastatic tumor cells led to endothelial hyperpermeability contrary to the exosomes released by non-metastatic tumor cells [60]. Under hypoxia, miR-23a upregulation in TEXs leads to the accumulation of hypoxia-inducible factor-1 α, enhancing angiogenesis, and inhibits tight junction protein ZO-1, increasing vascular permeability [61].

After the intravasation of tumor cells, TEXs traveling through the bloodstream develop "pre-metastatic niches" by modifying microenvironments in distant target organs and affecting organ-specific stromal cells [62]. Exosomes from highly metastatic melanomas reprogrammed bone marrow progenitor cells, resulting in exosome-mediated tyrosine-protein kinase Met signaling [63]. Uptake of exosomes derived from pancreatic ductal adenocarcinomas by Kupffer cells induced transforming growth factor-β secretion and fibronectin production, which initiated liver pre-metastatic niche formation [64]. Specific integrin expression patterns on TEXs were shown to correlate with the localization of TEXs and organotropic metastasis [65]. Targeting the exosomal integrin $\alpha_6\beta_4$ was associated with lung metastasis, whereas targeting the exosomal integrin $\alpha_v\beta_5$ was associated with liver metastasis. TEXs can also enter sentinel lymph nodes and influence lymph node distribution of cancer cells, which is driven by synchronized molecular signals that affect tumor metastasis [66]. Meanwhile, reports indicate that exosomes from non-metastatic cells inhibit metastasis. Exosomes isolated from non-metastatic patient sera suppressed experimental lung metastasis by increasing the number of patrolling monocytes in the lungs and inducing macrophage differentiation, leading to immune surveillance in the pre-metastatic niche [67].

As TEXs contain immunosuppressive ligands as well as immunostimulatory tumor-associated antigens (TAAs), they can play roles in mediating tumor immunity [68]. Binding immune-inhibitory ligands of TEXs to T cell receptors and IL-2 receptors leads to tolerogenic signals [69]. TEXs also induce apoptosis of $CD8^{+}$ T lymphocytes and differentiation of myeloid precursor cells and regulatory T cells. They can also inhibit the cytotoxic functions of natural killer (NK) cells via downregulation of NK group 2D, an NK-activating receptor that recognizes ligands on the surface of malignant cells as well as TEXs [70]. Meanwhile, DEXs pulsed with tumor peptides activate cytotoxic T lymphocytes [71]. Mast cell-derived exosomes associated with antigens induce maturation of DCs. Antigen presentation by DCs activates B and T cells [72].

Exosomes have also been reported to mediate cancer chemoresistance by cargo transfer [73]. Exosomes derived from drug-resistant breast cancer cells modulated the cell cycle and drug-induced apoptosis, which might be dependent on selective miRNA patterns [73]. Restoration of miR-151a via exosomes derived from temozolomide (TMZ)-resistant glioblastoma multiforme enhances chemosensitivity to TMZ, whereas miR-151a loss drives TMZ resistance [74]. In cisplatin-resistant tumor cells, acidic pH in the extracellular microenvironment reduces cisplatin uptake into tumor cells and increases cisplatin levels eliminated via TEXs [75]. Exosomes isolated from fibroblast-derived conditioned medium prime cancer stem cells and promote chemoresistance in colorectal cancer [76]. Transfer of exosomal RNA from stromal to breast cancer cells activates signal transducer and activator of transcription 1 and NOTCH3 signaling, which regulate the expansion of chemoresistant cancer cells [77].

3. Potential Therapeutic Applications of Exosomes in Cancer

3.1. Exosomes as Diagnostic Biomarkers for Cancer

3.1.1. Identification Techniques of Exosomes in Liquid Biopsy

As exosomes have the potential to be used as prognostic biomarkers, isolation and identification of exosomes and their contents are also critical issues. For exosome isolation/purification, various methods have been evaluated. The most common method is differential centrifugation at 300× *g* for 10 min, 2000× *g* for 10 min, and 10,000× *g* for 30 min, followed by ultracentrifugation at 100,000× *g* for 2 h. For higher purity of exosomes, gradient centrifugation using sucrose can also be used [78]. An immuno-isolation method using antibody-coated magnetic beads can also be used to obtain higher purity and recovery rates. Rapid surface protein characterization using flow cytometry provides additional benefits with this method [79]. However, only specific types of exosomes can be isolated using this method, which can be considered as one of the limitations of this method. Currently, exosome extraction kits such as the ExoSpin™ Exosome purification Kit (Cell Guidance Systems LLC; St. Louis, MO, USA) and Total Exosome Isolation Kit™ (Life Technologies; Waltham, MA, USA) are also available. Typically, these kits use polymers such as polyethylene glycol with centrifugation to induce exosome sedimentation [80]. Microfluidic technology enables rapid and precise isolation/purification of exosomes with a very small volume of samples using a micro-electromechanical system [81]. Turbidimetry-enabled particle purification liquid chromatography based on the size exclusion principle also has been proven superior in purification of EVs in biofluids [82]. Using biosensors is another technique to determine exosomes with higher sensitivity and automated analysis [83]. For identification of exosome contents, general methods such as polymerase chain reaction (PCR), next-generation sequencing (NGS), and proteomics can be used.

3.1.2. Applications of Exosomes as Diagnostic Biomarkers for Cancer

Liquid biopsy tools such as CTCs, ctDNA, and exosomes have advantages in non-invasive diagnosis and prognosis over traditional tissue biopsy strategies. Prognostic potential of CTCs has been already tested for monitoring epithelium-originating tumors in clinical trials [84]. However, CTCs shed from the primary tumor are found as only a few CTCs per mL of blood among millions of erythrocytes or leukocytes, and CTC enrichment techniques are needed for their detection [85]. Fragmented DNA shed from tumor cells may reflect the genetic signature of tumors [86]. Analysis of ctDNA in blood is challenging because there is a small fraction of ctDNA among cell-free DNAs from leukocytes in the blood sample [87]. In addition, the heterogeneity of tumor cells makes determining tumor-specific mutation in the ctDNA sample difficult.

Compared to the limited amounts of CTCs or ctDNA in the bloodstream, exosomes can be detected not only in blood but also in urine, cerebrospinal fluid, or lymphatic exudate [88,89]. Exosomes can be actively involved in cellular communication by delivering various signaling molecules. They serve as effective carriers, as the lipid bilayer can protect the contents and directly deliver them to the target cells. Exosome contents include nucleic acids, enzymes, and various signaling proteins. The contents can vary depending on the cells of origin. Therefore, the identification of exosomal contents can provide important clues regarding the cells of origin, which makes them ideal biomarkers for the diagnosis of diseases such as cancer, infection, metabolic, and neurodegenerative disorders [90,91].

As exosomes and their contents released from cancer cells display unique properties, many attempts have been made to use TEXs as cancer diagnostic biomarkers (Table 1) [92]. TEXs play important roles in facilitating tumor growth and are involved in every step of cancer development, including angiogenesis, proliferation, metastasis, and fostering the tumor microenvironment by delivering relevant genes, growth factors, and cell signaling molecules [93–95]. For example, exosomes isolated from urine samples can be used to diagnose prostate cancer, bladder cancer, and glioblastoma. Typically, exosomal proteins related to epidermal growth factor receptor (EGFR) pathways (resistin, α-subunit of Gs protein, retinoic acid-induced protein 3, EGFR variant III, etc.) are present at diagnostic levels and hence, can be used as reliable biomarkers [96–98]. Prostate-specific antigen, survivin, and prostate cancer antigen 3 in exosomal contents can be also used for detecting prostate cancer [97,99,100]. Nucleic acids present in cancer exosomes, including miRNA, messenger RNA (mRNA), and long non-coding RNA (lncRNA), can also be used as diagnostic markers. For example, unique nucleic acids from exosomes have been identified in patients with glioblastoma [98]. Specific lncRNA, LINC00152, was also identified in gastric cancer-derived exosomes, which makes it a useful diagnostic biomarker [101]. miRNAs such as miR-21, -141, -200a, etc. can be detected in ovarian cancer patients, and miR-17-3p, -21, etc. were identified in lung cancer patients [102,103]. The genetic mutation in cancer patients is detectable by using exosome samples instead of CTCs or ctDNA. Exosomal RNA/DNA demonstrated the diagnostic value for KRAS mutation in pancreatic cancer and EGFR mutation in NSCLC [104–106]. BRAF mutation in EVs from lymphatic exudate of melanoma patients was reported to be useful for the prognosis [107]. Therefore, identification of unique exosomal contents corresponding to various types of cancers can help develop reliable diagnostic biomarkers (Figure 4).

3.2. Exosomes as Drug Delivery Vehicles for Oncotherapy

Exosomes are attractive nanovehicles for targeting cancer (Figure 4). As exosomes originate from endogenous cells, they possess low immunogenicity and thus induce low toxicity and side effects [108]. Exosomes are stable under physiological conditions. Owing to the presence of a lipid bilayer, they can protect the contents from the immune system and various enzymes. Furthermore, they demonstrate a homing capability by cell/tissue tropism with a longer circulation period, and can also cross the blood-brain barrier [109]. Unlike liposomes or other synthetic drug delivery nanoparticles, exosomes have characteristic membrane proteins and lipids that promote efficient targeting of exosomes to the recipient cell [110]. Exosomes can also enhance the delivery of contents as they can be directly fused or internalized into target cells. CD47, an integrin-associated protein upregulated in mesenchymal stem cells (MSCs), interacts with signal-regulatory protein, which helps inhibition of phagocytosis [111]. Thus, exosomes derived from fibroblast-like MSCs show the enhanced retention in the circulation in mice [6]. As the average size of exosomes ranges from 30 to 200 nm, passive targeting of exosomes to tumor tissue with enhanced permeability and retention effect can also be expected. With these benefits, a number of clinical and preclinical trials have been conducted to utilize exosomes as delivery vehicles.

Table 1. Applications of exosomes as diagnostic biomarkers for cancer.

Exosome Contents	Associated Molecule	Target Disease	Reference
Proteins	Resistin, α-subunit of Gs protein, retinoic acid-induced protein 3	Prostate, bladder cancer	[96,97]
	Epidermal growth factor receptor (EGFR) variant III	Glioblastoma	[98]
	Prostate-specific antigen, survivin, prostate cancer antigen 3	Prostate cancer	[97,99,100]
RNAs	lncRNA, LINC00152	Gastric cancer	[101]
	miR-21, miR-141, miR-200a	Ovarian cancer	[102]
	miR-17-3p, miR-21	Non-small cell lung cancer (NSCLC)	[103]
DNAs	Mutant KRAS	Pancreatic cancer	[104]
	EGFR T790M mutation	NSCLC	[105]
	Mutant KRAS, TP53	Pancreatic cancer	[106]
	$BRAF^{V600E}$ mutation	Melanoma	[107]

EGFR: epidermal growth factor receptor; NSCLC: non-small cell lung cancer.

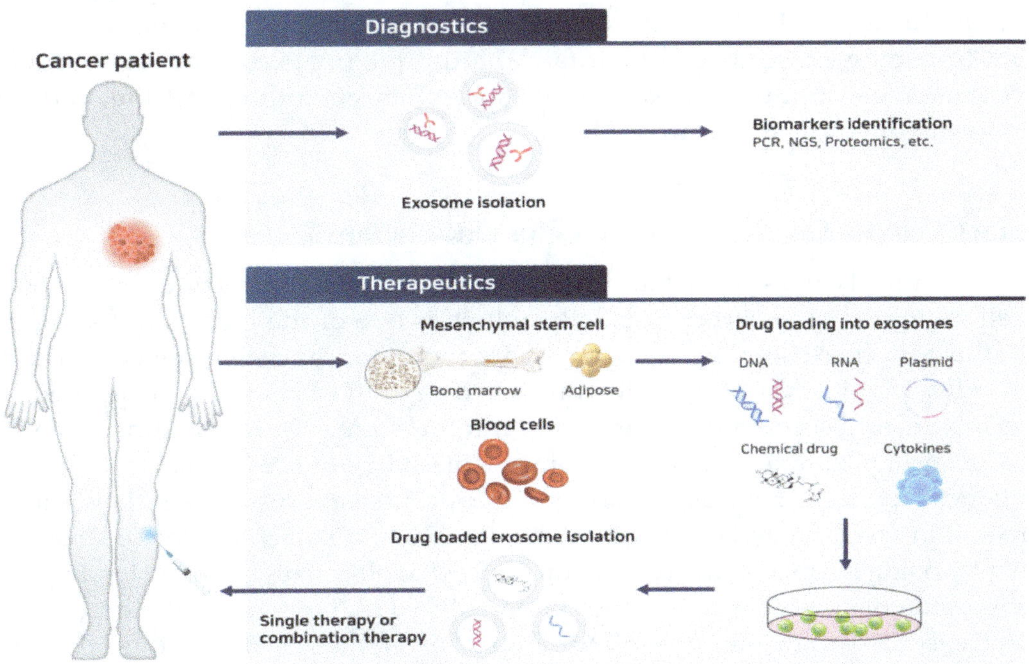

Figure 4. Potential therapeutic applications of exosomes in cancer. As exosomal components reflect the characteristics of the cells of origin, many attempts have been made to use tumor-derived exosomes (TEXs) as cancer diagnostic biomarkers. For identification of exosome contents, general methods such as polymerase chain reaction (PCR), next-generation sequencing (NGS), and proteomics can be used. Exosomes also have therapeutic potential as nanovehicles for drug delivery and personalized cancer immunotherapy. TEX: tumor-derived exosome; PCR: polymerase chain reaction; NGS: next-generation sequencing.

3.2.1. Methods for Loading Drugs into Exosomes

A number of studies have demonstrated that drug-loaded exosomes show better outcomes in inhibiting cancers, but the methods for loading drugs into exosomes also need to be explored further because they are closely related to the stability and loading efficiency of the drugs. To date, there are three types of drug loading methods for exosomes: exogenous loading, endogenous loading, and liposome fusion loading [108]. Exogenous loading refers to the method that directly entraps the drugs inside an isolated exosome with simple incubation, sonication, electroporation, repeated freeze/thaw, and extrusion [112]. Simple incubation can be easily used, but the average loading efficiency of paclitaxel into EVs was below 10%. Sonication can elevate the average loading efficiency up to 28.29%, but affecting the loading amount of hydrophobic drugs by altering the membrane of the exosome is an issue [113]. Typically, the exogenous loading method is advantageous in maintaining the aqueous stability of drug-loaded exosomes over one month at 4 °C and 37 °C, but it is limited by relatively low loading efficiency. Endogenous loading refers to a method that entraps desired molecules in exosomes by modifying the cells of origin before the isolation of exosomes. For instance, treating host cells with chemical drugs, such as paclitaxel, can induce the release of exosomes loaded with paclitaxel [114]. For protein or gene delivery, host cells can be transfected with desired genes, which facilitates the release of exosomes with desired proteins or genes [115]. Although the endogenous loading method demonstrates a relatively high loading efficiency, it is difficult to quantify the amount of content inside the exosome and maintain high purity. Exosomes used in membrane protein engineering approaches protect their cargo proteins, but can be degraded by proteinase [116]. The liposome fusion method uses the hybridization of drug-loaded liposomes and exosomes by the freeze/melting process. This fusion method exhibits higher loading efficiency, especially for loading large plasmids, including CRISPR-Cas9 expressing vectors [117]. However, it is unclear whether this hybridized liposome-exosome can maintain the unique properties of exosomes [118]. Hence, further comprehensive evaluation of parameters such as targetability, half-life, and side effects for hybridized liposome-exosome is required.

3.2.2. Delivering Chemical Drugs via Exosomes for Oncotherapy

As many of chemical drugs can act after being internalized into cancer cells, they need to diffuse through the cell membrane to exert cytotoxicity, which is one of the factors reducing the efficacy of drugs [108]. In this aspect, exosomes can be potential candidates for delivering chemical drugs directly into the target cells. Many trials have been conducted to deliver chemical drugs such as paclitaxel, doxorubicin, cisplatin, and curcumin by packaging into exosomes for the treatment of various cancers (Table 2). For instance, paclitaxel-loaded exosomes isolated from MSCs, macrophages, and prostate cancer cells enhanced antitumor efficacy against pancreatic, breast, prostate, and Lewis lung carcinomas both in in vitro and in vivo studies [113,114,119,120]. Similarly, doxorubicin-loaded exosomes were also examined either by using the mechanical extrusion method to obtain higher drug loading efficiency [121] or surface engineering of exosomes to enhance targetability [122] for the treatment of colon and breast cancers, respectively. Treatment with cisplatin-loaded exosomes could prolong the survival rate of mice with ovarian cancer compared to the free cisplatin-treated group [123]. Pancreatic cancer-derived exosomes containing curcumin also effectively induced apoptosis in pancreatic cancer cells [124]. These studies show that by using exosomes as delivery vehicles, chemical drugs can be delivered more efficiently to target cells, which results in better outcomes.

Table 2. Exosomes as drug delivery vehicles for oncotherapy in preclinical studies.

Therapeutic Molecules	Exosome Origin	Targeted Disease	Reference
Chemical drugs			
Paclitaxel	Macrophage	Lewis lung carcinoma	[113]
	MSC	Pancreatic, breast cancer	[114,119]
	Prostate cancer cell	Prostate cancer	[120]
Droxorubicin	U937 RAW264.7	Colon cancer	[121]
	DCs expressing iRGD	Breast cancer	[122]
Cisplatin	Hepatocarcinoma cell	Hepatocarcinoma	[123]
Curcumin	Pancreatic cancer cell	Pancreatic cancer	[124]
Proteins			
TRIM3	Gastric cancer cell	Gastric cancer	[125]
CD-UPRT fusion protein	HEK293T	Schwannoma	[126]
TRAIL	K562	Lymphoma	[127]
MHC class I/peptide complex	DC	Breast cancer	[128]
HSP70	Myeloma cell	Myeloma	[129]
EGFR nanobodies	Myeloid leukemia cell	Epidermal carcinoma	[130]
SIRPα	Embryonic kidney cell	Colon cancer	[131]
miRNA			
miR-145-5p	MSC	Pancreatic cancer	[115]
Let-7a	HEK293T expressing GE11	Breast cancer with EGFR	[132]
miR-146b	MSC	Glioma	[133]
miR-122	MSC	Hepatocellular carcinoma	[134]
miR-335-5p	Stellate cell	Hepatocellular carcinoma	[135]
miR-379	MSC	Breast cancer	[136]
miR-25-3p inhibitor	Colorectal cancer cell	Colorectal cancer	[137]
siRNA			
PLK-1 siRNA	HEK293T + MSC	Bladder cancer	[138]
GRP78 siRNA	MSC	Hepatocellular carcinoma	[139]
HSP27 siRNA	Neuroblastoma cell	Neuroblastoma	[140]
mRNA			
Cas9 mRNA	Red blood cell	Breast cancer	[141]
PTEN mRNA	Mouse embryonic fibroblast serum	Glioma	[142]
ECRG4 mRNA	Neuroblastoma cell	Tongue carcinoma	[143]

MSC: mesenchymal stem cell; DC: dendritic cell; TRIM: tripartite motif-containing protein; CD: cytosine deaminase; UPRT: uracil phosphoribosyltransferase; TRAIL: TNF-related apoptosis-inducing ligand; MHC: major histocompatibility complex; HSP: heat shock protein; EGFR: epidermal growth factor receptor; SIRP: signal-regulatory protein; PLK: polo-like kinase; Cas: CRISPR associated protein; PTEN: phosphatase and tensin homolog; ECRG: esophageal cancer related gene.

3.2.3. Delivering Therapeutic Proteins via Exosomes for Oncotherapy

As the efficacy of many proteins is limited due to several barriers such as short half-life, low delivery rate, and induction of resistance, the use of appropriate delivery vehicles is one of the best ways to achieve successful protein drug therapies. As exosomes can protect the contents from various enzymes and the immune system, they can act as effective delivery vehicles for proteins. Proteins can be loaded either inside or on the surface of the exosome, based on the mechanism of action of the drugs. However, as therapeutic proteins are macromolecules, it is difficult to directly incorporate them into exosomes. Therefore, genetic modification of the cells of origin leading to the expression of therapeutic proteins in exosomes is usually preferred to prepare protein-loaded exosomes. Several preclinical studies regarding protein-loaded exosomes are summarized in Table 2. For example, the delivery of tripartite motif-containing protein 3 (TRIM3) using gastric cancer-derived exosomes successfully suppressed the proliferation, migration, and metastasis of gastric cancer [125]. Similarly, apoptosis-inducing proteins such as suicide-inducing fusion protein or TNF-related apoptosis-inducing ligand (TRAIL) were loaded into exosomes, and this method could elicit substantially reduced tumor growth in in vivo tumor models [126,127]. Some signaling-related proteins can be expressed on the surface of exosomes to improve tumor immunity, such as the major histocompatibility (MHC) class I/peptide complex [128]. In other studies, immunogenic proteins such as HSP70 were loaded onto exosomes, which resulted in enhanced antitumor T cell activity [129]. Some studies have shown that EGFR nanobodies anchored on exosomes via glycosylphosphatidylinositol (GPI) could bind to EGFR-expressing tumor cells with higher affinity [130].

3.2.4. Delivering RNA Drugs via Exosomes for Oncotherapy

Similar to therapeutic proteins, delivery vehicles are an essential component of successful gene therapy. Exosomes can protect genes from various enzymes, such as DNases and RNases, and can also directly deliver genes inside the cells, which enhances their therapeutic efficacy. Different types of RNAs such as mRNA, miRNA, and siRNA are promising candidates for the treatment of cancers. Studies on exosome RNA delivery are summarized in Table 2. miRNAs are non-coding RNAs involved in the regulation of gene expression. Pathophysiological conditions such as cancer are usually characterized by abnormal expression of certain types of miRNAs, which suggests that targeting miRNAs could be an effective way to treat cancer [144]. Treatment with exosomes overexpressing miR-122 showed substantially elevated chemosensitivity in hepatocellular carcinoma [134]. As the downregulation of miR-335-5p in both hepatocellular carcinoma and stellate cells acts as a pro-tumorigenic factor, delivery of miR-335-5p-overexpressing exosome exhibited substantial tumor shrinkage in an in vivo tumor model [135]. Similarly, treatment with miR-379-overexpressing exosomes significantly suppressed tumor growth in a T47D breast tumor model [136]. In addition, miR-145-5p overexpression inhibited the proliferation of pancreatic ductal adenocarcinoma and induced tumor cell apoptosis in an in vivo model [115]. In addition, specific miRNA inhibitors could act as potential drug candidates. For example, miR-25-3p is known to play an important role in facilitating colorectal cancer metastasis and promoting angiogenesis by targeting KLF (Kruppel-like factor)-2 and KLF-4, which implies that miR-25-3p can be a promising target for treating colorectal cancer. One study showed that treatment with exosomes loaded with miR-25-3p inhibitor considerably attenuated the tumor metastasis of colorectal cancer by balancing the level of miR-25-3p [137]. Silencing target genes using siRNA is another way to inhibit tumor growth. As overexpression of polo-like kinase (PLK)-1 is associated with the development of bladder cancer, one study showed that treatment with PLK-1 siRNA containing exosomes inhibited bladder cancer growth [138]. Similarly, as GRP78 overexpression is implicated in the growth and metastasis of hepatocellular carcinoma, treatment with GRP78 siRNA expressing exosomes resulted in an efficacious antitumor response in a sorafenib-resistant hepatocellular carcinoma model [139]. HSP27, a member of the heat-shock protein family, is known to promote neuron maturation and can be involved in the development of neuroblastoma. Treatment with Hsp27 siRNA-tagged exosome showed a significant reduction in tumor growth of the neuroblastoma cell line SH-SY5Y [140]. mRNA can be another candidate for anticancer therapy using exosome vehicles. Transferring CRISPR-associated protein (Cas) 9 mRNA-expressing exosomes from red blood cells induced miRNA inhibition and Cas9 genome editing effects in a breast cancer model [141]. Phosphatase and tensin homolog (PTEN) and esophageal cancer-related gene (ECRG) 4 are classified as tumor suppressors and are generally mutated in cancer cells. Therefore, treatment with exosomes expressing PTEN or ECRG4 mRNA could inhibit the growth of glioma cells [145] and tongue squamous cell carcinoma cells, respectively [143]. These studies show that depending on the target, different types of RNA therapeutics can be chosen, and the therapeutic efficacy of these drugs can be substantially enhanced by using exosomes as delivery vehicles.

3.3. Exosomes into Personalized Cancer Immunotherapy Drug Design (Single or in Combination)

Exosomes can be used as cell-free vaccines owing to the fact that exosomes derived from various donor cells, such as immune cells and cancer cells, are involved in fostering antitumor immunity [146]. DEXs can perform important immunostimulatory functions, as DCs that act as sentinel antigen-presenting cells play a crucial role in orchestrating cancer-specific adaptive immunity [147]. The surface of DEXs is characterized by various functional molecules for priming T cells such as MHC class I/II and costimulatory molecules including CD40, CD80, and CD86 [148]. This can foster antitumor immunity by inducing the activation of both innate and adaptive immunity. In several preclinical tests, treating DEXs could elicit antitumor effects and prolong the survival rate of tumor-bearing mice by expanding the repertoire of tumor-specific cytotoxic T cells as well as activating naïve T cells [149]. DEXs are also known to induce NK cell-mediated cytotoxicity to inhibit tumor growth [150]. In order to potentiate the efficacy of DEX as a therapeutic cancer vaccine, choosing an appropriate TAA and a relevant adjuvant is essential. To date, both for human and preclinical studies, only MHC class I/II binding peptides such as Epstein-Barr virus, melanoma-associated antigen, and melanoma antigen recognized by T cells-1 have been used for DEX vaccine [13,14,151]. However, the use of peptide-based DEX vaccine was not very effective in inducing antitumor effects due to modest activation of antitumor immune responses [152]. Several studies have reported that a more intense adaptive immune response can be induced with a DEX vaccine loaded with protein antigen [152,153]. This effect might be attributable to the presence of a broad range of epitopes with protein antigens, which might be more effective in activating various repertoires of tumor-specific cytotoxic T cells. These studies demonstrate the ability of personalized DEX vaccines based on patient tumor lysates. Therefore, a more potent DEX vaccine that can evoke strong antigen-specific responses can be manufactured with the loading of an allogenic protein antigen. Based on other studies, B cells are required to boost antitumor immunity, suggesting that epitopes, which activate B cells, are also needed for a successful DEX vaccine [152,153]. These results also provide a rationale for utilizing personalized protein antigens as cargo in DEX vaccines. General adjuvants such as interferon-γ and toll-like receptor agonists, including polyinosinic:polycytidylic acid and CpG oligodeoxynucleotides can be used to potentiate the efficacy of DEX vaccine. It is also known that the use of these adjuvants can result in the maturation of DCs and ultimately produce more immunogenic DEXs [154]. DEXs derived from mature DCs are known to express more costimulatory surface molecules including CD40, CD80, CD86, and intercellular adhesion molecule -1 and MHC class I/II [149]. Even though the DEX vaccine seems to show promising outcomes, several challenges remain until it can be widely used in clinics, as the research on this therapy is still at an early stage. For example, it is not clear whether the mass production of personalized DEX vaccine possessing a homogenous quality is available. At present, there are no clear guidelines regarding the production of exosome-based therapeutics. Proper storage, maintenance of stability, and route of administration for exosomes are also the issues to be considered [155].

Besides, CAR exosomes derived from effector CAR-T cells showed cytotoxic effects on cancer cells [156]. Intravenous injection of CAR exosomes into a mouse xenograft model exerted potent tumor growth inhibition. A combination of exosomal therapy with other immunotherapies such as immune-checkpoint blockers, cytokines, adoptive T cell transfer, and cancer vaccines might be a good way to elicit synergistic anticancer effects. For the combined utilization of CAR exosomes and CAR-T cells, further clinical/preclinical studies are required, and the clinically applicable scheme should be proposed.

4. Conclusions and Future Perspectives

Recently, cancer therapeutics has made great strides, and various clinical trials for targeted cancer therapy or immunotherapy have been conducted singly or in combination. However, there remains an unmet need, because only a few types of cancer patients are restrictedly responsive to current immune checkpoint blockers. In order to elevate response rate for personalized immunotherapy, prognostic biomarkers need to be established. As mentioned above, exosomal components can be used as diagnostic biomarkers in liquid biopsy, nanovehicles for delivery of anticancer drugs, and mediators between cells affecting tumor immunity. Detection of marker proteins or nucleic acids in circulating exosomes shows potential for predicting patients' clinical response [157]. "TEXs-on-chip" techniques using patient-derived tumor spheroids obtained from liquid biopsy will be applicable in the near future.

In that vaccinating autologous exosomes obtained from patients can avoid allograft reaction and carry tumor antigens, patient-derived exosomes have received attention as good candidates for personalized immunotherapy. Using bone marrow aspirate from patients, MSCs are isolated and expanded ex vivo, and then large scale of MSCs engineered with anticancer genes can be transplanted to patients for personalized treatment [158]. As exosomes released from MSCs acquire tropism toward tumor locations and the corresponding receptors with the original MSC, they would mediate anticancer activity [159]. Engineering techniques for enhancing therapeutic efficacy of these cell-free vaccines are needed for the development of innovative personalized immunotherapy.

Author Contributions: Conceptualization, J.U.C. and S.R.H.; methodology and validation, I.-K.P.; investigation, J.U.C.; writing—Original draft preparation, J.U.C.; writing—Review and editing, S.R.H.; visualization, Y.-K.L.; supervision, I.-K.P.; funding acquisition, Y.-K.L. and S.R.H. All authors have read and agreed to the published version of the manuscript.

Acknowledgments: The authors acknowledge KB BIOMED Inc. for administrative and technical support.

Abbreviations

ADC	Antibody-drug conjugate
ALIX	Apoptosis-linked gene 2-interacting protein X
ALK	Anaplastic lymphoma kinase
CAF	Cancer-associated fibroblasts
CAR	Chimeric antigen receptor
Cas	CRISPR-associated protein
CD	Cytosine deaminase
CSF1R	Colony-stimulating factor-1 receptor
CTC	Circulating-tumor cell
ctDNA	Circulating tumor DNA
CTLA4	Cytotoxic T-lymphocyte-associated protein 4
DC	Dendritic cell
DEX	Dendritic cell-derived exosome
ECM	Extracellular matrix
ECRG	Esophageal cancer-related gene
EGFR	Epidermal growth factor receptor
EpCAM	Epithelial cell adhesion molecule
ER	Endoplasmic reticulum
ESCRT	Endosomal sorting complex required for transport
EV	Extracellular vesicle
FGFR	Fibroblast growth factor receptor
FLT3	FMS-like tyrosine kinase 3
GM-CSF	Granulocyte-macrophage colony-stimulating factor
GPI	Glycosylphosphatidylinositol
HRS	Hepatocyte growth factor-regulated tyrosine kinase substrate
HSP	Heat shock protein
ICAM	Intercellular adhesion molecule
ILV	Intraluminal vesicle
KLF	Kruppel-like factor
lncRNA	Long non-coding RNA
MHC	Major histocompatibility complex
miRNA	MicroRNA
MMP	Matrix metalloproteinase
mRNA	Messenger RNA
MSC	Mesenchymal stem cell
MV	Microvesicle
MVB	Multivesicular body
NGS	Next-generation sequencing
NK	Natural killer
NSCLC	Non-small cell lung cancer

NTRK	Neurotrophic tyrosine receptor kinase
PCR	Polymerase chain reaction
PD-1	Programmed cell death protein 1
PD-L1	Programmed cell death ligand 1
PLK	Polo-like kinase
PTEN	Phosphatase and tensin homolog
ROS1	C-ros oncogene 1
RTK	Receptor tyrosine kinase
siRNA	Small interfering RNA
SIRP	Signal-regulatory protein
TAA	Tumor-associated antigen
TEX	Tumor-derived exosome
TMZ	Temozolomide
TRAIL	TNF-related apoptosis-inducing ligand
TRIM	Tripartite motif-containing protein
UPRT	Uracil phosphoribosyltransferase

References

1. Mullard, A. 2019 FDA drug approvals. *Nat. Rev. Drug Discov.* **2020**, *19*, 79–84. [CrossRef] [PubMed]
2. Zhang, Y.; Li, M.; Gao, X.; Chen, Y.; Liu, T. Nanotechnology in cancer diagnosis: Progress, challenges and opportunities. *J. Hematol. Oncol.* **2019**, *12*, 137. [CrossRef] [PubMed]
3. Jeon, S.M.; Kwon, J.W.; Choi, S.H.; Park, H.Y. Economic burden of lung cancer: A retrospective cohort study in South Korea, 2002-2015. *PLoS ONE* **2019**, *14*, e0212878. [CrossRef] [PubMed]
4. Yang, Y.; Miller, C.R.; Lopez-Beltran, A.; Montironi, R.; Cheng, M.; Zhang, S.; Koch, M.O.; Kaimakliotis, H.Z.; Cheng, L. Liquid Biopsies in the Management of Bladder Cancer: Next-Generation Biomarkers for Diagnosis, Surveillance, and Treatment-Response Prediction. *Crit. Rev. Oncog.* **2017**, *22*, 389–401. [CrossRef]
5. Melo, S.A.; Sugimoto, H.; O'Connell, J.T.; Kato, N.; Villanueva, A.; Vidal, A.; Qiu, L.; Vitkin, E.; Perelman, L.T.; Melo, C.A.; et al. Cancer exosomes perform cell-independent microRNA biogenesis and promote tumorigenesis. *Cancer Cell* **2014**, *26*, 707–721. [CrossRef]
6. Kamerkar, S.; LeBleu, V.S.; Sugimoto, H.; Yang, S.; Ruivo, C.F.; Melo, S.A.; Lee, J.J.; Kalluri, R. Exosomes facilitate therapeutic targeting of oncogenic KRAS in pancreatic cancer. *Nature* **2017**, *546*, 498–503. [CrossRef]
7. Jan, M.; Majeti, R. Clonal evolution of acute leukemia genomes. *Oncogene* **2013**, *32*, 135–140. [CrossRef]
8. Turner, N.C.; Reis-Filho, J.S. Genetic heterogeneity and cancer drug resistance. *Lancet. Oncol.* **2012**, *13*, e178–e185. [CrossRef]
9. Waldmann, T.A. Immunotherapy: Past, present and future. *Nat. Med.* **2003**, *9*, 269–277. [CrossRef]
10. Chauhan, V.P.; Stylianopoulos, T.; Boucher, Y.; Jain, R.K. Delivery of molecular and nanoscale medicine to tumors: Transport barriers and strategies. *Annu. Rev. Chem. Biomol. Eng.* **2011**, *2*, 281–298. [CrossRef]
11. Au, J.L.; Yeung, B.Z.; Wientjes, M.G.; Lu, Z.; Wientjes, M.G. Delivery of cancer therapeutics to extracellular and intracellular targets: Determinants, barriers, challenges and opportunities. *Adv. Drug. Deliv. Rev.* **2016**, *97*, 280–301. [CrossRef] [PubMed]
12. Mendez-Ferrer, S.; Bonnet, D.; Steensma, D.P.; Hasserjian, R.P.; Ghobrial, I.M.; Gribben, J.G.; Andreeff, M.; Krause, D.S. Bone marrow niches in haematological malignancies. *Nat. Rev. Cancer* **2020**, *20*, 285–298. [CrossRef] [PubMed]
13. Escudier, B.; Dorval, T.; Chaput, N.; Andre, F.; Caby, M.P.; Novault, S.; Flament, C.; Leboulaire, C.; Borg, C.; Amigorena, S.; et al. Vaccination of metastatic melanoma patients with autologous dendritic cell (DC) derived-exosomes: Results of thefirst phase I clinical trial. *J Transl. Med.* **2005**, *3*, 10. [CrossRef] [PubMed]
14. Morse, M.A.; Garst, J.; Osada, T.; Khan, S.; Hobeika, A.; Clay, T.M.; Valente, N.; Shreeniwas, R.; Sutton, M.A.; Delcayre, A.; et al. A phase I study of dexosome immunotherapy in patients with advanced non-small cell lung cancer. *J. Transl. Med.* **2005**, *3*, 9. [CrossRef] [PubMed]
15. Dai, S.; Wei, D.; Wu, Z.; Zhou, X.; Wei, X.; Huang, H.; Li, G. Phase I clinical trial of autologous ascites-derived exosomes combined with GM-CSF for colorectal cancer. *Mol. Ther.* **2008**, *16*, 782–790. [CrossRef] [PubMed]

16. Hessvik, N.P.; Llorente, A. Current knowledge on exosome biogenesis and release. *Cell Mol. Life Sci.* **2018**, *75*, 193–208. [CrossRef]
17. Huotari, J.; Helenius, A. Endosome maturation. *EMBO J.* **2011**, *30*, 3481–3500. [CrossRef]
18. Raposo, G.; Stoorvogel, W. Extracellular vesicles: Exosomes, microvesicles, and friends. *J. Cell Biol.* **2013**, *200*, 373–383. [CrossRef]
19. Wollert, T.; Hurley, J.H. Molecular mechanism of multivesicular body biogenesis by ESCRT complexes. *Nature* **2010**, *464*, 864–869. [CrossRef]
20. Henne, W.M.; Stenmark, H.; Emr, S.D. Molecular mechanisms of the membrane sculpting ESCRT pathway. *Cold Spring Harb. Perspect. Biol.* **2013**, *5*, a016766. [CrossRef]
21. Glickman, M.H.; Ciechanover, A. The ubiquitin-proteasome proteolytic pathway: Destruction for the sake of construction. *Physiol. Rev.* **2002**, *82*, 373–428. [CrossRef]
22. Mani, A.; Gelmann, E.P. The ubiquitin-proteasome pathway and its role in cancer. *J. Clin. Oncol.* **2005**, *23*, 4776–4789. [CrossRef]
23. Clifford, S.C.; Cockman, M.E.; Smallwood, A.C.; Mole, D.R.; Woodward, E.R.; Maxwell, P.H.; Ratcliffe, P.J.; Maher, E.R. Contrasting effects on HIF-1alpha regulation by disease-causing pVHL mutations correlate with patterns of tumourigenesis in von Hippel-Lindau disease. *Hum. Mol. Genet.* **2001**, *10*, 1029–1038. [CrossRef]
24. Hashizume, R.; Fukuda, M.; Maeda, I.; Nishikawa, H.; Oyake, D.; Yabuki, Y.; Ogata, H.; Ohta, T. The RING heterodimer BRCA1-BARD1 is a ubiquitin ligase inactivated by a breast cancer-derived mutation. *J. Biol. Chem.* **2001**, *276*, 14537–14540. [CrossRef] [PubMed]
25. Stuffers, S.; Sem Wegner, C.; Stenmark, H.; Brech, A. Multivesicular endosome biogenesis in the absence of ESCRTs. *Traffic* **2009**, *10*, 925–937. [CrossRef] [PubMed]
26. Wubbolts, R.; Leckie, R.S.; Veenhuizen, P.T.; Schwarzmann, G.; Mobius, W.; Hoernschemeyer, J.; Slot, J.W.; Geuze, H.J.; Stoorvogel, W. Proteomic and biochemical analyses of human B cell-derived exosomes. Potential implications for their function and multivesicular body formation. *J. Biol. Chem.* **2003**, *278*, 10963–10972. [CrossRef]
27. Trajkovic, K.; Hsu, C.; Chiantia, S.; Rajendran, L.; Wenzel, D.; Wieland, F.; Schwille, P.; Brugger, B.; Simons, M. Ceramide triggers budding of exosome vesicles into multivesicular endosomes. *Science* **2008**, *319*, 1244–1247. [CrossRef] [PubMed]
28. Mobius, W.; Ohno-Iwashita, Y.; van Donselaar, E.G.; Oorschot, V.M.; Shimada, Y.; Fujimoto, T.; Heijnen, H.F.; Geuze, H.J.; Slot, J.W. Immunoelectron microscopic localization of cholesterol using biotinylated and non-cytolytic perfringolysin O. *J. Histochem. Cytochem.* **2002**, *50*, 43–55. [CrossRef]
29. Baietti, M.F.; Zhang, Z.; Mortier, E.; Melchior, A.; Degeest, G.; Geeraerts, A.; Ivarsson, Y.; Depoortere, F.; Coomans, C.; Vermeiren, E.; et al. Syndecan-syntenin-ALIX regulates the biogenesis of exosomes. *Nat. Cell Biol.* **2012**, *14*, 677–685. [CrossRef]
30. Geminard, C.; De Gassart, A.; Blanc, L.; Vidal, M. Degradation of AP2 during reticulocyte maturation enhances binding of hsc70 and Alix to a common site on TFR for sorting into exosomes. *Traffic* **2004**, *5*, 181–193. [CrossRef]
31. Tamai, K.; Tanaka, N.; Nakano, T.; Kakazu, E.; Kondo, Y.; Inoue, J.; Shiina, M.; Fukushima, K.; Hoshino, T.; Sano, K.; et al. Exosome secretion of dendritic cells is regulated by Hrs, an ESCRT-0 protein. *Biochem. Biophys. Res. Commun.* **2010**, *399*, 384–390. [CrossRef] [PubMed]
32. Batagov, A.O.; Kuznetsov, V.A.; Kurochkin, I.V. Identification of nucleotide patterns enriched in secreted RNAs as putative cis-acting elements targeting them to exosome nano-vesicles. *BMC Genomics.* **2011**, *12*, S3–S18. [CrossRef] [PubMed]
33. Gibbings, D.J.; Ciaudo, C.; Erhardt, M.; Voinnet, O. Multivesicular bodies associate with components of miRNA effector complexes and modulate miRNA activity. *Nat. Cell. Biol.* **2009**, *11*, 1143–1149. [CrossRef] [PubMed]
34. Spaull, R.; McPherson, B.; Gialeli, A.; Clayton, A.; Uney, J.; Heep, A.; Cordero-Llana, O. Exosomes populate the cerebrospinal fluid of preterm infants with post-haemorrhagic hydrocephalus. *Int J. Dev. Neurosci.* **2019**, *73*, 59–65. [CrossRef] [PubMed]
35. Huleihel, L.; Hussey, G.S.; Naranjo, J.D.; Zhang, L.; Dziki, J.L.; Turner, N.J.; Stolz, D.B.; Badylak, S.F. Matrix-bound nanovesicles within ECM bioscaffolds. *Sci. Adv.* **2016**, *2*, e1600502. [CrossRef]

36. Zhuang, X.; Xiang, X.; Grizzle, W.; Sun, D.; Zhang, S.; Axtell, R.C.; Ju, S.; Mu, J.; Zhang, L.; Steinman, L. Treatment of brain inflammatory diseases by delivering exosome encapsulated anti-inflammatory drugs from the nasal region to the brain. *Mol. Ther.* **2011**, *19*, 1769–1779. [CrossRef]
37. Sung, B.H.; von Lersner, A.; Guerrero, J.; Krystofiak, E.S.; Inman, D.; Pelletier, R.; Zijlstra, A.; Ponik, S.M.; Weaver, A.M. A live cell reporter of exosome secretion and uptake reveals pathfinding behavior of migrating cells. *Nat. Commun.* **2020**, *11*, 2092. [CrossRef]
38. Tian, T.; Zhu, Y.L.; Zhou, Y.Y.; Liang, G.F.; Wang, Y.Y.; Hu, F.H.; Xiao, Z.D. Exosome uptake through clathrin—mediated endocytosis and macropinocytosis and mediating miR-21 delivery. *J. Biol. Chem.* **2014**, *289*, 22258–22267. [CrossRef]
39. Fitzner, D.; Schnaars, M.; van Rossum, D.; Krishnamoorthy, G.; Dibaj, P.; Bakhti, M.; Regen, T.; Hanisch, U.K.; Simons, M. Selective transfer of exosomes from oligodendrocytes to microglia by macropinocytosis. *J. Cell Sci.* **2011**, *124*, 447–458. [CrossRef]
40. Nanbo, A.; Kawanishi, E.; Yoshida, R.; Yoshiyama, H. Exosomes derived from Epstein-Barr virus-infected cells are internalized via caveola-dependent endocytosis and promote phenotypic modulation in target cells. *J. Virol.* **2013**, *87*, 10334–10347. [CrossRef]
41. Valapala, M.; Vishwanatha, J.K. Lipid raft endocytosis and exosomal transport facilitate extracellular trafficking of annexin A2. *J. Biol. Chem.* **2011**, *286*, 30911–30925. [CrossRef] [PubMed]
42. Bonsergent, E.; Lavieu, G. Content release of extracellular vesicles in a cell-free extract. *FEBS Lett.* **2019**, *593*, 1983–1992. [CrossRef] [PubMed]
43. Mincheva-Nilsson, L.; Baranov, V. Cancer exosomes and NKG2D receptor-ligand interactions: Impairing NKG2D-mediated cytotoxicity and anti-tumour immune surveillance. *Semin. Cancer Biol.* **2014**, *28*, 24–30. [CrossRef]
44. Thery, C.; Zitvogel, L.; Amigorena, S. Exosomes: Composition, biogenesis and function. *Nat. Rev. Immunol.* **2002**, *2*, 569–579. [CrossRef] [PubMed]
45. Imai, T.; Takahashi, Y.; Nishikawa, M.; Kato, K.; Morishita, M.; Yamashita, T.; Matsumoto, A.; Charoenviriyakul, C.; Takakura, Y. Macrophage-dependent clearance of systemically administered B16BL6-derived exosomes from the blood circulation in mice. *J. Extracell. Vesicles* **2015**, *4*, 26238. [CrossRef]
46. Maas, S.L.N.; Breakefield, X.O.; Weaver, A.M. Extracellular Vesicles: Unique Intercellular Delivery Vehicles. *Trends. Cell Biol.* **2017**, *27*, 172–188. [CrossRef] [PubMed]
47. Oushy, S.; Hellwinkel, J.E.; Wang, M.; Nguyen, G.J.; Gunaydin, D.; Harland, T.A.; Anchordoquy, T.J.; Graner, M.W. Glioblastoma multiforme-derived extracellular vesicles drive normal astrocytes towards a tumour-enhancing phenotype. *Philos. Trans. R. Soc. Lond. B. Biol. Sci.* **2018**, *373*, 200160477. [CrossRef]
48. Syn, N.; Wang, L.; Sethi, G.; Thiery, J.P.; Goh, B.C. Exosome-Mediated Metastasis: From Epithelial-Mesenchymal Transition to Escape from Immunosurveillance. *Trends. Pharmacol. Sci.* **2016**, *37*, 606–617. [CrossRef]
49. Sato, S.; Weaver, A.M. Extracellular vesicles: Important collaborators in cancer progression. *Essays. Biochem.* **2018**, *62*, 149–163.
50. Hoshino, D.; Kirkbride, K.C.; Costello, K.; Clark, E.S.; Sinha, S.; Grega-Larson, N.; Tyska, M.J.; Weaver, A.M. Exosome secretion is enhanced by invadopodia and drives invasive behavior. *Cell Rep.* **2013**, *5*, 1159–1168. [CrossRef]
51. Baroni, S.; Romero-Cordoba, S.; Plantamura, I.; Dugo, M.; D'ippolito, E.; Cataldo, A.; Cosentino, G.; Angeloni, V.; Rossini, A.; Daidone, M. Exosome-mediated delivery of miR-9 induces cancer-associated fibroblast-like properties in human breast fibroblasts. *Cell Death Dis.* **2016**, *7*, e2312. [CrossRef] [PubMed]
52. Webber, J.P.; Spary, L.K.; Sanders, A.J.; Chowdhury, R.; Jiang, W.G.; Steadman, R.; Wymant, J.; Jones, A.T.; Kynaston, H.; Mason, M.D.; et al. Differentiation of tumour-promoting stromal myofibroblasts by cancer exosomes. *Oncogene* **2015**, *34*, 290–302. [CrossRef] [PubMed]
53. Webber, J.; Steadman, R.; Mason, M.D.; Tabi, Z.; Clayton, A. Cancer exosomes trigger fibroblast to myofibroblast differentiation. *Cancer Res.* **2010**, *70*, 9621–9630. [CrossRef] [PubMed]
54. Luga, V.; Zhang, L.; Viloria-Petit, A.M.; Ogunjimi, A.A.; Inanlou, M.R.; Chiu, E.; Buchanan, M.; Hosein, A.N.; Basik, M.; Wrana, J.L. Exosomes mediate stromal mobilization of autocrine Wnt-PCP signaling in breast cancer cell migration. *Cell* **2012**, *151*, 1542–1556. [CrossRef] [PubMed]

55. Shimoda, M.; Principe, S.; Jackson, H.W.; Luga, V.; Fang, H.; Molyneux, S.D.; Shao, Y.W.; Aiken, A.; Waterhouse, P.D.; Karamboulas, C. Loss of the Timp gene family is sufficient for the acquisition of the CAF-like cell state. *Nat. Cell Biol.* **2014**, *16*, 889–901. [CrossRef] [PubMed]
56. Lazar, I.; Clement, E.; Dauvillier, S.; Milhas, D.; Ducoux-Petit, M.; LeGonidec, S.; Moro, C.; Soldan, V.; Dalle, S.; Balor, S. Adipocyte exosomes promote melanoma aggressiveness through fatty acid oxidation: A novel mechanism linking obesity and cancer. *Cancer Res.* **2016**, *76*, 4051–4057. [CrossRef] [PubMed]
57. Ribeiro, M.F.; Zhu, H.; Millard, R.W.; Fan, G.C. Exosomes Function in Pro- and Anti-Angiogenesis. *Curr. Angiogenes* **2013**, *2*, 54–59.
58. Carrasco-Ramírez, P.; Greening, D.W.; Andrés, G.; Gopal, S.K.; Martín-Villar, E.; Renart, J.; Simpson, R.J.; Quintanilla, M. Podoplanin is a component of extracellular vesicles that reprograms cell-derived exosomal proteins and modulates lymphatic vessel formation. *Oncotarget* **2016**, *7*, 16070. [CrossRef]
59. Umezu, T.; Ohyashiki, K.; Kuroda, M.; Ohyashiki, J. Leukemia cell to endothelial cell communication via exosomal miRNAs. *Oncogene* **2013**, *32*, 2747–2755. [CrossRef]
60. Schillaci, O.; Fontana, S.; Monteleone, F.; Taverna, S.; Di Bella, M.A.; Di Vizio, D.; Alessandro, R. Exosomes from metastatic cancer cells transfer amoeboid phenotype to non-metastatic cells and increase endothelial permeability: Their emerging role in tumor heterogeneity. *Sci. Rep.* **2017**, *7*, 1–15. [CrossRef]
61. Hsu, Y.L.; Hung, J.Y.; Chang, W.A.; Lin, Y.S.; Pan, Y.C.; Tsai, P.H.; Wu, C.Y.; Kuo, P.L. Hypoxic lung cancer-secreted exosomal miR-23a increased angiogenesis and vascular permeability by targeting prolyl hydroxylase and tight junction protein ZO-1. *Oncogene* **2017**, *36*, 4929–4942. [CrossRef] [PubMed]
62. Liu, Y.; Cao, X. Characteristics and Significance of the Pre-metastatic Niche. *Cancer Cell* **2016**, *30*, 668–681. [CrossRef] [PubMed]
63. Peinado, H.; Alečković, M.; Lavotshkin, S.; Matei, I.; Costa-Silva, B.; Moreno-Bueno, G.; Hergueta-Redondo, M.; Williams, C.; García-Santos, G.; Ghajar, C.M. Melanoma exosomes educate bone marrow progenitor cells toward a pro-metastatic phenotype through MET. *Nat. Med.* **2012**, *18*, 883–891. [CrossRef] [PubMed]
64. Costa-Silva, B.; Aiello, N.M.; Ocean, A.J.; Singh, S.; Zhang, H.; Thakur, B.K.; Becker, A.; Hoshino, A.; Mark, M.T.; Molina, H. Pancreatic cancer exosomes initiate pre-metastatic niche formation in the liver. *Nat. cell Biol.* **2015**, *17*, 816–826. [CrossRef]
65. Hoshino, A.; Costa-Silva, B.; Shen, T.-L.; Rodrigues, G.; Hashimoto, A.; Mark, M.T.; Molina, H.; Kohsaka, S.; Di Giannatale, A.; Ceder, S. Tumour exosome integrins determine organotropic metastasis. *Nature* **2015**, *527*, 329–335. [CrossRef]
66. Hood, J.L.; San, R.S.; Wickline, S.A. Exosomes released by melanoma cells prepare sentinel lymph nodes for tumor metastasis. *Cancer Res.* **2011**, *71*, 3792–3801. [CrossRef]
67. Plebanek, M.P.; Angeloni, N.L.; Vinokour, E.; Li, J.; Henkin, A.; Martinez-Marin, D.; Filleur, S.; Bhowmick, R.; Henkin, J.; Miller, S.D.; et al. Pre-metastatic cancer exosomes induce immune surveillance by patrolling monocytes at the metastatic niche. *Nat. Commun.* **2017**, *8*, 1319. [CrossRef]
68. Whiteside, T.L. Exosomes and tumor-mediated immune suppression. *J. Clin. Investig.* **2016**, *126*, 1216–1223. [CrossRef]
69. Clayton, A.; Mitchell, J.P.; Court, J.; Mason, M.D.; Tabi, Z. Human tumor-derived exosomes selectively impair lymphocyte responses to interleukin-2. *Cancer Res.* **2007**, *67*, 7458–7466. [CrossRef]
70. Clayton, A.; Mitchell, J.P.; Court, J.; Linnane, S.; Mason, M.D.; Tabi, Z. Human tumor-derived exosomes down-modulate NKG2D expression. *J. Immunol.* **2008**, *180*, 7249–7258. [CrossRef]
71. Zitvogel, L.; Regnault, A.; Lozier, A.; Wolfers, J.; Flament, C.; Tenza, D.; Ricciardi-Castagnoli, P.; Raposo, G.; Amigorena, S. Eradication of established murine tumors using a novel cell-free vaccine: Dendritic cell-derived exosomes. *Nat. Med.* **1998**, *4*, 594–600. [CrossRef] [PubMed]
72. Skokos, D.; Botros, H.G.; Demeure, C.; Morin, J.; Peronet, R.; Birkenmeier, G.; Boudaly, S.; Mecheri, S. Mast cell-derived exosomes induce phenotypic and functional maturation of dendritic cells and elicit specific immune responses in vivo. *J. Immunol.* **2003**, *170*, 3037–3045. [CrossRef] [PubMed]
73. Chen, W.X.; Liu, X.M.; Lv, M.M.; Chen, L.; Zhao, J.H.; Zhong, S.L.; Ji, M.H.; Hu, Q.; Luo, Z.; Wu, J.Z.; et al. Exosomes from drug-resistant breast cancer cells transmit chemoresistance by a horizontal transfer of microRNAs. *PLoS ONE* **2014**, *9*, e95240. [CrossRef] [PubMed]

74. Zeng, A.; Wei, Z.; Yan, W.; Yin, J.; Huang, X.; Zhou, X.; Li, R.; Shen, F.; Wu, W.; Wang, X. Exosomal transfer of miR-151a enhances chemosensitivity to temozolomide in drug-resistant glioblastoma. *Cancer Lett.* **2018**, *436*, 10–21. [CrossRef] [PubMed]
75. Federici, C.; Petrucci, F.; Caimi, S.; Cesolini, A.; Logozzi, M.; Borghi, M.; D'Ilio, S.; Lugini, L.; Violante, N.; Azzarito, T.; et al. Exosome release and low pH belong to a framework of resistance of human melanoma cells to cisplatin. *PLoS ONE* **2014**, *9*, e88193. [CrossRef] [PubMed]
76. Hu, Y.; Yan, C.; Mu, L.; Huang, K.; Li, X.; Tao, D.; Wu, Y.; Qin, J. Fibroblast-Derived Exosomes Contribute to Chemoresistance through Priming Cancer Stem Cells in Colorectal Cancer. *PLoS ONE* **2015**, *10*, e0125625. [CrossRef]
77. Boelens, M.C.; Wu, T.J.; Nabet, B.Y.; Xu, B.; Qiu, Y.; Yoon, T.; Azzam, D.J.; Twyman-Saint Victor, C.; Wiemann, B.Z.; Ishwaran, H.; et al. Exosome transfer from stromal to breast cancer cells regulates therapy resistance pathways. *Cell* **2014**, *159*, 499–513. [CrossRef]
78. Thery, C.; Amigorena, S.; Raposo, G.; Clayton, A. Isolation and characterization of exosomes from cell culture supernatants and biological fluids. *Curr. Protoc. Cell Biol.* **2006**, *30*, 3.22.1–3.22.29. [CrossRef]
79. Tauro, B.J.; Greening, D.W.; Mathias, R.A.; Ji, H.; Mathivanan, S.; Scott, A.M.; Simpson, R.J. Comparison of ultracentrifugation, density gradient separation, and immunoaffinity capture methods for isolating human colon cancer cell line LIM1863-derived exosomes. *Methods* **2012**, *56*, 293–304. [CrossRef]
80. Taylor, D.D.; Shah, S. Methods of isolating extracellular vesicles impact down-stream analyses of their cargoes. *Methods* **2015**, *87*, 3–10. [CrossRef]
81. Riahi, R.; Shaegh, S.A.; Ghaderi, M.; Zhang, Y.S.; Shin, S.R.; Aleman, J.; Massa, S.; Kim, D.; Dokmeci, M.R.; Khademhosseini, A. Automated microfluidic platform of bead-based electrochemical immunosensor integrated with bioreactor for continual monitoring of cell secreted biomarkers. *Sci. Rep.* **2016**, *6*, 24598. [CrossRef] [PubMed]
82. Kaddour, H.; Lyu, Y.; Shouman, N.; Mohan, M.; Okeoma, C.M. Development of Novel High-Resolution Size-Guided Turbidimetry-Enabled Particle Purification Liquid Chromatography (PPLC): Extracellular Vesicles and Membraneless Condensates in Focus. *Int. J. Mol. Sci* **2020**, *21*, 5361. [CrossRef] [PubMed]
83. Doldan, X.; Fagundez, P.; Cayota, A.; Laiz, J.; Tosar, J.P. Electrochemical Sandwich Immunosensor for Determination of Exosomes Based on Surface Marker-Mediated Signal Amplification. *Anal. Chem.* **2016**, *88*, 10466–10473. [CrossRef] [PubMed]
84. Miller, M.C.; Doyle, G.V.; Terstappen, L.W. Significance of Circulating Tumor Cells Detected by the CellSearch System in Patients with Metastatic Breast Colorectal and Prostate Cancer. *J. Oncol.* **2010**, *2010*, 617421. [CrossRef] [PubMed]
85. Yu, M.; Stott, S.; Toner, M.; Maheswaran, S.; Haber, D.A. Circulating tumor cells: Approaches to isolation and characterization. *J. Cell Biol.* **2011**, *192*, 373–382. [CrossRef]
86. Wan, J.C.M.; Massie, C.; Garcia-Corbacho, J.; Mouliere, F.; Brenton, J.D.; Caldas, C.; Pacey, S.; Baird, R.; Rosenfeld, N. Liquid biopsies come of age: Towards implementation of circulating tumour DNA. *Nat. Rev. Cancer* **2017**, *17*, 223–238. [CrossRef] [PubMed]
87. Heitzer, E.; Ulz, P.; Geigl, J.B. Circulating tumor DNA as a liquid biopsy for cancer. *Clin. Chem.* **2015**, *61*, 112–123. [CrossRef]
88. Figueroa, J.M.; Skog, J.; Akers, J.; Li, H.; Komotar, R.; Jensen, R.; Ringel, F.; Yang, I.; Kalkanis, S.; Thompson, R.; et al. Detection of wild-type EGFR amplification and EGFRvIII mutation in CSF-derived extracellular vesicles of glioblastoma patients. *Neuro. Oncol.* **2017**, *19*, 1494–1502. [CrossRef]
89. Pisitkun, T.; Shen, R.F.; Knepper, M.A. Identification and proteomic profiling of exosomes in human urine. *Proc. Natl. Acad. Sci. USA* **2004**, *101*, 13368–13373. [CrossRef]
90. Chung, I.M.; Rajakumar, G.; Venkidasamy, B.; Subramanian, U.; Thiruvengadam, M. Exosomes: Current use and future applications. *Clin. Chim. Acta.* **2020**, *500*, 226–232. [CrossRef]
91. Jalalian, S.H.; Ramezani, M.; Jalalian, S.A.; Abnous, K.; Taghdisi, S.M. Exosomes, new biomarkers in early cancer detection. *Anal. Biochem.* **2019**, *571*, 1–13. [CrossRef] [PubMed]
92. Logozzi, M.; De Milito, A.; Lugini, L.; Borghi, M.; Calabro, L.; Spada, M.; Perdicchio, M.; Marino, M.L.; Federici, C.; Iessi, E.; et al. High levels of exosomes expressing CD63 and caveolin-1 in plasma of melanoma patients. *PLoS ONE* **2009**, *4*, e5219. [CrossRef] [PubMed]

93. Qu, J.L.; Qu, X.J.; Zhao, M.F.; Teng, Y.E.; Zhang, Y.; Hou, K.Z.; Jiang, Y.H.; Yang, X.H.; Liu, Y.P. Gastric cancer exosomes promote tumour cell proliferation through PI3K/Akt and MAPK/ERK activation. *Dig. Liver. Dis.* **2009**, *41*, 875–880. [CrossRef] [PubMed]
94. Yang, L.; Wu, X.H.; Wang, D.; Luo, C.L.; Chen, L.X. Bladder cancer cell-derived exosomes inhibit tumor cell apoptosis and induce cell proliferation in vitro. *Mol. Med. Rep.* **2013**, *8*, 1272–1278. [CrossRef] [PubMed]
95. Sento, S.; Sasabe, E.; Yamamoto, T. Application of a Persistent Heparin Treatment Inhibits the Malignant Potential of Oral Squamous Carcinoma Cells Induced by Tumor Cell-Derived Exosomes. *PLoS ONE* **2016**, *11*, e0148454. [CrossRef] [PubMed]
96. Smalley, D.M.; Sheman, N.E.; Nelson, K.; Theodorescu, D. Isolation and identification of potential urinary microparticle biomarkers of bladder cancer. *J. Proteome Res.* **2008**, *7*, 2088–2096. [CrossRef]
97. Nilsson, J.; Skog, J.; Nordstrand, A.; Baranov, V.; Mincheva-Nilsson, L.; Breakefield, X.O.; Widmark, A. Prostate cancer-derived urine exosomes: A novel approach to biomarkers for prostate cancer. *Br. J. Cancer* **2009**, *100*, 1603–1607. [CrossRef]
98. Skog, J.; Wurdinger, T.; van Rijn, S.; Meijer, D.H.; Gainche, L.; Sena-Esteves, M.; Curry, W.T., Jr.; Carter, B.S.; Krichevsky, A.M.; Breakefield, X.O. Glioblastoma microvesicles transport RNA and proteins that promote tumour growth and provide diagnostic biomarkers. *Nat. Cell Biol.* **2008**, *10*, 1470–1476. [CrossRef]
99. Khan, S.; Jutzy, J.M.; Valenzuela, M.M.; Turay, D.; Aspe, J.R.; Ashok, A.; Mirshahidi, S.; Mercola, D.; Lilly, M.B.; Wall, N.R. Plasma-derived exosomal survivin, a plausible biomarker for early detection of prostate cancer. *PLoS ONE* **2012**, *7*, e46737. [CrossRef]
100. McKiernan, J.; Donovan, M.J.; Margolis, E.; Partin, A.; Carter, B.; Brown, G.; Torkler, P.; Noerholm, M.; Skog, J.; Shore, N.; et al. A Prospective Adaptive Utility Trial to Validate Performance of a Novel Urine Exosome Gene Expression Assay to Predict High-grade Prostate Cancer in Patients with Prostate-specific Antigen 2–10 ng/mL at Initial Biopsy. *Eur. Urol.* **2018**, *74*, 731–738. [CrossRef]
101. Li, Q.; Shao, Y.; Zhang, X.; Zheng, T.; Miao, M.; Qin, L.; Wang, B.; Ye, G.; Xiao, B.; Guo, J. Plasma long noncoding RNA protected by exosomes as a potential stable biomarker for gastric cancer. *Tumour. Biol.* **2015**, *36*, 2007–2012. [CrossRef] [PubMed]
102. Taylor, D.D.; Gercel-Taylor, C. MicroRNA signatures of tumor-derived exosomes as diagnostic biomarkers of ovarian cancer. *Gynecol. Oncol.* **2008**, *110*, 13–21. [CrossRef] [PubMed]
103. Rabinowits, G.; Gercel-Taylor, C.; Day, J.M.; Taylor, D.D.; Kloecker, G.H. Exosomal microRNA: A diagnostic marker for lung cancer. *Clin. Lung Cancer* **2009**, *10*, 42–46. [CrossRef] [PubMed]
104. Allenson, K.; Castillo, J.; San Lucas, F.A.; Scelo, G.; Kim, D.U.; Bernard, V.; Davis, G.; Kumar, T.; Katz, M.; Overman, M.J.; et al. High prevalence of mutant KRAS in circulating exosome-derived DNA from early-stage pancreatic cancer patients. *Ann. Oncol.* **2017**, *28*, 741–747. [CrossRef]
105. Castellanos-Rizaldos, E.; Grimm, D.G.; Tadigotla, V.; Hurley, J.; Healy, J.; Neal, P.L.; Sher, M.; Venkatesan, R.; Karlovich, C.; Raponi, M.; et al. Exosome-Based Detection of EGFR T790M in Plasma from Non-Small Cell Lung Cancer Patients. *Clin. Cancer Res.* **2018**, *24*, 2944–2950. [CrossRef]
106. Yang, S.; Che, S.P.; Kurywchak, P.; Tavormina, J.L.; Gansmo, L.B.; Correa de Sampaio, P.; Tachezy, M.; Bockhorn, M.; Gebauer, F.; Haltom, A.R.; et al. Detection of mutant KRAS and TP53 DNA in circulating exosomes from healthy individuals and patients with pancreatic cancer. *Cancer Biol. Ther.* **2017**, *18*, 158–165. [CrossRef]
107. Garcia-Silva, S.; Benito-Martin, A.; Sanchez-Redondo, S.; Hernandez-Barranco, A.; Ximenez-Embun, P.; Nogues, L.; Mazariegos, M.S.; Brinkmann, K.; Amor Lopez, A.; Meyer, L.; et al. Use of extracellular vesicles from lymphatic drainage as surrogate markers of melanoma progression and BRAF (V600E) mutation. *J. Exp. Med.* **2019**, *216*, 1061–1070. [CrossRef]
108. Zhao, X.; Wu, D.; Ma, X.; Wang, J.; Hou, W.; Zhang, W. Exosomes as drug carriers for cancer therapy and challenges regarding exosome uptake. *Biomed. Pharmacother.* **2020**, *128*, 110237. [CrossRef]
109. Hu, Q.; Su, H.; Li, J.; Lyon, C.; Tang, W.; Wan, M.; Hu, T.Y. Clinical applications of exosome membrane proteins. *Precis. Clin. Med.* **2020**, *3*, 54–66. [CrossRef]
110. Johnsen, K.B.; Gudbergsson, J.M.; Skov, M.N.; Pilgaard, L.; Moos, T.; Duroux, M. A comprehensive overview of exosomes as drug delivery vehicles—endogenous nanocarriers for targeted cancer therapy. *Biochim. Biophys. Acta.* **2014**, *1846*, 75–87. [CrossRef]

111. Jaiswal, S.; Jamieson, C.H.; Pang, W.W.; Park, C.Y.; Chao, M.P.; Majeti, R.; Traver, D.; van Rooijen, N.; Weissman, I.L. CD47 is upregulated on circulating hematopoietic stem cells and leukemia cells to avoid phagocytosis. *Cell* **2009**, *138*, 271–285. [CrossRef] [PubMed]
112. Haney, M.J.; Klyachko, N.L.; Zhao, Y.; Gupta, R.; Plotnikova, E.G.; He, Z.; Patel, T.; Piroyan, A.; Sokolsky, M.; Kabanov, A.V.; et al. Exosomes as drug delivery vehicles for Parkinson's disease therapy. *J. Control. Release* **2015**, *207*, 18–30. [CrossRef] [PubMed]
113. Kim, M.S.; Haney, M.J.; Zhao, Y.; Mahajan, V.; Deygen, I.; Klyachko, N.L.; Inskoe, E.; Piroyan, A.; Sokolsky, M.; Okolie, O.; et al. Development of exosome-encapsulated paclitaxel to overcome MDR in cancer cells. *Nanomedicine* **2016**, *12*, 655–664. [CrossRef] [PubMed]
114. Pascucci, L.; Cocce, V.; Bonomi, A.; Ami, D.; Ceccarelli, P.; Ciusani, E.; Vigano, L.; Locatelli, A.; Sisto, F.; Doglia, S.M.; et al. Paclitaxel is incorporated by mesenchymal stromal cells and released in exosomes that inhibit in vitro tumor growth: A new approach for drug delivery. *J. Control. Release* **2014**, *192*, 262–270. [CrossRef] [PubMed]
115. Ding, Y.; Cao, F.; Sun, H.; Wang, Y.; Liu, S.; Wu, Y.; Cui, Q.; Mei, W.; Li, F. Exosomes derived from human umbilical cord mesenchymal stromal cells deliver exogenous miR-145-5p to inhibit pancreatic ductal adenocarcinoma progression. *Cancer Lett.* **2019**, *442*, 351–361. [CrossRef]
116. Sterzenbach, U.; Putz, U.; Low, L.H.; Silke, J.; Tan, S.S.; Howitt, J. Engineered Exosomes as Vehicles for Biologically Active Proteins. *Mol. Ther.* **2017**, *25*, 1269–1278. [CrossRef]
117. Sato, Y.T.; Umezaki, K.; Sawada, S.; Mukai, S.A.; Sasaki, Y.; Harada, N.; Shiku, H.; Akiyoshi, K. Engineering hybrid exosomes by membrane fusion with liposomes. *Sci Rep.* **2016**, *6*, 21933. [CrossRef]
118. Lee, J.; Kim, J.; Jeong, M.; Lee, H.; Goh, U.; Kim, H.; Kim, B.; Park, J.H. Liposome-based engineering of cells to package hydrophobic compounds in membrane vesicles for tumor penetration. *Nano Lett.* **2015**, *15*, 2938–2944. [CrossRef]
119. Kalimuthu, S.; Gangadaran, P.; Rajendran, R.L.; Zhu, L.; Oh, J.M.; Lee, H.W.; Gopal, A.; Baek, S.H.; Jeong, S.Y.; Lee, S.W.; et al. A New Approach for Loading Anticancer Drugs Into Mesenchymal Stem Cell-Derived Exosome Mimetics for Cancer Therapy. *Front. Pharmacol.* **2018**, *9*, 1116. [CrossRef]
120. Saari, H.; Lázaro-Ibáñez, E.; Viitala, T.; Vuorimaa-Laukkanen, E.; Siljander, P.; Yliperttula, M. Microvesicle-and exosome-mediated drug delivery enhances the cytotoxicity of Paclitaxel in autologous prostate cancer cells. *J. Control. Release* **2015**, *220*, 727–737. [CrossRef]
121. Jang, S.C.; Kim, O.Y.; Yoon, C.M.; Choi, D.S.; Roh, T.Y.; Park, J.; Nilsson, J.; Lotvall, J.; Kim, Y.K.; Gho, Y.S. Bioinspired exosome-mimetic nanovesicles for targeted delivery of chemotherapeutics to malignant tumors. *ACS Nano* **2013**, *7*, 7698–7710. [CrossRef] [PubMed]
122. Tian, Y.; Li, S.; Song, J.; Ji, T.; Zhu, M.; Anderson, G.J.; Wei, J.; Nie, G. A doxorubicin delivery platform using engineered natural membrane vesicle exosomes for targeted tumor therapy. *Biomaterials* **2014**, *35*, 2383–2390. [CrossRef] [PubMed]
123. Tang, K.; Zhang, Y.; Zhang, H.; Xu, P.; Liu, J.; Ma, J.; Lv, M.; Li, D.; Katirai, F.; Shen, G.X.; et al. Delivery of chemotherapeutic drugs in tumour cell-derived microparticles. *Nat. Commun.* **2012**, *3*, 1282. [CrossRef] [PubMed]
124. Osterman, C.J.; Lynch, J.C.; Leaf, P.; Gonda, A.; Ferguson Bennit, H.R.; Griffiths, D.; Wall, N.R. Curcumin Modulates Pancreatic Adenocarcinoma Cell-Derived Exosomal Function. *PLoS ONE* **2015**, *10*, e0132845. [CrossRef] [PubMed]
125. Fu, H.; Yang, H.; Zhang, X.; Wang, B.; Mao, J.; Li, X.; Wang, M.; Zhang, B.; Sun, Z.; Qian, H.; et al. Exosomal TRIM3 is a novel marker and therapy target for gastric cancer. *J. Exp. Clin. Cancer Res.* **2018**, *37*, 162. [CrossRef] [PubMed]
126. Mizrak, A.; Bolukbasi, M.F.; Ozdener, G.B.; Brenner, G.J.; Madlener, S.; Erkan, E.P.; Strobel, T.; Breakefield, X.O.; Saydam, O. Genetically engineered microvesicles carrying suicide mRNA/protein inhibit schwannoma tumor growth. *Mol. Ther.* **2013**, *21*, 101–108. [CrossRef]
127. Rivoltini, L.; Chiodoni, C.; Squarcina, P.; Tortoreto, M.; Villa, A.; Vergani, B.; Burdek, M.; Botti, L.; Arioli, I.; Cova, A.; et al. TNF-Related Apoptosis-Inducing Ligand (TRAIL)-Armed Exosomes Deliver Proapoptotic Signals to Tumor Site. *Clin. Cancer Res.* **2016**, *22*, 3499–3512. [CrossRef]

128. Andre, F.; Chaput, N.; Schartz, N.E.; Flament, C.; Aubert, N.; Bernard, J.; Lemonnier, F.; Raposo, G.; Escudier, B.; Hsu, D.H.; et al. Exosomes as potent cell-free peptide-based vaccine. I. Dendritic cell-derived exosomes transfer functional MHC class I/peptide complexes to dendritic cells. *J. Immunol.* **2004**, *172*, 2126–2136. [CrossRef]

129. Xie, Y.; Bai, O.; Zhang, H.; Yuan, J.; Zong, S.; Chibbar, R.; Slattery, K.; Qureshi, M.; Wei, Y.; Deng, Y.; et al. Membrane-bound HSP70-engineered myeloma cell-derived exosomes stimulate more efficient CD8(+) CTL- and NK-mediated antitumour immunity than exosomes released from heat-shocked tumour cells expressing cytoplasmic HSP70. *J. Cell Mol. Med.* **2010**, *14*, 2655–2666. [CrossRef]

130. Kooijmans, S.A.; Aleza, C.G.; Roffler, S.R.; van Solinge, W.W.; Vader, P.; Schiffelers, R.M. Display of GPI-anchored anti-EGFR nanobodies on extracellular vesicles promotes tumour cell targeting. *J. Extracell. Vesicles* **2016**, *5*, 31053. [CrossRef]

131. Koh, E.; Lee, E.J.; Nam, G.H.; Hong, Y.; Cho, E.; Yang, Y.; Kim, I.S. Exosome-SIRPalpha, a CD47 blockade increases cancer cell phagocytosis. *Biomaterials* **2017**, *121*, 121–129. [CrossRef] [PubMed]

132. Ohno, S.; Takanashi, M.; Sudo, K.; Ueda, S.; Ishikawa, A.; Matsuyama, N.; Fujita, K.; Mizutani, T.; Ohgi, T.; Ochiya, T.; et al. Systemically injected exosomes targeted to EGFR deliver antitumor microRNA to breast cancer cells. *Mol. Ther.* **2013**, *21*, 185–191. [CrossRef]

133. Katakowski, M.; Buller, B.; Zheng, X.; Lu, Y.; Rogers, T.; Osobamiro, O.; Shu, W.; Jiang, F.; Chopp, M. Exosomes from marrow stromal cells expressing miR-146b inhibit glioma growth. *Cancer Lett.* **2013**, *335*, 201–204. [CrossRef] [PubMed]

134. Lou, G.; Song, X.; Yang, F.; Wu, S.; Wang, J.; Chen, Z.; Liu, Y. Exosomes derived from miR-122-modified adipose tissue-derived MSCs increase chemosensitivity of hepatocellular carcinoma. *J. Hematol. Oncol.* **2015**, *8*, 122. [CrossRef] [PubMed]

135. Wang, F.; Li, L.; Piontek, K.; Sakaguchi, M.; Selaru, F.M. Exosome miR-335 as a novel therapeutic strategy in hepatocellular carcinoma. *Hepatology* **2018**, *67*, 940–954. [CrossRef] [PubMed]

136. O'Brien, K.P.; Khan, S.; Gilligan, K.E.; Zafar, H.; Lalor, P.; Glynn, C.; O'Flatharta, C.; Ingoldsby, H.; Dockery, P.; De Bhulbh, A.; et al. Employing mesenchymal stem cells to support tumor-targeted delivery of extracellular vesicle (EV)-encapsulated microRNA-379. *Oncogene* **2018**, *37*, 2137–2149. [CrossRef] [PubMed]

137. Zeng, Z.; Li, Y.; Pan, Y.; Lan, X.; Song, F.; Sun, J.; Zhou, K.; Liu, X.; Ren, X.; Wang, F.; et al. Cancer-derived exosomal miR-25-3p promotes pre-metastatic niche formation by inducing vascular permeability and angiogenesis. *Nat. Commun.* **2018**, *9*, 5395. [CrossRef]

138. Greco, K.A.; Franzen, C.A.; Foreman, K.E.; Flanigan, R.C.; Kuo, P.C.; Gupta, G.N. PLK-1 Silencing in Bladder Cancer by siRNA Delivered With Exosomes. *Urology* **2016**, *91*, e1–e7. [CrossRef]

139. Li, H.; Yang, C.; Shi, Y.; Zhao, L. Exosomes derived from siRNA against GRP78 modified bone-marrow-derived mesenchymal stem cells suppress Sorafenib resistance in hepatocellular carcinoma. *J. Nanobiotechnol.* **2018**, *16*, 103. [CrossRef]

140. Shokrollahi, E.; Nourazarian, A.; Rahbarghazi, R.; Salimi, L.; Karbasforush, S.; Khaksar, M.; Salarinasab, S.; Abhari, A.; Heidarzadeh, M. Treatment of human neuroblastoma cell line SH-SY5Y with HSP27 siRNA tagged-exosomes decreased differentiation rate into mature neurons. *J. Cell Physiol.* **2019**, *234*, 21005–21013. [CrossRef]

141. Usman, W.M.; Pham, T.C.; Kwok, Y.Y.; Vu, L.T.; Ma, V.; Peng, B.; Chan, Y.S.; Wei, L.; Chin, S.M.; Azad, A.; et al. Efficient RNA drug delivery using red blood cell extracellular vesicles. *Nat. Commun.* **2018**, *9*, 2359. [CrossRef] [PubMed]

142. Yang, Z.; Shi, J.; Xie, J.; Wang, Y.; Sun, J.; Liu, T.; Zhao, Y.; Zhao, X.; Wang, X.; Ma, Y.; et al. Large-scale generation of functional mRNA-encapsulating exosomes via cellular nanoporation. *Nat. Biomed. Eng.* **2020**, *4*, 69–83. [CrossRef] [PubMed]

143. Mao, L.; Li, X.; Gong, S.; Yuan, H.; Jiang, Y.; Huang, W.; Sun, X.; Dang, X. Serum exosomes contain ECRG4 mRNA that suppresses tumor growth via inhibition of genes involved in inflammation, cell proliferation, and angiogenesis. *Cancer Gene Ther.* **2018**, *25*, 248–259. [CrossRef] [PubMed]

144. Ell, B.; Mercatali, L.; Ibrahim, T.; Campbell, N.; Schwarzenbach, H.; Pantel, K.; Amadori, D.; Kang, Y. Tumor-induced osteoclast miRNA changes as regulators and biomarkers of osteolytic bone metastasis. *Cancer Cell* **2013**, *24*, 542–556. [CrossRef] [PubMed]

145. Yang, L.; Zhang, L.; Lu, L.; Wang, Y. lncRNA UCA1 Increases Proliferation and Multidrug Resistance of Retinoblastoma Cells Through Downregulating miR-513a-5p. *DNA Cell Biol.* **2020**, *39*, 69–77. [CrossRef] [PubMed]
146. Syn, N.L.; Wang, L.; Chow, E.K.; Lim, C.T.; Goh, B.C. Exosomes in Cancer Nanomedicine and Immunotherapy: Prospects and Challenges. *Trends Biotechnol.* **2017**, *35*, 665–676. [CrossRef] [PubMed]
147. Munich, S.; Sobo-Vujanovic, A.; Buchser, W.J.; Beer-Stolz, D.; Vujanovic, N.L. Dendritic cell exosomes directly kill tumor cells and activate natural killer cells via TNF superfamily ligands. *Oncoimmunology* **2012**, *1*, 1074–1083. [CrossRef]
148. Viaud, S.; Terme, M.; Flament, C.; Taieb, J.; Andre, F.; Novault, S.; Escudier, B.; Robert, C.; Caillat-Zucman, S.; Tursz, T.; et al. Dendritic cell-derived exosomes promote natural killer cell activation and proliferation: A role for NKG2D ligands and IL-15Ralpha. *PLoS ONE* **2009**, *4*, e4942. [CrossRef]
149. Viaud, S.; Ploix, S.; Lapierre, V.; Thery, C.; Commere, P.H.; Tramalloni, D.; Gorrichon, K.; Virault-Rocroy, P.; Tursz, T.; Lantz, O.; et al. Updated technology to produce highly immunogenic dendritic cell-derived exosomes of clinical grade: A critical role of interferon-gamma. *J. Immunother.* **2011**, *34*, 65–75. [CrossRef]
150. Damo, M.; Wilson, D.S.; Simeoni, E.; Hubbell, J.A. TLR-3 stimulation improves anti-tumor immunity elicited by dendritic cell exosome-based vaccines in a murine model of melanoma. *Sci. Rep.* **2015**, *5*, 17622. [CrossRef]
151. Besse, B.; Charrier, M.; Lapierre, V.; Dansin, E.; Lantz, O.; Planchard, D.; Le Chevalier, T.; Livartoski, A.; Barlesi, F.; Laplanche, A.; et al. Dendritic cell-derived exosomes as maintenance immunotherapy after first line chemotherapy in NSCLC. *Oncoimmunology* **2016**, *5*, e1071008. [CrossRef] [PubMed]
152. Naslund, T.I.; Gehrmann, U.; Qazi, K.R.; Karlsson, M.C.; Gabrielsson, S. Dendritic cell-derived exosomes need to activate both T and B cells to induce antitumor immunity. *J. Immunol.* **2013**, *190*, 2712–2719. [CrossRef] [PubMed]
153. Hiltbrunner, S.; Larssen, P.; Eldh, M.; Martinez-Bravo, M.J.; Wagner, A.K.; Karlsson, M.C.; Gabrielsson, S. Exosomal cancer immunotherapy is independent of MHC molecules on exosomes. *Oncotarget* **2016**, *7*, 38707–38717. [CrossRef] [PubMed]
154. Pitt, J.M.; Andre, F.; Amigorena, S.; Soria, J.C.; Eggermont, A.; Kroemer, G.; Zitvogel, L. Dendritic cell-derived exosomes for cancer therapy. *J. Clin. Investig.* **2016**, *126*, 1224–1232. [CrossRef] [PubMed]
155. Srinivasan, S.; Vannberg, F.O.; Dixon, J.B. Lymphatic transport of exosomes as a rapid route of information dissemination to the lymph node. *Sci. Rep.* **2016**, *6*, 24436. [CrossRef] [PubMed]
156. Fu, W.; Lei, C.; Liu, S.; Cui, Y.; Wang, C.; Qian, K.; Li, T.; Shen, Y.; Fan, X.; Lin, F.; et al. CAR exosomes derived from effector CAR-T cells have potent antitumour effects and low toxicity. *Nat. Commun.* **2019**, *10*, 4355. [CrossRef]
157. Vafaizadeh, V.; Barekati, Z. Immuno-Oncology Biomarkers for Personalized Immunotherapy in Breast Cancer. *Front. Cell Dev. Biol.* **2020**, *8*, 162. [CrossRef]
158. Dai, L.J.; Moniri, M.R.; Zeng, Z.R.; Zhou, J.X.; Rayat, J.; Warnock, G.L. Potential implications of mesenchymal stem cells in cancer therapy. *Cancer Lett.* **2011**, *305*, 8–20. [CrossRef]
159. Kim, H.S.; Choi, D.Y.; Yun, S.J.; Choi, S.M.; Kang, J.W.; Jung, J.W.; Hwang, D.; Kim, K.P.; Kim, D.W. Proteomic analysis of microvesicles derived from human mesenchymal stem cells. *J. Proteome Res.* **2012**, *11*, 839–849. [CrossRef]

13

Small Extracellular Vesicles Isolated from Serum May Serve as Signal-Enhancers for the Monitoring of CNS Tumors

Gabriella Dobra [1,2], Matyas Bukva [1,2], Zoltan Szabo [3], Bella Bruszel [3], Maria Harmati [1], Edina Gyukity-Sebestyen [1], Adrienn Jenei [4], Monika Szucs [5,6], Peter Horvath [1], Tamas Biro [7], Almos Klekner [4] and Krisztina Buzas [1,8,9,*]

1 Laboratory of Microscopic Image Analysis and Machine Learning, Institute of Biochemistry, Biological Research Centre, H-6726 Szeged, Hungary; dobragab@yahoo.co.uk (G.D.); bukvamatyas@gmail.com (M.B.); harmatimarcsi@gmail.com (M.H.); e.gyukity.sebestyen@gmail.com (E.G.-S.); horvath.peter@brc.hu (P.H.)

2 Department of Medical Genetics, Doctoral School of Interdisciplinary Medicine, University of Szeged, H-6720 Szeged, Hungary

3 Department of Medical Chemistry, Faculty of Medicine, University of Szeged, H-6720 Szeged, Hungary; szabo.zoltan@med.u-szeged.hu (Z.S.); bruszel.bella@med.u-szeged.hu (B.B.)

4 Department of Neurosurgery, Clinical Centre, University of Debrecen, H-4032 Debrecen, Hungary; jenei.adrienn@med.unideb.hu (A.J.); aklekner@yahoo.com (A.K.)

5 Department of Medical Physics and Informatics, Faculty of Medicine, University of Szeged, H-6720 Szeged, Hungary; szucs.monika@med.u-szeged.hu

6 Department of Medical Physics and Informatics, Faculty of Science and Informatics, University of Szeged, H-6720 Szeged, Hungary

7 Department of Immunology, Faculty of Medicine, University of Debrecen, H-4032 Debrecen, Hungary; biro.lcmp@gmail.com

8 Department of Immunology, Faculty of Medicine, University of Szeged, H-6720 Szeged, Hungary

9 Department of Immunology, Faculty of Science and Informatics, University of Szeged, H-6720 Szeged, Hungary

* Correspondence: kr.buzas@gmail.com;

Abstract: Liquid biopsy-based methods to test biomarkers (e.g., serum proteins and extracellular vesicles) may help to monitor brain tumors. In this proteomics-based study, we aimed to identify a characteristic protein fingerprint associated with central nervous system (CNS) tumors. Overall, 96 human serum samples were obtained from four patient groups, namely glioblastoma multiforme (GBM), non-small-cell lung cancer brain metastasis (BM), meningioma (M) and lumbar disc hernia patients (CTRL). After the isolation and characterization of small extracellular vesicles (sEVs) by nanoparticle tracking analysis (NTA) and atomic force microscopy (AFM), liquid chromatography -mass spectrometry (LC-MS) was performed on two different sample types (whole serum and serum sEVs). Statistical analyses (ratio, Cohen's d, receiver operating characteristic; ROC) were carried out to compare patient groups. To recognize differences between the two sample types, pairwise comparisons (Welch's test) and ingenuity pathway analysis (IPA) were performed. According to our knowledge, this is the first study that compares the proteome of whole serum and serum-derived sEVs. From the 311 proteins identified, 10 whole serum proteins and 17 sEV proteins showed the highest intergroup differences. Sixty-five proteins were significantly enriched in sEV samples, while 129 proteins were significantly depleted compared to whole serum. Based on principal component analysis (PCA) analyses, sEVs are more suitable to discriminate between the patient groups. Our results support that sEVs have greater potential to monitor CNS tumors, than whole serum.

Keywords: extracellular vesicles; cancer biomarker; proteomics

1. Introduction

According to the World Health Organization (WHO), cancer is the second leading cause of death, accounting for an estimated 9.6 million cases in 2018. Globally, 1 in 6 deaths is due to cancer [1]. The cancer burden continues to grow worldwide, exerting tremendous physical, emotional and financial strain on individuals, families, communities and on health systems [2].

The diagnosis of central nervous system (CNS) tumors is based on CT and MRI scans, as well as on the histopathological analysis of samples obtained by biopsy or via surgical resection. However, these procedures are highly invasive, uncomfortable for the patient, bear a considerable risk of complications and provide limited information on tumor status. Therefore, biomarkers appropriate for monitoring disease progression and response to treatment are eagerly required. While repeated MRI scans serve as the standard method to follow patients, it has little prognostic value for long-term recurrence [3]. Thus, neuro-oncological research aims to identify novel biomarkers suitable for monitoring CNS tumors in clinical practice [4].

Liquid biopsy is in the spotlight of biomarker-focused research, as body fluids are easily accessible sources of biomarkers and are available with minimally invasive and low cost sampling procedures. Also, multiple sampling allows the monitoring of disease progression and therapeutic response [5]. Every cell, including neoplastic cells, release molecular markers into the circulation. Tumor-derived biomarkers include proteins, nucleic acids, circulating tumor cells, platelets and tumor-derived extracellular vesicles that accumulate in urine, cerebrospinal fluid, saliva and blood [6].

Blood is the most easily accessible source for biomarkers, thus it is frequently used to assess disease status in malignancies such as prostate, liver and ovarian cancers based on the serum concentrations of PSA, AFP and CA125, respectively. In neuro-oncology, blood-based biomarkers are mainly used to evaluate toxicity and safety of treatments to guide patient management. For example, myelosuppression is a common risk associated with temozolomide treatment and radiotherapy, thus standard practice dictates weekly tests of complete blood count, including whole blood cell differential and platelet counts during definitive chemoradiotherapy [7]. Finding biomarkers for blood-based CNS tumor monitoring is more challenging, as the blood-brain barrier (BBB) prevents the release of tumor-related biomarkers into peripheral blood. However, it would have outstanding benefits in clinical patient management, thus efforts to identify blood based biomarkers, including proteins, nucleic acids, circulating tumor cells and extracellular vesicles are currently in the forefront of neuro-oncological research [8].

Extracellular vesicles (EVs) are promising cancer biomarkers accessible via liquid biopsy, because they are cell-secreted, nano-sized and stably exist in all types of body fluids. EVs contain a sample of the cytosolic milieu, including an abundance of DNA, RNA, proteins and other analytes, while externally they also resemble their cell of origin [9]. EVs are small, lipid bilayer-enclosed vesicles released by both cancer and non-cancerous cells into the extracellular space [10].

EVs secreted by cancer cells communicate with neighboring stromal cells or even with cells at distant sites, inducing an alteration of the cell program [11,12]. Pre-metastatic niche formation has been shown in several tumors, for example, in pancreatic, lung, colorectal and ovarian cancers [13–16]. Also, EVs may be taken up by immune cells, leading to immunosuppression [17]. More recently, EVs have even gained a role in cancer diagnosis and therapy [18–20] as biomarker molecules that may be identified in different primary tumors with high sensitivity and specificity [21]. Regarding pancreatic cancer, Kalluri and colleagues found that glypican-1 (GPC1), a cell surface proteoglycan, is specifically enriched in circulating exosomes (30–200 nm endosome-derived EVs). GPC1 is suitable to differentiate early- and late-stage pancreatic cancer from benign diseases of the pancreas, with an accuracy of 100% [22]. The available evidence also supports that tumor-derived EVs can cross the BBB [23,24], however, currently no clinically relevant EV biomarkers are accepted for the monitoring of CNS tumors.

Several studies report on gene or protein expression analyses of CNS tumor tissue (specifically, glioblastoma), allowing to identify biomarkers that could be secreted into the blood and thus could be detected from serum samples. Recent studies have aimed to identify one and two specific biomarkers for the reliable evaluation of actual tumor status [25–27] but none of these proteins alone was found to be sufficiently specific and sensitive to serve as a monitoring marker.

Regarding that previous attempts to find surrogate serum markers for brain tumors have failed when based on a single or only few candidate factors, we made an attempt to identify a characteristic protein fingerprint of 10–20 candidate markers associated with CNS tumors.

For this purpose, 96 serum samples were collected from four patient groups according to the criteria of the National Ethical Committee and proteomics analysis was performed using liquid chromatography and mass spectrometry (LC-MS). The serum samples were obtained from patients diagnosed with the two most common types of brain tumors [28], namely malignant glioblastoma multiforme (GBM) and benign meningioma (M), as well as from patients with a prevalent brain metastasis [29] originating from non-small-cell lung cancer (BM). Patients with lumbar disc herniation served as controls (CTRL). Following a statistical selection, these four patient groups were compared with respect to the identified proteins. In parallel, small extracellular vesicles (sEVs) were isolated from the serum samples by differential centrifugation and proteomics and statistical analyses were also performed on these sEV samples, allowing to compare the suitability of these two different sample types. According to the best of our knowledge, this is the first study that compares the proteome of whole serum and serum-derived sEV samples. Results from the proteomics analysis indicate that using a protein fingerprint of serum-derived sEVs instead of analyzing whole serum increases the accuracy of distinguishing between the clinical samples, that is, between the patient groups. Our results support that sEVs have a greater potential for the proteomics-based monitoring of CNS tumors compared to whole serum analysis.

2. Results

2.1. EV Samples Show sEV Properties with Similar Concentration and Size Distribution in the Different Patient Groups

To verify the value of circulating extracellular vesicles as potential biomarkers for CNS tumors, EVs were isolated from the serum of patients with glioblastoma multiforme (GBM), single brain metastasis originating from non-small-cell lung cancer (BM) and meningioma (M), as well as from control patients with lumbar disc herniation (CTRL). Each group included 24 individuals of both genders with various ages. Extracellular vesicles were isolated from the sera by differential centrifugation and were characterized by atomic force microscopy (AFM) and nanoparticle tracking analysis (NTA). Pools of 6 samples were formed in all groups, allowing four parallel samples to be tested per group (see in Section 4.1). Western blot analyses were also performed to demonstrate the EV nature (Figure S1).

EVs were divided into subtypes based on their size range, separating small EVs (sEVs) and medium/large EVs (m/lEVs) [30]. AFM analysis revealed that the small EV subtype includes various structures. Mean and mode diameters of the particles, represented by an average of the 16 sample pools, were measured as 112 nm and 86 nm by NTA, respectively (Figure 1A).

The quantitative characterization of serum sEVs by NTA (Figure 1B) revealed no significant differences between the four patient groups regarding the size and concentration of circulating sEVs. However, within the groups high individual differences were observed in the measured parameters of the sEVs.

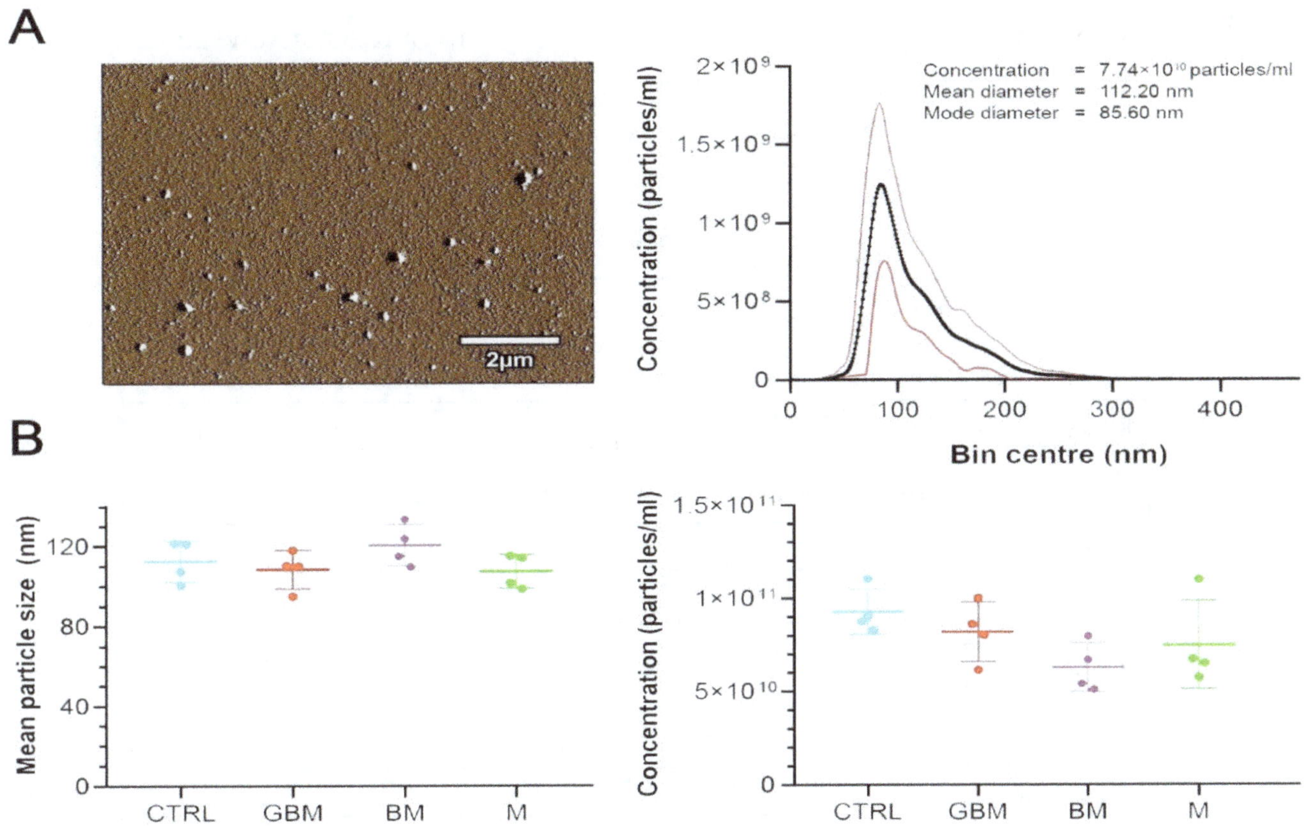

Figure 1. Characterization and quantitative properties of the small extracellular vesicle (sEV) samples. (**A**) Atomic force microscopy (AFM) image of sEV isolates displays vesicles with diameters within the range of 50–140 nm. The diagram shows the size distribution of the 96 sEV samples isolated from the serum, presenting the mean +/−95% CI values measured by nanoparticle tracking analysis (NTA). (**B**) Dot plots show the number and size distribution of small extracellular vesicles (sEVs) displayed in mean size (left) and concentration (right) values for each sample pools (4 samples/group).

2.2. Statistical Analysis of LC-MS Data Reveals Characteristic Proteomic Fingerprints for Each Patient Group and Informs on the Suitability of the Two Different Sample Types in Distinguishing CNS Tumors

We aimed to identify the differences between the four patient groups to reveal the characteristic protein profiles associated with the CNS tumors in point. Using an intensity ratio of >2 or <0.5 with Cohen's d effect size of 2 as a cut-off, we investigated which proteins show reliable intensity difference and which proteins can separate at least one group from the others based on a receiver operating characteristic (ROC) analysis. Moreover, utilizing principal component analysis (PCA) with k-means clustering, we were able to compare the suitability of the two different sample types to distinguish between the CNS tumors in point. Figure 2 shows the flowchart of LC-MS data processing and the results of the statistical analyses.

Proteomics analyses by LC-MS (Step 1) were performed on whole serum and sEV samples obtained from patients with GBM, BM, M and CTRL. Individual samples (n = 24) in each group were arranged into 4 pools (see in Section 4.1) to eliminate individual variances, reduce sample number, shorten the time of LC-MS measurements and reduce the need for materials. The Data independent acquisition (DIA) mode constructed spectral library revealed 311 proteins (see Table S1). Based on Pearson's correlation analyses (Step 2), one of the sEV control samples had to be excluded from further statistical analyses (Table S2). After excluding unreliable proteins, as well as proteins with missing values (Step 3), a total of 262 proteins remained for the final analysis.

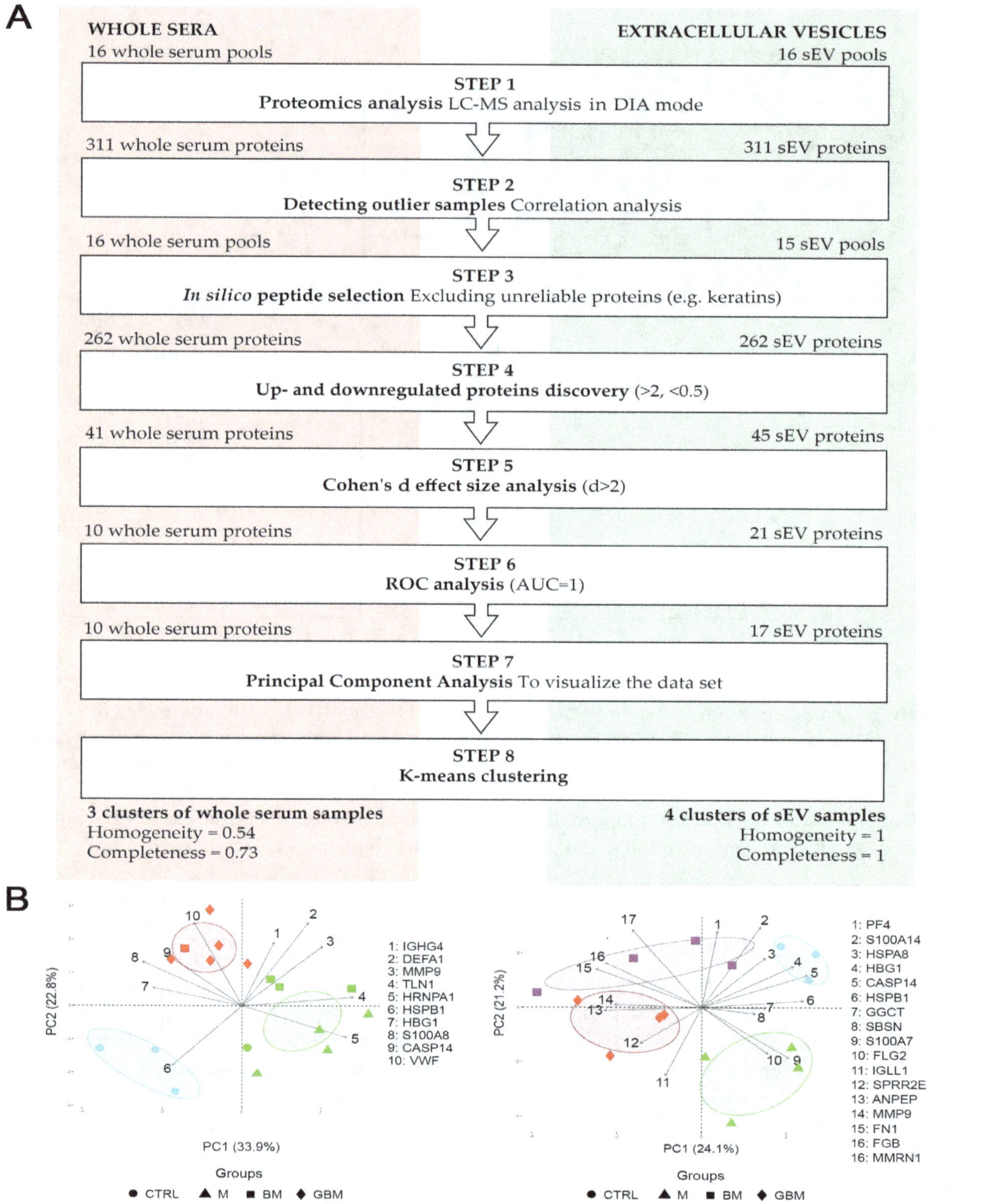

Figure 2. Statistical analysis of the proteome of whole serum (left) and sEV samples (right). (**A**) The flowchart shows the steps of selecting the proteins revealed by liquid chromatography and mass spectrometry (LC-MS) (**B**) The diagrams visualize the results of the principal component analysis (PCA) and k-means clustering. X and Y axes of PCA biplots show principal component 1 (PC1) and principal component 2 (PC2) with explained variances. Arrows represent the coefficients of each protein for PC1 versus the coefficients for PC2, showing the significance of each protein in influencing PCs. Different dots represent the 4 patient groups. Colors indicate the clusters formed by k-means clustering; ellipses indicating the 95% confidence interval were constructed around the barycenters of the clusters.

Following basic processing, up- and down-regulated protein discovery (Step 4) resulted in 41 whole serum proteins and 45 sEV proteins. In addition to comparing each CNS tumor group to CTRL, between-group differences among the CNS tumor groups were also assessed in the protein selection process. As clinically relevant incidence is an important consideration for selecting the proteins identified, Cohen's d effect size was adopted as an indicator of between-groups difference. The Cohen's d effect size analysis (Step 5) with a threshold of $d > 2$ yielded 10 and 21 proteins in the whole serum and sEV samples, respectively. In the ROC analyses (Step 6) 10 whole serum proteins (MMP9, HSPB1, CASP14, HBG1, IGHG4, DEFA1, VWF, HNRNPA1, S100A8, TLN1) and 17 sEV proteins (MMP9, HSPB1, CASP14, HBG1, FGB, GGCT, PF4, S100A7, FN1, ANPEP, FLG2, HSPA8, IGLL1, MMRN1, S100A14, SBSN, SPRR2E) were found to meet the AUC = 1 selection criteria. Table S3 includes the UniProt ID, Gene symbol, ratio of intensity means > 2 or < 0.5 and Cohen's d effect size > 2 parameters for the selected proteins. The two sample groups shared four significantly altered proteins (highlighted in Table S3), namely MMP9, CASP14, HBG1 and HSPB1.

Following protein selection, PCA (Step 7) was performed to visualize the dataset, where several potentially correlated proteins were projected into a smaller number of variables. K-means clustering (Step 8) on the whole serum PCA biplot resulted in 3 inhomogeneous or incomplete clusters. Calculated cluster homogeneity and completeness scores are 0.56 and 0.73, respectively. In contrast to whole serum samples, the clustering of sEV samples formed homogeneous and complete clusters, with homogeneity and completeness scores of 1. The results of the PCA analyses and k-means clustering indicate considerable differences between the whole serum and sEV samples (Figure 2B). We found that the accuracy of distinguishing between various CNS tumors can be increased using a protein panel from serum-derived sEVs, compared to analyzing whole serum samples.

2.3. Statistical Evaluation and IPA of LC-MS Data Revealed the Background of Suitability Differences between Whole Serum and sEV Samples

2.3.1. Quantitative Changes of the Proteome May Affect the Suitability of sEV Samples to Provide Biomarkers for CNS Tumor Status Monitoring

Statistical comparison of the proteome of sEV and whole serum samples was performed to reveal quantitative differences affecting the suitability of different sample types to provide biomarkers for CNS tumor status monitoring. Pairwise statistical comparison (Welch's test) was used to identify proteins significantly enriched or depleted in sEV samples compared to whole serum samples (Figure 3). Sixty-five proteins were found to be significantly enriched in sEV samples, while 129 proteins were significantly depleted ($p < 0.05$). Using our sEV purification protocol detailed in the Section 4, we obtained a uniform particle size range of sEVs but the magnitude of quantitative changes in the sEV versus whole serum proteome suggested the possible presence of lipoprotein and serum protein contaminations. The level of apolipoproteins was decreased in sEV enriched samples (sEV/serum mean ratio is 0.66), however this fraction could not be completely eliminated. Besides, well known high abundance serum proteins (e.g., ALB) dominated the protein content of sEV enriched samples too. However, the enrichment of non-tissue specific (ITGA2B, ITGB3, LGALS3BP), epithelial cell (CD5L) and platelet related (STOM, TSPAN9) EV marker proteins [31] confirms sEV enrichment (sEV/serum mean ratio is 26.58), while it also demonstrates the presence of sEVs produced during clotting.

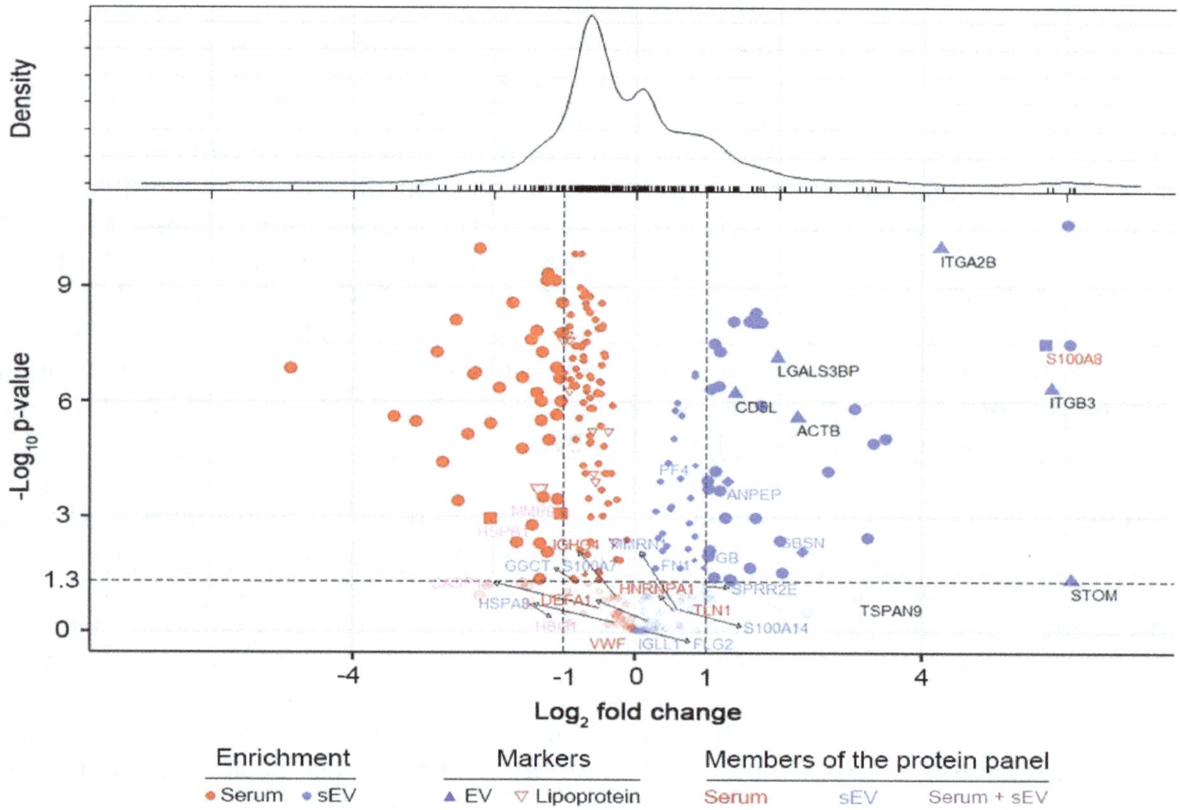

Figure 3. Quantitative comparison of the proteome of sEV and whole serum samples. Volcano plot represents the observed changes in average MS intensities in paired sEV vs. serum comparisons. Protein enrichment is marked with red and blue colored symbols in whole serum and sEVs, respectively. Lipoproteins (empty red upside-down triangles), elements of our whole serum protein panel (red letters, square symbols), sEV protein panel (blue letters, diamond symbols) and common members of the two protein panels (purple letters) are highlighted. Values of –log (p) were obtained from paired Welch's test in sEV/serum comparisons. Density estimation of log2 (fold change) values is shown on top.

Among the 17 proteins of the sEV marker panel described in Section 2.2 only 6 were significantly enriched in the sEV samples and 5 of the 10 proteins comprising the specific serum panel had higher abundance in whole serum (Figure 3). These findings suggest that the better suitability of sEV enriched samples to serve a biomarker source is not explained by a total increase in the abundance of specific proteins. (Detailed proteomics findings, protein annotation and sEV enrichment data are available in Table S1).

Additional sample processing (sEV isolation) may introduce higher technical variance in case of sEV samples, thus it may reduce the analytical suitability of this sample type. Our analysis revealed a similar level of variance for proteins quantified in each sample type (excluding contaminants)—median coefficients of variation within each patient group were in the ranges of 20.78–23.87% for sEV and 20.21–24.45% for serum samples (see Figure S2 for CV distributions).

2.3.2. Biological Background Might Be Responsible for the Increased Suitability of sEV Samples to Provide Biomarkers for CNS Tumor Status Monitoring

To gain insight into the biological background of the obtained proteomics data, IPAwas applied. We performed 'Core Analyses' for whole serum and sEV data separately, yielding a list of significantly influenced 'Diseases and Functions' in each patient group ($p < 0.05$). Using 'Comparison Analysis,' we were able to develop heatmaps covering the relevant systemic and tumor-related functions, as well as the activated or inhibited immune functions (Figure 4A). Regarding whole serum samples, many of the significantly influenced functions identified are related to CNS involvement and active immune regulatory processes but the patient groups are not clearly distinguished on the heatmaps (Figure 4A, left panels). In contrast, on two of the three sEV proteome-based heatmaps M was evidently separated from the malignant tumor groups (Figure 4A, right panels), where tumor progression-related functions (e.g., angiogenesis, proliferation and migration of tumor cells) were detected to be highly activated and the activated immune functions (e.g., cell movement or activation of myeloid cells) predominate over inhibited immune functions (e.g., phagocytosis).

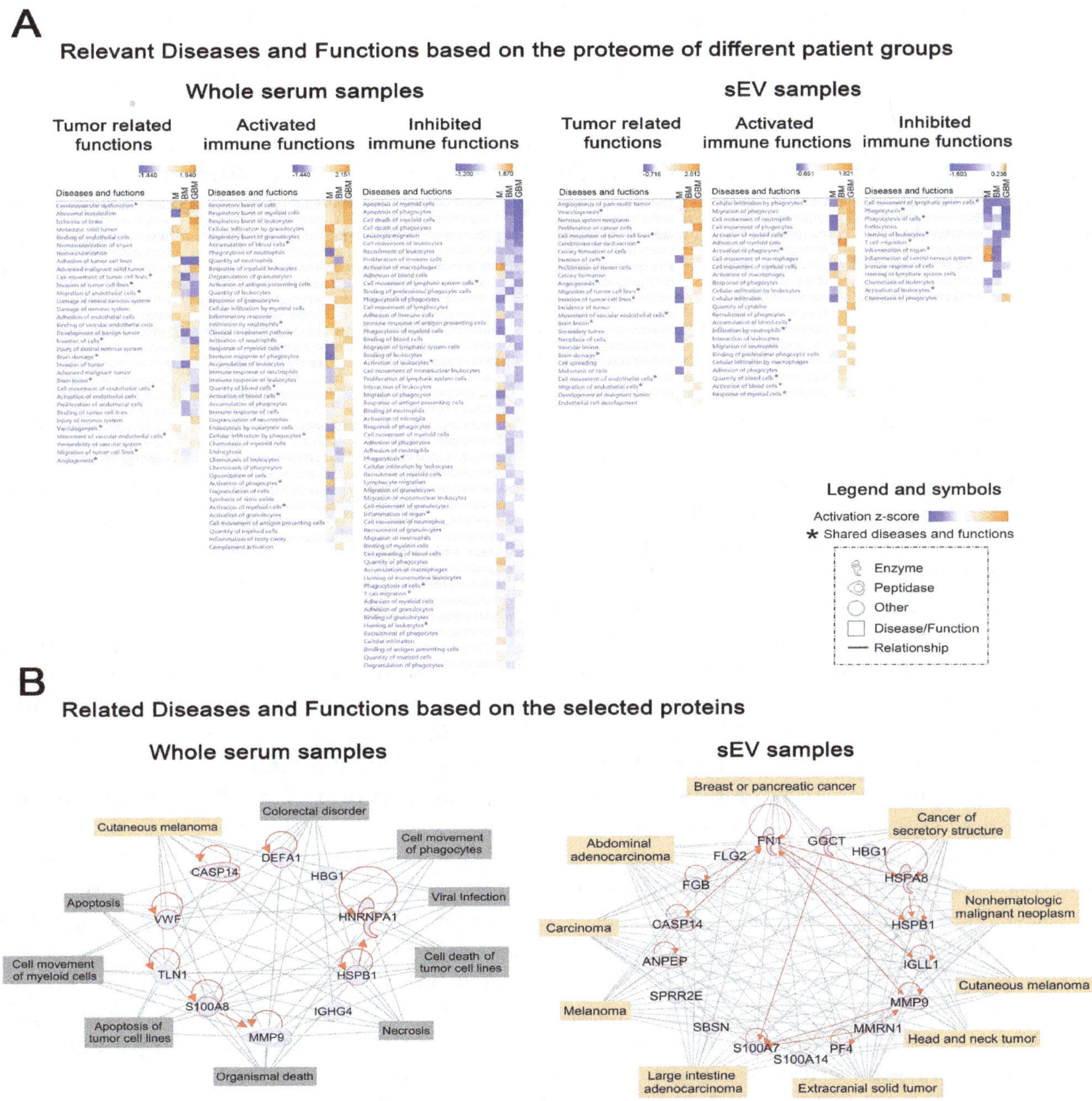

Figure 4. Ingenuity Pathway Analysis (IPA) analyses of whole serum (left) and sEV (right) data derived from the LC-MS analysis. (**A**) Heatmaps show relevant 'Diseases and Functions' in three separated panels related to systemic and tumor-related functions, as well as activated and inhibited immune functions. Z-score indicates activation or inhibition rates of the relevant 'Diseases and Functions' in the three tumorous patient groups compared to the control group. * symbol indicates the shared diseases and functions in whole serum and sEVs. (**B**) Networks display the selected 10 whole serum or 17 sEV proteins (blue symbols) and their relationships (red lines). Top ten related 'Diseases (highlighted in orange symbols) and Functions (highlighted in grey)' are connected by grey lines.

Next, we attempted to specify the common biological role of the characteristic protein profiles identified. Therefore, we elaborated two networks containing the selected 10 and 17 proteins identified based on whole serum and the sEV data, respectively (Figure 4B). Using the 'Grow tool,' the top ten influenced 'Diseases and Functions' were integrated into the networks. In case of the whole serum network (Figure 4B, left panel), nine different related 'Diseases and Functions' were identified, including viral infection, apoptosis, necrosis or cell movement of phagocytes and myeloid cells and only one was cancer-related. In contrast, the top ten influenced diseases identified on the sEV network based on the identified 17 proteins (Figure 4B, right panel) were all tumor-associated, suggesting their potential involvement in the pathophysiology of cancers.

3. Discussion

Non-invasive diagnostic tests are of outstanding clinical importance because of their minimal burden and risk to the patient, their repeatability, low cost, high information content and easy accessibility. In CNS tumors, a minimally invasive technique for describing the actual tumor status should be particularly important. Conventional MRI tests commonly used for the monitoring of CNS tumors are not absolutely appropriate for discriminating between various tumor types (e.g., cannot differentiate between glioblastomas and solitary metastases, CNS lymphomas or other glioma grades [32]) and cannot distinguish recurrence from pseudoprogression. Brain biopsies, as another option, are highly challenging and risky, especially when multiple sampling is required for long-term follow-up [33]. For several cancer types, blood-based tumor markers, such as PSA, AFP and CA125 have been introduced into clinical practice and research for the identification of further noninvasive biomarkers applicable for monitoring a wider scale of malignant diseases is ongoing [34]. However, regarding CNS tumors these studies have generally failed, presumably explained by several reasons, including (1) the barrier function of BBB (releasing less tumor 'information' into the systemic circulation), (2) the presence of molecules released into the blood from other sources and (3) possibly because of the complexity of tumor tissues (such as glioblastoma multiforme). These issues hamper attempts to use a single or only a few biomarkers to diagnose and monitor CNS tumors.

Based on these considerations, we aimed to detect the characteristic protein fingerprint of some common CNS tumors, trying to amplify the signal/information that brain tumors release into the circulation. For this purpose, the protein content of 96 clinical serum samples and related sEV samples isolated from the whole serum was measured by LC-MS. Serum samples were collected from three brain tumor groups considered as the most common malignant, benign and metastatic brain tumors (glioblastoma multiforme, meningioma [28] and brain metastasis of non-small-cell lung cancer [29]) and a control group (lumbar disc herniation).

To examine whether the proteomes of serum and sEV samples are suitable for differentiating between the CNS tumors in point, that is, whether they are applicable to diagnose and monitor the disease, the proteomes of these four patient groups were compared. The effectiveness of tumor type distinction may be increased if the analysis is restricted to proteins which exhibit significant between-group differences. Protein selection was carried out as described in literature [35] (using ratio of intensity means; Cohen's d effect size; ROC) but much stricter thresholds were applied (>2, <0.5; $d > 2$; $AUC = 1$, respectively). Statistical selection yielded a collection of proteins whose intensity showed significant between-group differences and thus these proteins could be reckoned as the most suitable molecules for distinguishing between the tumor types examined. Specifically, protein selection yielded a 10- and 17-membered protein panel for whole serum and sEV samples, respectively. While none of these proteins appeared to be able to distinguish between the patient groups individually, their combination was found to reliably discriminate between the different patient groups suggesting that instead of a few candidates, a specific protein panel is required for a perfect differentiation between various tumor types.

To evaluate group distinction efficiency, PCA with k-means clustering was carried out according to literature [36]. Homogeneity and completeness scores of the clusters were calculated to measure the performance of k-means clustering. Cluster homogeneity and completeness mean that each cluster contains only samples from the same group and all samples of a given group are assigned to the same cluster. Both scores are bounded below by 0 and above by 1. A score of 1 indicates perfect homogeneity or completeness. PCA revealed that sEV samples were more suitable for group distinction. Despite carefully selected and perfectly identical statistical analyses for the two sample types, the homogeneity and completeness scores for the whole serum analysis were 0.56 and 0.73, respectively, compared to scores 1 and 1 for the analysis of sEV samples. The explanation for these findings is illustrated on a PCA biplot (Figure 2). Regarding serum samples, the proteins that can separate two given groups by the appropriate ratio and effect size may have similar intensities in other groups as well. For example, DEFA1 is important in distinguishing the CTRL group from the BM and GBM groups, however, it shows similar intensities in the GBM and BM groups, hampering the separation of these groups (see DEFA1 arrow pointing between the BM and GBM groups in the whole serum PCA plot). Still, DEFA1 cannot be removed, because it plays a key role in separating the CTRL group from malignant tumors. In contrast, the majority of the proteins identified in the sEV samples were able to separate any given group from all the others.

To check whether the poorer performance of whole serum proteins in distinguishing between the patient groups is attributed to the number of the proteins considered, we performed another PCA analysis including only up- and down-regulated proteins selected from the whole identified panel (see Figure 2, Step 4), yielding a similar number of proteins for the two sample types. The PCA analysis of these 41 whole serum and 45 sEV proteins yielded similar results as the previous analysis of carefully selected proteins only and the sEV sample type proved to perform better again. Although the sEV sample was far from being perfect in this case (4 groups were recognized with a homogeneity score of 0.66 and a completeness score of 0.66), the results for the whole serum analysis indicated that not even the sample groups can be recognized based on these proteins only 2 groups were recognized, homogeneity—0.07, completeness: 0.40) (Figure S3). These findings support that sEVs have a better efficiency in distinguishing between various patient groups, irrespective of the order of magnitude of proteins analyzed for the comparison of sEV and whole serum samples.

To investigate the background of our observations, we performed a quantitative proteomics comparison of the two sample types. A quantitative evaluation of sEV purification protocols was suggested based on quantitative LC-MS based proteomics approach, using enrichment analysis of carefully selected sEV markers along with medium specific contamination marker proteins (e.g., lipoproteins and serum). To the best of our knowledge, we are the first group to quantitatively compare the proteome of serum derived small extracellular vesicles with that of the original whole serum samples. sEV enrichment may increase the relative abundance of proteins present in higher concentration within sEVs and the increased signal-to-noise ratio may be beneficial for the quantitative LC-MS analysis of such proteins. On the other hand, proteins in serum are originating from different sources of the human body. Any fractionation (e.g., enrichment of a specific sEV population) may decrease the suppressing effect of the uninformative protein fraction released from sources not specific for the target disease. No association was revealed between being a sEV marker and sEV enrichment, suggesting that it not the overall enrichment process that should be responsible for the increased suitability of sEV samples to provide biomarkers for CNS tumor monitoring. Instead, the removal of an uninformative protein fraction, providing a more specific sample, may explain why the sEV sample is more applicable for distinguishing between various CNS cancer patient groups. Compared to whole serum samples, EVs may be more suitable for investigating tumor related molecular patterns, as the characteristic fingerprint molecules are present in higher concentrations in sEV samples and are accompanied by less contaminating molecules that may bias the analytical findings.

To understand the biological background for our proteomics-based data, IPA was used for the separate analyses of whole serum and sEV data. 'Core Analyses' were performed, yielding a list of significantly influenced 'Diseases and Functions' comprising tumor-related functions as well as activated or inhibited immune functions in each patient group ($p < 0.05$). 'Comparison Analysis' was carried out to compare the affected 'Diseases and Functions' in the different patient groups. Regarding whole serum samples, many of the significantly influenced functions identified were associated with CNS involvement and active immune regulatory processes but the patient groups were not clearly distinguished on the heatmaps. In contrast, on the sEV proteome-based heatmaps the benign M was clearly separated from the malignant tumors, for which numerous tumor progression-related functions (e.g., angiogenesis, proliferation and migration of tumor cells) were found to be highly activated. The generated IPA heatmaps also revealed that the proteome of sEV samples may provide more specific information on the immune reactions characteristic to the patient groups. We assume that activated immune functions (e.g., cellular infiltration and migration of phagocytes) may play a crucial role in the development of an immune-suppressive microenvironment, while antitumoral immune responses (e.g., phagocytosis, inflammation) might be inhibited.

Serum is a dual source of biomolecular information on cancer, as it contains the molecules released by cancer cells, as well as those released during the immune system's tumor-specific responses [37]. Therefore, the differences observed in the serum vesicles isolated from different patient groups may not only mirror tumor-specific processes but also those related to the associated immune responses [38,39]. Samples enriched in sEVs can offer an amplified source of relevant information, representing not only the specific tumor tissue but also the associated immune responses. Thus, an appropriate protein panel, covering both sources, may have improved efficiency for CNS tumor classification and monitoring.

In addition, the networks developed based on the IPA 'Grow tool' demonstrated that the biological background of the sEV-based characteristic protein profile is more specifically associated with the tumor types compared with the whole serum based protein profile. The role of some of the proteins included in the sEV-based characteristic protein profile has already been described, for example in GBM biology, making these proteins promising targets for extracellular vesicle-based biomarker development [25].

In addition to the proteomics-based comparison of EV samples, we also examined the EV concentration of individual serum samples. Interestingly, no significant differences were detected between the four patient groups regarding the concentration of serum sEVs, with a mean size of 112.2 nm. Osti et al. observed higher EV plasma levels in GBM patients, brain metastases and extra-axial brain tumors compared to healthy controls. Other researchers also demonstrated higher EV concentration in tumorous patients, when unfractionated EV isolates [40] or a wider spectra of EVs were analyzed [41,42]. However, other non-neoplastic diseases of the central nervous system may also increase the number of small-sized circulating EVs, as it was demonstrated in acute ischemic stroke [43] or multiple sclerosis patients [44]. Our vesicle number measurement results, as well as the findings detailed above suggest that the elevated sEV concentration cannot be clearly attributed to the presence of the tumor as immune responses or other systemic responses also contribute to the circulating EV population. Our proteomics-based findings, coupled with the available literature data, suggest that circulating small-sized EVs show important qualitative but not quantitative differences between benign or malignant brain tumors and spinal disc herniation.

Liu and colleagues highlighted that serum is not the perfect choice for a representative sampling of circulating EVs [45], as a high fraction of EVs may be lost during coagulation and also blood components (e.g., platelets) may release microvesicles (MV) during clotting, altering the original MV content of blood samples. However, serum is still the preferred sample form for blood-based clinical diagnoses and it is a practical choice for future clinical developments. It should be noted that co-purification of proteins [46] and lipoprotein particles [47] in EV isolation methods is a common and well known challenge [31]. The presence of protein aggregates [48] and lipoproteins in sEV isolates may provide additional explanation for the lack of increase in the concentration of enriched sEV particles in cancer patients' serum, contrary to literature data on plasma [49] or serum samples [40]. Efforts to eliminate lipoproteins are described in numerous papers reporting on attempts to introduce more sophisticated methods (e.g., combination of ultracentrifugation and size-exclusion chromatography) [50]. In fact, these laborious and instrumentation demanding methods are of high importance in scientific research of the molecular contents of EVs but they may not be applicable in routine clinical practice. Our sEV isolation protocol has several advantages, as it does not require expensive equipment or highly trained professionals and the entire procedure (along with characterization) is performed within 4 h, therefore, this technique could be easily incorporated into clinical practice.

Our quantitative proteomics results demonstrate that even a simple sEV enrichment protocol can increase the diagnostic potential of serum samples for the identification and classification of patients with different CNS cancers. This finding also supports that even a low-efficacy sEV enrichment/purification method may be appropriate to enhance the analytical applicability of serum samples for CNS cancer monitoring, however, in such cases a quantitative description of enrichment efficiency is definitely required for the right interpretation of the analytical results [51].

In conclusion, our findings support that extracellular vesicles have a greater potential for the monitoring of CNS tumors compared to whole serum samples. Using EV samples is a possible way to amplify the signals released by brain tumors into the circulation. Given the easy-to-implement isolation and enrichment protocol established in this study, the introduction of EV analysis would be beneficial in clinical practice.

4. Materials and Methods

4.1. Patients

Blood samples of 96 patients treated between March 2015 and January 2018 in the Department of Neurosurgery, University of Debrecen were analyzed. Samples were obtained from patients with primary glioblastoma multiforme (GBM), meningioma (M) and single brain metastasis originating from non-small-cell lung cancer (BM). Control samples (CTRL) were collected from patients with spinal disc herniation without evidence of cancer. This non-tumor patient group served as control group in comparison to the patients having different intracranial tumor to distinguish the effects of tumorous processes from the CNS involvement on circulating sEVs. Each group contained 24 individuals with mixed ages and genders. As shown in the Table 1, six-sample-pools were created from the individuals, allowing four parallel samples to be tested per group. Blood samples were collected one day prior to neurosurgical procedure in each tumor case. None of the patients received radio- or chemotherapy before tumor resection. Blood samples were stored by the Neurosurgical Brain Tumor and Tissue Bank of Debrecen according to the criteria of the National Research Ethics Committee. An informed consent form was signed by each patient; the study was conducted in accordance with the Declaration of Helsinki. This study was carried out according to two ethical approvals, namely 51450-2/2015/EKU (0411/15), Medical Research Council, Scientific and Research Ethics Committee, Budapest, October 30, 2015 and 121/2019-SZTE, University of Szeged, Human Investigation Review Board, Albert Szent-Györgyi Clinical Centre, Szeged, 19 July 2019.

Table 1. Patient cohort [1].

♦ **Glioblastoma Multiforme**	**GBM**	**GBM1**	**GBM2**	**GBM3**	**GBM4**
Total No. of patients	***n* = 24**	$n = 6$	$n = 6$	$n = 6$	$n = 6$
Age, Median (range)	**67 (33–82)**	64.5 (38–82)	69.5 (33–76)	67.5 (49–74)	66.5 (63–77)
Mean	**64.9**	62.7	63.8	64.7	68.5
Sex (%), Male	**13 (54.2)**	3 (50)	3 (50)	5 (83.3)	2 (33.3)
Female	**11 (45.8)**	3 (50)	3 (50)	1 (16.7)	4 (66.7)
■ **Brain Metastasis**	**BM**	**BM1**	**BM2**	**BM3**	**BM4**
Total No. of patients	***n* = 24**	$n = 6$	$n = 6$	$n = 6$	$n = 6$
Age, Median (range)	**64 (42–82)**	66.5 (51–82)	68 (62–71)	63.5 (42–81)	59.5 (53–64)
Mean	**64**	67.7	67.5	59.7	59.5
Sex (%), Male	**13 (54.2)**	2 (33.3)	3 (50)	4 (66.7)	4 (66.7)
Female	**11 (45.8)**	4 (66.7)	3 (50)	2 (33.3)	2 (33.3)
▲ **Meningioma**	**M**	**M1**	**M2**	**M3**	**M4**

Table 1. *Cont.*

♦ **Glioblastoma Multiforme**	**GBM**	**GBM1**	**GBM2**	**GBM3**	**GBM4**
Total No. of patients	***n* = 24**	*n* = 6	*n* = 6	*n* = 6	*n* = 6
Age, Median (range)	**60 (30–79)**	54.5 (39–69)	62 (30–66)	61.5 (44–75)	66.5 (52–79)
Mean	**58.0**	53.5	53	59.3	66
Sex (%), Male	**4 (16.7)**	0 (0)	0 (0)	1 (16.7)	3 (50)
Female	**20 (83.3)**	6 (100)	6 (100)	5 (83.3)	3 (50)
• **Control**	**CTRL**	**CTRL1**	**CTRL2**	**CTRL3**	**CTRL4**
Total No. of patients	***n* = 24**	*n* = 6	*n* = 6	*n* = 6	*n* = 6
Age, Median (range)	**50.5 (20–81)**	46.5 (26–71)	47 (20–62)	70.5 (49–81)	52.5 (41–69)
Mean	**52.9**	46.5	45	67.2	53
Sex (%), Male	**9 (37.5)**	2 (33.3)	4 (66.7)	4 (66.7)	4 (66.7)
Female	**15 (62.5)**	4 (66.7)	2 (33.3)	2 (33.3)	2 (33.3)

[1] The table summarizes the main characteristics of the patient groups examined. Each group (average values highlighted in bald) included 24 individuals, converted into six-sample-pools to yield four samples per group for further analysis.

4.2. Preparation of Serum Samples, sEV Isolation and Characterization

Blood samples were collected into BD Vacutainer SST II Advance Tubes (Becton, Dickinson and Company, Franklin Lakes, NJ, USA), allowed to clot for at least 1 h at room temperature and centrifuged for 20 min at 3000× *g*, 10 °C to remove cells. Following the 3000× *g* centrifugation, the supernatant serum was transferred to new Eppendorf tubes and centrifuged for 30 min at 10,000× *g*, 4 °C to remove debris and large vesicles. One milliliter serum aliquot was diluted with DPBS (Ca^{2+}-free, Mg^{2+}-free, Lonza Group Ltd., Basel, Switzerland) to 8 mL and ultracentrifuged for 70 min, at 100,000× *g*, 4 °C (polycarbonate tubes, fixed angle T-1270 rotor, Thermo Fisher Scientific, Waltham, MA, USA). The pellet was resuspended in 100 uL DPBS and stored at −80 °C until further processing. This sEV isolation protocol served to reach intermediate recovery and intermediate specificity according to MISEV2018 [31].

SEVs were characterized by AFM (Oxford Instruments Asylum Research, Abingdon, UK), as described previously [52] and NTA using a NanoSight NS300 instrument (Malvern Panalytical Ltd., Malvern, UK) as it described below. Classical EV markers were presented by Western blot analyses using NuPAGE reagents and an XCell SureLock Mini-Cell System (Thermo Fisher Scientific, Waltham, MA, USA) according to manufacturer's protocols. For detection of the CD81 and Alix markers, we used rabbit anti-human CD81 (1:1000, Sigma-Aldrich, St. Louis, MO, USA) and rabbit anti-human Alix (1:1000, Sigma-Aldrich, St. Louis, MO, USA) primary antibody and HRP-conjugated anti-rabbit IgG (1:1000, R&D Systems, Minneapolis, MN, USA) secondary antibody.4.3. Quantitative analysis of sEVs by NTA.

As suggested by a recent study [53], sEVs were diluted in particle free DPBS and analyzed using a NanoSight NS300 instrument with 532 nm laser (Malvern Panalytical Ltd., Malvern, UK). The measurements were performed on the 16 sEV sample pools (described in 4.1). Six videos of 60 s were recorded for each sample under constant settings (Camera level: 15; Treshold: 4, 25 °C; 60–80 particles/frame) and analyzed to obtain data on size distribution and particle concentration.

4.3. Proteomic Analysis by LC-MS

4.3.1. 'In Solution' Digestion

Individual samples containing 20 µg protein were diluted to 10 µL with 0.1 M NH_4HCO_3 (pH = 8.0) buffer; 12 µL 0.1% RapiGest SF (Waters, Milford, MA, USA) and 2 µL 55 mM dithioeritritol solution was added and kept at 60 °C for 30 min to unfold and reduce proteins. A volume of 2 µL 200 mM iodo acetamide solution was added to alkylate the proteins which were kept for an additional 30 min in the dark at room temperature. The samples were digested overnight at 37 °C with trypsin (Thermo

Scientific, Waltham, MA, USA, enzyme/protein ratio: 0.4 to 1). The digestion was stopped by addition of 1 μL of concentrated formic acid.

4.3.2. LC-MS

The separation of the digested samples was carried out on a nanoAcquity UPLC, (Waters, Milford, MA, USA) using Waters ACQUITY UPLC M-Class Peptide C18 (130 Å, 1.78 μm, 75 μm × 250 mm) column with a nonlinear 90 min gradient. Eluents were water (A) and acetonitrile (B) containing 0.1 *V*/*V*% formic acid and the separation of the peptide mixture was performed at 45 °C with 0.35 μL/min flow rate using an optimized nonlinear LC gradient (3–40% B). The LC was coupled to a high-resolution Q Exactive Plus quadrupole-orbitrap hybrid mass spectrometer (Thermo Scientific, Waltham, MA, USA). The quantitative measurements of digested individual samples were performed in DIA mode. The survey scan for DIA method operated with 35,000 resolution. The full scan was performed between 380 to 1020 *m*/*z*. The AGC target was set to 5 × 106 or 120 ms maximum injection time. In the 400–1000 *m*/*z* region 22 *m*/*z* wide overlapping windows were acquired at 17,500 resolution (AGC target: 3 × 106 or 100 ms injection time, normalized collision energy: 27 for charge 2). The quantitative analysis was performed in Encyclopedia 0.81 [54] using default settings after deconvolution, peak picking and conversion of raw MS files to mzML format in Proteowizard [55]. A comprehensive spectral library [56] of 10,000 human proteins was used for peptide identification. Protein quantities calculated by the Encyclopedia software based on summed intensities of the automatically filtered peptides were used in further statistical evaluations.

4.4. Statistical Analysis

The collected data about the whole serum and extracellular vesicles were reduced and analyzed using statistical methods. Pearson's correlation analysis was used to investigate the outlier samples [57], contaminating proteins (cytokeratins) and proteins with missing values were excluded from the proteomic data [58]. Data were log-transformed to reduce skewness and increase linearity [59]. Cohen's d effect size was calculated to measure the difference between the protein intensity means, outcomes in two different groups [60,61]. Pairwise ROC analysis allowed us to find those proteins which can separate at least one group to the others [62]. The calculated ROC AUC (area under the ROC curve) values are accepted if it equals to 1. In order to transform several (potentially) correlated proteins into a (smaller) number of uncorrelated variables and visualize the dataset, PCA with k-means clustering was performed [63–66]. Homogeneity and completeness scores of the clusters were calculated to measure the performance of k-means clustering [67]. Two-tailed Welch's *t*-test was performed to identify the significantly enriched or depleted proteins in sEV samples. The statistical analyses were performed using R statistical program (version 3.6.3 with pROC, FactoMineR, factoextra and ggplot2 packages; Vienna, Austria), Python programming language (version 3.8, Scotts Valley, CA, USA) and Perseus (MaxQuant, Munich, Germany). Values of $p < 0.05$ were considered significant (see in Appendix A more detailed). GraphPad Prism 8 (San Diego, CA, USA) was used for further data visualization.

4.5. In Silico Analysis of LC-MS Data

Protein data derived from the LC-MS were analyzed by the IPA (Qiagen Bioinformatics, Hilden, Germany). Using fold change values, 'Core Analysis' were performed for whole serum and sEV data separately to identify 'Diseases and Functions,' which can be significantly influenced by the described proteomes ($p < 0.05$). After 'Comparison Analysis,' we created heatmaps of the relevant 'Diseases and Functions,' that is, tumor-related and immunological functions showing regulatory differences between the three CNS tumor groups. Activation z-score calculated by IPA indicates the extent and direction of the effect that given proteins have on function/disease.

The selected 10 whole serum proteins and 17 sEV proteins were introduced to custom pathways as well. Then, 'Connect tool' of IPA was used to reveal the relationships between these molecules and

'Grow tool' was applied to search the top ten 'Diseases and Functions' assigned to the 10 whole serum proteins or 17 sEV proteins. Results are displayed in two networks created by the IPA Path Designer.

Confidence level was set to 'Experimentally observed' for all IPA procedures, which enables literature data-based analysis but excludes unproven predictions.

4.6. Data Availability

All datasets generated during the current study are available from the corresponding author upon reasonable request.

We have submitted all relevant data of our experiments to the EV-TRACK [68] knowledgebase EV-TRACK ID: EV200080.

5. Conclusions

Our study aimed to detect the characteristic protein fingerprint of the most common CNS tumors. Intending to amplify the signal that brain tumors release into the circulation, in addition to the whole serum's, the protein content of the small extracellular vesicles isolated from the serum was also examined.

Comparative proteomic analysis suggests that sEVs may be more suitable for investigating tumor related molecular patterns, because these molecules are present in higher concentrations in sEV samples compared to whole serum samples and have less 'noise' that may bias the analytical findings. In silico analyses revealed that the biological background of the sEV-based characteristic protein profile of the samples is more specifically associated with the tumor types compared with the whole serum based protein profile. Samples enriched in sEVs can offer an amplified source of relevant information, representing not only the specific tumor tissue but also the associated immune responses.

These findings revealed that circulating small-sized extracellular vesicles were more suitable for separating different patient groups. The number of proteins applied for monitoring cannot be reduced to a few individual molecules, instead, a specific protein panel is required for perfect differentiation. To the best of our knowledge, we are the first group to quantitatively compare the proteome of serum derived small extracellular vesicles with that of the original whole serum samples.

In conclusion, our findings support that extracellular vesicles have a greater potential for monitoring CNS tumors, compared to whole serum samples. Considering that analyzing sEVs can be performed easily, incorporating our method into clinical practice would be of great benefit.

Supplementary Materials
Figure S1: Western blot analyses of classical EV markers, Figure S2: Intragroup Coefficients of variation (CV) distributions, Figure S3: PCA dotplot constructed after statistical selection based on the means of intensity ratio; Table S1: 311 membered protein table of DIA mode constructed spectral library; Table S2: Sample correlation matrix; Table S3: List of the selected proteins.

Author Contributions: Conceptualization, K.B.; methodology, G.D. and K.B.; validation, G.D., B.B. and A.J.; formal analysis G.D., M.B., Z.S., M.H., M.S.; investigation, G.D., M.B. and E.G.-S.; resources, A.K., P.H., T.B. and K.B.; writing—original draft preparation, G.D., M.B., Z.S. and M.H.; writing—review and editing, G.D. and K.B.; visualization, G.D, M.B., Z.S. and M.H.; supervision, A.K. and K.B.; project administration, G.D. and K.B.; funding acquisition, K.B. All authors have read and agreed to publish the final version of the manuscript.

Acknowledgments: The authors thank Lilla Pinter for her technical assistance and Zsolt Szegletes for taking the AFM images. The authors thank Dora Bokor, PharmD, for proofreading the manuscript.

Abbreviations

AFM	Atomic force microscopy
BBB	Blood brain barrier
BM	Brain metastasis originating from non-small-cell lung cancer
CNS	Central nervous system
CTRL	Controls – lumbar disc hernia patients
DIA	Data independent acquisition
EVs	Extracellular Vesicles
GBM	Primary glioblastoma multiforme
IPA	Ingenuity Pathway Analysis
LC-MS	Liquid chromatography - mass spectrometry
M	Meningioma
NTA	Nanoparticle tracking analysis
PCA	Principal component analysis
ROC	Receiver operating characteristic
sEVs	small extracellular vesicles

Appendix A. Detailed Description of the Statistical Analyses.

The collected data about the whole serum and extracellular vesicles were reduced and analyzed using statistical methods. Pearson's correlation analysis was used to investigate the outlier samples [57]. Contaminating proteins (cytokeratins) and proteins with missing values were excluded from the proteomic data [58]. Data were log-transformed to reduce skewness and increase linearity [59].

Cohen's d effect size was calculated to measure the difference between the protein intensity mean, outcomes in two different groups. The formula of the Cohen's d effect size

$$d = \frac{|\overline{X}_1 - \overline{X}_2|}{\sqrt{\frac{(n_1-1)SD_1^2+(n_2-1)SD_2^2}{n_1+n_2-2}}}$$

where X is the mean protein intensity in a given group, SD is standard deviation and n is sample size [60]. In this study we say at least 2 effect size is necessary. It indicates that the mean of group 1 is at the 97.7 percentile of group 2, and the nonoverlapping area of the two distributions at least is 81.1% [61].

Pairwise ROC analysis allowed us to find those proteins which can separate at least one group to the others [62]. The ROC analysis use the true positive rate (sensitivity) and the true negative rate (specificity) at various threshold settings. Plotting the sensitivity against the 1-specificity we get the ROC curve, under the area under this curve measure the separability of the given variable (protein). AUC = 0.5 represents an unsuitable variable to the separate two groups. If AUC = 1, the separation using the actual variable is error-free. In our study the calculated AUC (area under the curve) values are accepted if it equals to 1.

In order to transform several (potentially) correlated proteins into a (smaller) number of uncorrelated variables, and visualize the dataset, principal component analysis (PCA) with k-means clustering was performed. The goal of PCA is to reduce a large number of correlated variables with a set of uncorrelated principal components. These components can be thought of as linear combinations of the original variables that are optimally weighted and derived from the correlation matrix of the data [63].

K-means clustering was performed on the obtained PCA score plots by Hartigan-Wong algorithm [64,65]. Optimal numbers of clusters were determined with Silhouette method and these recommended values were used for clustering [66].

Homogeneity and completeness scores of the clusters were calculated to measure the performance of k-means clustering Cluster homogeneity and completeness mean that each cluster contains only samples from the same group, and all samples of a given group are assigned to the same cluster. Both scores are bounded below by 0 and above by 1. A score of 1 indicates the perfect homogeneity or completeness. [67].

Two-tailed Welch's t-test was performed to identify significantly enriched or depleted proteins in sEV samples.

The statistical analyses were performed using R statistical program (version 3.6.3 with pROC, FactoMineR, factoextra and ggplot2 packages), Python programming language (version 3.8) and Perseus (MaxQuant). Values of $p < 0.05$ were considered significant. GraphPad Prism 8 was used for further data visualization.

References

1. World Health Organization. *WHO Guidelines for the Pharmacological and Radiotherapeutic Management of Cancer Pain in Adults and Adolescents*; World Health Organization, 2018; License: CC BY-NC-SA 3.0 IGO. Available online: https://apps.who.int/iris/handle/10665/279700 (accessed on 1 June 2020).
2. World Health Organization. *Guide to Cancer Early Diagnosis*; World Health Organization, 2017; License: CC BY-NC-SA 3.0 IGO. Available online: https://apps.who.int/iris/handle/10665/254500 (accessed on 1 June 2020).
3. Garden, G.A.; Campbell, B.M. Glial biomarkers in human central nervous system disease. *Glia* **2016**, *64*, 1755–1771. [CrossRef] [PubMed]
4. Staedtke, V.; Dzaye, O.; Holdhoff, M. Actionable molecular biomarkers in primary brain tumors. *Trends Cancer* **2016**, *2*, 338–349. [CrossRef] [PubMed]
5. Good, D.M.; Thongboonkerd, V.; Novak, J.; Bascands, J.L.; Schanstra, J.P.; Coon, J.J.; Dominiczak, A.; Mischak, H. Body fluid proteomics for biomarker discovery: Lessons from the past hold the key to success in the future. *J. Proteome Res.* **2007**, *6*, 4549–4555. [CrossRef] [PubMed]
6. Best, M.G.; Sol, N.; Zijl, S.; Reijneveld, J.C.; Wesseling, P.; Wurdinger, T. Liquid biopsies in patients with diffuse glioma. *Acta Neuropathol.* **2015**, *129*, 849–865. [CrossRef] [PubMed]
7. Gerber, D.E.; Grossman, S.A.; Zeltzman, M.; Parisi, M.A.; Kleinberg, L. The impact of thrombocytopenia from temozolomide and radiation in newly diagnosed adults with high-grade gliomas. *Neuro Oncol.* **2007**, *9*, 47–52. [CrossRef]
8. Cagney, D.N.; Sul, J.; Huang, R.Y.; Ligon, K.L.; Wen, P.Y.; Alexander, B.M. The FDA NIH Biomarkers, EndpointS and other Tools (BEST) resource in neuro-oncology. *Neuro Oncol.* **2018**, *20*, 1162–1172. [CrossRef]
9. Sheridan, C. Exosome cancer diagnostic reaches market. *Nat. Biotechnol.* **2016**, *34*, 359–360. [CrossRef]
10. Colombo, M.; Raposo, G.; Théry, C. Biogenesis, secretion and intercellular interactions of exosomes and other extracellular vesicles. *Annu. Rev. Cell Dev. Biol* **2014**, *30*, 255–289. [CrossRef] [PubMed]
11. Nogués, L.; Benito-Martin, A.; Hergueta-Redondo, M.; Peinado, H. The influence of tumour-derived extracellular vesicles on local and distal metastatic dissemination. *Mol. Asp. Med.* **2018**, *60*, 15–26. [CrossRef]
12. Hoshino, A.; Costa-Silva, B.; Shen, T.L.; Rodrigues, G.; Hashimoto, A.; Tesic Mark, M.; Molina, H.; Kohsaka, S.; Di Giannatale, A.; Ceder, S.; et al. Tumour exosome integrins determine organotropic metastasis. *Nature* **2015**, *527*, 329–335. [CrossRef] [PubMed]
13. Costa-Silva, B.; Aiello, N.M.; Ocean, A.J.; Singh, S.; Zhang, H.; Thakur, B.K.; Becker, A.; Hoshino, A.; Mark, M.T.; Molina, H.; et al. Pancreatic cancer exosomes initiate pre-metastatic niche formation in the liver. *Nat. Cell Biol.* **2015**, *17*, 816–826. [CrossRef] [PubMed]
14. Liu, Y.; Gu, Y.; Han, Y.; Zhang, Q.; Jiang, Z.; Zhang, X.; Huang, B.; Xu, X.; Zheng, J.; Cao, X. Tumor Exosomal RNAs Promote Lung Pre-metastatic Niche Formation by Activating Alveolar Epithelial TLR3 to Recruit Neutrophils. *Cancer Cell* **2016**, *30*, 243–256. [CrossRef] [PubMed]
15. Zeng, Z.; Li, Y.; Pan, Y.; Lan, X.; Song, F.; Sun, J.; Zhou, K.; Liu, X.; Ren, X.; Wang, F.; et al. Cancer-derived exosomal miR-25-3p promotes pre-metastatic niche formation by inducing vascular permeability and angiogenesis. *Nat. Commun.* **2018**, *9*, 5395. [CrossRef] [PubMed]
16. Feng, W.; Dean, D.C.; Hornicek, F.J.; Shi, H.; Duan, Z. Exosomes promote pre-metastatic niche formation in ovarian cancer. *Mol. Cancer* **2019**, *18*, 124. [CrossRef] [PubMed]
17. Chen, G.; Huang, A.C.; Zhang, W.; Zhang, G.; Wu, M.; Xu, W.; Yu, Z.; Yang, J.; Wang, B.; Sun, H.; et al. Exosomal PD-L1 contributes to immunosuppression and is associated with anti-PD-1 response. *Nature* **2018**, *560*, 382–386. [CrossRef] [PubMed]

18. Scavo, M.P.; Depalo, N.; Tutino, V.; De Nunzio, V.; Ingrosso, C.; Rizzi, F.; Notarnicola, M.; Curri, M.L.; Giannelli, G. Exosomes for Diagnosis and Therapy in Gastrointestinal Cancers. *Int. J. Mol. Sci.* **2020**, *21*, 367. [CrossRef] [PubMed]
19. Basu, B.; Ghosh, M.K. Extracellular Vesicles in Glioma: From Diagnosis to Therapy. *Bioessays* **2019**, *41*, e1800245. [CrossRef]
20. Kosaka, N.; Kogure, A.; Yamamoto, T.; Urabe, F.; Usuba, W.; Prieto-Vila, M.; Ochiya, T. Exploiting the message from cancer: The diagnostic value of extracellular vesicles for clinical applications. *Exp. Mol. Med.* **2019**, *51*, 31. [CrossRef]
21. Möhrmann, L.; Huang, H.J.; Hong, D.S.; Tsimberidou, A.M.; Fu, S.; Piha-Paul, S.A.; Subbiah, V.; Karp, D.D.; Naing, A.; Krug, A. Liquid biopsies using plasma exosomal nucleic acids and plasma cell-free DNA compared with clinical outcomes of patients with advanced cancers. *Clin. Cancer Res.* **2018**, *24*, 181–188. [CrossRef]
22. Melo, S.A.; Luecke, L.B.; Kahlert, C.; Fernandez, A.F.; Gammon, S.T.; Kaye, J.; LeBleu, V.S.; Mittendorf, E.A.; Weitz, J.; Rahbari, N.; et al. Glypican-1 identifies cancer exosomes and detects early pancreatic cancer. *Nature* **2015**, *523*, 177–182. [CrossRef]
23. Choy, C.; Jandial, R. Breast Cancer Exosomes Breach the Blood-Brain Barrier. *Neurosurgery* **2016**, *78*, N10-1. [CrossRef] [PubMed]
24. García-Romero, N.; Carrión-Navarro, J.; Esteban-Rubio, S.; Lázaro-Ibáñez, E.; Peris-Celda, M.; Alonso, M.M.; Guzmán-De-Villoria, J.; Fernández-Carballal, C.; de Mendivil, A.O.; García-Duque, S.; et al. DNA sequences within glioma-derived extracellular vesicles can cross the intact blood-brain barrier and be detected in peripheral blood of patients. *Oncotarget* **2017**, *8*, 1416–1428. [CrossRef] [PubMed]
25. Gollapalli, K.; Ray, S.; Srivastava, R.; Renu, D.; Singh, P.; Dhali, S.; Bajpai Dikshit, J.; Srikanth, R.; Moiyadi, A.; Srivastava, S. Investigation of serum proteome alterations in human glioblastoma multiforme. *Proteomics* **2012**, *12*, 2378–2390. [CrossRef] [PubMed]
26. Gállego Pérez-Larraya, J.; Paris, S.; Idbaih, A.; Dehais, C.; Laigle-Donadey, F.; Navarro, S.; Capelle, L.; Mokhtari, K.; Marie, Y.; Sanson, M.; et al. Diagnostic and prognostic value of preoperative combined GFAP, IGFBP-2 and YKL-40 plasma levels in patients with glioblastoma. *Cancer* **2014**, *120*, 3972–3980. [CrossRef] [PubMed]
27. Figueroa, J.M.; Carter, B.S. Detection of glioblastoma in biofluids. *J. Neurosurg.* **2018**, *129*, 334–340. [CrossRef] [PubMed]
28. Ostrom, Q.T.; Gittleman, H.; Truitt, G.; Boscia, A.; Kruchko, C.; Barnholtz-Sloan, J.S. CBTRUS Statistical Report: Primary Brain and Other Central Nervous System Tumors Diagnosed in the United States in 2011-2015. *Neuro Oncol.* **2018**, *20*, iv1–iv86. [CrossRef]
29. Fox, B.D.; Cheung, V.J.; Patel, A.J.; Suki, D.; Rao, G. Epidemiology of Metastatic Brain Tumors. *Neurosurg. Clin. N. Am.* **2011**, *22*, 1–6. [CrossRef]
30. Théry, C.; Witwer, K.W.; Aikawa, E.; Alcaraz, M.J.; Anderson, J.D.; Andriantsitohaina, R.; Antoniou, A.; Arab, T.; Archer, F.; Atkin-Smith, G.K.; et al. Minimal information for studies of extracellular vesicles 2018 (MISEV2018): A position statement of the International Society for Extracellular Vesicles and update of the MISEV2014 guidelines. *J. Extracell. Vesicles* **2018**, *7*, 1535750. [CrossRef]
31. de Menezes-Neto, A.; Sáez, M.J.; Lozano-Ramos, I.; Segui-Barber, J.; Martin-Jaular, L.; Ullate, J.M.; Fernandez-Becerra, C.; Borrás, F.E.; Del Portillo, H.A. Size-exclusion chromatography as a stand-alone methodology identifies novel markers in mass spectrometry analyses of plasma-derived vesicles from healthy individuals. *J. Extracell. Vesicles* **2015**, *4*, 27378. [CrossRef]
32. Weber, M.A.; Zoubaa, S.; Schlieter, M.; Jüttler, E.; Huttner, H.B.; Geletneky, K.; Ittrich, C.; Lichy, M.P.; Kroll, A.; Debus, J.; et al. Essig Diagnostic performance of spectroscopic and perfusion MRI for distinction of brain tumors. *Neurology* **2006**, *66*, 1899–1906. [CrossRef]
33. Shankar, G.M.; Balaj, L.; Stott, S.L.; Nahed, B.; Carter, B.S. Liquid biopsy for brain tumors. *Expert Rev. Mol. Diagn.* **2017**, *17*, 943–947. [CrossRef] [PubMed]
34. Marrugo-Ramírez, J.; Mir, M.; Samitier, J. Blood-Based Cancer Biomarkers in Liquid Biopsy: A Promising Non-Invasive Alternative to Tissue Biopsy. *Int. J. Mol. Sci.* **2018**, *21*, 2877. [CrossRef] [PubMed]
35. Miyauchi, E.; Furuta, T.; Ohtsuki, S.; Tachikawa, M.; Uchida, Y.; Sabit, H.; Obuchi, W.; Baba, T.; Watanabe, M.; Terasaki, T.; et al. Identification of blood biomarkers in glioblastoma by SWATH mass spectrometry and quantitative targeted absolute proteomics. *PLoS ONE* **2018**, *13*, e0193799. [CrossRef] [PubMed]

36. Rhie, S.K.; Perez, A.A.; Lay, F.D.; Schreiner, S.; Shi, J.; Polin, J.; Farnham, P.J. A high-resolution 3D epigenomic map reveals insights into the creation of the prostate cancer transcriptome. *Nat. Commun.* **2019**, *10*, 4154. [CrossRef] [PubMed]
37. Anderson, K.S.; LaBaer, J. The sentinel within: Exploiting the immune system for cancer biomarkers. *J. Proteome Res.* **2005**, *4*, 1123–1133. [CrossRef] [PubMed]
38. Wen, C.; Seeger, R.C.; Fabbri, M.; Wang, L.; Wayne, A.S.; Jong, A.Y. Biological roles and potential applications of immune cell-derived extracellular vesicles. *J. Extracell. Vesicles* **2017**, *6*, 1400370. [CrossRef] [PubMed]
39. Veerman, R.E.; Güçlüler Akpinar, G.; Eldh, M.; Gabrielsson, S. Immune Cell-Derived Extracellular Vesicles –Functions and Therapeutic Applications. *Trends Mol. Med.* **2019**, *25*, 382–394. [CrossRef]
40. Gercel-Taylor, C.; Atay, S.; Tullis, R.H.; Kesimer, M.; Taylor, D.D. Nanoparticle analysis of circulating cell-derived vesicles in ovarian cancer patients. *Anal. Biochem.* **2012**, *428*, 44–53. [CrossRef]
41. Lázaro-Ibáñez, E.; Sanz-Garcia, A.; Visakorpi, T.; Escobedo-Lucea, C.; Siljander, P.; Ayuso-Sacido, A.; Yliperttula, M. Different gDNA content in the subpopulations of prostate cancer extracellular vesicles: Apoptotic bodies, microvesicles and exosomes. *Prostate* **2014**, *74*, 1379–1390. [CrossRef]
42. König, L.; Kasimir-Bauer, S.; Bittner, A.K.; Hoffmann, O.; Wagner, B.; Santos Manvailer, L.F.; Kimmig, R.; Horn, P.A.; Rebmann, V. Elevated levels of extracellular vesicles are associated with therapy failure and disease progression in breast cancer patients undergoing neoadjuvant chemotherapy. *Oncoimmunology* **2017**, *27*, e1376153. [CrossRef]
43. Ji, Q.; Ji, Y.; Peng, J.; Zhou, X.; Chen, X.; Zhao, H.; Xu, T.; Chen, L.; Xu, Y. Increased Brain-Specific MiR-9 and MiR-124 in the Serum Exosomes of Acute Ischemic Stroke Patients. *PLoS ONE* **2016**, *11*, e0163645. [CrossRef] [PubMed]
44. Galazka, G.; Mycko, M.P.; Selmaj, I.; Raine, C.S.; Selmaj, K.W. Multiple sclerosis: Serum-derived exosomes express myelin proteins. *Mult. Scler.* **2018**, *24*, 449–458. [CrossRef] [PubMed]
45. Liu, M.-L.; Werth, V.P.; Williams, K.J. Blood plasma versus serum: Which is right for sampling circulating membrane microvesicles in human subjects? *Ann. Rheum. Dis.* **2020**, *79*, e73. [CrossRef] [PubMed]
46. Smolarz, M.; Pietrowska, M.; Matysiak, N.; Mielańczyk, Ł.; Widłak, P. Proteome Profiling of Exosomes Purified from a Small Amount of Human Serum: The Problem of Co-Purified Serum Components. *Proteomes* **2019**, *7*, 18. [CrossRef]
47. Sódar, B.W.; Kittel, Á.; Pálóczi, K.; Vukman, K.V.; Osteikoetxea, X.; Szabó-Taylor, K.; Németh, A.; Sperlágh, B.; Baranyai, T.; Giricz, Z.; et al. Low-density lipoprotein mimics blood plasma-derived exosomes and microvesicles during isolation and detection. *Sci. Rep.* **2016**, *6*, 24316. [CrossRef] [PubMed]
48. Filipe, V.; Hawe, A.; Jiskoot, W. Critical evaluation of Nanoparticle Tracking Analysis (NTA) by NanoSight for the measurement of nanoparticles and protein aggregates. *Pharm. Res.* **2010**, *27*, 796–810. [CrossRef] [PubMed]
49. Osti, D.; Del Bene, M.; Rappa, G.; Santos, M.; Matafora, V.; Richichi, C.; Faletti, S.; Beznoussenko, G.V.; Mironov, A.; Bachi, A.; et al. Clinical Significance of Extracellular Vesicles in Plasma from Glioblastoma Patients. *Clin. Cancer Res.* **2019**, *25*, 266–276. [CrossRef]
50. Karimi, N.; Cvjetkovic, A.; Jang, S.C.; Crescitelli, R.; Hosseinpour Feizi, M.A.; Nieuwland, R.; Lötvall, J.; Lässer, C. Detailed analysis of the plasma extracellular vesicle proteome after separation from lipoproteins. *Cell Mol. Life Sci.* **2018**, *75*, 2873–2886. [CrossRef]
51. Xu, R.; Greening, D.W.; Zhu, H.J.; Takahashi, N.; Simpson, R.J. Extracellular vesicle isolation and characterization: Toward clinical application. *J. Clin. Investig.* **2016**, *126*, 1152–1162. [CrossRef]
52. Harmati, M.; Tarnai, Z.; Decsi, G.; Kormondi, S.; Szegletes, Z.; Janovak, L.; Dekany, I.; Saydam, O.; Gyukity-Sebestyen, E.; Dobra, G.; et al. Stressors alter intercellular communication and exosome profile of nasopharyngeal carcinoma cells. *J. Oral Pathol. Med.* **2017**, *46*, 259–266. [CrossRef]
53. Parsons, M.E.M.; McParland, D.; Szklanna, P.B.; Guang, M.H.Z.; O'Connell, K.; O'Connor, H.D.; McGuigan, C.; Ní Áinle, F.; McCann, A.; Maguire, P.B. A Protocol for Improved Precision and Increased Confidence in Nanoparticle Tracking Analysis Concentration Measurements between 50 and 120 nm in Biological Fluids. *Front. Cardiovasc. Med.* **2017**, *4*, 68. [CrossRef] [PubMed]
54. Searle, B.C.; Pino, L.K.; Egertson, J.D.; Ting, Y.S.; Lawrence, R.T.; MacLean, B.X.; Villén, J.; MacCoss, M.J. Chromatogram libraries improve peptide detection and quantification by data independent acquisition mass spectrometry. *Nat. Commun.* **2018**, *9*, 5128. [CrossRef] [PubMed]

55. Chambers, M.; Maclean, B.; Burke, R.; Amodei, D.; Ruderman, D.L.; Neumann, S.; Gatto, L.; Fischer, B.; Pratt, B.; Egertson, J.; et al. A cross-platform toolkit for mass spectrometry and proteomics. *Nat. Biotechnol.* **2012**, *30*, 918–920. [CrossRef] [PubMed]
56. Rosenberger, G.; Koh, C.C.; Guo, T.; Röst, H.L.; Kouvonen, P.; Collins, B.C.; Heusel, M.; Liu, Y.; Caron, E.; Vichalkovski, A.; et al. A repository of assays to quantify 10,000 human proteins by SWATH-MS. *Sci. Data* **2014**, *1*, 140031. [CrossRef] [PubMed]
57. Metz, T.O.; Qian, W.J.; Jacobs, J.M.; Gritsenko, M.A.; Moore, R.J.; Polpitiya, A.D.; Monroe, M.E.; Camp, D.G., 2nd; Mueller, P.W.; Smith, R.D. Application of proteomics in the discovery of candidate protein biomarkers in a diabetes autoantibody standardization program sample subset. *J. Proteome Res.* **2008**, *7*, 698–707. [CrossRef] [PubMed]
58. Hodge, K.; Have, S.T.; Hutton, L.; Lamond, A.I. Cleaning up the masses: Exclusion lists to reduce contamination with HPLC-MS/MS. *J. Proteom.* **2013**, *88*, 92–103. [CrossRef] [PubMed]
59. Curran-Everett, D. Explorations in statistics: The log transformation. *Adv. Physiol. Educ.* **2018**, *42*, 343–347. [CrossRef]
60. Lakens, D. Calculating and reporting effect sizes to facilitate cumulative science: A practical primer for t-tests and ANOVAs. *Front. Psychol.* **2013**, *4*, 863. [CrossRef]
61. Cohen, J. *Statistical Power Analysis for the Behavioral Sciences*, 2nd ed.; Routledge: New York, NY, USA, 1988. [CrossRef]
62. Fawcett, T. An introduction to ROC analysis. *Pattern Recognit. Lett.* **2006**, *27*, 861–874. [CrossRef]
63. Husson, F.; Le, S.; Pagès, J. *Exploratory Multivariate Analysis by Example Using R*, 2nd ed.; Chapman and Hall/CRC: New York, NY, USA, 2017. [CrossRef]
64. Hartigan, J.A.; Wong, M.A. Algorithm AS 136: A K-Means Clustering Algorithm. *J. Appl. Stat.* **1979**, *28*, 100. [CrossRef]
65. Ding, C.; He, X. K-means clustering via principal component analysis. In Proceedings of the Twenty-First International Conference on Machine Learning, Banff, AB, Canada, 4–8 July 2004; Association for Computing Machinery: New York, NY, USA, 2004. [CrossRef]
66. Rousseeuw, P.J. Silhouettes: A graphical aid to the interpretation and validation of cluster analysis. *J. Comput. Appl. Math.* **1987**, *20*, 53–65. [CrossRef]
67. Pedregosa, F.; Varoquaux, G.; Gramfort, A.; Michel, V.; Thirion, B.; Grisel, O.; Blondel, M.; Prettenhofer, P.; Weiss, R.; Dubourg, V.; et al. Scikit-learn: Machine Learning in Python. *J. Mach. Learn. Res.* **2012**, *12*, 2825–2830.
68. Van Deun, J.; Mestdagh, P.; Agostinis, P.; Akay, Ö.; Anand, S.; Anckaert, J.; Martinez, Z.A.; Baetens, T.; Beghein, E.; Bertier, L.; et al. EV-TRACK: Transparent reporting and centralizing knowledge in extracellular vesicle research. *Nat. Methods* **2017**, *14*, 228–232. [CrossRef] [PubMed]

Permissions

All chapters in this book were first published by MDPI; hereby published with permission under the Creative Commons Attribution License or equivalent. Every chapter published in this book has been scrutinized by our experts. Their significance has been extensively debated. The topics covered herein carry significant findings which will fuel the growth of the discipline. They may even be implemented as practical applications or may be referred to as a beginning point for another development.

The contributors of this book come from diverse backgrounds, making this book a truly international effort. This book will bring forth new frontiers with its revolutionizing research information and detailed analysis of the nascent developments around the world.

We would like to thank all the contributing authors for lending their expertise to make the book truly unique. They have played a crucial role in the development of this book. Without their invaluable contributions this book wouldn't have been possible. They have made vital efforts to compile up to date information on the varied aspects of this subject to make this book a valuable addition to the collection of many professionals and students.

This book was conceptualized with the vision of imparting up-to-date information and advanced data in this field. To ensure the same, a matchless editorial board was set up. Every individual on the board went through rigorous rounds of assessment to prove their worth. After which they invested a large part of their time researching and compiling the most relevant data for our readers.

The editorial board has been involved in producing this book since its inception. They have spent rigorous hours researching and exploring the diverse topics which have resulted in the successful publishing of this book. They have passed on their knowledge of decades through this book. To expedite this challenging task, the publisher supported the team at every step. A small team of assistant editors was also appointed to further simplify the editing procedure and attain best results for the readers.

Apart from the editorial board, the designing team has also invested a significant amount of their time in understanding the subject and creating the most relevant covers. They scrutinized every image to scout for the most suitable representation of the subject and create an appropriate cover for the book.

The publishing team has been an ardent support to the editorial, designing and production team. Their endless efforts to recruit the best for this project, has resulted in the accomplishment of this book. They are a veteran in the field of academics and their pool of knowledge is as vast as their experience in printing. Their expertise and guidance has proved useful at every step. Their uncompromising quality standards have made this book an exceptional effort. Their encouragement from time to time has been an inspiration for everyone.

The publisher and the editorial board hope that this book will prove to be a valuable piece of knowledge for researchers, students, practitioners and scholars across the globe.

List of Contributors

Luigi Pasini and Paola Ulivi
Biosciences Laboratory, Istituto Scientifico Romagnolo per lo Studio e la Cura dei Tumori (IRST) IRCCS, 47014 Meldola, Italy

Carolina Herrero, Carlos Casas-Arozamena, Alicia Abalo and Ana Vilar
Translational Medical Oncology Group (Oncomet), Health Research Institute of Santiago de Compostela (IDIS), University Hospital of Santiago de Compostela (SERGAS), Trav. Choupana s/n, 15706 Santiago de Compostela, Spain

Alexandre de la Fuente
Nasasbiotech, S.L., Canton Grande 3, 15003 A Coruña, Spain

Victor Sebastian, Martin Prieto and Manuel Arruebo
Department of Chemical Engineering, Aragon Institute of Nanoscience (INA), University of Zaragoza, 50018 Zaragoza, Spain
Networking Research Center on Bioengineering, Biomaterials and Nanomedicine (CIBER-BBN), Monforte de Lemos 3–5, 28029 Madrid, Spain

Eva Colás and Antonio Gil-Moreno
Biomedical Research Group in Gynecology, Vall Hebron Research Institute (VHIR), Universitat Autonoma de Barcelona, 119–129 Pg. Vall d'Hebron, 08035 Barcelona, Spain
Centro de Investigación Biomédica en Red de Cáncer (CIBERONC), Monforte de Lemos 3–5, 28029 Madrid, Spain

Juan Cueva, Miguel Abal and Laura Muinelo-Romay
Translational Medical Oncology Group (Oncomet), Health Research Institute of Santiago de Compostela (IDIS), University Hospital of Santiago de Compostela (SERGAS), Trav. Choupana s/n, 15706 Santiago de Compostela, Spain
Centro de Investigación Biomédica en Red de Cáncer (CIBERONC), Monforte de Lemos 3–5, 28029 Madrid, Spain

Gema Moreno-Bueno
Centro de Investigación Biomédica en Red de Cáncer (CIBERONC), Monforte de Lemos 3–5, 28029 Madrid, Spain
Department of Biochemistry, Universidad Autonoma de Madrid (UAM), Instituto de Investigaciones Biomedicas "Alberto Sols" (CSIC-UAM), IdiPaz, Pedro Rico, 6, 28029 Madrid, Spain
Fundacion MD Anderson Internacional, C/Gómez Hemans 2, 28033 Madrid, Spain

Sophie K. Horrevorts, Dorian A. Stolk, Myrthe Hulst, Bert van Het Hof, Sanne Duinkerken, Marieke H. Heineke, Sophie A. Dusoswa, Juan J. Garcia-Vallejo, Sandra J. van Vliet and Yvette van Kooyk
Amsterdam UMC, Department of Molecular Cell Biology and Immunology, Amsterdam Infection and Immunity Institute, Cancer Center Amsterdam, Vrije Universiteit Amsterdam, 1081 HV Amsterdam, The Netherlands

Tanja D. de Gruijl
Amsterdam UMC, Department of Medical Oncology, Cancer Center Amsterdam, Vrije Universiteit Amsterdam, 1081 HV Amsterdam, The Netherlands

Rieneke van de Ven
Amsterdam UMC, Department of Otolaryngology/Head and Neck Surgery, Cancer Center Amsterdam, Vrije Universiteit Amsterdam, 1081 HV Amsterdam, The Netherlands
Amsterdam UMC, Department of Medical Oncology, Cancer Center Amsterdam, Vrije Universiteit Amsterdam, 1081 HV Amsterdam, The Netherlands

Wenbin Ma
Ludwig Institute for Cancer Research, B-1200 Brussels, Belgium
Walloon Excellence in Life Sciences and Biotechnology, B-1200 Brussels, Belgium
De Duve Institute, Université Catholique de Louvain, B-1200 Brussels, Belgium

Rienk Nieuwland
Amsterdam UMC, Laboratory of Experimental Clinical Chemistry and Vesicle Observation Centre, Academic Medical Centre, 1081 HV Amsterdam, The Netherlands

Arjan A. van de Loosdrecht
Amsterdam UMC, Department of Hematology, Cancer Center Amsterdam, Vrije Universiteit Amsterdam, 1081 HV Amsterdam, The Netherlands

Thomas Simon, Anish Kumaran, Diana-Florentina Veselu and Georgios Giamas
Department of Biochemistry and Biomedicine, School of Life Sciences, University of Sussex, Brighton BN1 9QG, UK

Aneta Zebrowska, Piotr Widlak and Monika Pietrowska
Maria Sklodowska-Curie National Research Institute of Oncology, Gliwice Branch, 44-100 Gliwice, Poland

Theresa Whiteside
UPMC Hillman Cancer Center, University of Pittsburgh, Pittsburgh, PA 15213, USA
Department of Pathology, University of Pittsburgh School of Medicine Pittsburgh, Pittsburgh, PA 15261, USA

Stephanie Burrola and Wei-Qun Ding
Department of Pathology, University of Oklahoma Health Science Center, Oklahoma City, OK 73104, USA

Haoyao Sun
Department of Pathology, University of Oklahoma Health Science Center, Oklahoma City, OK 73104, USA
Department of Radiation Oncology, The Affiliated Suzhou Hospital of Nanjing Medical University, Suzhou 215001, China

Jinchang Wu
Department of Radiation Oncology, The Affiliated Suzhou Hospital of Nanjing Medical University, Suzhou 215001, China
Section of Oncology, The Second Affiliated Hospital of Xuzhou Medical University, Xuzhou 221006, China

Seda Tuncay Cagatay, Ammar Mayah and Munira Kadhim
Department of Biological and Medical Sciences, Oxford Brookes University, Oxford OX3 0BP, UK

Mariateresa Mancuso, Simonetta Pazzaglia and Anna Saran
Laboratory of Biomedical Technologies, Italian National Agency for New Technologies, Energy and Sustainable Economic Development (ENEA), 00123 Rome, Italy

Paola Giardullo
Laboratory of Biomedical Technologies, Italian National Agency for New Technologies, Energy and Sustainable Economic Development (ENEA), 00123 Rome, Italy
Department of Radiation Physics, Guglielmo Marconi University, 00193 Rome, Italy

Amuthachelvi Daniel, Damien Traynor, Aidan D. Meade and Fiona Lyng
Centre for Radiation and Environmental Science, FOCAS Research Institute, Technological University Dublin, D08 NF82 Dublin, Ireland

Soile Tapio
Institute of Radiation Biology, Helmholtz Zentrum München, 85764 Munich, Germany

Katrin S Reiners and Gunther Hartmann
Institute of Clinical Chemistry and Clinical Pharmacology, University of Bonn, 53127 Bonn, Germany

Johannes Weller, Niklas Schäfer, Christina Schaub and Ulrich Herrlinger
Division of Clinical Neurooncology, Department of Neurology, Center of Integrated Oncology Aachen-Bonn-Cologne-Düsseldorf, Partner Site Bonn, University Hospital Bonn, 53127 Bonn, Germany

Theophilos Tzaridis
Institute of Clinical Chemistry and Clinical Pharmacology, University of Bonn, 53127 Bonn, Germany
Division of Clinical Neurooncology, Department of Neurology, Center of Integrated Oncology Aachen-Bonn-Cologne-Düsseldorf, Partner Site Bonn, University Hospital Bonn, 53127 Bonn, Germany
Tumor Initiation & Maintenance Program, NCI-Designated Cancer Center, Sanford Burnham Prebys Medical Discovery Institute, La Jolla, CA 92037, USA

Daniel Bachurski and Michael Hallek
Department I of Internal Medicine, Center for Integrated Oncology Aachen-Bonn-Cologne-Düsseldorf, Partner Site Cologne, CECAD Center of Excellence on "Cellular Stress Responses in Aging-Associated Diseases", Center for Molecular Medicine Cologne, University of Cologne, 50937 Cologne, Germany

Björn Scheffler
DKFZ-Division Translational Neurooncology at the West German Cancer Center (WTZ), German Cancer Consortium (DKTK), DKFZ Heidelberg & Partner Site Univ Hospital Essen, 45147 Essen, Germany

Martin Glas
Division of Clinical Neurooncology, Department of Neurology and West German Cancer Center (WTZ), German Cancer Consortium, University Hospital Essen, 45147 Essen, Germany

Stefan Wild
Miltenyi Biotec & Biomedicine GmbH, 51429 Bergisch Gladbach, Germany

Christoph Coch
Institute of Clinical Chemistry and Clinical Pharmacology, University of Bonn, 53127 Bonn, Germany
Miltenyi Biotec & Biomedicine GmbH, 51429 Bergisch Gladbach, Germany

Tomasz Lorenc
1st Department of Clinical Radiology, Medical University of Warsaw, 02-004 Warsaw, Poland

Katarzyna Klimczyk, Izabela Michalczewska, Monika Słomka, Grażyna Kubiak-Tomaszewska and Wioletta Olejarz
Department of Biochemistry and Pharmacogenomics, Faculty of Pharmacy, Medical University of Warsaw, 02-097 Warsaw, Poland
Centre for Preclinical Research, Medical University of Warsaw, 02-097 Warsaw, Poland

Magdalena Żmigrodzka, Olga Witkowska-Piłaszewicz and Anna Winnicka
Department of Pathology and Veterinary Diagnostics, Institute of Veterinary Medicine, Warsaw University of Life Sciences (WULS-SGGW), Nowoursynowska 159c, 02-787 Warsaw, Poland

Donatella Lucchetti, Claudio Ricciardi-Tenore, Filomena Colella, Luigi Perelli and Ruggero De Maria
Department of Translational Medicine and Surgery, Università Cattolica del Sacro Cuore, 00168 Rome, Italy

Alessandra Battaglia
Department of Life Science and Public Health, Università Cattolica del Sacro Cuore, 00168 Rome, Italy

Giovanni Scambia and Andrea Fattorossi
Laboratory of Cytometry and Immunology, Department of Obstetrics and Gynecology, Università Cattolica del Sacro Cuore, Fondazione Policlinico Universitario A. Gemelli IRCCS, 00168 Rome, Italy

Alessandro Sgambato
Centro di Riferimento Oncologico della Basilicata (IRCCS-CROB), Rionero in Vulture (PZ), 85028 Potenza, Italy

Jeong Uk Choi
College of Pharmacy, Chonnam National University, 77 Yongbong-ro, Buk-gu, Gwangju 61186, Korea

In-Kyu Park
Department of Biomedical Sciences, Chonnam National University Medical School, 322 Seoyang-ro, Hwasun 58128, Korea

Yong-Kyu Lee
Department of Chemical and Biological Engineering, Korea National University of Transportation, 50 Daehak-ro, Chungju, Chungbuk 27469, Korea

Seung Rim Hwang
College of Pharmacy, Chosun University, 309 Pilmun-daero, Dong-gu, Gwangju 61452, Korea
Department of Biomedical Sciences, Graduate School, Chosun University, 309 Pilmun-daero, Dong-gu, Gwangju 61452, Korea

Maria Harmati, Edina Gyukity-Sebestyen and Peter Horvath
Laboratory of Microscopic Image Analysis and Machine Learning, Institute of Biochemistry, Biological Research Centre, H-6726 Szeged, Hungary

Gabriella Dobra and Matyas Bukva
Laboratory of Microscopic Image Analysis and Machine Learning, Institute of Biochemistry, Biological Research Centre, H-6726 Szeged, Hungary
Department of Medical Genetics, Doctoral School of Interdisciplinary Medicine, University of Szeged, H-6720 Szeged, Hungary

Zoltan Szabo and Bella Bruszel
Department of Medical Chemistry, Faculty of Medicine, University of Szeged, H-6720 Szeged, Hungary

Adrienn Jenei and Almos Klekner
Department of Neurosurgery, Clinical Centre, University of Debrecen, H-4032 Debrecen, Hungary

Monika Szucs
Department of Medical Physics and Informatics, Faculty of Medicine, University of Szeged, H-6720 Szeged, Hungary
Department of Medical Physics and Informatics, Faculty of Science and Informatics, University of Szeged, H-6720 Szeged, Hungary

Tamas Biro
Department of Immunology, Faculty of Medicine, University of Debrecen, H-4032 Debrecen, Hungary

Krisztina Buzas
Laboratory of Microscopic Image Analysis and Machine Learning, Institute of Biochemistry, Biological Research Centre, H-6726 Szeged, Hungary
Department of Immunology, Faculty of Medicine, University of Szeged, H-6720 Szeged, Hungary
Department of Immunology, Faculty of Science and Informatics, University of Szeged, H-6720 Szeged, Hungary

Index

www.ingramcontent.com/pod-product-compliance
Lightning Source LLC
Chambersburg PA
CBHW080109230426
43735CB00005B/586
* 9 7 8 1 6 3 9 8 7 7 8 3 6 *